Problem Books in Mathematics

For further volumes:
http://www.springer.com/series/714

Dmytro Gusak · Alexander Kukush · Alexey Kulik·
Yuliya Mishura · Andrey Pilipenko

Theory of Stochastic Processes

With Applications to Financial Mathematics and Risk Theory

Springer

Dmytro Gusak
Institute of Mathematics
of Ukrainian National
Academy of Sciences
Kyiv 01601
Ukraine
random@imath.kiev.ua

Alexander Kukush
Department of Mathematical Analysis
Faculty of Mechanics and Mathematics
National Taras Shevchenko University of Kyiv
Kyiv 01033
Ukraine
alexander_kukush@univ.kiev.ua

Alexey Kulik
Institute of Mathematics
of Ukrainian National
Academy of Sciences
Kyiv 01601
Ukraine
kulik@imath.kiev.ua

Yuliya Mishura
Department of Probability Theory
and Mathematical Statistics
Faculty of Mechanics and Mathematics
National Taras Shevchencko University of Kyiv
Kyiv 01033
Ukraine
myus@univ.kiev.ua

Andrey Pilipenko
Institute of Mathematics
of Ukrainian National
Academy of Sciences
Kyiv 01601
Ukraine
apilip@imath.kiev.ua

Series Editor
Peter Winkler
Department of Mathematics
Dartmouth College
Hanover, NH 03755-3551
USA
peter.winkler@dartmouth.edu

ISSN 0941-3502
ISBN 978-1-4614-2506-9 e-ISBN 978-0-387-87862-1
DOI 10.1007/978-0-387-87862-1
Springer New York Dordrecht Heidelberg London

Mathematics Subject Classification (2000): 60-xx:60Gxx 60G07 60H10 91B30

Softcover reprint of the hardcover 1st edition 2009

Printed on acid-free paper

Springer is part of Springer Science+Business Media (www.springer.com)

To our families

Preface

This collection of problems is planned as a textbook for university courses in the theory of stochastic processes and related special courses. The problems in the book have a wide spectrum of the level of difficulty and can be useful for readers with various levels of mastering in the theory of stochastic processes. Together with technical and illustrative problems intended for beginners, the book contains a number of problems of theoretical nature that can be useful for students and undergraduate students that pursue advanced studies in the theory of stochastic processes and its applications. Among others, the important aim of the book is to provide a teaching staff an efficient tool for preparing seminar studies, tests, and exams concerning university courses in the theory of stochastic processes and related topics. While composing the book, the authors have partially used the collections of problems in probability theory [16, 65, 75, 83]. Also, some exercises and problems from the monographs and textbooks [4, 9, 19, 22, 82] were used. At the same time, a large part of our problem book contains original material.

The book is organized as follows. The problems are collected into chapters, each chapter being devoted to a certain topic. At the beginning of each chapter, the theoretical grounds for the corresponding topic are given briefly together with the list of bibliography, which the reader can use in order to study this topic in more detail. For the most of the problems, either hints or complete solutions (or answers) are given, and some of the problems are provided with both hints and solutions (answers). However, the authors do not recommend that a reader use the hints systematically, because solving a problem without assistance is much more useful than using a ready-made idea. Some statements that have a particular theoretical interest are formulated on theoretical grounds, and their proofs are formulated as problems for the reader. Such problems are supplied with either complete solutions or detailed hints.

In order to work with the problem book efficiently, a reader should be acquainted with probability theory, calculus, and measure theory within the scope of respective university courses. Standard notions, such as random variable, measurability, independence, Lebesgue measure and integral, and so on are used without additional discussion. All the new notions and statements required for solving the problems are given either on theoretical grounds or in the formulations of the problems straightforwardly. However, sometimes a notion is used in the text before its formal definition. For instance, the Wiener and Poisson processes are processes with independent increments and thus are formally introduced in a Theoretical grounds for Chapter 5, but these processes are used widely in the problems of Chapters 2 to 4. The authors recommend that a reader who comes to an unknown notion or object use the Index in order to find the corresponding formal definition. The same recommendation concerns some standard abbreviations and symbols listed at the end of the book.

Some problems in the book form cycles: solutions to one of them are grounded on statements of others or on auxiliary constructions described in some preceding solutions. Sometimes, on the contrary, it is proposed to prove the same statement within different problems using essentially different techniques. The authors recommend a reader pay specific attention to these fruitful internal links between various topics of the theory of stochastic processes.

Every part of the book was composed substantially by one author. Chapters 1–6, and 16 are composed by A. Kulik, Chapters 7, 12–15, 18, and 19 by Yu. Mishura, Chapters 8–10 by A. Pilipenko, Chapter 17 by A. Kukush, and Chapter 20 by D. Gusak. Chapter 11 was prepared jointly by D. Gusak and A. Pilipenko. At the same time, every author has made a contribution to other parts of the book by proposing separate problems or cycles of problems, improving preliminary versions of theoretical grounds, and editing the final text.

The authors would like to express their deep gratitude to M. Portenko and A. Ivanov for their careful reading of a preliminary version of the book and valuable comments that led to significant improvement of the text. The authors are also grateful to T. Yakovenko, G. Shevchenko, O. Soloveyko, Yu. Kartashov, Yu. Klimenko, A. Malenko, and N. Ryabova for their assistance in translation, preparing files and pictures, and composing the subject index and references.

The theory of stochastic processes is an extended discipline, and the authors understand that the problem book in its current form may cause critical remarks from readers, concerning either the structure of the book or the content of separate chapters. While publishing the problem book in its current form, the authors are open for remarks, comments, and propositions, and express in advance their gratitude to all their correspondents.

Kyiv
December 2008

Dmytro Gusak
Alexander Kukush
Alexey Kulik
Yuliya Mishura
Andrey Pilipenko

Contents

1

Definition of stochastic process. Cylinder σ-algebra, finite-dimensional distributions, the Kolmogorov theorem

Theoretical grounds

Let $(\Omega, \mathcal{F}, \mathsf{P})$ be a probability space, $(\mathbb{X}, \mathcal{X})$ be a measurable space, and $\mathbb{T}$ be some set.

Definition 1.1. A random function X *with phase space* $\mathbb{X}$ *and parameter set* $\mathbb{T}$ *is a function* $X : \mathbb{T} \times \Omega \ni (t, \omega) \mapsto X(t, \omega) \in \mathbb{X}$ *such that for any* $t \in \mathbb{T}$ *the mapping* $X(t, \cdot) : \omega \mapsto X(t, \omega)$ *is* $\mathcal{F} - \mathcal{X}$*-measurable.*

Hereinafter the mapping $X(t, \cdot)$ is denoted as $X(t)$. According to commonly accepted terminology it is a *random element* taking values in $\mathbb{X}$. The definition introduced above is obviously equivalent to the following one.

Definition 1.2. A random function X *with phase state* $\mathbb{X}$ *and parameter set* $\mathbb{T}$ *is a family of random elements* $\{X(t), t \in \mathbb{T}\}$ *with values in* $\mathbb{X}$ *indexed by points of* $\mathbb{T}$.

A random function, as defined in Definitions 1.1 and 1.2, sometimes is also called a *stochastic (random) process*, but usually the term *stochastic process* is reserved for the case where $\mathbb{T}$ is an interval or a ray on the real axis $\mathbb{R}$. *A random sequence* (or *stochastic process with discrete time*) is a random function defined on $\mathbb{T} \subset \mathbb{Z}$. *A random field* is a random function defined on $\mathbb{T} \subset \mathbb{R}^d, d > 1$.

For a fixed $\omega \in \Omega$, function $X(\cdot, \omega) : \mathbb{T} \ni t \mapsto X(t, \omega)$ is called *a trajectory* or *a realization* of the random function X.

Denote by $\mathcal{X}^{\otimes m} = \mathcal{X} \otimes \cdots \otimes \mathcal{X}$ the product of m copies of σ-algebra $\mathcal{X}$, that is, the least σ-algebra of a subsets of $\mathbb{X}^m$ that contains every set of the form

$$A_1 \times \cdots \times A_m, \quad A_1, \ldots, A_m \in \mathcal{X}.$$

Definition 1.3. *For given values* $m \geq 1$ *and* $t_1, \ldots, t_m \in \mathbb{T}$, ($m$-dimensional) finite-dimensional distribution $\mathsf{P}^X_{t_1,\ldots,t_m}$ *of random function* X *is the joint distribution of random elements* $X(t_1), \ldots, X(t_m)$ *or, equivalently, the distribution of the vector* $(X(t_1), \ldots, X(t_m))$ *considered as a random element with values in* $(\mathbb{X}^m, \mathcal{X}^{\otimes m})$. *The set* $\{\mathsf{P}^X_{t_1,\ldots,t_m}, t_1, \ldots, t_m \in \mathbb{T}, m \geq 1\}$ *is called* the set *(or* the family*)* of finite-dimensional distributions *of the random function* X.

D. Gusak et al., *Theory of Stochastic Processes*, Problem Books in Mathematics,
DOI 10.1007/978-0-387-87862-1_1, © Springer Science+Business Media, LLC 2010

Theorem 1.1. *(The Kolmogorov theorem on finite-dimensional distributions) Let* $\mathbb{X}$ *be a complete separable metric space and* $\mathcal{X}$ *be its Borel* σ*-algebra. Suppose a family* $\{\mathsf{P}_{t_1,\ldots,t_m}, t_1,\ldots,t_m \in \mathbb{T}, m \geq 1\}$ *be given, such that, for any* $m \geq 1$ *and* $t_1,\ldots,t_m \in \mathbb{T}$, $\mathsf{P}_{t_1,\ldots,t_m}$ *is a probability measure on* $(\mathbb{X}^m, \mathcal{X}^{\otimes m})$*. The following* consistency conditions *are necessary and sufficient for a random function to exist, such that the family* $\{\mathsf{P}_{t_1,\ldots,t_m}, t_1,\ldots,t_m \in \mathbb{T}, m \geq 1\}$ *is the family of finite-dimensional distributions for this function.*

(1) For any $m \geq 1, t_1,\ldots,t_m \in \mathbb{T}, B_1,\ldots,B_m \in \mathcal{X}$ *and arbitrary permutation* $\pi : \{1,\ldots,m\} \to \{1,\ldots,m\}$,

$$\mathsf{P}_{t_1,\ldots,t_m}(B_1 \times \cdots \times B_m) = \mathsf{P}_{t_{\pi(1)},\ldots,t_{\pi(m)}}(B_{\pi(1)} \times \cdots \times B_{\pi(m)})$$

(permutation invariance).

(2) For any $m > 1, t_1,\ldots,t_m \in \mathbb{T}, B_1,\ldots,B_{m-1} \in \mathcal{X}$,

$$\mathsf{P}_{t_1,\ldots,t_m}(B_1 \times \cdots \times B_{m-1} \times \mathbb{X}) = \mathsf{P}_{t_1,\ldots,t_{m-1}}(B_1 \times \cdots \times B_{m-1})$$

(projection invariance).

The random function provided by the Kolmogorov theorem can be constructed on a special probability space. Further on, we describe the construction of this probability space.

Let $\Omega = \mathbb{X}^{\mathbb{T}}$ be a space of all functions $\omega : \mathbb{T} \to \mathbb{X}$.

Definition 1.4. *A set* $A \subset \Omega$ *is called* a cylinder set, *if it has the following representation*

$$A = \{\omega \in \Omega \mid (\omega(t_1),\ldots,\omega(t_m)) \in B\} \tag{1.1}$$

for some $m \geq 1, t_1,\ldots,t_m \in \mathbb{T}, B \in \mathcal{X}^{\otimes m}$.

A cylinder set has many representations of the form (1.1). A set B in any representation (1.1) of the cylinder set A is called a base (or basis) of A.

A class of all cylinder sets is denoted by $\mathcal{C}(\mathbb{X},\mathbb{T})$ or simply by $\mathcal{C}$. This class is an algebra, but, in general, it is not a σ-algebra (see Problem 1.35). A minimal σ-algebra $\sigma(\mathcal{C})$ that contains this class is called a *σ-algebra generated by cylinder sets* or *cylinder σ-algebra.* The random function in the Kolmogorov theorem can be defined on the space $\Omega = \mathbb{X}^{\mathbb{T}}$ with the σ-algebra $\mathcal{F} = \sigma(\mathcal{C})$, $X(t,\omega) = \omega(t), t \in \mathbb{T}, \omega \in \Omega$, and probability P constructed in some special way (see, for instance, [79], Chapter 2 or [9], Appendix 1).

The cylinder σ-algebra has the following useful characterization (see Problem 1.29).

Theorem 1.2. *A set* $A \subset \mathbb{X}^{\mathbb{T}}$ *belongs to* $\sigma(\mathcal{C}(\mathbb{X},\mathbb{T}))$ *if and only if there exists a sequence* $(t_n)_{n=1}^{\infty} \subset \mathbb{T}$ *and a set* $B \in \sigma(\mathcal{C}(\mathbb{X},\mathbb{N}))$ *such that*

$$A = \{\omega \mid (\omega(t_n))_{n=1}^{\infty} \in B\}.$$

Bibliography

[9], Chapter I; [24], Volume 1, Chapter I, §4; [25], Chapter II, §2; [15], Chapter II, §§1,2; [79], Chapters 1,2.

Problems

1.1. Let η be a random variable with the distribution function F. Prove that $X(t)$ is a stochastic process, if (a) $X(t) = \eta t$; (b) $X(t) = \min(\eta, t)$; (c) $X(t) = \max(\eta, t^2)$; (d) $X(t) = \operatorname{sign}(\eta + t)$, where

$$\operatorname{sign} x = \begin{cases} 1, & x \geq 0, \\ -1, & x < 0. \end{cases}$$

Draw the trajectories of the process X. Find one-dimensional distributions of the process X.

1.2. Let τ be a random variable with uniform distribution on $[0,1]$ and $\{X(t), t \in [0,1]\}$ be a *waiting process* corresponding to this variable; that is, $X(t) = 1\!\!\mathrm{I}_{t \geq \tau}, t \in [0,1]$. Find all (a) one-dimensional; (b) two-dimensional; (c) m-dimensional distributions of the process X.

1.3. Two devices start to operate at the instant of time $t = 0$. They operate independently of each other for random periods of time and after that they shut down. The operating time of the ith device has a distribution function F_i, $i = 1, 2$. Let $X(t)$ be the number of operating devices at the instant t. Find one- and two-dimensional distributions of the process $\{X(t), t \in \mathbb{R}^+\}$.

1.4. Let $\xi_1, \ldots, \xi_n$ be independent identically distributed random variables with distribution function F, and

$$X(x) = \frac{1}{n} \#\{k \mid \xi_k \leq x\} = \frac{1}{n} \sum_{k=1}^{n} 1\!\!\mathrm{I}_{\xi_k \leq x}, \quad x \in \mathbb{R}$$

(remark that $X(\cdot) \equiv F_n^*(\cdot)$ is the empirical distribution function based on the sample $\xi_1, \ldots, \xi_n$). Find all (a) one-dimensional; (b) two-dimensional; (c) m-dimensional distributions of the process X.

Here and below, # denotes the number of elements of a set.

1.5. Is it possible for stochastic processes $\{X_1(t), X_2(t), t \geq 0\}$ to have (a) identical one-dimensional distributions, but different two-dimensional ones; (b) identical two-dimensional distributions, but different three-dimensional ones?

1.6. Let $\Omega = \mathbb{T} = \mathbb{R}^+, \mathcal{F} = \mathcal{B}(\mathbb{R}^+), A \subset \mathbb{R}^+$. Here and below, $\mathcal{B}(\mathbb{X})$ denotes the Borel σ-algebra in $\mathbb{X}$. Prove that:
(1) $X_A(t,\omega) = 1\!\!\mathrm{I}_{t=\omega} \cdot 1\!\!\mathrm{I}_{\omega \in A}$ is a stochastic process for an arbitrary set A.
(2) $Y_A(t,\omega) = 1\!\!\mathrm{I}_{t \geq \omega} \cdot 1\!\!\mathrm{I}_{\omega \in A}$ is a random process if and only if $A \in \mathcal{B}(\mathbb{R}^+)$.
Depict all possible realizations of the processes X_A, Y_A.

1.7. At the instant of failure of some unit in a device, this unit is immediately replaced by a reserve one. Nonfailure operating times for each unit are random variables, jointly independent and exponentially distributed with parameter $\alpha > 0$. Let $X(t)$ be the number of failures up to the time moment t. Find finite-dimensional distributions of the process $X(t)$, if there are (a) n reserve units; (b) an infinite number of reserve units.

1.8. Let random variable ξ have distribution function F. Denote

$$F^{[-1]}(x) = \inf\{y|\ F(y) > x\}, \quad x \in [0,1]$$

(the function $F^{[-1]}$ is called *the generalized inverse* function for F or *the quantile transformation* of F), and set $\zeta = F^{[-1]}(\varepsilon)$, where ε is a random variable uniformly distributed on $[0,1]$. Prove that ζ has the same distribution with ξ.

1.9. Prove that it is possible to construct a sequence of independent identically distributed random variables defined on the probability space $\Omega = [0,1], \mathcal{F} = \mathcal{B}([0,1])$, $\mathsf{P} = \lambda^1|_{[0,1]}$, which
(a) take the values 0 and 1 with probabilities $\frac{1}{2}$;
(b) are uniformly distributed on $[0,1]$;
(c) have an arbitrary distribution function F.

Here and below, $\lambda^1|_{[0,1]}$ denotes the restriction of the Lebesgue measure λ^1 to $[0,1]$.

1.10. Prove that it is impossible to construct on the probability space $\Omega = [0,1], \mathcal{F} = \mathcal{B}([0,1])$, $\mathsf{P} = \lambda^1|_{[0,1]}$ a family of independent identically distributed random variables $\{\xi_t, t \in [0,1]\}$ with a nondegenerate distribution.

1.11. Let μ, ν be such distributions on $\mathbb{R}^2$ that $\mu(\mathbb{R} \times A) = \nu(A \times \mathbb{R})$ for every $A \in \mathcal{B}(\mathbb{R})$. Prove that it is possible to construct random variables ξ, η, ζ defined on some probability space in such a way that the joint distribution of ξ and η equals μ, and the joint distribution of η and ζ equals ν.

1.12. Assume a two-parameter family of distributions $\{\mu_{m,n},\ m,n \geq 1\}$ on $\mathbb{R}^2$ is given, consistent in the sense that
(a) for any $A, B \in \mathcal{B}(\mathbb{R})$ and $m,n \geq 1$, $\mu_{m,n}(A \times B) = \mu_{n,m}(B \times A)$;
(b) for any $A \in \mathcal{B}(\mathbb{R})$ and $l,m,n \geq 1$, $\mu_{n,m}(A \times \mathbb{R}) = \mu_{n,l}(A \times \mathbb{R})$.

Is it true that for any such family there exists a sequence of random variables $\{\xi_n, n \geq 1\}$ satisfying the relations $\mu_{m,n}(C) = \mathsf{P}((\xi_m, \xi_n) \in C)$ for any $m,n \geq 1$ and $C \in \mathcal{B}(\mathbb{R}^2)$?

1.13. Let $\{X_n, n \geq 1\}$ be a random sequence. Prove that the following extended random variables (i.e., the variables with possible values $+\infty$ and $-\infty$) are measurable with respect to the σ-algebra generated by cylinder sets: (a) $\sup_n X_n$; (b) $\limsup_n X_n$; (c) number of partial limits of the sequence $\{X_n\}$.

1.14. In the previous problem, let random variables in the sequence $\{X_n, n \geq 1\}$ be independent. Which ones among extended variables presented in the items a) — c) are degenerate, that is, take some value from $\mathbb{R} \cup \{-\infty, +\infty\}$ with probability 1?

1.15. Suppose that in Problem 1.13 random variables $\{X_n, n \geq 1\}$ may be dependent, but for some $m \geq 2$ and for any $l = 1, \ldots, m$ random variables $\{X_{nm+l}, n \geq 0\}$ are independent. Which ones among extended variables presented in the items a) — c) of Problem 1.13 are degenerate?

1.16. Let $\{\xi_n, n \geq 1\}$ be a sequence of i.i.d. random variables. Indicate such a sequence $\{a_n, n \geq 1\} \subset \mathbb{R}^+$ that $\limsup_{n\to+\infty} \xi_n/a_n = 1$ almost surely, if (a) $\xi_n \sim \mathcal{N}(0, \sigma^2)$; (b) $\xi_n \sim \text{Exp}(\lambda)$; (c) $\xi_n \sim \text{Pois}(\lambda)$.

1.17. Let $\{X(t), t \in \mathbb{R}^+\}$ be a stochastic process with right continuous trajectories, $a \in C(\mathbb{R}^+)$ be some deterministic function. Prove that the following variables are extended random variables (that is, measurable functions from Ω to $\mathbb{R} \cup \{-\infty, +\infty\}$):
(a) $\sup_{t\in\mathbb{R}^+} X(t)/a(t)$;
(b) $\limsup_{t\to+\infty} X(t)/a(t)$;
(c) the number of partial limits of the function $X(\cdot)/a(\cdot)$ as $t \to +\infty$;
(d) $\text{Var}(X(\cdot), [a,b])$ (variation of $X(\cdot)$ on the interval $[a,b]$).

1.18. Are the random variables presented in the previous problem measurable for an arbitrary stochastic process without any additional conditions on its trajectories?

1.19. Suppose that a random process $\{X(t), t \in [0;1]\}$ has continuous trajectories. Prove that the following sets are measurable.

$$A = \{\omega \in \Omega |\, X(t,\omega), t \in [0;1] \text{ satisfies Lipschitz condition}\}.$$
$$B = \{\omega \in \Omega |\, \min_{t\in[0;1]} X(t,\omega) < 7\}.$$
$$C = \{\omega \in \Omega |\, \int_0^1 X^2(s,\omega)ds > 3\max_{s\in[0;1]} X(s,\omega)\}.$$
$$D = \{\omega \in \Omega |\, \exists t \in [0;1) : X(t,\omega) = 1\}.$$
$$E = \{\omega \in \Omega |\, X(1/2,\omega) + 3\sin X(1,\omega) \leq 0\}.$$
$$F = \{\omega \in \Omega |\, X(t,\omega), t \in [0,1] \text{ is monotonically nondecreasing}\}.$$
$$G = \{\omega \in \Omega |\, \exists t_1, t_2 \in [0,1], t_1 \neq t_2 : X(t_1,\omega) = X(t_2,\omega) = 0\}.$$
$$H = \{\omega \in \Omega |\, X(t,\omega), t \in [0,1) \text{ is monotonically increasing}\}.$$

$I = \{\omega \in \Omega |$ at some point trajectory $X(\cdot,\omega)$ is tangent from above to the axis Ox; that is, there exists such an interval $[\tau_1, \tau_2] \subset [0,1]$, that $X(t,\omega) \geq 0$ as $t \in [\tau_1, \tau_2]$ and $\min_{t\in[\tau_1,\tau_2]} X(t,\omega) = 0\}$.

1.20. Let $\{X_n(t), n \geq 1, t \in [0,1]\}$ be a sequence of random processes with continuous trajectories. Prove that the set $\{\omega \in \Omega |\, \sum_n X_n(t,\omega)$ uniformly converges on $[0,1]\}$ is a random event.

1.21. Let $\Gamma \subset \mathbb{R}$ be an open set and suppose that the trajectories of a process $\{X(t), t \in \mathbb{R}^+\}$ are right continuous and have left limits.
(1) Are the following functions extended random variables: (a) $\tau^\Gamma \equiv \sup\{t : \forall s \leq t, X(s) \notin \Gamma\}$; (b) $\tau_\Gamma \equiv \inf\{t : X(t) \in \Gamma\}$?
(2) Prove that $\tau^\Gamma = \tau_\Gamma$ (this value is called the *hitting time* of the set Γ by the process X).

1.22. Solve the previous problem assuming Γ is a closed set.

1.23. Let $\Gamma \subset \mathbb{R}$ be some set and $\tau_\Gamma \equiv \inf\{t : X(t) \in \Gamma\}$ be a hitting time of Γ by a process $\{X(t), t \in \mathbb{R}^+\}$ with right continuous trajectories that have left limits. Is the variable τ_Γ an extended random variable?

1.24. Solve the previous problem assuming Γ is a closed set and trajectories of the process $\{X(t), t \in \mathbb{R}^+\}$ do not satisfy any continuity conditions.

1.25. Suppose that trajectories of the process $\{X(t), t \in [0,1]\}$ are right continuous and have left limits. Prove that for any $\omega \in \Omega$ the trajectory $X(\cdot,\omega)$ is Riemann integrable on $[0,1]$, and $\int_0^1 X(t)\,dt$ is a random variable.

1.26. Suppose that trajectories of the process $\{X(t), t \in [0,1]\}$ are right continuous. Is it true that for any $\omega \in \Omega$ trajectory $X(\cdot,\omega)$ is Riemann integrable on $[0,1]$?

1.27. Let trajectories of the process $\{X(t), t \in [0,1]\}$ be right continuous, and τ be a random variable with values in $[0,1]$. Prove that the function $X(\tau) : \Omega \ni \omega \mapsto X(\tau(\omega),\omega)$ is a measurable function; that is, $X(\tau)$ is a random variable.

1.28. Present an example of random process $\{X(t), t \in [0,1]\}$ and random variable τ taking values in $[0,1]$ such that $X(\tau)$ is not a random variable.

1.29. Prove Theorem 1.2.

1.30. Prove that the following subsets of $\mathbb{R}^{[0,1]}$ do not belong to the σ-algebra generated by cylinder sets.
(a) The set of all continuous functions
(b) The set of all bounded functions
(c) The set of all Borel functions

1.31. Construct a random process $\{X(t), t \in [0,1]\}$ defined on probability space $\Omega = [0,1], \mathcal{F} = \mathcal{B}([0,1]), \mathsf{P} = \lambda^1|_{[0,1]}$ in such a way that the set $\{\omega \in \Omega |$ function $X(\cdot,\omega)$ is continuous$\}$ is not a random event.

1.32. Construct a random process $\{X(t), t \in [0,1]\}$ defined on probability space $\Omega = [0,1], \mathcal{F} = \mathcal{B}([0,1]), \mathsf{P} = \lambda^1|_{[0,1]}$ in such a way that the set $\{\omega \in \Omega |$ function $X(\cdot,\omega)$ is bounded$\}$ is not a random event.

1.33. Construct a random process $\{X(t), t \in [0,1]\}$ defined on probability space $\Omega = [0,1], \mathcal{F} = \mathcal{B}([0,1]), \mathsf{P} = \lambda^1|_{[0,1]}$ in such a way that the set $\{\omega \in \Omega |$function $X(\cdot,\omega)$ is measurable$\}$ is not a random event.

1.34. Prove that there exist subsets of the set $\{0,1\}^{\mathbb{N}}$ that do not belong to the σ-algebra generated by cylinder sets (suppose that $\mathcal{X} = 2^{\{0,1\}}$).

1.35. Prove that the class $\mathcal{C}(\mathbb{X}, \mathbb{T})$ of cylinder sets is an algebra; if $\mathbb{T}$ is an infinite set and $\mathbb{X}$ contains at least two points, then the class is not a σ-algebra.

1.36. Let $X = \{X(t), t \in \mathbb{T}\}$ be a random function defined on some probability space $(\Omega, \mathcal{F}, \mathsf{P})$ with phase space $\mathbb{X}$. Prove that for any subset $A \subset \mathbb{X}^{\mathbb{T}}$ that belongs to the cylinder σ-algebra we have: $\{\omega \in \Omega | X(\cdot, \omega) \in A\} \in \mathcal{F}$.

1.37. Let $X = \{X(t), t \in \mathbb{T}\}$ be a random function defined on some probability space $(\Omega, \mathcal{F}, \mathsf{P})$ with phase space $\mathbb{X}$ and let $\{\omega \in \Omega | X(\cdot, \omega) \in A\} \in \mathcal{F}$ for some subset $A \subset \mathbb{X}^{\mathbb{T}}$. Can we assert that A belongs to cylinder σ-algebra? Compare with the previous problem.

Hints

1.2. For any $t_1 < \cdots < t_m$ the random variables $X(t_1), \ldots, X(t_m)$ can take values 0 and 1, only. Moreover, if $X(t_j) = 1$, then $X(t_k) = 1$ for $k = j+1, \ldots, m$. Therefore the joint distribution of $X(t_1), \ldots, X(t_m)$ is concentrated at the points $z_0 = (1, \ldots, 1), z_1 = (0, 1, \ldots, 1), \ldots, z_{m-1} = (0, \ldots, 0, 1), z_m = (0, \ldots, 0)$ (there are $m+1$ such points). The fact that $(X(t_1), \ldots, X(t_m)) = z_j$ $(j = 2, \ldots, m-1)$ means that $X(t_{j-1}) = 0$ and $X(t_j) = 1$, which gives $\tau \in (t_{j-1}, t_j]$.

1.6. (1) $X(t, \cdot) \equiv 0$ if $t \notin A$ and $X(t, \cdot) = \mathbb{I}_{\{t\}}(\cdot)$ if $t \in A$; that is, in both these cases we have a measurable function.
(2) $\{\omega | \, X(t, \omega) = 1\} = \{\omega \le t, \omega \in A\}$.

1.8. $\{F_{-1}(\varepsilon) \le x\} = \{\inf\{y | \, F(y) > \varepsilon\} \le x\} = \bigcap_{n=1}^{\infty} \{\inf\{y | \, F(y) > \varepsilon\} < x + 1/n\} = \bigcap_{n=1}^{\infty} \bigcup_{y \in \mathbb{Q}, y < x+1/n} \{F(y) > \varepsilon\} = \bigcap_{n=1} \{\varepsilon < F((x+1/n)-)\} = \{\varepsilon \le F(x)\}$.

1.9. (a) Let $\varepsilon_k(\omega)$ be equal to the kth digit after the point in the binary notation for the number $\omega \in [0,1]$. Then $\{\varepsilon_k, k \in \mathbb{N}\}$ are i.i.d. random variables, which take on values 0 and 1 with probabilities $\frac{1}{2}$ (prove it!). (b) Sets $\mathbb{N}$ and $\mathbb{N}^2$ are equinumerous, therefore on $[0,1]$ there exists a double sequence $\{\varepsilon_{k,j}, k, j \in \mathbb{N}^2\}$ of i.i.d. random variables that take on values 0 and 1 with probabilities $\frac{1}{2}$. Take $\xi_k = \sum_{j=1}^{\infty} 2^{-j} \varepsilon_{k,j}$.
(c) Use item (b) and Problem 1.8.

1.10. Suppose that the set $A \subset \mathcal{B}(\mathbb{R})$ is such that $\mathsf{P}(\xi_t \in A) \in (0,1)$; then for any $t \neq s$ the distance in $L_2(\Omega, \mathcal{F}, \mathsf{P})$ between $\mathbb{I}_{\xi_t \in A}$ and $\mathbb{I}_{\xi_s \in A}$ is equal to some constant $c_A > 0$; that is, the space $L_2(\Omega, \mathcal{F}, \mathsf{P})$ is not separable. Compare with properties of the space $L_2([0,1])$.

1.11. Let $\varepsilon_1, \varepsilon_2$ be independent variables with uniform distribution on $[0,1]$; $\tilde{\xi}, \tilde{\eta}$ be random variables with joint distribution μ; $\{F(y), y \in \mathbb{R}\}$ be distribution function of $\tilde{\eta}$; and $\{F(x/y), x, y \in \mathbb{R}\}$ be conditional distribution function of $\tilde{\xi}$ under condition that $\{\eta = y\}$ (i.e., $P(\tilde{\xi} \le a, \tilde{\eta} \le b) = \int_{-\infty}^{b} F(a/y)\, dF(y), a, b \in \mathbb{R}$). Denote by $F^{[-1]}$ and $F^{[-1]}(\cdot/y)$ the generalized inverse functions of F, $F(\cdot/y), y \in \mathbb{R}$, take $\xi = F^{[-1]}(\varepsilon_1, \varepsilon_2), \eta = F^{[-1]}(\varepsilon_2)$, and prove that ξ, η have joint distribution μ (see Problem 1.8). After that repeat the same procedure (with the same $\varepsilon_1, \varepsilon_2$) for variables $\tilde{\zeta}, \tilde{\eta}$ such that the joint distribution of $\tilde{\eta}, \tilde{\zeta}$ equals to ν.

1.12. Find a symmetric matrix of the size 3×3, which is not a nonnegatively defined one, but such that all three matrices obtained by obliteration of the ith row and the ith column ($i = 1,2,3$) of this matrix are nonnegatively defined. Does there exist a Gaussian three-dimensional vector with such a covariance matrix?

1.13. (c) Let $N(\omega)$ be the number of partial limits of the sequence $\{X_n(\omega), n \geq 1\}$, then $\{\omega \in \Omega | \ N(\omega) \geq k\} = \bigcup_{\alpha_1,\dots,\alpha_{2k} \in \mathbb{Q}, \alpha_1 < \dots < \alpha_{2k}} \bigcap_{j=1}^{k} \{\omega \in \Omega | \ \{X_n(\omega), n \in \mathbb{N}\} \cap (\alpha_{2j-1}, \alpha_{2j})$ is an infinite set$\}$. Prove this and use it for solving the problem.

1.16. The relation $\limsup_{n \to +\infty} \xi_n / a_n = 1$ a.s. is equivalent to the following. For every $\varepsilon > 0$, $\mathsf{P}(\limsup_n \{\xi_n \geq (1+\varepsilon) a_n\}) = 0$ and $\mathsf{P}(\limsup_n \{\xi_n \geq (1-\varepsilon) a_n\}) = 1$ (prove it). Use the Borel–Cantelli lemma. In the item (c), for estimation of probability of the event $\{\xi_n \geq m\}$ prove that $\sum_{k \geq m} \lambda^k / k! \sim \lambda^m / m!$, $m \to +\infty$ and use Stirling formula.

1.17. This problem can be reduced to Problem 1.13 by using discretization of time; for example,

$$\sup_{t \in \mathbb{R}^+} \frac{X(t)}{a(t)} = \sup_{n \geq 1} \sup_{m \geq 1} \frac{X(\frac{m}{n})}{a(\frac{m}{n})}.$$

1.21. $X(t, \omega) \in \Gamma$ if and only if there exists such an $\varepsilon(\omega) > 0$ that $X(s, \omega) \in \Gamma, s \in [t, t + \varepsilon(\omega))$. Therefore $\{\tau^{\Gamma} < a\} = \bigcap_{b \in \mathbb{Q}, b < a} \{X(b) \in \Gamma\} = \{\tau_{\Gamma} < a\}$.

1.22. Denote by $d(x, \Gamma) = \inf_{y \in \Gamma} \{|x - y|\}$ the distance between x and the set Γ; $d(\cdot, \Gamma)$ is a Borel function (prove it!). It is obvious that $X(t, \omega) \in \Gamma$ if and only if $d(X(s, \omega), \Gamma) \to 0, s \to t+$, therefore $\{\tau^{\Gamma} < a\} = \bigcap_{m \geq 1} \{\inf_{b \in \mathbb{Q}, b < a - 1/m} d(X(b), \Gamma) = 0\} = \{\tau_{\Gamma} < a\}$.

1.23. Let $\Gamma \subset (0, +\infty)$ be a nonmeasurable set, $\Omega = \mathbb{R}^+, \mathcal{F} = \mathcal{B}(\mathbb{R}^+)$, and $X(t, \omega) = \omega 1\!\!\mathrm{I}_{t \geq 1}, t, \omega \in \mathbb{R}^+$ (check that it is a stochastic process). Then the function τ_{Γ} is not measurable.

1.24. Consider the process presented in item (a) of Problem 1.6 with a nonmeasurable set A such that that $0 \notin A, 1 \in A$. Show that the function τ_{Γ} with $\Gamma = \{1\}$ is not measurable.

1.25. Integrability follows from the criterion of Riemann integrability and the properties of the functions, which are right continuous and have left limits (see Problem 3.29). The integral $\int_0^1 X(t)\,dt$ is a random variable because $\int_0^1 X(t, \omega)\,dt = \lim_n n^{-1} \sum_{k=1}^{n} X(k/n, \omega)$ for every $\omega \in \Omega$.

1.26. Under given conditions the trajectory is not necessarily bounded (give an example!).

1.27. If τ takes values in countable set $\{t_i\} \subset \mathbb{T}$, then $X(\tau) = \sum_i X(t_i) 1\!\!\mathrm{I}_{\tau = t_i}$ is a measurable function. In general, $X(\tau) = \lim_n X(]n\tau[/n)$, where $]x[= \min\{m \in \mathbb{N} | \ m \geq x\}$.

1.28. Use the process from item (a) of Problem 1.6, suppose that $\tau(\omega) = \omega$, and select a set A in the proper way.

1.30. Check that for arbitrary set $\mathbb{T}' \subset [0,1]$ with no more then countable complement, the continuous (bounded, measurable) function x may be modified on the set $\mathbb{T}'$ in such a way that it will not be continuous (correspondingly, bounded or measurable) anymore. Use Theorem 1.2.

1.31. Use the process from item (1) of Problem 1.6.

1.32,1.33. Modify the process given in item (1) of Problem 1.6, assuming $X(t,\omega) = \sum_{k\geq 1} \mathbb{I}_{t=\omega/k}\mathbb{I}_{\omega\in A_k}$ and choosing the sets A_k in the proper way.

1.34. Sequences of 0 and 1 that contain an infinite number of zeros and units can be put into one-to-one correspondence with the numbers from the interval $[0,1]$, which are not binary-rational. This correspondence is given by $[0,1] \ni x = \sum_k 2^{-k}\varepsilon_k, \{\varepsilon_k\} \in \{0,1\}^{\mathbb{N}}$ and is measurable together with inverse mapping with respect to $\mathcal{B}([0,1])$ and $\sigma(\mathcal{C}(\{0,1\},\mathbb{N}))$. Now one can use the fact that there exist subsets of $[0,1]$ which are not Borel sets.

1.35. In order to prove that $\mathcal{C}(\mathbb{X},\mathbb{T})$ is an algebra, use the considerations presented in the solution of Problem 1.29. For proving that $\mathcal{C}(\mathbb{X},\mathbb{T})$ is not a σ-algebra, use the same considerations as in the hint of Problem 1.30 with a finite set $\mathbb{T}'$.

1.36. Let $\mathcal{K}$ be a class of such $A \subset \mathbb{X}^{\mathbb{T}}$, that $\{\omega \in \Omega \mid X(\cdot,\omega) \in A\} \in \mathcal{F}$. Then $\mathcal{K} \supset \mathcal{C}(\mathbb{X},\mathbb{T})$ and $\mathcal{K}$ is a σ-algebra. Thus, $\mathcal{K} \supset \sigma(\mathcal{C}(\mathbb{X},\mathbb{T}))$.

Answers and Solutions

1.2. For every $m \geq 1$, $0 \leq t_1 < \cdots < t_m \leq 1$, the m-dimensional distribution $\mathsf{P}_{t_1,\ldots,t_m}$ assigns to the points $(1,\ldots,1)$, $(0,1,\ldots,1)$, $(0,\ldots,0,1)$, $(0,\ldots,0)$ the weights $t_1, (t_2-t_1), \ldots, (t_m-t_{m-1}), (1-t_m)$ correspondingly.

1.4. For every $m \geq 1$, $0 \leq t_1 < \cdots < t_m \leq 1$, the m-dimensional distribution $\mathsf{P}_{t_1,\ldots,t_m}$ is concentrated at the points $(l_1/n,\ldots,l_m/m)$, where $l_1 \leq l_2 \leq \cdots \leq l_m$ and $l_1,\ldots l_m \in \{0,\ldots,n\}$ (we suppose $0 \leq t_1 < \cdots < t_m \leq 1$). The weight of the point $(l_1/n,\ldots,l_m/m)$ equals

$$\frac{n![F(t_1)]^{l_1}[F(t_2)-F(t_1)]^{l_2}\ldots[F(t_m)-F(t_{m-1})]^{l_m}[1-F(t_m)]^{n-l_1-\cdots-l_m}}{l_1!l_2!\ldots l_m!(n-l_1-\cdots-l_m)!}.$$

1.5. (a) Yes; (b) yes.

1.12. Not for any.

1.14. Variables given in items (b), (c).

1.15. Variables given in items (b), (c).

1.16. (a) $a_n = \sigma\sqrt{2\ln n}$. (b) $a_n = \lambda^{-1}\ln n$. (c) $a_n = \phi^{-1}(\ln n)$, where $\phi(t) = t\ln t$, $t \in [1,+\infty)$.

1.18. No.

1.23. No.

1.26. No.

1.29. Use the "principle of the fitting sets". That is, consider the class $\mathcal{K}$ of the sets $A \in \sigma(\mathcal{C}(\mathbb{X},\mathbb{T}))$, for which there exist such $(t_n)_{n=1}^{\infty}$ and $B \in \sigma(\mathcal{C}(\mathbb{X},\mathbb{N}))$ that $A = \{\omega|\ (\omega(t_n))_{n=1}^{\infty} \in B\}$. If we prove that $\mathcal{K}$ is a σ-algebra which contains a class that generates $\sigma(\mathcal{C}(\mathbb{X},\mathbb{T}))$, then we prove that $\mathcal{K} = \sigma(\mathcal{C}(\mathbb{X},\mathbb{T}))$. It is obvious that $\mathcal{K} \supset \mathcal{C}(\mathbb{X},\mathbb{T})$, thus the second requirement is met. Next, $\varnothing, \mathbb{X}^{\mathbb{T}} \in \mathcal{K}$ because for representation of these sets it is sufficient to choose an arbitrary sequence $(t_n)_{n=1}^{\infty}$ and put $B = \varnothing$ or $B = \mathbb{X}^{\mathbb{N}}$, correspondingly. Let $A = \{\omega|\ (\omega(t_n))_{n=1}^{\infty} \in B\}$, then $\mathbb{X}^{\mathbb{T}}\backslash A = \{\omega|\ (\omega(t_n))_{n=1}^{\infty} \in \mathbb{X}^{\mathbb{N}}\backslash B\}$. Finally, if $A = \{\omega|\ (\omega(t_n))_{n=1}^{\infty} \in B\}$, and $(s_j, j \in J)$ is some at most countable family of points from the parameter set, then A may be presented in the form $A = \{\omega|\ ((\omega(t_n))_{n=1}^{\infty}, (\omega(s_j))_{j\in J}) \in \tilde{B}\}$ with $\tilde{B} = B \times \mathbb{X}^J$. Thus, if $\{A^k = \{\omega|\ (\omega(t_n^k))_{n=1}^{\infty} \in B^k\}, k \geq 1\}$ is a countable collection of sets from the class $\mathcal{K}$, then assuming $\{s_n, n \geq 1\} = \bigcup_k \{t_n^k, n \geq 1\}$ and writing the sets A^k in the form $A^k = \{\omega|\ (\omega(s_n))_{n=1}^{\infty} \in \tilde{B}^k\}, k \geq 1$, we obtain $\bigcup_k A^k = \{\omega|\ (\omega(s_n))_{n=1}^{\infty} \in \bigcup_k \tilde{B}^k\} \in \mathcal{K}$.

1.37. No, we can not. For example, a trajectory of the waiting process is bounded for every ω (Problem 1.2), but the set of bounded functions does not belong to $\sigma(\mathcal{C}(\mathbb{R},[0,1]))$ (Problem 1.30).

2

Characteristics of a stochastic process. Mean and covariance functions. Characteristic functions

Theoretical grounds

In this chapter we consider random functions with the phase space being either real line $\mathbb{R}$ or complex plane $\mathbb{C}$.

Definition 2.1. *Assume that* $\mathsf{E}|X(t)| < +\infty$, $t \in \mathbb{T}$. *Function* $\{a_X(t) = \mathsf{E}X(t), t \in \mathbb{T}\}$ *is called* the mean function *(or simply* the mean*) of the random function* X. *Function* $\tilde{X}(t) = X(t) - a_X(t), t \in \mathbb{T}$ *is called the* centered *(or* compensated*) function, corresponding to function* X.

Recall that *covariance* of two real-valued random variables ξ and η, both having the second moment, is defined as $\operatorname{cov}(\xi,\eta) = \mathsf{E}(\xi - \mathsf{E}\xi)(\eta - \mathsf{E}\eta) = \mathsf{E}\xi\eta - \mathsf{E}\xi\mathsf{E}\eta$. If ξ,η are complex-valued and $\mathsf{E}|\xi|^2 < +\infty, \mathsf{E}|\eta|^2 < +\infty$ then $\operatorname{cov}(\xi,\eta) = \mathsf{E}(\xi - \mathsf{E}\xi)\overline{(\eta - \mathsf{E}\eta)} = \mathsf{E}\xi\overline{\eta} - \mathsf{E}\xi\mathsf{E}\overline{\eta}$ (here "—", the overbar, is a sign of complex conjugation).

Definition 2.2. *Assume that* $\mathsf{E}|X(t)|^2 < +\infty$, $t \in \mathbb{T}$. *Function*

$$R_X(t,s) = \operatorname{cov}(X(t),X(s)), \quad t,s \in \mathbb{T}$$

is called the covariance function *(or simply* the covariance*) of the random function* X. *If* X,Y *are two functions with* $\mathsf{E}|X(t)|^2 < +\infty, \mathsf{E}|Y(t)|^2 < +\infty, t \in \mathbb{T}$, *then* $\{R_{X,Y}(t,s) = \operatorname{cov}(X(t),Y(s)), t,s \in \mathbb{T}\}$ *is called* the mutual covariance function *for the functions* X,Y.

Definition 2.3. *Let* $\mathbb{T}$ *be some set, function* K *be defined on* $\mathbb{T} \times \mathbb{T}$, *and take values in* $\mathbb{C}$. *Function* K *is* nonnegatively defined *if*

$$\sum_{j,k=1}^{m} K(t_j,t_k)c_j\overline{c}_k \geq 0$$

for any $m \in \mathbb{N}$ *and any* $t_1,\dots,t_m \in \mathbb{T}, c_1,\dots,c_m \in \mathbb{C}$.

D. Gusak et al., *Theory of Stochastic Processes*, Problem Books in Mathematics, DOI 10.1007/978-0-387-87862-1_2,

This definition is equivalent to the following one.

Definition 2.4. *Function $K : \mathbb{T} \times \mathbb{T} \to \mathbb{C}$ is* nonnegatively defined *if for any $m \in \mathbb{N}$ and any $t_1, \dots, t_m \in \mathbb{T}$ the matrix $K_{t_1 \dots t_m} = \{K(t_j, t_k)\}_{j,k=1}^m$ is nonnegatively defined.*

Proposition 2.1. *Covariance R_X of an arbitrary stochastic process X is nonnegatively defined. And vice versa, if $a : \mathbb{T} \to \mathbb{C}$ and $K : \mathbb{T} \times \mathbb{T} \to \mathbb{C}$ are some functions and K is nonnegatively defined, then on some probability space there exists random function X such that $a = a_X, K = R_X$.*

Remark 2.1. Recall that the mean vector and covariance matrix for a random vector $\xi = (\xi_1, \dots, \xi_m)$ are $a_\xi = (\mathsf{E}\xi_j)_{j=1}^m$ and $R_\xi = (\mathrm{cov}(\xi_j, \xi_k))_{j,k=1}^m$, respectively. If the conditions of Proposition 2.1 hold, then for any $m \in \mathbb{N}, t_1, \dots, t_m \in \mathbb{T}$ the covariance matrix for the vector $(X(t_1), \dots, X(t_m))$ is equal to $K_{t_1 \dots t_m}$ (see Definition 2.4) and the mean vector is equal to $a_{t_1 \dots t_m} = (a(t_j))_{j=1}^m$.

Recall that for a random vector $\xi = (\xi_1, \dots, \xi_m)$ with real-valued components, its *characteristic function* (or equivalently, *common characteristic function* of the random variables $\xi_1, \dots, \xi_m$) is defined by

$$\phi_\xi(z) = \mathsf{E}e^{i(\xi, z)_{\mathbb{R}^m}} = \mathsf{E}e^{i\sum_{j=1}^m \xi_j z_j}, \quad z = (z_1, \dots, z_m) \in \mathbb{R}^m.$$

Theorem 2.1. *(The Bochner theorem) An arbitrary function $\phi : \mathbb{R}^m \to \mathbb{C}$ is a characteristic function of some random vector if and only if the following three conditions are satisfied.*
(1) $\phi(0) = 1$.
(2) ϕ is continuous in the neighborhood of 0.
(3) For any $m \in \mathbb{N}$ and $z_1, \dots, z_m \in \mathbb{R}, c_1, \dots, c_m \in \mathbb{C}$

$$\sum_{j,k=1}^m \phi(z_j - z_k) c_j \bar{c}_k \ge 0.$$

Definition 2.5. *Let X be a real-valued random function. For a fixed $m \ge 1$ and $t_1, \dots, t_m \in \mathbb{T}$, the common characteristic function of $X(t_1), \dots, X(t_m)$ is denoted by $\phi^X_{t_1, \dots, t_m}$ and is called the* (m-dimensional) characteristic function *of the random function X. The set $\{\phi^X_{t_1, \dots, t_m}, t_1, \dots, t_m \in \mathbb{T}, m \ge 1\}$ is called* the set *(or* the family*) of finite-dimensional characteristic functions of the random function X.*

Mean and covariance functions of a random function do not determine the finite-dimensional distributions of this function uniquely (e.g., see Problem 6.7). On the other hand, the family of finite-dimensional characteristic functions of the random function X has unique correspondence to its finite-dimensional characteristics because the characteristic function of a random vector determines the distribution of this vector uniquely. The following theorem is the reformulation of the Kolmogorov theorem (Theorem 1.1) in terms of characteristic functions.

Theorem 2.2. *Consider a family* $\{\phi_{t_1,\ldots,t_m} : \mathbb{R}^m \to \mathbb{C}, t_1,\ldots,t_m \in \mathbb{T}, m \geq 1\}$ *such that for any* $m \geq 1, t_1,\ldots,t_m \in \mathbb{T}$ *the function* $\phi_{t_1,\ldots,t_m}$ *satisfies the conditions of the Bochner theorem. The following* consistency conditions *are necessary and sufficient for such a random function X to exist that the family* $\{\phi_{t_1,\ldots,t_m} : \mathbb{R}^m \to \mathbb{C}, t_1,\ldots,t_m \in \mathbb{T}, m \geq 1\}$ *is the family of its finite-dimensional characteristic functions.*

(1) For any $m \geq 1, t_1,\ldots,t_m \in \mathbb{T}, z_1,\ldots,z_m \in \mathbb{R}$ *and any permutation* $\pi : \{1,\ldots,m\} \to \{1,\ldots,m\}$,

$$\phi_{t_1,\ldots,t_m}(z_1,\ldots,z_m) = \phi_{t_{\pi(1)},\ldots,t_{\pi(m)}}\left(z_{\pi(1)},\ldots,z_{\pi(m)}\right).$$

(2) For any $m > 1, t_1,\ldots,t_m \in \mathbb{T}, z_1,\ldots,z_{m-1} \in \mathbb{R}$,

$$\phi_{t_1,\ldots,t_m}(z_1,\ldots,z_{m-1},0) = \phi_{t_1,\ldots,t_{m-1}}(z_1,\ldots,z_{m-1}).$$

Bibliography

[9], Chapter II; [24], Volume 1, Chapter IV, §1; [25], Chapter I, §1; [79], Chapter 16.

Problems

2.1. Find the covariance function for (a) the Wiener process; (b) the Poisson process.

2.2. Let W be the Wiener process. Find the mean and covariance functions for the process $X(t) = W^2(t), t \geq 0$.

2.3. Let W be the Wiener process. Find the covariance function for the process X if
(a) $X(t) = W(1/t)$, $t > 0$.
(b) $X(t) = W(e^t)$, $t \in \mathbb{R}$.
(c) $X(t) = W\left(1 - t^2\right)$, $t \in [-1,1]$.

2.4. Let W be the Wiener process. Find the characteristic function for $W(2) + 2W(1)$.

2.5. Let N be the Poisson process with intensity λ. Find the characteristic function for $N(2) + 2N(1)$.

2.6. Let W be the Wiener process. Find:
(a) $\mathsf{E}(W(t))^m, m \in \mathbb{N}$.
(b) $\mathsf{E}\exp(2W(1) + W(2))$.
(c) $\mathsf{E}\cos(2W(1) + W(2))$.

2.7. Let N be the Poisson process with intensity λ. Find:
(a) $\mathsf{P}(N(1) = 2, N(2) = 3, N(3) = 5)$.
(b) $\mathsf{P}(N(1) \leq 2, N(2) = 3, N(3) \geq 5)$.
(c) $\mathsf{E}(N(t) + 1)^{-1}$.
(d) $\mathsf{E}N(t)(N(t) - 1) \cdot \cdots \cdot (N(t) - k), k \in \mathbb{Z}^+$.

2.8. Let W be the Wiener process and $f \in C([0,1])$. Find the characteristic function for random variable $\int_0^1 f(s)W(s)\,ds$ (the integral is defined for every ω in the Riemann sense; see Problem 1.25). Prove that this random variable is normally distributed.

2.9. Let W be the Wiener process, $f \in C([0,1])$, $X(t) = \int_0^t f(s)W(s)\,ds$, $t \in [0,1]$. Find $R_{W,X}$.

2.10. Let N be the Poisson process, $f \in C([0,1])$. Find the characteristic functions of random variables: (a) $\int_0^1 f(s)N(s)\,ds$; (b) $\int_0^1 f(s)dN(s) \equiv \sum f(s)$, where summation is taken over all $s \in [0,1]$ such that $N(s) \neq N(s-)$.

2.11. Let N be the Poisson process, $f,g \in C([0,1])$, $X(t) = \int_0^t f(s)N(s)\,ds$, $Y(t) = \int_0^t g(s)dN(s)$, $t \in [0,1]$. Find: (a) $R_{N,X}$; (b) $R_{N,Y}$; (c) $R_{X,Y}$.

2.12. Find all one-dimensional and m-dimensional characteristic functions: (a) for the process introduced in Problem 1.2; (b) for the process introduced in Problem 1.4.

2.13. Find the covariance function of the process $X(t) = \xi_1 f_1(t) + \cdots + \xi_n f_n(t)$, $t \in \mathbb{R}$, where $f_1, \dots, f_n$ are nonrandom functions, and $\xi_1, \dots, \xi_n$ are noncorrelated random variables with variances $\sigma_1^2, \dots, \sigma_n^2$.

2.14. Let $\{\xi_n, n \geq 1\}$ be the sequence of independent square integrable random variables. Denote $a_n = \mathsf{E}\xi_n, \sigma_n^2 = \operatorname{Var}\xi_n$.
(1) Prove that series $\sum_n \xi_n$ converges in the mean square sense if and only if the series $\sum_n a_n$ and $\sum_n \sigma_n^2$ are convergent.
(2) Let $\{f_n(t),\ t \in \mathbb{R}\}_{n\in\mathbb{N}}$ be the sequence of nonrandom functions. Formulate the necessary and sufficient conditions for the series $X(t) = \sum_n \xi_n f_n(t)$ to converge in the mean square for every $t \in \mathbb{R}$. Find the mean and covariance functions of the process X.

2.15. Are the following functions nonnegatively defined: (a) $K(t,s) = \sin t \sin s$; (b) $K(t,s) = \sin(t+s)$; (c) $K(t,s) = t^2 + s^2$ $(t,s \in \mathbb{R})$?

2.16. Prove that for $\alpha > 2$ the function $K(t,s) = \frac{1}{2}\left(t^\alpha + s^\alpha - |t-s|^\alpha\right)$, $t,s \in \mathbb{R}^m$ is not a covariance function.

2.17. (1) Let $\{X(t),\ t \in \mathbb{R}^+\}$ be a stochastic process with independent increments and $\mathsf{E}|X(t)|^2 < +\infty, t \in \mathbb{R}^+$. Prove that its covariance function is equal to $R_X(t,s) = F(t \wedge s)$, $t,s \in \mathbb{R}^+$, where F is some nondecreasing function.
(2) Let $\{X(t), t \in \mathbb{R}^+\}$ be a stochastic process with $R_X(t,s) = F(t \wedge s), t,s \in \mathbb{R}^+$, where F is some nondecreasing function. Does it imply that X is a process with independent increments?

2.18. Let N be the Poisson process with intensity λ. Let $X(t) = 0$ when $N(t)$ is odd and $X(t) = 1$ when $N(t)$ is even.
(1) Find the mean and covariance of the process X.
(2) Find $R_{N,X}$.

2.19. Let W and N be the independent Wiener process and Poisson process with intensity λ, respectively. Find the mean and covariance of the process $X(t) = W(N(t))$. Is X a process with independent increments?

2.20. Find $R_{X,W}$ and $R_{X,N}$ for the process from the previous problem.

2.21. Let N_1, N_2 be two independent Poisson processes with intensities λ_1, λ_2, respectively. Define $X(t) = (N_1(t))^{N_2(t)}$, $t \in \mathbb{R}^+$ if at least one of the values $N_1(t)$, $N_2(t)$ is nonzero and $X(t) = 1$ if $N_1(t) = N_2(t) = 0$. Find:
(a) The mean function of the process X
(b) The covariance function of the process X

2.22. Let X, Y be two independent and centered processes and $c > 0$ be a constant. Prove that $R_{X+Y} = R_X + R_Y$, $R_{\sqrt{c}X} = cR_X$, $R_{XY} = R_X R_Y$.

2.23. Let K_1, K_2 be two nonnegatively defined functions and $c > 0$. Prove that the following functions are nonnegatively defined: (a) $R = K_1 + K_2$; (b) $R = cK_1$; (c) $R = K_1 \cdot K_2$.

2.24. Let K be a nonnegatively defined function on $\mathbb{T} \times \mathbb{T}$.
(1) Prove that for every polynomial $P(\cdot)$ with nonnegative coefficients the function $R = P(K)$ is nonnegatively defined.
(2) Prove that the function $R = e^K$ is nonnegatively defined.
(3) When it is additionally assumed that for some $p \in (0,1)$ $K(t,t) < p^{-1}$, $t \in \mathbb{T}$, prove that the function $R = (1 - pK)^{-1}$ is nonnegatively defined.

2.25. Give the probabilistic interpretation of items (1)–(3) of the previous problem; that is, construct the stochastic process for which R is the covariance function.

2.26. Let $K(t,s) = ts, t, s \in \mathbb{R}^+$. Prove that for an arbitrary polynomial P the function $R = P(K)$ is nonnegatively defined if and only if all coefficients of the polynomial P are nonnegative. Compare with item (1) of Problem 2.24.

2.27. Which of the following functions are nonnegatively defined: (a) $K(t,s) = \sin(t-s)$; (b) $K(t,s) = \cos(t-s)$; (c) $K(t,s) = e^{-(t-s)}$; (d) $K(t,s) = e^{-|t-s|}$; (e) $K(t,s) = e^{-(t-s)^2}$; (f) $K(t,s) = e^{-(t-s)^4}$?

2.28. Let $K \in C([a,b] \times [a,b])$. Prove that K is nonnegatively defined if and only if the integral operator $A_K : L_2([a,b]) \to L_2([a,b])$, defined by

$$A_K f(t) = \int_a^b K(t,s) f(s)\, ds, \quad f \in L_2([a,b]),$$

is nonnegative.

2.29. Let A_K be the operator from the previous problem. Check the following statements.
(a) The set of eigenvalues of the operator A_K is at most countable.
(b) The function K is nonnegatively defined if and only if every eigenvalue of the operator A_K is nonnegative.

2.30. Let $K(s,t) = F(t-s)$, $t,s \in \mathbb{R}$, where the function F is periodic with period 2π and $F(x) = \pi - |x|$ for $|x| \le \pi$. Construct the Gaussian process with covariance K of the form $\sum_n \varepsilon_n f_n(t)$, where $\{\varepsilon_n, n \ge 1\}$ is a sequence of the independent normally distributed random variables.

2.31. Solve the previous problem assuming that F has period 2 and $F(x) = (1-x)^2$, $x \in [0,1]$.

2.32. Denote $\{\tau_n, n \ge 1\}$ the jump moments for the Poisson process $N(t), \tau_0 = 0$. Let $\{\varepsilon_n, n \ge 0\}$ be i.i.d. random variables that have expectation a and variance σ^2. Consider the stochastic processes $X(t) = \sum_{k=0}^{n} \varepsilon_k$, $t \in [\tau_n, \tau_{n+1})$, $Y(t) = \varepsilon_n$, $t \in [\tau_n, \tau_{n+1})$, $n \ge 0$. Find the mean and covariance functions of the processes X, Y. Exemplify the models that lead to such processes.

2.33. A radiation measuring instrument accumulates radiation with the rate that equals a Roentgen per hour, right up to the failing moment. Let $X(t)$ be the reading at point of time $t \ge 0$. Find the mean and covariance functions for the process X if $X(0) = 0$, the failing moment has distribution function F, and after the failure the measuring instrument is fixed (a) at zero point; (b) at the last reading.

2.34. The device registers a Poisson flow of particles with intensity $\lambda > 0$. Energies of different particles are independent random variables. Expectation of every particle's energy is equal to a and variance is equal to σ^2. Let $X(t)$ be the readings of the device at point of time $t \ge 0$. Find the mean and covariance functions of the process X if the device shows
(a) Total energy of the particles have arrived during the time interval $[0,t]$.
(b) The energy of the last particle.
(c) The sum of the energies of the last K particles.

2.35. A Poisson flow of claims with intensity $\lambda > 0$ is observed. Let $X(t), t \in \mathbb{R}$ be the time between t and the moment of the last claim coming before t. Find the mean and covariance functions for the process X.

Hints

2.1. See the hint to Problem 2.17.

2.4. Because the variables $(W(1), W(2))$ are jointly Gaussian, the variable $W(2) + 2W(1)$ is normally distributed. Calculate its mean and variance and use the formula for the characteristic function of the Gaussian distribution. Another method is proposed in the following hint.

2.5. $N(2) + 2N(1) = N(2) - N(1) + 3N(1)$. The values $N(2) - N(1)$ and $N(1)$ are Poisson-distributed random variables and thus their characteristic functions are known. These values are independent, that is, the required function can be obtained as a product.

2.6. (a) If $\eta \sim \mathcal{N}(0,1)$, then $\mathsf{E}\eta^{2k-1} = 0, \mathsf{E}\eta^{2k} = (2k-1)!! = (2k-1)(2k-3)\cdots 1$ for $k \in \mathbb{N}$. Prove and use this for the calculations.
(b) Use the explicit formula for the Gaussian density.
(c) Use formula $\cos x = \frac{1}{2}\left(e^{ix} + e^{-ix}\right)$ and Problem 2.4.

2.10. (a) Make calculations similar to those of Problem 2.8.
(b) Obtain the characteristic functions of the integrals of piecewise constant functions f and then uniformly approximate the continuous function by piecewise constant ones.

2.17. (1) Let $s \le t$; then values $X(t) - X(s)$ and $X(s)$ are independent which means that they are uncorrelated. Therefore $\operatorname{cov}(X(t), X(s)) = \operatorname{cov}(X(t) - X(s), X(s)) + \operatorname{cov}(X(s), X(s)) = \operatorname{cov}(X(t \wedge s), X(t \wedge s))$. The case $t \le s$ can be treated similarly.

2.23. Items (a) and (b) can be proved using the definition. In item (c) you can use the previous problem.

2.24. Proof of item (1) can be directly obtained from the previous problem. For the proof of items (2) and (3) use item (1), Taylor decomposition of the functions $x \mapsto e^x, x \mapsto (1-px)^{-1}$ and a fact that the pointwise limit of a sequence of nonnegatively defined functions is also a nonnegatively defined function. (Prove this fact!).

Answers and Solutions

2.1. $R_W(t,s) = t \wedge s, R_N(t,s) = \lambda(t \wedge s)$.

2.2. $a_X(t) = t$, $R_X(t,s) = 2(t \wedge s)^2$.

2.3. For arbitrary $f : \mathbb{R}^+ \to \mathbb{R}^+$, the covariance function for the process $X(t) = W(f(t)), t \in \mathbb{R}^+$ is equal to $R_X(t,s) = R_W(f(t), f(s)) = f(t) \wedge f(s)$.

2.8. Let $I_n = n^{-1}\sum_{k=1}^n f(k/n)W(k/n)$. Because the process W a.s. has continuous trajectories and the function f is continuous, the Riemann integral sum I_n converges to $I = \int_0^1 f(t)W(t)\,dt$ a.s. Therefore $\phi_{I_n}(z) \to \phi_I(z), n \to +\infty, z \in \mathbb{R}$. Hence,

$$\mathsf{E}e^{izI_n} = \mathsf{E}e^{izn^{-1}\sum_{k=1}^n f(k/n)W(k/n)} = \mathsf{E}e^{i\sum_{k=1}^n \left[zn^{-1}\sum_{j=k}^n f(j/n)\right](W(k/n)-W((k-1)/n))}$$
$$= \prod_{k=1}^n e^{-(2n)^{-1}\left[zn^{-1}\sum_{j=k}^n f(j/n)\right]^2} \to e^{-(z^2/2)\int_0^1\left(\int_t^1 f(s)\,ds\right)^2 dt}, \quad n \to \infty.$$

Thus I is a Gaussian random variable with zero mean and variance $\int_0^1\left(\int_t^1 f(s)\,ds\right)^2 dt$.

2.9. $R_{W,X}(t,s) = \int_0^s f(r)(t \wedge r)\,dr$.

2.10. (a) $\phi(z) = \exp\left(\lambda \int_0^1 \left[e^{iz\int_t^1 f(s)\,ds} - 1\right] dt\right)$.
(b) $\phi(z) = \exp\left(\lambda \int_0^1 \left[e^{izf(t)} - 1\right] dt\right)$.

2.11. $R_{N,X}(t,s)=\lambda^2\int_0^s f(r)(t\wedge r)\,dr$, $R_{N,Y}(t,s)=\lambda^2\int_0^{t\wedge s} g(r)\,dr$, $R_{X,Y}(t,s)=\lambda^2\times\int_0^t f(u)\left[\int_0^{u\wedge s} g(r)\,dr\right]du$.

2.12. (a) Let $0\le t_1<\cdots<t_n\le 1$; then $\phi_{t_1,\dots,t_m}(z_1,\dots,z_m)=t_1e^{iz_1+\dots+iz_m}+(t_2-t_1)e^{iz_2+\dots+iz_m}+\dots+(t_m-t_{m-1})e^{iz_m}+(1-t_m)$.
(b) Let $0\le t_1<\cdots<t_n\le 1$, then

$$\phi_{t_1,\dots,t_m}(z_1,\dots,z_m)=\Big[F(t_1)e^{iz_1n^{-1}+\dots+iz_mn^{-1}}+(F(t_2)-F(t_1))e^{iz_2n^{-1}+\dots+iz_mn^{-1}}+\cdots$$
$$+(F(t_m)-F(t_{m-1}))e^{iz_mn^{-1}}+(1-F(t_m))\Big]^n.$$

2.13. $R_X(t,s)=\sum_{k=1}^n\sigma_k^2f_k(t)f_k(s)$.

2.15. (a) Yes; (b) no; (c) no.

2.17. (2) No, it does not.

2.18. (1) $a_X(t)=\frac{1}{2}\left(1+e^{-2\lambda t}\right), R_X(t,s)=\frac{1}{4}\left(e^{-2\lambda|t-s|}-e^{-2\lambda(t+s)}\right)$.
(2) $R_{N,X}(t,s)=-\lambda(t\wedge s)e^{-2\lambda s}$.

2.19. $a_X\equiv 0, R_X(t,s)=\lambda(t\wedge s)$. X is the process with independent increments.

2.20.

$$R_{X,W}(t,s)=\mathsf{E}[N(t)\wedge s]=e^{-\lambda t}\left[\sum_{k<s}\frac{k(\lambda t)^k}{k!}+s\cdot\sum_{k\ge s}\frac{(\lambda t)^k}{k!}\right],\quad R_{X,N}\equiv 0.$$

2.21. $a_X(t)=\exp\left[\lambda_1te^{\lambda_2t}-(\lambda_1+\lambda_2)t\right]$; function R_X is not defined because $\mathsf{E}X^2(t)=+\infty, t>0$.

2.25. There exist several interpretations, let us give two of them.

The first one: let $R=f(K)$ and $f(x)=\sum_{m=0}^\infty c_mx^m$ with $c_m\ge 0, m\in\mathbb{Z}^+$. Let the radius of convergence of the series be equal to $r_f>0$ and $K(t,t)<r_f, t\in\mathbb{R}^+$. Consider a triangular array $\{X_{m,k}, 1\le k\le m\}$ of independent centered identically distributed processes with the covariance function K. In addition, let random variable ξ be independent of $\{X_{m,k}\}$ and $\mathsf{E}\xi=0, \mathsf{D}\xi=1$. Then the series $X(t)=\sqrt{c_0}\xi+\sum_{m=1}^\infty\sqrt{c_m}\prod_{k=1}^m X_{m,k}(t)$ converges in the mean square for any t and the covariance function of the process X is equal to R.

The second one: using the same notations, denote $c=\sum_{k=0}^\infty c_k, p_k=c_k/c$, $k\ge 0$. Let $\{X_m, m\ge 1\}$ be a sequence of independent identically distributed centered processes with the covariance function K, and ξ be as above. Let η be the random variable, independent both on ξ and the processes $\{X_m, m\ge 1\}$, with $P(\eta=k)=p_k, k\in\mathbb{Z}^+$. Consider the process $X(t)=\sqrt{c}\prod_{k=1}^\eta X_k(t)$ assuming that $\prod_{k=1}^0 X_k(t)=\xi$. Then the covariance function of the process X is equal to R. In particular, the random variable η should have a Poisson distribution in item (2) and a geometric distribution in item (3).

2.26. Consider the functions $R_k = (\partial^{2k}/\partial t^k \partial s^k)R, k \geq 0$. These functions are nonnegatively defined (one can obtain this fact by using either Definition 2.3 or Theorem 4.2). Function R_k can be represented in the form $R_k = P_k(K)$, where the absolute term of the polynomial P_k equals the kth coefficient of the polynomial P multiplied by $(k!)^2$. Now, the required statement follows from the fact that $Q(t,t) \geq 0$ for any nonnegatively defined function Q.

2.27. Functions from the items (b), (d), (e) are nonnegatively defined; the others are not.

2.28. Let K be nonnegatively defined. Then for any $f \in C([a,b])$,

$$(A_K f, f)_{L_2([a,b])} = \int_a^b \int_a^b K(t,s) f(t) f(s)\, ds dt$$
$$= \lim_{n\to\infty} \sum_{j,k=1}^{n} \left(\frac{b-a}{n}\right)^2 K\left(a + \frac{j(b-a)}{n}, a + \frac{k(b-a)}{n}\right) \geq 0$$

because every sum under the limit sign is nonnegative. Because $C([a,b])$ is a dense subset in $L_2([a,b])$ the above inequality yields that $(A_K f, f)_{L_2([a,b])} \geq 0$, $f \in L_2([a,b])$. On the other hand, let $(A_K f, f)_{L_2([a,b])} \geq 0$ for every $f \in L_2([a,b])$, and let points $t_1, \dots, t_m$ and constants $z_1, \dots, z_m$ be fixed. Choose m sequences of continuous functions $\{f_n^1, n \geq 1\}, \dots, \{f_n^m, n \geq 1\}$ such that, for arbitrary function $\phi \in C([a,b])$, $\int_a^b \phi(t) f_n^j(t)\, dt \to \phi(t_j), n \to \infty, j = 1, \dots, m$. Putting $f_n = \sum_{j=1}^m z_j f_n^j$, we obtain that $\sum_{j,k=1}^m z_j z_k K(t_j, t_k) = \lim_{n\to\infty} \int_a^b \int_a^b K(t,s) f_n(t) f_n(s)\, ds dt = \lim_{n\to\infty} (A_K f_n, f_n) \geq 0$.

2.29. Statement (a) is a particular case of the theorem on the spectrum of a compact operator. Statement (b) follows from the previous problem and theorem on spectral decomposition of a compact self-adjoint operator.

3

Trajectories. Modifications. Filtrations

Theoretical grounds

Definition 3.1. *Random functions* $\{X(t), t \in \mathbb{T}\}$ *and* $\{Y(t), t \in \mathbb{T}\}$*, defined on the same probability space, are called* equivalent *(or* stochastically equivalent*), if* $\mathsf{P}(X(t) = Y(t)) = 1$ *for any* $t \in \mathbb{T}$*. Random functions* $\{X(t), t \in \mathbb{T}\}$ *and* $\{Y(t), t \in \mathbb{T}\}$*, possibly defined on different probability spaces, are called* stochastically equivalent in a wide sense *if their corresponding finite-dimensional distributions coincide.*

A random function Y equivalent to X is called a *modification* of the random function X.

Definition 3.2. *Let* $\mathbb{T}$ *be a linearly ordered set. A* filtration *(or a* flow of σ-algebras*) on a probability space* $(\Omega, \mathcal{F}, P)$ *is a family of* σ*-algebras* $\mathbb{F} = \{\mathcal{F}_t, t \in \mathbb{T}\}$ *that satisfies the condition* $\mathcal{F}_s \subset \mathcal{F}_t \subset \mathcal{F}, s, t \in \mathbb{T}, s \le t$*. Filtration is called* complete *if every* σ*-algebra* $\mathcal{F}_t$ *includes all null probability sets from* $\mathcal{F}$.

Denote $\mathcal{F}_{t+} = \bigcap_{s>t} \mathcal{F}_s, \mathcal{F}_{t-} = \bigvee_{s<t} \mathcal{F}_s$ (recall that $\bigvee_{\alpha \in \mathcal{A}} \mathcal{G}_\alpha$ denotes the least σ-algebra that contains every σ-algebra $\mathcal{G}_\alpha, \alpha \in \mathcal{A}$).

Definition 3.3. *Filtration* $\mathbb{F} = \{\mathcal{F}_t, t \in \mathbb{T}\}$ *is* left-hand continuous *if* $\mathcal{F}_{t-} = \mathcal{F}_t, t \in \mathbb{T}$ *and* right-hand continuous *if* $\mathcal{F}_{t+} = \mathcal{F}_t, t \in \mathbb{T}$*. Filtration that is both left-hand and right-hand continuous is* continuous.

An important class of filtrations contains filtrations generated by stochastic processes (or, more generally, by random functions). Let $\{X(t), t \in \mathbb{T}\}$ be a random function and a set $\mathbb{T}$ be linearly ordered. For every $t \in \mathbb{T}$ denote $\mathcal{F}_t^{X,0} = \sigma(X(s), s \le t)$ and define σ-algebra $\mathcal{F}_t^X$ as the *augmentation* of $\mathcal{F}_t^{X,0}$ w.r.t. the measure P; that is, $\mathcal{F}_t^X = \sigma(\mathcal{F}_t^{X,0} \cup \mathcal{N}_{\mathsf{P}})$ with $\mathcal{N}_{\mathsf{P}} = \{A \in \mathcal{F} \mid \mathsf{P}(A) = 0\}$.

Definition 3.4. *Filtration* $\mathbb{F}^X = \{\mathcal{F}_t^X, t \in \mathbb{T}\}$ *is called* the filtration generated by the random function X, *or* the natural filtration for the function X.

D. Gusak et al., *Theory of Stochastic Processes*, Problem Books in Mathematics,
DOI 10.1007/978-0-387-87862-1_3,

Sometimes, in the definition of the natural filtration, the augmentation operation is not involved, and one calls *natural filtration* the family $\mathbb{F}^{X,0} = \{\mathcal{F}_t^{X,0}, t \in \mathbb{T}\}$. The Problems 3.37, 3.44 explain the role of the augmentation operation. We remark that the augmentation of a σ-algebra $\mathcal{G}$ differs slightly from the *completion* of $\mathcal{G}$ (the latter is defined as the least σ-algebra that contains all sets from $\mathcal{G}$ and all subsets of the sets from $\mathcal{N}_{\mathsf{P}}$), but if the probability P is complete then these two σ-algebras coincide.

Definition 3.5. *Let the parametric set* $\mathbb{T}$ *be endowed by a* σ*-algebra* $\mathcal{T}$. *A random function* $\{X(t), t \in \mathbb{T}\}$ *is called* measurable *if* X *is measurable as a function* $\mathbb{T} \times \Omega \to \mathbb{X}$*; that is,*

$$\{(t,\omega) | X(t,\omega) \in B\} \in \mathcal{T} \otimes \mathcal{F}, \quad B \in \mathcal{X}.$$

Further on we assume $\mathbb{T}, \mathbb{X}$ to be metric spaces with the metrics d and ρ, respectively, and $\mathcal{T}, \mathcal{X}$ to be corresponding Borel σ-algebras.

Theorem 3.1. *Assume that the space* $\mathbb{T}$ *is separable, the space* $\mathbb{X}$ *is complete, and random function* $\{X(t), t \in \mathbb{T}\}$ *is* continuous in probability*; that is, for any* $t \in \mathbb{T}$ *and* $\varepsilon > 0$,

$$\mathsf{P}(\rho(X(t), X(s)) > \varepsilon) \to 0, \quad s \to t.$$

Then there exists a modification of the function X *that is measurable (i.e., a* measurable modification*).*

Remark 3.1. In most textbooks the statement of Theorem 3.1 is formulated and proved under the condition that the space $\mathbb{T}$ is compact or σ-compact (e.g., $\mathbb{T} = \mathbb{R}, \mathbb{T} = \mathbb{R}^d$, etc.). For the separable space $\mathbb{T}$, this statement still holds true. This follows from the result of Problem 3.45, published primarily in the paper [86] (see also Problem 3.46).

Definition 3.6. *A random function* X *is called* separable *if there exist a countable dense subset* $\mathbb{T}_0 \subset \mathbb{T}$ *and a set* $N \in \mathcal{F}$ *with* $\mathsf{P}(N) = 0$ *such that, for every open set* $G \subset \mathbb{T}$ *and closed set* $F \subset \mathbb{X}$,

$$\{\omega | X(t,\omega) \in F, \, t \in G\} \triangle \{\omega | X(t,\omega) \in F, \, t \in G \cap \mathbb{T}_0\} \subset N.$$

The set $\mathbb{T}_0$ *is called the* set of separability *for the function* X.

Theorem 3.2. *Let the space* $\mathbb{T}$ *be separable and the space* $\mathbb{X}$ *be compact. Then every random function* $\{X(t), t \in \mathbb{T}\}$ *has a modification being a separable random function (i.e., a* separable modification*).*

Let us consider the question of existence of a modification of a random function with all its trajectories being continuous functions (i.e., a *continuous modification*).

Theorem 3.3. *Assume that the space* $\mathbb{T}$ *is compact, the space* $\mathbb{X}$ *is complete, and a random function* $\{X(t), t \in \mathbb{T}\}$ *is continuous in probability. Then the continuous modification for the function* X *exists if and only if the following condition holds true,*

$$\mathsf{P}\left(\bigcap_{n=1}^{\infty}\bigcup_{m=1}^{\infty}\bigcap_{s,t\in\mathbb{T}_0,d(t,s)<\frac{1}{m}}\left\{\rho(X(t),X(s))<\frac{1}{n}\right\}\right)=1,$$

where $\mathbb{T}_0$ is an arbitrary countable dense subset of $\mathbb{T}$.

In what follows, we assume the space $\mathbb{X}$ to be complete. The following theorem gives a sufficient condition for a stochastic process to possess a continuous modification, formulated in terms of two-dimensional distributions of the process.

Theorem 3.4. *Let $\{X(t),t\in[0,T]\}$ be a stochastic process. Suppose there exist a nondecreasing function $\{g(h),h\in[0,T]\}$ and some function $\{q(c,h),c\in\mathbb{R}^+,h\in[0,T]\}$ such that*

$$\mathsf{P}(\rho(X(t),X(t+h))>cg(h))\le q(c,h),\quad h>0,\quad t\in[h,T-h],$$

$$\sum_{n=0}^{\infty}g(2^{-n}T)<+\infty,\quad \sum_{n=1}^{\infty}2^n q(c,2^{-n}T)<+\infty,!\quad c\in\mathbb{R}^+. \tag{3.1}$$

Then the process X has a continuous modification.

As a corollary of Theorem 3.4, one can obtain the well-known *sufficient Kolmogorov condition* for existence of a continuous modification.

Theorem 3.5. *Let $\{X(t),t\in[0,T]\}$ be a stochastic process satisfying*

$$\mathsf{E}\rho^{\alpha}(X(t),X(s))\le C|t-s|^{1+\beta},\quad t,s\in[0,T]$$

with some positive constants α,β,C. Then the process X possesses a continuous modification.

Under the sufficient Kolmogorov condition, the properties of the trajectories of the process X can be specified in more detail. Recall that function $f(t),t\in[0,T]$ is said to satisfy *the Hölder condition with index γ* ($\gamma>0$) if

$$\sup_{t\neq s}|t-s|^{-\gamma}\rho(f(t),f(s))<+\infty.$$

Theorem 3.6. *Under conditions of Theorem 3.5, for arbitrary $\gamma<\beta/\alpha$ the process X has a modification with the trajectories satisfying the Hölder condition with index γ.*

Analogues of the sufficient Kolmogorov condition are available for random functions defined on parametric sets that may have more complicated structure than an interval. Let us give a version of this condition for random fields.

Theorem 3.7. *Let $\{\xi(x),\ x=(x_1,\dots,x_d)\in D\subset\mathbb{R}^d\}$ be a random field such that*

$$\mathsf{E}\rho^{\alpha}(\xi(x),\xi(y))\le C\|x-y\|^{d+\beta},\quad x,y\in D$$

with some positive constants α,β,C. Then the field ξ possesses a continuous modification.

Remark 3.2. In numerous models and examples, there arises a wide class of stochastic processes that do not possess continuous modification because of the jump discontinuities of their trajectories. The most typical and important example here is the Poisson process. This leads to the following definitions and notation.

A function $f:[a,b]\to\mathbb{X}$ is called *càdlàg* if it is right continuous and has left-hand limits in every point of $[a,b]$. This notation is the abbreviation for the French phrase *continue à droite, limite à gauche*. Similarly, a function that is left continuous and has right-hand limits in every point is called *càglàd*. Analogous English abbreviations *rcll* and *rllc* are used less frequently. The set of all càdlàg functions $f:[a,b]\to\mathbb{X}$ is denoted $\mathbb{D}([a,b],\mathbb{X})$ and is called *the Skorohod space*. The short notation for $\mathbb{D}([a,b],\mathbb{R})$ is $\mathbb{D}([a,b])$.

The following theorem gives sufficient conditions for a stochastic process to possess a càdlàg modification, formulated in terms of three-dimensional distributions of the process.

Theorem 3.8. *Let $\{X(t),t\in[0,T]\}$ be a continuous in probability stochastic process. Suppose that there exist a nondecreasing function $\{g(h),h\in[0,T]\}$ and a function $\{q(c,h),c\in\mathbb{R}^+,h\in[0,T]\}$ such that* (3.1) *holds true and*

$$\mathsf{P}(\{\rho(X(t-h),X(t))>cg(h)\}\cap\{\rho(X(t),X(t+h))>cg(h)\})\le q(c,h),$$

$h>0,t\in[h,T-h]$. Then the process X has càdlàg modification.

Also, for existence of either càdlàg or continuous modifications sufficient conditions are available, formulated in terms of conditional probabilities.

Theorem 3.9. *Let $\{X(t),\ t\in[0,T]\}$ be a stochastic process, and $\{\alpha(\varepsilon,\delta),\ \varepsilon,\delta>0\}$ be a family of constants such that*

$$\mathsf{P}(\rho(X(t),X(s))>\varepsilon/\mathcal{F}_s^X)\le\alpha(\varepsilon,\delta)\quad a.s.,\ 0\le s\le t\le s+\delta\le T,\quad \varepsilon>0.$$

Then

(1) If $\lim_{\delta\to 0+}\alpha(\varepsilon,\delta)=0$ for any $\varepsilon>0$, then the process X has càdlàg modification.

(2) If $\lim_{\delta\to 0+}\delta^{-1}\alpha(\varepsilon,\delta)=0$ for any $\varepsilon>0$, then the process X has continuous modification.

Bibliography

[9], Chapter I; [24], Volume 1, Chapter III, §2–5; [25], Chapter IV, §2–5; [15], Chapter II, §2; [79], Chapters 8–11.

Problems

3.1. Prove that if the domain $\mathbb{T}$ of a random function X is countable and σ-algebra $\mathcal{T}$ includes all one-point sets, then the random function X is measurable.

3.2. (1) Prove that if a process is measurable then each of its trajectories is a measurable function.
(2) Give an example of a nonmeasurable stochastic process with all its trajectories being measurable functions.
(3) Give an example of a stochastic process with all its trajectories being nonmeasurable functions.

3.3. Let $\Omega = [0,1]$, and σ-algebra $\mathcal{F}$ consist of all the subsets of $[0,1]$ having their Lebesgue measure equal either 0 or 1. Let $X(t,\omega) = \mathbb{I}_{t=\omega}, \omega \in [0,1], t \in [0,1]$. Prove that (a) X is a stochastic process; (b) X is not measurable.

3.4. Prove that stochastic process $\{X(t), t \in \mathbb{R}^+\}$ is measurable assuming its trajectories are: (a) right continuous; (b) left continuous.

3.5. Assume it is known that every trajectory of the process $\{X(t), t \in \mathbb{R}^+\}$ is either right continuous or left continuous. Does it imply that this process is measurable? Compare with the previous problem.

3.6. Prove that if a process $\{X(t),\ t \in \mathbb{R}^+\}$ is measurable and a random variable τ possesses its values in $\mathbb{R}^+$, then $X(\tau)$ is random variable. Compare this problem and Problem 3.4 with Problems 1.27 and 1.28.

3.7. A process $\{X(t),\ t \in \mathbb{R}^+\}$ is called *progressively measurable* if for every $T > 0$ the restriction of the function X to $[0,T] \times \Omega$ is $\mathcal{B}([0,T]) \otimes \mathcal{F}_T^X - \mathcal{X}$-measurable. Construct a process X that is measurable, but not progressively measurable.

3.8. Let stochastic process $\{X(t),\ t \in \mathbb{R}^+\}$ be continuous in probability. Prove that this process has: (a) measurable modification; (b) progressively measurable modification.

3.9. Let all values of a process $\{X(t),\ t \in \mathbb{R}^+\}$ be independent and uniformly distributed on $[0,1]$.
(a) Does this process have a càdlàg modification?
(b) Does this process have a measurable modification?

3.10. Let $\mathbb{T} = \mathbb{R}^+, (\Omega, \mathcal{F}, \mathsf{P}) = (\mathbb{R}^+, \mathcal{B}(\mathbb{R}^+), \mu)$, where μ is a probability measure on $\mathbb{R}^+$ that does not have any atoms. Introduce the processes X, Y by $X(t,\omega) = \mathbb{I}_{\{t=\omega\}}$, $Y(t,\omega) = 0$, $t \in \mathbb{T}$, $\omega \in \Omega$.
(1) Prove that X and Y are stochastically equivalent; that is, Y is a modification of X.
(2) Check that all the trajectories of Y are continuous and all the trajectories of X are discontinuous.

3.11. Prove that if a stochastic process $\{X(t),\ t \in \mathbb{R}\}$ has a continuous modification, then every process Y, stochastically equivalent to X in a wide sense, has a continuous modification too.

3.12. Assume that the random field $\{\xi(x),\ x \in \mathbb{R}^d\}$ is such that, for every $n \in \mathbb{N}$, the field $\{\xi(x), \|x\| \le n\}$ has a continuous modification. Prove that the field $\{\xi(x),\ x \in \mathbb{R}^d\}$ itself has a continuous modification.

3.13. Let $\{X(t), t \in \mathbb{R}^+\}$ be a Gaussian process with zero mean and covariance function $R_X(t,s)$ that is equal to

(1) $\exp[-|t-s|]$ (Ornstein–Uhlenbeck process).
(2) $\frac{1}{2}(t^{2H} + s^{2H} - |t-s|^{2H})$, $H \in (0,1]$ (fractional Brownian motion).

Prove that the process $X(\cdot)$ has a continuous modification. Find values of γ such that these processes have modifications with their trajectories satisfying the Hölder condition with index γ.

3.14. Prove that if a function $f : [0,1] \to \mathbb{R}$ satisfies the Hölder condition with index $\gamma > 1$, then f is a constant function. Derive that $K(t,s) = \frac{1}{2}(t^{2H} + s^{2H} - |t-s|^{2H})$ is not a covariance function for $H > 1$. Compare with Problem 2.16.

3.15. Let Gaussian process $\{X(t), t \in \mathbb{R}\}$ be stochastically continuous, and $t_0 \in \mathbb{R}$ be fixed. Prove that the process $Y(t) = \mathsf{E}[X(t)/X(t_0)], t \in \mathbb{R}$ has a continuous modification.

3.16. Prove that the Wiener process has a modification with every trajectory satisfying the Hölder condition with arbitrary index $\gamma < \frac{1}{2}$.

3.17. Let W be the Wiener process. Prove that $\limsup_{t \to 0+} W(t)/\sqrt{t} = +\infty$ with probability one. In particular, almost every trajectory of the Wiener process does not satisfy the Hölder condition with index $\gamma = \frac{1}{2}$.

3.18. Prove that for any $\alpha > \frac{1}{2}$

$$\mathsf{P}\left(\lim_{t \to +\infty} \frac{W(t)}{t^\alpha} = 0 \right) = 1.$$

3.19. Let $\{W(t), t \ge 0\}$ be the Wiener process. Prove that there exists the limit in probability

$$\operatorname*{P-lim}_{n \to \infty} \sum_{k=0}^{n} \left(W\left(\frac{k+1}{n}\right) - W\left(\frac{k}{n}\right) \right)^2,$$

and find this limit. Prove that

$$\operatorname*{P-lim}_{n \to \infty} \sum_{k=0}^{n} \left| W\left(\frac{k+1}{n}\right) - W\left(\frac{k}{n}\right) \right| = \infty.$$

3.20. Prove that almost all trajectories of the Wiener process have unbounded variation on $[0,1]$.

3.21. Let $\{W(t),\ t \in \mathbb{R}^+\}$ be the Wiener process. Prove that, with probability one, $\lambda^1\{t \ge 0 |\ W(t) = 0\} = 0$.

3.22. Prove that, with probability one, the Wiener process attains its maximum value on $[0,1]$ only once.

3.23. Let W be the Wiener process. For $a,b \in \mathbb{R}^+, a<b$ denote

$$C_{ab} = \{x \in C(\mathbb{R}^+) \mid x(t) = x(a),\ t \in [a,b]\}.$$

Prove that the set $\{\omega \mid W(\cdot,\omega) \in \bigcup_{a<b} C_{ab}\}$ is a random event that has zero probability; that is, almost every trajectory of W is nonconstant on every interval.

3.24. Let process X be continuous in probability. Assume the probability space $(\Omega, \mathcal{F}, \mathsf{P})$ is such that Ω is countable and every point $\omega \in \Omega$ has positive probability. Prove that all trajectories of the process X are continuous.

3.25. Let process $\{X(t), t \in [a,b]\}$ be continuous in probability. Prove that if X has a càdlàg modification then it has a càglàd modification too, and vice versa.

3.26. Let $\{X(t),\ t \in \mathbb{R}^+\}$ be a stochastic process that has càdlàg trajectories. Define $Y(t) = X(t-) = \lim_{s \to t-} X(s), t>0, Y(0) = X(0)$. Prove that Y is a stochastic process with càglàd trajectories.

3.27. Let τ be a random variable taking values in $\mathbb{R}^+$ and $X(t) = \mathbb{I}_{t>\tau},\ t \geq 0$. Find a condition on the distribution of τ that would be necessary and sufficient for the process X to have a càdlàg modification.

3.28. Let $\{X(t),\ t \in [0,T]\}$ be a continuous in probability stochastic process with independent increments. Prove that X has a càdlàg modification.

3.29. Let $f \in \mathbb{D}([a,b])$. Prove that
(a) Function f is bounded.
(b) The set of discontinuities of the function f is at most countable.

3.30. Let the trajectories of a stochastic process X belong to the space $\mathbb{D}([0,T])$ and let $c>0$ be given. Prove that $\tau_c = \inf\{t \mid |X(t) - X(t-)| \geq c\}$ is a random variable.

3.31. Let the trajectories of a stochastic process X belong to the space $\mathbb{D}([0,T])$ and let $\Gamma \in \mathcal{B}(\mathbb{R})$ be given. Prove that $\tau = \inf\{t \mid X(t) - X(t-) \in \Gamma\}$ is a random variable.

3.32. Let $\{X(t),\ t \in \mathbb{R}\}$ be a separable stochastic process taking its values in a complete metric space. Prove that if the process X has a continuous modification then there exists a set $\tilde{N} \in \mathcal{F}$ with $\mathsf{P}(\tilde{N}) = 0$ such that the trajectory $X(\cdot,\omega)$ is continuous for any $\omega \notin \tilde{N}$.

3.33. Let $\{X(t),\ t \in \mathbb{R}\}$ be a separable stochastic process taking its values in a complete metric space. Prove that if the process X has a càdlàg modification then there exists a set $\tilde{N} \in \mathcal{F}$ with $\mathsf{P}(\tilde{N}) = 0$ such that the trajectory $X(\cdot,\omega)$ is càdlàg for any $\omega \notin \tilde{N}$.

3.34. Let $\{X(t),\ t\in\mathbb{R}\}$ be a separable stochastic process taking its values in a complete metric space. Assume that the process X has a measurable modification. Does this imply existence of such a set $\tilde{N}\in\mathcal{F}$ with $\mathsf{P}(\tilde{N})=0$ that restriction of the function $X(\cdot,\cdot)$ to $\mathbb{R}\times(\Omega\backslash\tilde{N})$ is a measurable function?

3.35. Let $\{X(t),\ t\in\mathbb{R}\}$ be a stochastic process taking its values in $\mathbb{X}=[0,1]$. Does separability of X imply separability of the subspace $L_2(\Omega,\mathcal{F},\mathsf{P})$ generated by the family $\{X(t),\ t\in\mathbb{R}\}$ of the values of this process?

3.36. Prove the following characterization of the σ-algebra $\mathcal{F}_t^X$: it includes all $A\in\mathcal{F}$ for which there exists a set $A_0\in\mathcal{F}_t^{X,0}$ such that $A\triangle A_0\in\mathcal{N}_{\mathsf{P}}$.

3.37. (1) Let X,Y be two stochastically equivalent processes. Prove they generate the same filtration.
(2) Make an example of two stochastically equivalent processes X,Y such that the corresponding filtrations $\{\mathcal{F}_t^{X,0},t\in\mathbb{T}\}$ and $\{\mathcal{F}_t^{Y,0},t\in\mathbb{T}\}$ do not coincide.

3.38. Let

$$X(t)=\begin{cases}0, & t\in[0,\frac{1}{2}],\\ (t-\frac{1}{2})\eta & t\in(\frac{1}{2},1]\end{cases},$$

where $\mathsf{P}(\eta=\pm1)=\frac{1}{2}$. Describe explicitly the natural filtration for the process X. Is this filtration: (a) right continuous? (b) left continuous?

3.39. Let τ be a random variable uniformly distributed on $[0,1]$, $\mathcal{F}=\sigma(\tau)$ (σ-algebra, generated by random variable τ), $X(t)=\mathbb{I}_{t>\tau}$, $t\in[0,1]$. Describe explicitly the natural filtration for the process X. Is this filtration: a) right continuous? b) left continuous?

3.40. Let stochastic process $\{X(t),t\in\mathbb{R}^+\}$ have continuous trajectories.
(1) Prove that its natural filtration is left continuous.
(2) Provide an example that this filtration is not necessarily right continuous.

3.41. Provide an example of a process having càdlàg trajectories and generating filtration that is neither left continuous nor right continuous.

3.42. Is a filtration generated by a Wiener process: (a) left continuous? (b) right continuous?

3.43. Is a filtration generated by a Poisson process: (a) left continuous? (b) right continuous?

3.44. Let $\{W(t),\ t\in\mathbb{R}^+\}$ be a Wiener process and assume that all its trajectories are continuous.
(1) Is it necessary for a filtration $\{\mathcal{F}_t^{W,0},\ t\in\mathbb{R}^+\}$ to be: (a) left continuous? (b) right continuous?
(2) Answer the same questions for the filtration $\{\mathcal{F}_t^{N,0},\ t\in\mathbb{R}^+\}$, where $\{N(t),\ t\in\mathbb{R}^+\}$ is the Poisson process with càdlàg trajectories.

3.45. Let $(\Omega, \mathcal{F}, \mathsf{P})$ be a probability space, Y be a separable metric space, and $(A, \mathfrak{A})$ be a measurable space. Consider the sequence of $\mathcal{F} \otimes \mathfrak{A}$-measurable random elements $\{X_n = X_n(a),\ a \in A, n \geq 1\}$ taking their values in Y. Assume that for every $a \in A$ there exists a limit in probability of the sequence $\{X_n(a), n \geq 1\}$.

Prove that there exists an $\mathcal{F} \otimes \mathfrak{A}$-measurable random element $X(a),\ a \in A$, taking its values in Y, such that $X_n(a) \xrightarrow{\mathsf{P}} X(a),\ n \to \infty$ for every $a \in A$.

3.46. Prove Theorem 3.1.

Hints

3.1. $\{(t,\omega) | X(t,\omega) \in B\} = \bigcup_{t \in \mathbb{T}} \{t\} \times \{\omega | X(t,\omega) \in B\}$; the union is at most countable. Every set $\{t\} \times \{\omega | X(t,\omega) \in B\}$ belongs to $\mathfrak{T} \otimes \mathfrak{F}$ because $X(t)$ is a random element by the definition of a random function and therefore $\{\omega | X(t,\omega) \in B\} \in \mathfrak{F}$.

3.4. You can use Problem 3.1 and relation $X(t,\omega) = \lim_{n\to\infty} X\left(([tn]+1)/n, \omega\right)$ in item (a) or $X(t,\omega) = \lim_{n\to\infty} X\left([tn]/n, \omega\right)$ in item (b).

3.5. Consider a sum of a process X_A from item (1) of Problem 1.6 and a process $Y(t,\omega) = \mathbb{I}_{t>\omega}$.

3.11. Use Theorem 3.3.

3.13, 3.16. Use Theorem 3.6 with $\alpha = 2n, n \in \mathbb{N}$.

3.15. If $\mathsf{D}X(t_0) > 0$, then

$$\mathsf{E}[X(t)/X(t_0)] = \mathsf{E}X(t) + (X(t_0) - \mathsf{E}X(t_0)) \times \frac{\operatorname{cov}(X(t), X(t_0))}{\mathsf{D}X(t_0)}.$$

If $\mathsf{D}X(t_0) = 0$, then $Y(t) = \mathsf{E}X(t)$.

3.18. Use Problems 3.16 and 6.5 (d).

3.17. Consider a sequence of random variables $\xi_k = 2^{-(k+1)/2}(W(2^{-k}) - W(2^{-k-1}))$, $k \in \mathbb{N}$ and prove that $\mathsf{P}(\limsup_{k\to\infty} |\xi_k| = +\infty) = 1$. Use this identity.

3.21. Introduce a function $\mathbb{I}_{\{(t,\omega)|W(t,\omega)=0\}}$ and use the Fubini theorem.

3.22. It is sufficient to prove that for any $a < b$ probability of the event $\{\max_{t\in[0,a]} W(t) = \max_{t\in[b,1]} W(t)\}$ is equal to zero. One has

$$\max_{t\in[b,1]} W(t) = W(a) + (W(b) - W(a)) + \max_{t\in[b,1]} (W(t) - W(b))$$

and the variables $W(b) - W(a)$ and $\max_{t\in[b,1]}(W(t) - W(b))$ are jointly independent with the σ-algebra $\mathfrak{F}_a^W$. In addition, the distribution of $W(b) - W(a)$ is absolutely continuous. Therefore, the conditional distribution of $\max_{t\in[b,1]} W(t)$ w.r.t. $\mathfrak{F}_a^W$ is absolutely continuous. This implies the needed statement.

3.23. The fact that $\mathsf{P}(W \in C_{ab}) = 0$ for fixed $a < b$, can be proved similarly to the previous hint. Then you can use that $\bigcup_{a<b} C_{ab} = \bigcup_{a<b,a,b\in\mathbb{Q}} C_{ab}$.

3.24. Assume that there exist $\omega \in \Omega, t \in \mathbb{T}$ and a sequence $t_n \to t$ such that $X(t_n,\omega) \not\to X(t,\omega)$ and prove that $X(t_n)$ does not converge in probability to $X(t)$.

3.28. Use Theorem 3.9 and item (b) of Problem 4.4.

3.30. $\tau_c < t \Leftrightarrow \exists \varepsilon > 0 \forall \delta > 0 \exists u, v \in [0, t-\varepsilon] \cap \mathbb{Q} : |u - v| < \delta, |X(u) - X(v)| > c - \delta.$

3.31. Consider a sequence $\theta_k = \tau_{1/k}, k \geq 1$ (see Problem 3.30); then $\tau = \min\{\theta_k | X(\theta_k) - X(\theta_k -) \in \Gamma\}$.

3.34. Put $\Omega = [0,1]$, $\mathcal{F}$ is the σ-algebra of Lebesgue measurable sets, P is the Lebesgue measure, $X(t,\omega) = 1\!\!1_{t=\omega} 1\!\!1_{t\in A} + 1\!\!1_{t>\omega}$, where A is a Lebesgue nonmeasurable set.

3.35. Consider a stochastic process $\{X(t), t \in \mathbb{R}\}$ with i.i.d. values uniformly distributed on $[0,1]$ (see Problem 1.10 and corresponding hint). This process has a separable modification due to Theorem 3.2.

3.36. Use the "principle of the fitting sets": prove that the class given in the formulation of the problem is a σ-algebra containing $\mathcal{F}_t^{X,0}$ and $\mathcal{N}_{\mathsf{P}}$. Prove that every σ-algebra containing both $\mathcal{F}_t^{X,0}$ and $\mathcal{N}_{\mathsf{P}}$ should also contain this class.

3.42. (a) Use Problem 3.40.
(b) Use Problem 5.54.

3.43. (a) For any $t > 0, B \in \mathcal{B}(\mathbb{R})$, one has $\{N(t) \in B\} \triangle \{N(t-) \in B\} \subset \{N(t-) \neq N(t)\} \in \mathcal{N}_{\mathsf{P}}$ (assuming that N has càdlàg trajectories). Therefore $\{N(t) \in B\} \in \sigma\left(\mathcal{N}_{\mathsf{P}} \cup \bigvee_{s<t} \mathcal{F}_s^{N,0}\right) \subset \mathcal{F}_{t-}^N$.
(b) Use Problem 5.54.

3.44. (1b) On a null measure set, construct a process similar to the one given in Problem 3.38.

3.45. Prove that there exists an $\mathcal{F} \otimes \mathfrak{A}$-measurable random sequence $\{n_k(a) | \ a \in A, \ k \geq 1\}$ such that for all $k \in \mathbb{N}, \ a \in A : \mathsf{P}\left(\rho(X_{n_{k+1}(a)}(a), X_{n_k(a)}(a)) > 2^{-k}\right) < 2^{-k}$. Check that there exists an $\mathcal{F} \otimes \mathfrak{A}$-measurable random element $X(a), a \in A$ such that $X_{n_k(a)}(a) \to X(a), \ k \to \infty$ almost surely for every $a \in A$.

3.46. Choose a sequence of measurable subsets $\{U_{n,k}\}$ of the space $\mathbb{T}$ such that for any $n \geq 1$:
(1) $\mathbb{T} = \cup_k U_{n,k}$.
(2) The sets $U_{n,k}$ and $U_{n,j}$ are disjoint for $k \neq j$ and their diameters do not exceed $1/n$. Let $t_{n,k} \in U_{n,k}$. Then $X_n(t) = \sum_k X(t_{n_k}) 1\!\!1_{U_{n,k}}(t)$ is measurable. Prove that for any t sequence $\{X_n(t), n \geq 1\}$ converges to $X(t)$ in probability as $n \to \infty$ and use Problem 3.45.

Answers and Solutions

3.2. Item (1) immediately follows from the Fubini theorem. As an example for item (2), one can use the process from Problem 1.6 with a Borel nonmeasurable set. In item (3), one of the possible examples is: $\Omega = \mathbb{T} = [0,1], \mathcal{T} = \mathcal{F} = \mathcal{B}([0,1]), X(t,\omega) = \mathbb{I}_{t\in A}$, where A is a Borel nonmeasurable set.

3.3. Every variable $X(t,\cdot)$ is an indicator function for a one-point set and, obviously, is measurable w.r.t. $\mathcal{F}$. This proves (a). In order to prove that the process X is not measurable, let us show that the set $\{(t,\omega) | \, t=\omega\}$ does not belong to $\mathcal{B}([0,1]) \otimes \mathcal{F}$. Denote $\mathcal{K}$ the class of sets $C \subset [0,1]^2$ satisfying the following condition. There exists a set Δ_C with its Lebesgue measure equal to 0 such that for arbitrary $t, \omega_1, \omega_2 \in [0,1]$ it follows from $(t,\omega_1) \in C, (t,\omega_2) \notin C$ that at least one of the points ω_1, ω_2 belongs to Δ_C. Then $\mathcal{K}$ is a σ-algebra (prove this!). In addition, $\mathcal{K}$ contains all the sets of the type $C = A \times B, B \in \mathcal{F}$. Then, by the "principle of the fitting sets", class $\mathcal{K}$ contains $\mathcal{B}([0,1]) \otimes \mathcal{F}$. On the other hand, the set $\{(t,\omega) | \, t=\omega\}$ does not belong to $\mathcal{K}$ (verify this!).

3.5. It does not.

3.6. Because a superposition of two measurable mappings is also measurable, it is sufficient to prove that the mapping $\Omega \ni \omega \mapsto (\tau(\omega),\omega) \in \mathbb{R}^+ \times \Omega$ is $\mathcal{F} - \mathcal{B}(\mathbb{R}^+) \otimes \mathcal{F}$-measurable. For $C = A \times B, A \in \mathcal{B}(\mathbb{R}^+), B \in \mathcal{F}$ we have that $\{\omega | \, (\tau(\omega),\omega) \in C\} = \{\omega | \, \tau(\omega) \in A\} \cap B \in \mathcal{F}$. Because "rectangles" $C = A \times B$ generate $\mathcal{B}(\mathbb{R}^+) \otimes \mathcal{F}$, this proves the needed measurability.

3.7. Take $\Omega = [0,1], \mathcal{F} = \mathcal{B}([0,1])$,

$$X(t,\omega) = \begin{cases} \mathbb{I}_{t=\omega}, & t \in [0,1] \\ \omega, & t \in (1,2] \end{cases}.$$

Then the process X is measurable (verify this!). On the other hand, the σ-algebra $\mathcal{F}_1^X$ is degenerate, that is, contains only the sets of Lebesgue measure 0 or 1. Then the process X is not progressively measurable since its restriction to the time interval $[0,1]$ is not measurable (see Problem 3.3).

3.9. (a) No, because it would be right continuous in probability.
(b) No. Assume that such modification exists. Then for any $t \in \mathbb{R}^+, \omega \in \Omega$ there exists $Y(t,\omega) = \int_0^t \left(X(s,\omega) - \frac{1}{2}\right) ds$ (Lebesgue integral). By the Fubini theorem, $\mathsf{E}Y^2(t) = \int_0^t \int_0^t \mathsf{E}\left(X(s_1) - \frac{1}{2}\right)\left(X(s_2) - \frac{1}{2}\right) ds_1 ds_2 = \int_0^t \int_0^t \mathbb{I}_{s_1=s_2} ds_1 ds_2 = 0$, and then $Y(t) = 0$ a.s. Every trajectory of the process Y is continuous as a Lebesgue integral with varying upper bound, therefore $Y(\cdot,\omega) \equiv 0$ for almost all ω. Then $X(\cdot,\omega) \equiv \frac{1}{2}$ for the same ω. It is impossible.

3.13. a) $\gamma < \frac{1}{2}$; b) $\gamma < H$.

3.27. Denote by $\mathcal{K}_\tau$ the set of atoms of distribution for τ (i.e., the points $t \in \mathbb{R}^+$ for which $\mathsf{P}(\tau = t) > 0$). The required condition is as follows. For any $t \in \mathcal{K}_\tau$ there exists $\varepsilon > 0$ such that $\mathsf{P}(\tau \in (t, t+\varepsilon)) = 0$.

3.31. According to Problem 3.30, $\theta_k = \tau_{1/k}, k \in \mathbb{N}$ are random variables. Both the process X and the process $Y(t) = X(t-)$ are measurable (see Problems 3.4 and 3.26), therefore $Z_k = X(\theta_k) - X(\theta_k -), k \in \mathbb{N}$ are random variables (see Problem 3.6). Thus $\tau = \sum_{k=1}^{\infty} \theta_k \cdot \mathbb{I}_{Z_k \in \Gamma} \cdot \prod_{m>k} \mathbb{I}_{Z_m \notin \Gamma}$ is a random variable too.

3.34. Not true.

3.35. Not true.

3.37. (1) For any $t \in \mathbb{T}, B \in \mathcal{X}$ one has $\{X(t) \in B\} \triangle \{Y(t) \in B\} \subset \{X(t) \neq Y(t)\}$, and then $\{X(s) \in B\} \in \mathcal{F}_t^Y, s \le t$ (see Problem 3.36). Because $\mathcal{N}_\mathsf{P} \subset \mathcal{F}_t^Y$, one has $\mathcal{F}_t^X \subset \mathcal{F}_t^Y$. Similarly, it can be proved that $\mathcal{F}_t^Y \subset \mathcal{F}_t^X$.
(2) Let $\Omega = [0,1], \mathcal{F} = \mathcal{B}([0,1]), \mathsf{P}$ is the Lebesgue measure, $\mathbb{T} = \{t\}$, $X(t,\omega) \equiv 0, Y(t,\omega) = \mathbb{I}_{\omega=1}$. Then $\mathcal{F}_t^{X,0} = \{\varnothing, [0,1]\}$, $\mathcal{F}_t^{Y,0} = \{\varnothing, \{1\}, [0,1), [0,1]\}$.

3.38. (a) No; (b) yes.

3.39. $\mathcal{F}_t^X$ contains all sets of the form $\{\omega : \tau(\omega) \in A\}$, where A is an arbitrary Borel set such that either $\mathsf{P}(A \cap (t,1]) = 0$ or $\mathsf{P}(A \cap (t,1]) = 1-t$. Filtration $\mathbb{F}^X$ is both right and left continuous.

3.42. (a) Yes; (b) yes.

3.43. (a) Yes; (b) yes.

3.44. (1) (a) Yes; (b) no.
(2) (a) No; (b) no.

4

Continuity. Differentiability. Integrability

Theoretical grounds

Definition 4.1. *Let $\mathbb{T}$ and $\mathbb{X}$ be metric spaces with the metrics d and ρ, respectively. A random function $\{X(t), t \in \mathbb{T}\}$ taking values in $\mathbb{X}$ is said to be*

(1) Stochastically continuous *(or* continuous in probability*) at point $t \in \mathbb{T}$ if $\rho(X(t), X(s)) \to 0$ in probability as $s \to t$.*

(2) Continuous with probability one *(or* continuous almost surely, a.s.*) at point $t \in \mathbb{T}$ if $\rho(X(t), X(s)) \to 0$ with probability one as $s \to t$.*

(3) continuous in the L_p sense, $p > 0$ *(or* mean continuous of the power p *) at point $t \in \mathbb{T}$ if $\mathsf{E}\rho^p(X(t), X(s)) \to 0$ as $s \to t$.*

If a random function is stochastically continuous (continuous with probability one, continuous in the L_p sense) at every point of the parametric set $\mathbb{T}$ then it is said to be stochastically continuous *(respectively,* continuous with probability one *or* continuous in the L_p sense*) on this set.*

Note that sometimes, while dealing with continuity of the function X in the L_p sense, one assumes the additional condition $\mathsf{E}\rho^p(X(t), x) < +\infty, t \in \mathbb{T}$, where $x \in \mathbb{X}$ is a fixed point. Continuity in the L_1 sense is called *mean* continuity, and continuity in the L_2 sense is called *mean square* continuity.

Theorem 4.1. *Let $\{X(t),\ t \in [a,b]\}$ be a real-valued stochastic process with $\mathsf{E}X^2(t) < +\infty$ for $t \in [a,b]$. The process X is mean square continuous if and only if $a_X \in C([a,b]), R_X \in C([a,b] \times [a,b])$.*

Definition 4.2. *A real-valued stochastic process $\{X(t),\ t \in [a,b]\}$ is said to be* stochastically differentiable (differentiable with probability one, differentiable in the L_p sense) *at point $t \in [a,b]$ if there exists a random variable η such that*

$$\frac{X(t) - X(s)}{t - s} \to \eta, \quad s \to t$$

in probability (with probability one or in L_p, respectively). The random variable η is called the derivative of the process X at the point t and is denoted by $X'(t)$.

D. Gusak et al., *Theory of Stochastic Processes*, Problem Books in Mathematics,
DOI 10.1007/978-0-387-87862-1_4,

If a stochastic process is differentiable (in any sense introduced above) at every point t of the parametric set $\mathbb{T}$ then it is said to be differentiable (in this sense) on this set. If, in addition, its derivative $X' = \{X'(t), t \in \mathbb{T}\}$ is continuous (in the same sense), then the process X is said to be continuously differentiable (in this sense).

Theorem 4.2. *Let $\{X(t),\ t \in [a,b]\}$ be a real-valued stochastic process with $\mathsf{E}X^2(t) < +\infty$ for $t \in [a,b]$. The process X is continuously differentiable in the mean square if and only if $a_X \in C^1([a,b])$, $R_X \in C^1([a,b] \times [a,b])$ and there exists the continuous derivative $(\partial^2/\partial t \partial s) R_X$. In this case, $a_{X'}(t) = a'_X(t), R_{X'}(t,s) = (\partial^2/\partial t \partial s) R_X(t,s)$, $R_{X,X'}(t,s) = (\partial/\partial s) R_X(t,s)$.*

Definition 4.3. *Let $\{X(t),\ t \in [a,b]\}$ be a real-valued stochastic process. Assume there exists the random variable η such that, for any partition sequence $\{\lambda^n = \{a = t_0^n < t_1^n < \cdots < t_n^n = b\}, n \in \mathbb{N}\}$ with $\max_k (t_k^n - t_{k-1}^n) \to 0$ and for any sequence of suites $\{\theta_n = \{\theta_1^n, \ldots, \theta_n^n\}, n \in \mathbb{N}\}$ with $\theta_k^n \in [t_{k-1}^n, t_k^n], k \le n, n \in \mathbb{N}$, the following convergence takes place,*

$$\sum_{k=1}^{n} X(\theta_k^n)\left(t_k^n - t_{k-1}^n\right) \to \eta, \quad n \to \infty$$

either in probability, with probability one, or in the L_p sense.

Then the process X is said to be integrable (in probability, with probability one or in the L_p sense, *respectively) on $[a,b]$. The random variable η is denoted $\int_a^b X(t)\,dt$ and called the* integral *of the process X over $[a,b]$.*

Theorem 4.3. *Let $\{X(t), t \in [a,b]\}$ be a real-valued stochastic process with $\mathsf{E}X^2(t) < +\infty$ for $t \in [a,b]$. The process X is mean square integrable on $[a,b]$ if and only if the functions a_X and R_X are Riemann integrable on $[a,b]$ and $[a,b] \times [a,b]$, respectively. In this case,*

$$\mathsf{E}\int_a^b X(t)\,dt = \int_a^b a_X(t)\,dt, \quad \mathsf{D}\int_a^b X(t)\,dt = \int_a^b \int_a^b R_X(t,s)\,dt\,ds.$$

Bibliography

[24], Volume 1, Chapter IV, §3; [25], Chapter V, §1; [79], Chapter 14.

Problems

4.1. Let $\{X(t),\ t \in [0,1]\}$ be a waiting process; that is, $X(t) = \mathbb{I}_{t \ge \tau}, t \in [0,1]$, where τ is a random variable taking its values in $[0,1]$. Prove that the process X is continuous (in any sense: in probability, L_p, a.s.) if and only if the distribution function of τ is continuous on $[0,1]$.

4.2. Give an example of a process that is
(a) Stochastically continuous but not continuous a.s.
(b) Continuous a.s. but not mean square continuous.
(c) For given $p_1 < p_2$, continuous in L_{p_1} sense but not continuous in L_{p_2} sense.

4.3. Suppose that all values of a stochastic process $\{X(t),\ t \in \mathbb{R}^+\}$ are independent and uniformly distributed on $[0,1]$. Prove that the process is not continuous in probability.

4.4. Let $\{X(t),\ t \in [a,b]\}$ be continuous in probability process. Prove that this process is
(a) Bounded in probability:

$$\forall \varepsilon > 0 \exists C \forall t \in [a,b]: \quad \mathsf{P}(|X(t)| \geq C) < \varepsilon;$$

(b) Uniformly continuous in probability:

$$\forall \varepsilon > 0 \exists \delta > 0 \forall t,s \in [a,b], |t-s| < \delta: \quad \mathsf{P}(|X(t) - X(s)| \geq \varepsilon) < \varepsilon.$$

4.5. Prove that the necessary and sufficient condition for stochastic continuity of a real-valued process X at a point t is the following. For any $x, y \in \mathbb{R}, x < y$,

$$\mathsf{P}(X(t) \leq x,\ X(s) \geq y) + \mathsf{P}(X(t) \geq y,\ X(s) \leq x) \to 0, \quad s \to t.$$

4.6. Prove that the following condition is sufficient for a real-valued process $\{X(t),\ t \in \mathbb{R}\}$ to be stochastically continuous: for any points $t_0, s_0 \in \mathbb{R},\ x, y \in \mathbb{R}$,

$$\mathsf{P}(X(t) \leq x,\ X(s) \leq y) \to \mathsf{P}(X(t_0) \leq x,\ X(s_0) \leq y), \quad t \to t_0,\ s \to s_0.$$

4.7. (1) Let stochastic process $\{X(t),\ t \in [a,b]\}$ be mean square differentiable at the point $t_0 \in [a,b]$. Prove that $\operatorname{cov}(\eta, X'(t_0)) = (d/dt)|_{t=t_0} \operatorname{cov}(\eta, X(t))$ for any $\eta \in L_2(\Omega, \mathcal{F}, \mathsf{P})$.

(2) Let $\{X(t), t \in [a,b]\}$ be a mean square integrable stochastic process and $\eta \in L_2(\Omega, \mathcal{F}, \mathsf{P})$. Prove that $\operatorname{cov}\left(\eta, \int_a^b X(s)\,ds\right) = \int_a^b \operatorname{cov}(\eta, X(s))\,ds$.

4.8. Let $\{X(t), t \in \mathbb{R}\}$ be a real-valued centered process with independent increments and $\mathsf{E}X^2(t) < +\infty, t \in \mathbb{R}$. Prove that for every $t \in \mathbb{R}$ there exist the following mean square limits: $X(t-) = \lim_{s \to t-} X(s)$, $X(t+) = \lim_{s \to t+} X(s)$.

4.9. Let $\{N(t), t \in \mathbb{R}^+\}$ be the Poisson process.

(1) Prove that N is not differentiable in any point $t \in \mathbb{R}^+$ in the L_p sense for any $p \geq 1$.

(2) Prove that N is differentiable in an arbitrary point $t \in \mathbb{R}^+$ in the L_p sense for every $p \in (0,1)$.

(3) Prove that N is differentiable in an arbitrary point $t \in \mathbb{R}^+$ in probability and with probability 1.

Find the derivatives in items (2) and (3).

4.10. Let $\{X(t),\ t \geq 0\}$ be a real-valued homogeneous process with independent increments and $\mathsf{D}[X(1) - X(0)] > 0$. Prove that the processes $X(t)$ and $Y(t) := X(t+1) - X(t)$ are mean square continuous but not mean square differentiable.

4.11. Consider a mean square continuous process $\{X(t), t \in \mathbb{R}\}$ such that $\mathsf{D}(X(t) - X(s)) = F(t-s), t,s \in \mathbb{R}$ for some function F and $F''(0) = 0$. Describe all such processes.

4.12. Let the mean and covariance functions of the process X be equal $a_X(t) = t^2$, $R_X(t,s) = e^{ts}$. Prove that X is mean square differentiable and find :
(a) $\mathsf{E}[X(1) + X(2)]^2$.
(b) $\mathsf{E}X(1)X'(2)$.
(c) $\mathsf{E}\left[X(1) + \int_0^1 X(s)\,ds\right]^2$.
(d) $\operatorname{cov}\left(X'(1), \int_0^1 X(s)\,ds\right)$.
(e) $\mathsf{E}\left[X(1) + \int_0^1 X(s)\,ds\right]^4$ assuming additionally that the process X is Gaussian.

4.13. Let N be the Poisson process with intensity $\lambda = 1$ and $\widetilde{N}(t) = N(t) - t$ be the corresponding compensated process. Find:
(a) $\mathsf{E}\int_1^3 N(t)\,dt \cdot \int_2^4 N(t)\,dt$.
(b) $\mathsf{E}\int_0^1 \widetilde{N}(t)\,dt \cdot \int_1^2 \widetilde{N}(t)\,dt \cdot \int_2^3 \widetilde{N}(t)\,dt$.

4.14. Let $X(t) = \alpha \sin(t + \beta), t \in \mathbb{R}$, where α and β are independent random variables, α is exponentially distributed with parameter 2, and β is uniformly distributed on $[-\pi, \pi]$.

(1) Is the process X continuous or differentiable in the mean square?

(2) Prove using the definition that the mean square derivative $X'(t)$ is equal to $\alpha \cos(t + \beta)$.

(3) Find $R_X, R_{X,X'}, R_{X'}, a_X, a_{X'}$.

4.15. Suppose that the trajectories of a process $\{X(t),\ t \in \mathbb{R}\}$ are differentiable at the point t_0, and the process X is mean square differentiable at this point. Prove that the derivative of the process at the point t_0 in the mean square sense coincides almost surely with the ordinary derivative of the trajectory of the process.

4.16. Consider a mean square continuously differentiable stochastic process $\{X(t),\ t \in \mathbb{R}\}$.

(1) Prove that the processes X and X' have measurable modifications.
Further on, assume X and X' be measurable.

(2) Prove that

$$X(t) = X(0) + \int_0^t X'(s)ds \text{ a.s.}, \quad t \in \mathbb{R}. \tag{4.1}$$

Consider both the mean square integral and the Lebesgue integral for a fixed ω (by convention, $\int_0^t$ is equal to $-\int_t^0$ for $t < 0$).

(3) Prove that X has a continuous modification.

(4) Assume

$$\int_{-\infty}^{\infty} \mathsf{E}X^2(t)\,dt < \infty, \quad \int_{-\infty}^{\infty} \mathsf{E}(X'(t))^2\,dt < \infty.$$

Prove that for almost all ω the function $t \mapsto X(t,\omega)$ belongs to the Sobolev space $W_2^1(\mathbb{R})$, that is, a space of square-integrable functions with their weak derivatives being also square-integrable. Prove that, for P-almost all ω and λ^1-almost all t, the Sobolev derivative coincides with the mean square derivative.

(5) Will the previous statements still hold true if X is supposed to be mean square differentiable but not necessarily continuously mean square differentiable?

4.17. (1) Verify whether a process $\{X(t), t \in \mathbb{R}\}$ is either mean square continuous or mean square differentiable if
(a) $X(t) = \min(t, \tau)$, where the random variable τ is exponentially distributed.
(b) X has periodic trajectories with period 1 and $X(t) = t - \tau$, $t \in [\tau, \tau+1]$, where the random variable τ is uniformly distributed on $[0;1]$.

(2) Find R_X. If X is mean square differentiable then find X' and $R_{X'}, R_{X',X}$.

4.18. Let $X(t) = f(t-\tau)$, where τ is exponentially distributed and

$$f(x) = \begin{cases} 1-|x|, & |x| \le 1, \\ 0, & |x| > 1. \end{cases}$$

Is the process X mean square differentiable? If so, find X' and $\mathsf{E}(X'(1))^2$.

4.19. Let random variable τ have the distribution density

$$p(x) = \begin{cases} \frac{(1-x)^\alpha}{\alpha+1}, & x \in [0,1] \\ 0, & x \notin [0,1] \end{cases}$$

with $\alpha > 0$. Is the stochastic process

$$X(t) = \begin{cases} 0, & t \le \tau, \\ 1, & t \ge 1, \\ \frac{t-\tau}{1-\tau}, & t \in (\tau;1) \end{cases}$$

either mean square continuous or mean square differentiable (a) on $\mathbb{R}$. (b) On $[0,1]$?

4.20. Assume that function f is bounded and satisfies the Lipschitz condition. Let random variable τ be continuously distributed. Prove that the stochastic process $X(t) = f(t-\tau), t \in \mathbb{R}$ is mean square differentiable.

4.21. (*The fractional effect process*). Let $\{\varepsilon_n, n \ge 1\}$ be nonnegative nondegenerate i.i.d. random variables. Define $S_n = \sum_{k=1}^n \xi_k, f(x) = \left(1+x^4\right)^{-1}$. Prove that the stochastic process $X(t) = \sum_{n=1}^\infty f(t+S_n)$, $t \in \mathbb{R}$ is mean square continuously differentiable.

4.22. Is it possible that a stochastic process
(a) Has continuously differentiable trajectories but is not mean square continuously differentiable?
(b) Is mean square continuously differentiable but its trajectories are not continuously differentiable?

4.23. Prove that if $X(t),\ t\in\mathbb{R}^+$ is a mean square continuous process then the process $Y(t)=\int_0^t X(s)ds$ is mean square differentiable and $Y'(t)=X(t)$.

4.24. Let $\{W(t),\ t\in\mathbb{R}^+\}$ be the Wiener process.

(1) Prove that, for a given $\delta>0$, stochastic process $W_\delta(t)=(1/\delta)\int_t^{t+\delta}W(s)ds$, $t\in\mathbb{R}^+$ is mean square continuously differentiable. Find its derivative.

(2) Prove that $\text{l.i.m.}_{\delta\to0+}W_\delta(t)=W(t)$, where l.i.m. denotes the limit in the mean square sense.

Hints

4.4. Item (a) follows from item (b). Assume that the statement of item (b) is not true and prove that there exist $\varepsilon>0$ and sequences $\{t_n\}$ and $\{s_n\}$ converging to some point $t\in[a,b]$ such that $\mathsf{P}(|X(t_n)-X(s_n)|\geq\varepsilon)\geq\varepsilon$. Show that this contradicts the stochastic continuity of X at the point t.

4.6. Use Problem 4.5.

4.8. Prove and use the following fact. If H is a Hilbert space and $\{h_k,k\in\mathbb{N}\}$ is an orthogonal system of elements from this space with $\sum_k\|h_k\|_H^2<+\infty$, then there exists $h=\sum_k h_k\in H$, where the series converges in norm in H. In this problem, one should consider the increments of the process as elements in the Hilbert space $H=L_2(\Omega,\mathcal{F},\mathsf{P})$.

4.9. (1), (2) Show that

$$\mathsf{E}\left|\frac{N(t)-N(s)}{t-s}\right|^p\sim\lambda|t-s|^{1-p},s\to t.$$

(3)

$$\left\{\frac{N(t)-N(s)}{t-s}\to0,\quad s\to t\right\}\subset\{N(t-)=N(t)\}.$$

Derivatives of N in probability, a.s. and in the $L_p, p\in(0,1)$ sense are zero.

4.10. Use Problem 2.17 and Theorem 4.2.

4.12. (d) Use Problem 4.7.
(e) If $\xi\sim\mathrm{N}(a,\sigma^2)$ then $\mathsf{E}\xi^4=a^4+6a^2\sigma^2+3\sigma^4$. Prove this formula and use it.

4.16. (1) Both X and X' are stochastically continuous (prove this!), thus existence of their measurable modifications is provided by Theorem 3.1.

(2) Let $\varphi \in C_0^\infty(\mathbb{R})$ be a compactly supported nonnegative infinitely differentiable function with $\int_{-\infty}^{\infty}\varphi(x)dx = 1$. Define $\varphi_n(x) = n\varphi(nx)$, $X_n(t) := \int_{-\infty}^{\infty} X(t-s)\varphi_n(s)\, ds = \int_{-\infty}^{\infty} X(s)\varphi_n(t-s)ds$. Check that the trajectories of the process X_n are infinitely differentiable a.s., $X_n'(t) = \int_{-\infty}^{\infty} X'(s)\varphi_n(t-s)ds = \int_{-\infty}^{\infty} X(s)\varphi_n'(t-s)ds$, and for any t

$$\lim_{n\to\infty} X_n(t) = X(t), \quad \lim_{n\to\infty} X_n'(t) = X'(t) \tag{4.2}$$

in the mean square. Because the trajectories of X_n are continuously differentiable a.s., the Newton–Leibnitz formula implies

$$X_n(t) = X_n(0) + \int_0^t X_n'(s)ds \text{ a.s.} \tag{4.3}$$

One has

$$\mathsf{E}\left|\int_0^t X_n'(s)\,ds - \int_0^t X'(s)\,ds\right| \le \int_0^t \mathsf{E}\left|X_n'(s) - X'(s)\right| ds$$

and $\mathsf{E}|X_n'(s) - X'(s)| \le \mathsf{E}|X'(s)| + \int_{-\infty}^{\infty} \mathsf{E}|X'(s_1)|\varphi_n(s-s_1)ds_1$. Pass to the limit in (4.3) using (4.2) and the Lebesgue dominated convergence theorem.

(3) Follows immediately from the item (2).

(4) It is well known that if a function $f \in L_2(\mathbb{R})$ has the form

$$f(t) = c + \int_0^t g(s)ds, \quad t \in \mathbb{R}, \tag{4.4}$$

with some $c \in \mathbb{R}$, $g \in L_2(\mathbb{R})$, then $f \in W_2^1(\mathbb{R})$ and g is equal to its Sobolev derivative. Now use the result from the item (2). Note that, moreover, the function f appears to be absolutely continuous. In particular, f is differentiable for almost all t and g is its ordinary derivative.

(5) Statement (1) still holds true. In order to show this, consider the sequence of measurable processes $Y_n(t) := n\left(X\left(t+n^{-1}\right) - X(t)\right)$ (X is mean square continuous and thus can be supposed to be measurable). Then $Y_n(t) \to X'(t), n \to \infty$ in probability. Use Problem 3.45.

Statement (2) still holds true assuming that

$$\int_{-T}^{T} \mathsf{E}(X'(t))^2 dt < \infty, \quad T > 0. \tag{4.5}$$

This condition is needed in order to construct the approximating sequence $\{X_n\}$ and justify passing to the limit in (4.3). Note that under condition (4.5) the integral in the right-hand side of (4.1) is well defined as the Lebesque integral for a.s. ω but not necessarily as the mean square integral.

Statement (4) still holds true and statement (3) holds true assuming (4.5).

4.17. (1) (a) Use the Lebesgue dominated convergence theorem and check that $X'(t) = \mathbb{I}_{t\le\tau}$. Furthermore,

$$\mathsf{E}X(t) = \int_0^\infty \min(t,x)e^{-x}dx = \int_0^t xe^{-x}dx + \int_t^\infty te^{-x}dx = 1 - e^{-t}.$$

Let $s \le t$. Then

$$\mathsf{E}X(s)X(t) = \int_0^\infty \min(s,x)\min(t,x)e^{-x}dx =$$

$$\int_0^s x^2 e^{-x}dx + \int_s^t sxe^{-x}dx + \int_t^\infty ste^{-x}dx = -e^{-s}(s+2) + 2 - se^{-t}.$$

In order to obtain the covariance function of the derivative, use Theorem 4.2.
(b) Similarly to item (a), the process $X(t)$ is mean square continuous. If there exists $X'(t)$, then $X'(t) = 1$ a.s. (Problem 4.15) and the process $Y(t) = X(t) - t$ must be mean square continuously differentiable with zero derivative. That is, $Y(t) = Y(0)$ a.s. (Problem 4.16). But this is not correct.

4.19. The process $X(t)$ is mean square differentiable at every point t except possibly $t = 1$. At the same time, $X'(t) = 0$ when $t \le \tau$ or $t > 1$, and $X'(t) = 1/(1-\tau)$ when $t \in (\tau, 1]$. Because

$$\lim_{t \to 1-} \frac{X(t) - X(1)}{t-1} = \frac{1}{1-\tau} \ne 0 = \lim_{t \to 1+} \frac{X(t) - X(1)}{t-1} \quad \text{a.s.},$$

the process $\{X(t), t \in \mathbb{R}\}$ is not differentiable at $t = 1$.

Now, consider the restriction of $X(t)$ to $[0,1]$. Check that the mean square derivative $X'(1)$ exists if and only if $\mathsf{E}(1-\tau)^{-2} < \infty$; that is, $\alpha > 1$. Check that $X'(1) = (1-\tau)^{-1}$.

4.24. Use the considerations from the solution to Problem 4.23.

Answers and Solutions

4.5. For any $x < y$, $\mathsf{P}(X(t) \le x,\ X(s) \ge y) + \mathsf{P}(X(t) \ge y,\ X(s) \le x) \le \mathsf{P}(|X(t) - X(s)| > y - x) \to 0$ as $s \to t$ in the case when X is stochastically continuous at point t. On the other hand, let $\varepsilon, \delta > 0$ be fixed and we choose C so that $\mathsf{P}(|X(t)| > C) < \delta$. Consider the sets

$$A_k = \left\{(x,y) \mid x \le \frac{k\varepsilon}{2},\ y \ge \frac{(k+1)\varepsilon}{2}\right\}, \quad B_k = \left\{(x,y) \mid x \ge \frac{k\varepsilon}{2}, y \le \frac{(k-1)\varepsilon}{2}\right\},$$

$k \in \mathbb{Z}$. Let us assume that $m = [2C/\varepsilon] + 1$, then $\{(x,y) \mid |x-y| > \varepsilon, |x| \le C\} \subset \bigcup_{k=-m}^m (A_k \cup B_k)$. Under the assumptions made above $\mathsf{P}((X(t), X(s)) \in A_k \cup B_k) \to 0$ as $s \to t$ for every k. Therefore $\limsup_{s \to t} \mathsf{P}(|X(t) - X(s)| > \varepsilon) \le \mathsf{P}(|X(t)| > C) < \delta$. Because δ is arbitrary, we have $\mathsf{P}(|X(t) - X(s)| > \varepsilon) \to 0, s \to t$.

4.11. We get from the independence of increments that $\mathsf{D}(X(t) - X(0)) = nF(t/n)$ for every $n \in \mathbb{N}$, $t \in \mathbb{R}$. It follows from the equality $F(z) = F(-z)$ that $F'(0) = 0$. Obviously, $F(0) = \mathsf{D}(X(t) - X(t)) = 0$. Thus $\mathsf{D}(X(t) - X(s)) = o(|t-s|^2)$ as $|t - s| \to 0$. Therefore $\mathsf{D}(X(t) - X(0)) = 0$, $t \in \mathbb{R}$ and thus $X(t) = X(0) + a_X(t) - a_X(0)$, $t \in \mathbb{R}$.

4.12. (a) e^4+2e^2+e+25.
(b) e^2+4.
(c) $3e+\frac{2}{9}+\int_0^1 \frac{e^t-1}{t}\,dt$.
(d) 1.
(e) $\left(\frac{4}{3}\right)^4+6\left(\frac{4}{3}\right)^2\left(3e-2+\int_0^1 \frac{e^t-1}{t}\,dt\right)^2+3\left(3e-2+\int_0^1 \frac{e^t-1}{t}\,dt\right)^4$.

4.13. (a) $34+\frac{1}{3}$; (b) $\frac{1}{2}$.

4.14. (1) Yes, it is.
(2) By the Lebesgue dominated convergence theorem

$$\lim_{\varepsilon\to 0}\mathsf{E}\left(\frac{\alpha\sin(t+\varepsilon+\beta)-\alpha\sin(t+\beta)}{\varepsilon}-\alpha\cos(t+\beta)\right)^2=0.$$

The Lagrange theorem implies that the expression in parentheses is dominated by the quantity α^2.
(3) $R_X(s,t)=\frac{1}{2}\cos(t-s)$; $R_{X,X'}(s,t)=-\frac{1}{2}\sin(t-s)$; $R_{X'}(s,t)=\frac{1}{2}\cos(t-s)$, $a_X=a_{X'}\equiv 0$.

4.15. Mean square convergence of $(X(t)-X(t_0))/(t-t_0)$ to some random variable ξ as $t\to t_0$ implies convergence in probability. Thus, we can select a sequence $t_k\to t_0$, $k\to\infty$ such that

$$\frac{X(t_k)-X(t_0)}{t_k-t_0}\to\xi \text{ a.s.},\quad k\to\infty.$$

Since the trajectories of $X(t)$ are differentiable at the point t_0, the limit ξ coincides a.s. with ordinary derivative.

4.18. Similarly to item (a) of Problem 4.17, the process X is continuously differentiable in m.s. and

$$X'(t)=\begin{cases}1, & t\in[\tau-1,\tau],\\ -1, & t\in(\tau,\tau+1],\\ 0, & t\notin[\tau-1,\tau+1].\end{cases}$$

$$\mathsf{E}\left(X'(1)\right)^2=\mathsf{E}\mathbb{I}_{1\in[\tau-1;\tau+1]}=\mathsf{P}(\tau\in[0,2])=1-e^{2\alpha}.$$

4.20. The Lipschitz continuity of the function f implies its absolute continuity and therefore its differentiability at almost every point w.r.t. the Lebesgue measure. Let U be the set of the points where the derivative of the function f exists. Then, for every $t_0\in\mathbb{R}$

$$\mathsf{P}\left(\exists\lim_{t\to t_0}\frac{X(t)-X(t_0)}{t-t_0}\right)\geq\mathsf{P}(t_0-\tau\in U)=1.$$

On the other hand, for every $t,t_0,t\neq t_0$, the absolute value of the fraction $(X(t)-X(t_0))/(t-t_0)$ does not exceed the Lipschitz constant of the function f. Therefore, by the Lebesgue theorem on dominated convergence, the process $\{X(t),\ t\in\mathbb{R}\}$ is mean square differentiable and $X'(t)=g(t-\tau)$, where

$$g(t) = \begin{cases} f'(t), & t \in U \\ 0, & t \notin U \end{cases}.$$

4.22. (a) Yes, if $\mathsf{E}(X(t))^2 = +\infty$.
(b) Yes. Look at the process from Problem 4.17.

4.23. Let $\varepsilon > 0$ be arbitrary. Choose $\delta > 0$ such that $\mathsf{E}(X(t) - X(s))^2 < \varepsilon$ as soon as $|t - s| < \delta$. Thus, for every $s \in (t - \delta, t + \delta), s \neq t$:

$$\mathsf{E}\left(\frac{\int_s^t X(z)dz}{t-s} - X(t)\right)^2 = \mathsf{E}\left(\frac{\int_s^t (X(z) - X(t))dz}{t-s}\right)^2$$

$$\leq \frac{\mathsf{E}\int_s^t (X(z) - X(t))^2 dz}{t-s} < \varepsilon.$$

5

Stochastic processes with independent increments. Wiener and Poisson processes. Poisson point measures

Theoretical grounds

Definition 5.1. *Let* $\mathbb{T} \subset \mathbb{R}$ *be an interval. A stochastic process* $\{X(t), t \in \mathbb{T}\}$ *taking values in* $\mathbb{R}^d$ *is a* process with independent increments *if for any* $m \geq 1$ *and* $t_0, \ldots, t_m \in \mathbb{T}$, $t_0 < \cdots < t_m$ *the random vectors* $X(t_0), X(t_1) - X(t_0), \ldots, X(t_m) - X(t_{m-1})$ *are jointly independent. A process with independent increments is said to be* homogeneous *if, for any* $t, s, v, u \in \mathbb{T}$ *such that* $t - s = v - u$, *the increments* $X(t) - X(s)$ *and* $X(v) - X(u)$ *have the same distribution.*

The following theorem shows that all the finite-dimensional distributions for the process with independent increments on $\mathbb{T} = \mathbb{R}^+$ are uniquely determined by the starting distribution (i.e., the distribution of $X(0)$) and distributions of the increments (i.e., the distributions of $X(t) - X(s)$, $t > s \geq 0$).

Theorem 5.1. *The finite-dimensional distributions of the process with independent increments* $\{X(t), t \in \mathbb{R}^+\}$ *taking values in* $\mathbb{R}^d$ *are uniquely determined by the family of the characteristic functions*

$$\left\{ \phi_0(\cdot) = \mathsf{E} \exp\{i(\cdot, X(0))_{\mathbb{R}^d}\}, \ \phi_{s,t}(\cdot) = \mathsf{E} \exp\{i(\cdot, X(t) - X(s))_{\mathbb{R}^d}\}, \ 0 \leq s < t \right\}.$$

On the other hand, in order for a family of the functions $\{\phi_0, \phi_{s,t}, 0 \leq s < t\}$ *to determine a process with independent increments, it is necessary and sufficient that*
 (1) Every function ϕ_0, $\phi_{s,t}$, $0 \leq s < t$ *is a characteristics function of a random vector in* $\mathbb{R}^d$.
 (2) The following consistency condition is fulfilled

$$\phi_{s,t} \phi_{t,u} = \phi_{s,u}, \quad 0 \leq s < t < u.$$

Theorem 5.1 is a version of the Kolmogorov theorem on finite-dimensional distributions (see Theorems 1.1, 2.2 and Problem 5.2).

D. Gusak et al., *Theory of Stochastic Processes*, Problem Books in Mathematics, DOI 10.1007/978-0-387-87862-L5, © Springer Science+Business Media, LLC 2010

Definition 5.2. The (one-dimensional) Wiener process W *is the real-valued homogeneous process with independent increments on* $\mathbb{R}^+$ *such that* $W(0)=0$ *and for any* $t>s$ *the increment* $W(t)-W(s)$ *has the distribution* $\mathcal{N}(0,t-s)$. *The* multidimensional Wiener process *is the m-dimensional process* $W(t)=(W^1(t),\dots,W^m(t))$, *where* $\{W^i(t),t\geq 0\}$ *are jointly independent Wiener processes.*

Let $\mathbb{T}\subset\mathbb{R}$ be an interval and $\varkappa$ be a locally finite measure on $\mathcal{B}(\mathbb{T})$ (i.e., $\varkappa$ possesses finite values on bounded intervals).

Definition 5.3. The Poisson process X with intensity measure $\varkappa$ *on* $\mathbb{T}$ *is the process with independent increments such that the increment* $X(t)-X(s)$ *for any* $t>s$ *has the distribution* $\mathrm{Pois}(\varkappa((s,t]))$.

The Poisson process N with parameter $\lambda>0$ *is a homogeneous process with independent increments defined on* $\mathbb{T}=\mathbb{R}^+$, *such that* $N(0)=0$ *and for any* $t>s$ *the increment* $N(t)-N(s)$ *has the distribution* $\mathrm{Pois}(\lambda(t-s))$.

The Poisson process with parameter λ is, obviously, the Poisson process with intensity measure $\varkappa=\lambda\cdot\lambda^1|_{\mathbb{R}^+}$ on $\mathbb{T}=\mathbb{R}^+$ (λ^1 is a Lebesgue measure, $\lambda^1|_{\mathbb{R}^+}$ is its restriction to $\mathbb{R}^+$). Note that this form of the measure $\varkappa$ is implied by homogeneity of the process. Further on, we use the short name *the Poisson process* for the Poisson process with some parameter λ.

Let $0\leq\tau_1\leq\tau_2\leq\cdots$ be a sequence of random variables and $X(t)=\sum_{k=1}^{\infty}\mathbb{I}_{\tau_k\leq t}, t\in\mathbb{R}^+$. The process X is called the *registration process* associated with the sequence $\{\tau_k\}$. The terminology is due to the following widely used model. Assume that some sequence of events may happen at random (e.g., particles are registered by a device, claims are coming into a telephone exchange or a server, etc.). If the variable τ_k is interpreted as the time moment when the kth event happens, then $X(t)$ counts the number of events that happened until time t.

Proposition 5.1. *The Poisson process with parameter* λ *is the registration process associated with the sequence* $\{\tau_k\}$ *such that* $\tau_1,\tau_2-\tau_1,\tau_3-\tau_2,\dots$ *are i.i.d. random variables with the distribution* $\mathrm{Exp}(\lambda)$.

Let us give the full description of the characteristic functions for the increments of stochastically continuous homogeneous processes with independent increments (such processes are called *Lévy processes*).

Theorem 5.2. *(The Lévy–Khinchin formula) Let* $\{X(t),t\geq 0\}$ *be a stochastically continuous homogeneous processes with independent increments taking values in* $\mathbb{R}^d$. *Then*

$$\phi_{s,t}(z)=\exp[(t-s)\psi(z)],\quad s\leq t,\quad z\in\mathbb{R}^d,$$

$$\psi(z)=i(z,a)_{\mathbb{R}^d}-\frac{1}{2}(Bz,z)_{\mathbb{R}^d}+\int_{\mathbb{R}^d}\left[e^{i(z,u)_{\mathbb{R}^d}}-1-i(z,u)_{\mathbb{R}^d}\mathbb{I}_{\|u\|_{\mathbb{R}^d}\leq 1}\right]\Pi(du),\quad (5.1)$$

where $a\in\mathbb{R}^d$, *the matrix* $B\in\mathbb{R}^{d\times d}$ *is nonnegatively defined, and the measure* Π *satisfies the relation*

$$\int_{\mathbb{R}^d} (\|u\|^2_{\mathbb{R}^d} \wedge 1)\Pi(du) < +\infty.$$

The function ψ *is called the* cumulant *of the process* X; *the measure* Π *is called the* Lévy measure *of the process* X.

Consider a "multidimensional segment" $\mathbb{T} = \mathbb{T}_1 \times \cdots \times \mathbb{T}_d \subset \mathbb{R}^d$ with some segments $\mathbb{T}_i \subset \mathbb{R}, i = 1, \ldots, d$. On the set $\mathbb{T}$, the partial order is naturally defined: $t = (t^1, \ldots, t^d) \geq s = (s^1, \ldots, s^d) \Leftrightarrow t^i \geq s^i, i = 1, \ldots, d$. For a function $X : \mathbb{T} \to \mathbb{R}$ and $s \leq t$, we denote by $\Delta_{s,t}(X)$ the increment of X on the "segment" $(s,t] = (s^1,t^1] \times \cdots \times (s^d,t^d]$; that is,

$$\Delta_{s,t}(X) = \sum_{\varepsilon_1,\ldots,\varepsilon_d \in \{0,1\}} (-1)^{\varepsilon_1+\cdots+\varepsilon_d} X\Big(t^1 - \varepsilon_1(t^1 - s^1), \ldots, t^d - \varepsilon_d(t^d - s^d)\Big).$$

Definition 5.4. *Let* $\varkappa$ *be a locally finite measure on* $\mathcal{B}(\mathbb{T})$. *A real-valued random field* $\{X(t),\, t \in \mathbb{T}\}$ *is called the* Poisson random field with intensity measure $\varkappa$, *if for any* $t_0, \ldots, t_m \in \mathbb{T}$, $t_0 \leq t_1 \leq \cdots \leq t_m$ *the variables* $X(t_0)$, $\Delta_{t_0,t_1}(X), \ldots, \Delta_{t_{m-1},t_m}(X)$ *are jointly independent and for every* $s \leq t$ *the increment* $\Delta_{s,t}(X)$ *has the distribution* $\mathrm{Pois}(\varkappa((s,t]))$.

Let $(E, \mathcal{E}, \mu)$ be a space with σ-finite measure. Denote $\mathcal{E}_\mu = \{A \in \mathcal{E} \mid \mu(A) < +\infty\}$.

Definition 5.5. *A* random point measure *on a ring* $\mathcal{K} \subset \mathcal{E}_\mu$ *is the mapping* ν *that associates with every set* $A \in \mathcal{K}$ *a* $\mathbb{Z}^+$*-valued random variable* $\nu(A)$, *and satisfies the following condition. For any* $A_i \in \mathcal{K}, i \in \mathbb{N}$ *such that* $A_i \cap A_j = \varnothing, i \neq j, \bigcup_i A_i \in \mathcal{K}$,

$$\nu\Big(\bigcup_i A_i\Big) = \sum_i \nu(A_i) \quad a.s.$$

The random point measure is called the Poisson point measure with intensity measure μ, *if for any* $A_1, \ldots, A_m \in \mathcal{E}_\mu, A_i \cap A_j = \varnothing, i \neq j$ *the values* $\nu(A_1), \ldots, \nu(A_m)$ *are jointly independent and for every* $A \in \mathcal{E}_\mu$ *the value* $\nu(A)$ *has the distribution* $\mathrm{Pois}(\mu(A))$.

The term "random point measure" can be explained by the following result (see Problem 5.41) which shows that this object has a natural interpretation as a collection of measures indexed by $\omega \in \Omega$ and concentrated on the countable subsets of E ("point measures").

Proposition 5.2. *Let* ν *be a point measure on a ring* $\mathcal{K}$. *Assume that* $\sigma(\mathcal{K})$ *contains all one-point sets and there exists a countable ring* $\mathcal{K}_0 \subset \mathcal{K}$ *such that* $\sigma(\mathcal{K}_0) \supset \mathcal{K}$. *Then there exists the mapping* $\hat{\nu} : \Omega \times \mathcal{K} \to \mathbb{R}^+$ *such that:*

(1) For every $A \in \sigma(\mathcal{K})$ *the function* $\hat{\nu}(\cdot, A)$ *is an extended random variable.*

(2) For every $\omega \in \Omega$ *the mapping* $A \mapsto \hat{\nu}(\omega, A)$ *is the* σ*-finite measure concentrated at some enumerable set and taking natural values on the points of this set;*

(3) $\hat{\nu}(\cdot, A)$ *is equal to* $\nu(A)$ *a.s. for every* $A \in \mathcal{K}$.

If $\sigma(\mathcal{K}) = \mathcal{E}$ then the Poisson point measure with intensity measure μ defined on $\mathcal{K}$ can be extended to the Poisson point measure defined on $\mathcal{E}_\mu$, and such an extension is unique. For a given Poisson point measure ν with intensity μ, the corresponding *centered* (or *compensated*) Poisson point measure is defined as $\tilde{\nu}(A) = \nu(A) - \mu(A)$, $A \in \mathcal{E}_\mu$.

Let $f : E \to \mathbb{R}$ be measurable w.r.t. $\sigma(\mathcal{K})$ and $\{f \neq 0\} \subset A$ for some $A \in \mathcal{K}$. Then the integral of f over the random point measure ν is naturally defined by

$$\left[\int_E f(z)\nu(dz)\right](\omega) = \int_E f(z)\hat{\nu}(\omega, dz), \quad \omega \in \Omega, \tag{5.2}$$

where $\hat{\nu}$ is the collection of measures given by Proposition 5.2 (see Problem 5.42).

If ν is a Poisson measure, $\tilde{\nu}$ is the corresponding compensated measure, and $f \in L_2(E,\mu)$, then the integral $\int_E f(z)\tilde{\nu}(dz)$ is well defined as the stochastic integral over an orthogonal random measure (see Chapter 8 and Problem 5.43). Two definitions of the integrals mentioned above are adjusted in the sense that if $f \in L_2(E, d\mu)$ and $\{f \neq 0\} \in \mathcal{E}_\mu$ then

$$\int_E f(z)\nu(dz) - \int_E f(z)\mu(dz) = \int_E f(z)\tilde{\nu}(dz) \tag{5.3}$$

(see Problem 5.44).

Frequently, one needs to consider Poisson point measures defined on a product of spaces, for instance, $E = \mathbb{R}^+ \times \mathbb{R}^d, \mathcal{E} = \mathcal{B}(E)$. In this case, the above-defined integrals of a function $(s,u) \mapsto f(s,u)1\!\!1_{s \le t, u \in A}$ are denoted

$$\int_0^t \int_A f(s,u)\nu(ds,du), \quad \int_0^t \int_A f(s,u)\tilde{\nu}(ds,du).$$

Theorem 5.3. *Let $\{X(t), t \geq 0\}$ be a stochastically continuous homogeneous process with independent increments taking its values in $\mathbb{R}^d$. Let a, B, Π be, respectively, the vector, the matrix, and the measure appearing in the Lévy–Khinchin formula for the cumulant of this process (see Theorem* 5.2*). Then there exist the independent d-dimensional Wiener process W and Poisson point measure on $E = \mathbb{R}^+ \times \mathbb{R}^d$ with the intensity measure $\mu = \lambda^1|_{\mathbb{R}^+} \times \Pi$ such that*

$$X(t) = at + B^{1/2}W(t) + \int_0^t \int_{\{\|u\|_{\mathbb{R}^d} > 1\}} u\nu(ds,du) + \int_0^t \int_{\{\|u\|_{\mathbb{R}^d} \le 1\}} u\tilde{\nu}(ds,du), \tag{5.4}$$

$t \in \mathbb{R}^+$.

And vice versa, let X be determined by the equality (5.4) *with arbitrary a, B, and independent d-dimensional Wiener process W and Poisson point measure ν with the intensity measure $\mu = \lambda^1|_{\mathbb{R}^+} \times \Pi$. Then X is a Lévy process and its cumulant is determined by the equality* (5.1).

Bibliography

[9], Chapter II; [24], Volume 1, Chapter III, §1; [25], Chapter VI; [15], Chapter VIII; [78]; [79], Chapters 2, 27, 28.

Problems

5.1. Verify that the consistency condition (condition 2 of Theorem 5.1) holds true for characteristic functions of the increments for:
(a) Wiener process;
(b) Poisson process with intensity measure $\varkappa$.
Such verification is necessary for Definitions 5.2 and 5.3 to be formally correct.

5.2. Let the family of functions $\{\psi_0, \psi_{s,t}, 0 \le s < t\}$ be given, and let every function of the family be a characteristic function of some random variable. For any $m \ge 1, t_1, \dots, t_m \in \mathbb{R}^+, z_1, \dots, z_m \in \mathbb{R}$, put

$$\phi_{t_1,\dots,t_m}(z_1,\dots,z_m) = \psi_0(z_{\pi(1)} + \dots + z_{\pi(m)})\psi_{0,\pi(t_1)}(z_{\pi(1)} + \dots + z_{\pi(m)})$$

$$\times \psi_{\pi(t_1),\pi(t_2)}(z_{\pi(2)} + \dots + z_{\pi(m)}) \dots \psi_{\pi(t_{m-1}),\pi(t_m)}(z_{\pi(m)}),$$

where the permutation π is such that $t_{\pi(1)} \le \dots \le t_{\pi(m)}$.

Prove that the necessary and sufficient condition for the family

$$\{\phi_{t_1,\dots,t_m}, t_1,\dots,t_m \in \mathbb{T}, m \ge 1\}$$

to satisfy consistency conditions of Theorem 2.2 is that the family $\{\psi_0, \psi_{s,t}, 0 \le s < t\}$ satisfies the consistency condition of Theorem 5.1. Prove also that if these conditions hold then the process X with the finite-dimensional characteristic functions $\{\phi_{t_1,\dots,t_m}\}$ is a process with independent increments.

5.3. Let $\{\mu_t, t > 0\}$ be a family of probability measures on $\mathbb{R}$ such that, for any $s,t > 0$, μ_{t+s} equals the convolution of μ_t and μ_s. Prove that there exists homogeneous process with independent increments $\{X(t), t > 0\}$ such that for every t the distribution of $X(t)$ equals μ_t. Describe the finite-dimensional distributions of the process $\{X(t), t > 0\}$.

5.4. Let N be the Poisson process with intensity λ. Find:
(a) $\mathsf{P}(N(1) = 2, N(2) = 3, N(3) = 5)$.
(b) $\mathsf{P}(N(1) \le 2, N(2) = 3, N(3) \ge 5)$.
(c) $\mathsf{P}(N(\sqrt{2}) = 3)$.
(d) $\mathsf{P}(N(3) = \sqrt{2})$.
(e) $\mathsf{P}(N(4) = 3, N(1) = 2)$.
(f) $\mathsf{P}(N(2)N(3) = 2)$.
(g) $\mathsf{P}(N^2(2) \ge 3N(2) - 2)$.
(h) $\mathsf{P}(N(2) + N(3) = 1)$.

5.5. Find

$$\mathsf{E}(X(1) + 2X(2) + 3X(3))^2; \quad \mathsf{E}(X(1) + 2X(2))^3; \quad \mathsf{E}(X(1) + 2X(2) + 1)^3,$$

where X is:
(a) the Wiener process;
(b) the Poisson process.

5.6. Specify the finite-dimensional distributions for the Poisson process.

5.7. Assume that stochastic process $\{X(t), t \geq 0\}$ satisfies the conditions:
(a) X takes values in $\mathbb{Z}^+$ and $X(0) = 0$.
(b) X has independent increments.
(c) $\mathsf{P}(|X(t+h) - X(t)| > 1) = o(h)$, $\mathsf{P}(X(t+h) - X(t) = 1) = \lambda h + o(h)$, $h \to 0$ with some given $\lambda > 0$.

Prove that X is the Poisson process with intensity λ.

5.8. Assume that stochastic process $\{X(t),\ t \geq 0\}$ satisfies conditions (a),(b) from the previous problem and condition
(c′) $\mathsf{P}(|X(t+h)-X(t)| > 1) = o(h)$, $\mathsf{P}(X(t+h)-X(t) = 1) = \lambda(t)h + o(h)$, $h \to 0$, where $\lambda(\cdot)$ is some continuous nonnegative function.

Find the finite-dimensional distributions for the process X.

5.9. Let $\{N(t),\ t \in \mathbb{R}^+\}$ be the Poisson process, τ_1 be the time moment of its first jump. Find the conditional distribution of τ_1 given that the process has on $[0,1]$:
(a) Exactly one jump
(b) At most one jump
(c) At least m jumps ($m \in \mathbb{N}$).

5.10. Prove that the distribution of the Poisson process defined on $[0,1]$ conditioned that the process has m jumps ($m \in \mathbb{N}$) on $[0,1]$ is equal to the distribution of the process $X(t) = \sum_{k=1}^{m} \mathbb{I}_{\xi_k \leq t}$, $t \in [0,1]$, where $\xi_1, \ldots, \xi_m$ are i.i.d. random variables uniformly distributed on $[0,1]$.

5.11. Let τ be a random variable, and $X(t) = \mathbb{I}_{t \geq \tau}, t \in \mathbb{R}$ be the corresponding waiting process. What distribution should the random variable τ follow in order for X to be a process with independent increments?

5.12. Prove Proposition 5.1.

5.13. Let τ_n be the time moment of the nth jump for the Poisson process. Prove that the distribution density of τ_n equals

$$\frac{\lambda^n x^{n-1}}{n!} e^{-\lambda x}, \quad x \geq 0.$$

5.14. Let N be the Poisson process, τ_n be the time moment of its nth jump, and

$$X(t) = \begin{cases} N(t), & t \in [\tau_{2n}, \tau_{2n+1}), \\ N(t) - 1, & t \in [\tau_{2n-1}, \tau_{2n}). \end{cases}$$

Draw the trajectories of the process X. Calculate $\mathsf{P}(X(3) = 2)$, $\mathsf{P}(X(3) = 2, X(5) = 4)$. Is the process X a process with independent increments?

5.15. Let the number of signals transmitted via a communication channel during time $[0,t]$ be the Poisson process N with intensity $\lambda > 0$. Every signal is successfully received with probability $p \in (0;1)$, independently of the process N and other signals. Let $\{X(t), t \geq 0\}$ be the number of a signals received succesfully. Find: (a) one-dimensional; (b) multidimensional distributions of X.

5.16. The numbers of failures in work for plants A and B during the time period $[0,t]$ are characterized by two independent Poisson processes with intensities λ_1 and λ_2, respectively. Find: (a) one-dimensional; (b) multidimensional distributions of the total number of failures for both plants A and B during the time period $[0,t]$.

5.17. Let $\{\xi_n,\ n \geq 1\}$ be i.i.d. random variables with the distribution function F and ν be a Poisson distributed random variable with parameter λ independent of $\{\xi_n\}$. Prove that the process $\{X_\nu(x) = \sum_{n=1}^{\nu} \mathrm{1\!I}_{\xi_n \leq t}, t \in \mathbb{R}\}$ is the Poisson process with the intensity measure $\varkappa$ determined by the relation $\varkappa((a,b]) = \lambda(F(b) - F(a)), a \leq b$.

5.18. Within the conditions of the previous problem, prove that all multidimensional distributions of the process $\lambda^{-1}X_\nu$ weakly converge as $\lambda \to +\infty$ to the corresponding finite-dimensional distributions of a (nonrandom) process that is equal to the determinate function F.

5.19. Let $\{\xi_n, n \geq 1\}$ be i.i.d. random variables with the distribution function F and $\{N(t), t \in \mathbb{R}^+\}$ be the Poisson process with intensity λ independent of $\{\xi_n\}$. Prove that $\{X(x,t) = \sum_{n=1}^{N(t)} \mathrm{1\!I}_{\xi_n \leq x}, (x,t) \in \mathbb{R} \times \mathbb{R}^+\}$ is the Poisson random field with intensity $\varkappa \times \lambda^1|_{\mathbb{R}^+}$, where $\varkappa$ is the measure defined in Problem 5.17.

5.20. *The compound Poisson process* is a process of the form $X(t) = \sum_{k=1}^{N(t)} \xi_k, t \in \mathbb{R}^+$, where $\xi_k, k \in \mathbb{N}$ are i.i.d. random variables and $\{N(t),\ t \in \mathbb{R}^+\}$ is a Poisson process independent of $\{\xi_n\}$. Prove that the compound Poisson process is a homogeneous process with independent increments.

5.21. Prove that the sum of two processes with independent increments, which are independent of each other, is again a process with independent increments.

5.22. Let W be the Wiener process, $N_1, \ldots, N_m$ be the Poisson processes with parameters $\lambda_1, \ldots, \lambda_m$ and the processes $W, N_1, \ldots, N_m$ be jointly independent. Let also $c_0, \ldots, c_m \in \mathbb{R}$. Prove that $X = c_0 W + c_1 N_1 + \cdots + c_m N_m$ is a homogeneous process with independent increments and find the parameters a, B, Π in the Lévy–Khinchin formula for X (Theorem 5.2).

5.23. Let W be the Wiener process, N^λ and N^μ be two Poisson processes with intensities λ and μ, respectively. Let also $\{\eta_i, i \geq 1\}$ be i.i.d. random variables exponentially distributed with parameter α and $\{\zeta_k, k \geq 1\}$ be i.i.d. random variables exponentially distributed with parameter β. The processes W, N^λ, N^μ and sequences $\{\eta_i\}, \{\zeta_k\}$ are assumed to be jointly independent. Prove that

$$X(t) = bW(t) + \sum_{i=1}^{N^\lambda(t)} \eta_i - \sum_{i=1}^{N^\mu(t)} \zeta_i, \quad t \in \mathbb{R}^+$$

is a homogeneous process with independent increments and find the parameters a, B, Π in the Lévy–Khinchin formula for X. Find the characteristic function of the variable $X(t), t > 0$ and express the distribution density for this variable in integral form.

5.24. Prove that the m-dimensional Wiener process is a process with independent increments and each of its increments $W(t)-W(s)$, $t>s$ has the distribution $\mathcal{N}(0,(t-s)I_{\mathbb{R}^m})$.

5.25. Let $\{W(t),\ t\geq 0\}$ be the two-dimensional Wiener process, $B(0,r)=\{x\in\mathbb{R}^2|\ \|x\|\leq r\}$, $r>0$. Find $\mathsf{P}(W(t)\in B(0,r))$.

5.26. Let $\{W(t)=(W_1(t),\ldots,W_m(t)),\ t\in\mathbb{R}^+\}$ be an m-dimensional Wiener process and let set $A\subset\mathbb{R}^m$ have zero Lebesgue measure. Prove that the total time spent by W in the set A equals zero a.s. (compare with Problem 3.21).

5.27. Let $\{W(t)=(W_1(t),\ldots,W_m(t)),\ t\in\mathbb{R}^+\}$ be an m-dimensional Wiener process and $x=(x_1,\ldots,x_m)\in\mathbb{R}^m$ be a point such that $\sum_{i=1}^m x_i^2=1$. Prove that $Y(t):=\sum_{i=1}^m x_iW_i(t),\ t\in\mathbb{R}^+$ is the Wiener process.

5.28. Let $\{W(t)=(W_1(t),\ldots,W_m(t)),\ t\in\mathbb{R}^+\}$ be an m-dimensional Wiener process and an $(m\times m)$-matrix U have real-valued entries and be orthogonal (i.e., $UU^T=E$). Prove that $\{\widetilde{W}(t)=UW(t),\ t\geq 0\}$ is again an m-dimensional Wiener process.

5.29. Prove that there exists a homogeneous process $\{X(t),\ t>0\}$ with independent increments and distribution density

$$p_t(x)=\frac{1}{\Gamma(t)}x^{t-1}e^{-x}\mathbb{I}_{x>0}.$$

5.30. Prove that there exists homogeneous process $\{X(t),\ t>0\}$ with independent increments and the characteristic function $\mathsf{E}e^{izX(t)}=e^{-t|z|},\ t>0$.

5.31. Find the parameters a,B,Π in the Lévy–Khinchin formula for the process from:
(a) Problem 5.29.
(b) Problem 5.30.

5.32. Prove that there does not exist a process with independent increments X such that:
(a) $X(0)$ has a continuous distribution but the distribution of $X(1)$ has an atom.
(b) The distribution of $X(0)$ is absolutely continuous but the distribution of $X(1)$ is not.

5.33. Prove that there does not exist a process with independent increments X such that $X(0)$ is uniformly distributed on $[0,1]$ and $X(1)$ is exponentially distributed.

5.34. Prove that there does not exist a homogeneous process with independent increments X with $\mathsf{P}(X(1)-X(0)=\pm 1)=\frac{1}{2}$.

5.35. Give examples of homogeneous processes X with independent increments defined on $[0,1]$ for which $X(0)=0$ and the distribution of $X(1)$ is:
(a) discrete
(b) absolutely continuous
(c) continuous singular.

5.36. Prove that the process with independent increments $\{X(t),\ t \in \mathbb{R}^+\}$ is stochastically continuous if and only if its one-dimensional characteristic function $\phi_t(z) = \mathsf{E}e^{izX(t)}$, $t \in \mathbb{R}^+$, $z \in \mathbb{R}$ is a continuous function w.r.t. t for every fixed z.

5.37. Assume $\{X(t),\ t \in [0,1]\}$ is a process with independent increments. Does it imply that the process $Y(t) = X(-t)$, $t \in [-1,0]$ has independent increments? Compare with Problem 12.20.

5.38. Let $\{X(t),\ t \in \mathbb{R}^+\}$ be a nondegenerate homogeneous process with independent increments. Prove that $\mathsf{P}(|X(t)| > A) > 0$ for any $t > 0$ and $A > 0$.

5.39. Let $\{X(t),\ t \in \mathbb{R}^+\}$ be a nondegenerate homogeneous process with independent increments. Prove that for every $a > 0,\ b > 0$ there exist random variables $\tau_n, \sigma_n, n \geq 1$ such that almost surely $\tau_n < \sigma_n < \tau_{n+1},\ \sigma_n - \tau_n \leq a$ and $|X(\sigma_n) - X(\tau_n)| > b$ for every $n \geq 1$.

5.40. Let $\{X(t),\ t \in \mathbb{R}^+\}$ be a homogeneous process with independent increments and piecewise constant trajectories. Prove that X is a compound Poisson process (see Problem 5.20).

5.41. Prove Proposition 5.2.

5.42. Prove that the formula (5.2) defines a random variable (i.e., a measurable function of ω).

5.43. Prove that the compensated Poisson point measure $\tilde{\nu} = \nu - \mu$ is the centered orthogonal measure with a structural measure μ (here ν is a Poisson point measure with the intensity measure μ).

5.44. Prove equality (5.3).

5.45. Let ν be the Poisson point measure with intensity measure μ. Prove that

(1) The characteristic function of a random variable $\int_E f(z)\nu(dz)$ equals

$$\phi(z) = \exp\left[\int_E (e^{itf(z)} - 1)\mu(du)\right].$$

(2) The characteristic function of a random variable $\int_E g(z)\tilde{\nu}(dz)$ equals

$$\tilde{\phi}(z) = \exp\left[\int_E (e^{itg(z)} - 1 - iz)\mu(du)\right]$$

(the functions f, g are such that the corresponding integrals are correctly defined).

5.46. Let $\{X(t),\ t \in \mathbb{T}_1 \times \cdots \times \mathbb{T}_d\}$ be the Poisson random field with intensity measure $\varkappa$. Define the mapping ν on a ring $\mathcal{K} = \{\bigcup_{i=1}^m (s^i, t^i],\ s^i, t^i \in \mathbb{T},\ s^i \leq t^i, i = 1, \ldots, m,\ m \in \mathbb{N}\}$ by the equality

$$\nu\left(\bigcup_{i=1}^m (s^i, t^i]\right) = \sum_{i=1}^m \Delta_{s^i, t^i}(X), \text{ if } (s^i, t^i] \cap (s^j, t^j] = \varnothing, \quad i \neq j.$$

Prove that ν is the Poisson point measure with intensity measure $\varkappa$.

5.47. Let $\mathbb{T} \subset \mathbb{R}$ be an interval and let ν be a Poisson point measure with intensity measure $\varkappa$ defined on the ring $\mathcal{K} = \{\bigcup_{i=1}^m (s^i, t^i],\ s^i, t^i \in \mathbb{T},\ s^i \le t^i,\ i = 1, \dots, m,\ m \in \mathbb{N}\}$. Assume that $\{X(t),\ t \in \mathbb{T}\}$ is a stochastic process such that $X(t) - X(s) = \nu((s,t])$ for any $s, t \in \mathbb{T},\ s \le t$. Does it imply that X is the Poisson process with intensity measure $\varkappa$? Compare with the previous problem.

5.48. Let $\{X(t) = \sum_{k=1}^{N(t)} \xi_k, t \in \mathbb{R}^+\}$ be a compound Poisson process. Define the point measure ν on the ring $\mathcal{K} = \{\bigcup_{i=1}^m (s^i, t^i],\ s^i, t^i \in \mathbb{R}^+,\ s^i \le t^i, i = 1, \dots, m,\ m \in \mathbb{N}\}$ by equality $\nu(\bigcup_{i=1}^m (s^i, t^i]) = \sum_{i=1}^m \Delta_{s^i, t^i}(X)$, as $(s^i, t^i] \cap (s^j, t^j] = \varnothing,\ i \ne j$. What distribution should the random variables $\{\xi_k\}$ follow in order for ν to be a Poisson point measure? What is its intensity measure in that case?

5.49. Let $\{X(t) = \sum_{k=1}^{N(t)} \xi_k,\ t \in \mathbb{R}^+\}$ be a compound Poisson process, $E = \mathbb{R}^+ \times \mathbb{R}$, $\mathcal{K} = \{\bigcup_{i=1}^m (a^i, b^i],\ a^i, b^i \in E,\ a^i \le b^i,\ i = 1, \dots, m,\ m \in \mathbb{N}\}$. For $a = (s,x), b = (t,y) \in E, a \le b$ we define $\nu((a,b]) = \sum_{k=N(s)}^{N(t)} \mathbb{I}_{\xi_k \in (x,y]}$ and put $\nu(\bigcup_{i=1}^m (a^i, b^i]) = \sum_{i=1}^m \nu((a^i, b^i])$, as $(a^i, b^i] \cap (a^j, b^j] = \varnothing, i \ne j$. Prove that ν is a Poisson point measure with intensity measure $\lambda(\lambda^1 \times \mu)$, where λ is the parameter of the process N, λ^1 is the Lebesgue measure, and μ is the distribution of the variable ξ_1.

5.50. Let $(E, \mathcal{E})$ be a measurable space, $\{X_n, n \ge 1\}$ be a sequence of i.i.d random elements with values in E, and ζ be a random variable following the distribution $\mathrm{Pois}(\lambda)$ and independent of $\{X_n, n \ge 1\}$. Prove that the mapping $\nu : \mathcal{E} \ni A \mapsto \nu(A) = \sum_{k=1}^{\zeta} \mathbb{I}_{X_k \in A}$ is the Poisson point measure with the intensity measure $\lambda\mu$, where μ is the distribution of X_1.

5.51. Let $\{X(t) = \sum_{k=1}^{N(t)} \xi_k, t \in \mathbb{R}^+\}$ be a compound Poisson process, and $\alpha > 0$ be a fixed number. Prove that $\mathsf{E}|X(t)|^\alpha < +\infty$ for any $t > 0$ if and only if $\mathsf{E}|\xi_1|^\alpha < +\infty$.

5.52. Let X be a Lévy process with the Lévy measure Π, and $\alpha > 0$ be a fixed number. Prove that $\mathsf{E}\|X(t)\|_{\mathbb{R}^d}^\alpha < +\infty$ for any $t > 0$ if and only if $\int_{\|u\|_{\mathbb{R}^d} > 1} \|u\|_{\mathbb{R}^d}^\alpha \Pi(du) < +\infty$.

5.53. (General "0 and 1" rule) Let $\{\mathcal{G}_\alpha, \alpha \in \mathcal{A}\}$ be independent σ-algebras, $\mathcal{A} \supset \mathcal{A}_1 \supset \mathcal{A}_2 \supset \cdots, \bigcap_{n=1}^\infty \mathcal{A}_n = \varnothing$, $\mathcal{B}_k = \sigma\left(\bigcup_{\alpha \in \mathcal{A}_k} \mathcal{G}_\alpha\right)$, $k \in \mathbb{N}$. Prove: if $A \in \bigcap_{k=1}^\infty \mathcal{B}_k$ then $\mathsf{P}(A) = 0$ or 1.

5.54. Let the process $\{X(t),\ t > 0\}$ with independent increments have right-hand continuous trajectories. Prove that every random variable, measurable w.r.t. σ-algebra $\bigcap_{t>0} \sigma(X(s), s \le t)$, is degenerate, that is, possesses a nonrandom value with probability one.

5.55. Describe all centered continuous Gaussian processes with independent increments whose trajectories have bounded variation on any segment.

Hints

5.1. Write down the explicit expressions for the characteristic functions of the increments.

5.3. Use Theorem 5.1.

5.4. Express the events in the terms of increments of the process N. For example, $\{N(2)+N(3)=1\}=\{N(2)=0,N(3)=1\}=\{N(2)-N(0)=0,N(3)-N(2)=1\}$. This implies that $\mathsf{P}(N(2)+N(3)=1)=\mathsf{P}(N(2)-N(0)=0)\mathsf{P}(N(3)-N(2)=1)=e^{-2\lambda}\cdot[e^{-\lambda}\lambda]=\lambda e^{-3\lambda}$.

5.7. Write down the differential equations for the functions $f_k(t)=\mathsf{P}(X(t)=k)$, $k\in\mathbb{Z}^+$. Prove that the solution to this system of differential equations with $f_0(0)=1$, $f_k(0)=0$, $k\geq 1$ is unique. Verify that the corresponding probabilities for Poisson process satisfy this system.

5.9. The corresponding conditional distribution functions equal:
(a) $F_1(y)=\mathsf{P}(\tau_1\leq y/\tau_1\leq 1,\tau_2>1)$.
(b) $F_2(y)=\mathsf{P}(\tau_1\leq y/\tau_2>1)$.
(c) $F_3(y)=\mathsf{P}(\tau_1\leq y/\tau_m\leq 1)$.

Calculate these conditional probabilities using the identity $\{\tau_m\leq y\}=\{N(y)\geq m\}$.

5.10. Use Problem 1.2.

5.12. For a given $m\geq 1$ and $0<a_1<b_1<a_2<\cdots<b_{m-1}<a_m$, calculate the probability

$$\mathsf{P}(\tau_1\in(a_1,b_1],\dots,\tau_{m-1}\in(a_{m-1},b_{m-1}],\tau_m>a_m).$$

5.13. Differentiate by x the equality

$$\mathsf{P}(\tau_n\leq x)=\mathsf{P}(N(x)\geq n)=1-\sum_{k=0}^{n-1}((\lambda x)^k/k!)e^{-\lambda x},\quad x\geq 0.$$

5.17 — 5.20. Calculate the common characteristic functions for the increments.

5.21. Use the following general fact: if $\{\eta_\alpha,\alpha\in\mathcal{A}\}$ are jointly independent random variables, and $\mathcal{A}_1,\dots,\mathcal{A}_n$ are some disjoint subsets of $\mathcal{A}$ and $\zeta_i\in\sigma(\eta_\alpha,\alpha\in\mathcal{A}_i)$, $i=1,\dots,n$, then the variables $\zeta_1,\dots\zeta_n$ are jointly independent.

5.23. In order to obtain the characteristic function $\phi_{X(t)}$, use considerations similar to those used in the proof of Problem 5.17. In order to express the distribution density of $X(t)$, use the *inversion formula* for the characteristic function: $p_{X(t)}(x)=(2\pi)^{-1}\int_{\mathbb{R}}e^{-izx}\phi_{X(t)}(z)\,dz$.

5.28. Verify that the process $UW(t)$ is also a process with independent increments and $UW(t)-UW(s)$ follows the distribution $\mathcal{N}(0,(t-s)I_{\mathbb{R}^m})$.

5.29,5.30. Use Theorem 5.1.

5.32. Assuming $\zeta = \xi + \eta$ and variables ξ and η are independent, prove that:
(a) If the distribution of ξ does not have atoms then the distribution of ζ also does not have them. (b) If the distribution of ξ has the density then the distribution of ζ has it, too.

5.33. Write down the characteristic function ϕ for uniform distribution and the characteristic function ψ for exponential distribution. Check that ψ/ϕ is not a characteristic function of a random variable.

5.38. If $\mathsf{P}(|X(t)| > A) = 0$ then $\mathsf{P}(|X(t/2)| > \frac{A}{2}) = 0$ (Prove it!). Conclude that $\mathsf{D}X(t/2^k) \leq A^2/2^{2k}$. After that, deduce that $\mathsf{D}X(t) = 0$.

5.42. For a simple function f, the integral in the right-hand side of (5.2) is equal to the sum of the values $\hat{\nu}$ on a finite collection of sets $A \in \sigma(\mathcal{K})$ with some nonrandom weights, and thus it is a random variable according to statement (1) of Proposition 5.2. Any nonnegative measurable function can be monotonically approximated by a simple ones.

5.44. First prove the formula (5.3) for simple functions and then approximate (both pointwisely and in the $L_2(\mu)$ sense) an arbitrary function by a sequence of simple ones.

5.45. For a simple function f, the integrals are the sums of (independent) values of the Poisson measure ν or compensated Poisson measure $\tilde{\nu}$ with some nonrandom weights. Use the explicit formula for the characteristic function of the Poisson random variable. Approximate a measurable function by a sequence of simple ones.

5.49,5.50. Calculate the common characteristic functions for the values of ν on disjoint sets $A_1, \dots, A_n$.

5.54. Use Problem 5.53, putting $\mathcal{A} = \mathbb{N}$, $\mathcal{G}_\alpha = \sigma(X(t) - X(s), 2^{-\alpha-1} \leq s < t \leq 2^{-\alpha})$, $\alpha \in \mathbb{N}$, $\mathcal{A}_k = \{k, k+1, \dots\}$.

5.55. The process is identical to zero almost everywhere. In order to show this, make the appropriate change of time variable and use Problem 3.19.

Answers and Solutions

5.6. $\mathsf{P}^N_{t_1,\dots,t_m}(A) = \sum_{(u_1,\dots,u_m)\in A} \mathsf{P}(N(t_1) = u_1, \dots, N(t_m) = u_m)$. For $0 < t_1 < \dots < t_m$ and $u_1, \dots, u_m \in \mathbb{Z}^+$ such that $u_1 \leq \dots \leq u_m$,

$$\mathsf{P}(N(t_1) = u_1, \dots, N(t_m) = u_m) = \mathsf{P}(N(t_1) = u_1)\mathsf{P}(N(t_2) - N(t_1) = u_2 - u_1) \times \cdots$$

$$\times \mathsf{P}(N(t_m) - N(t_{m-1}) = u_m - u_{m-1}) = e^{-\lambda t_m} \frac{(\lambda t_1)^{u_1} \dots (\lambda t_m - \lambda t_{m-1})^{u_m - u_{m-1}}}{u_1! \dots (u_m - u_{m-1})!}.$$

5.8. $\mathsf{P}(X(t_i)=k_i, i=1,\ldots,m)=\mathsf{P}(N(\Lambda(t_i))=k_i, i=1,\ldots,m)$ where $\Lambda(t)=\int_0^t\lambda(s)\,ds, t\geq 0$ and N is the Poisson process with parameter $\lambda=1$.

5.10. Let $0\leq t_1<\cdots<t_n\leq 1, u_1\leq\cdots\leq u_n\leq m$ then

$$\begin{aligned}&\mathsf{P}(N(t_1)=u_1,\ldots,N(t_n)=u_n/N(1)=m)\\&=\left[e^{-\lambda}\frac{\lambda^m}{m!}\right]^{-1}\cdot\left[e^{-\lambda t_1}\frac{(\lambda t_1)^{u_1}}{u_1!}\right]\cdot\left[e^{-\lambda(t_2-t_1)}\frac{(\lambda(t_2-t_1))^{u_2-u_1}}{(u_2-u_1)!}\right]\\&\times\cdots\times\left[e^{-\lambda(1-t_n)}\frac{(\lambda(1-t_n))^{m-u_n}}{(m-u_n)!}\right].\end{aligned}$$

We finish the proof by using Problem 1.2.

5.11. The distribution should be degenerate.

5.12. Let $m\geq 1$ be fixed. For $0<a_1<b_1<a_2<\cdots<b_{m-1}<a_m$ we have that

$$\begin{aligned}&\mathsf{P}(\tau_1\in(a_1,b_1],\ldots,\tau_{m-1}\in(a_{m-1},b_{m-1}],\tau_m>a_m)\\&=\mathsf{P}(N(a_1)=0,N(b_1)-N(a_1)=1,\ldots,N(a_m)-N(b_{m-1})=0)\\&=e^{-\lambda a_1}\left[\lambda(b_1-a_1)e^{-\lambda(b_1-a_1)}\right]\ldots\left[\lambda(b_{m-1}-a_{m-1})e^{-\lambda(b_{m-1}-a_{m-1})}\right]\\&\times e^{-\lambda(a_m-b_{m-1})}=\int_{(a_1,b_1]\times\cdots\times(a_{m-1},b_{m-1}]\times(a_m,+\infty)}\lambda^m e^{-\lambda x_m}\,dx_1\ldots dx_m.\end{aligned}$$

From the same formula with a_m replaced by arbitrary $b_m>a_m$, we get

$$\mathsf{P}\Big((\tau_1,\ldots,\tau_m)\in A\Big)=\int_A\lambda^m e^{-\lambda x_m}\,dx_1\ldots dx_m \tag{5.5}$$

for every set A of the form

$$A=(a_1,b_1]\times\cdots\times(a_m,b_m],\quad a_1<b_1<\cdots<a_m<b_m. \tag{5.6}$$

The joint distribution of the variables $\tau_1,\ldots,\tau_m$ is concentrated on the set $\Delta_m:=\Big\{(x_1,\ldots,x_m)|\,0\leq x_1\leq\cdots\leq x_m\Big\}$. Because the family of sets of the type (5.6) is a semiring that generates a Borel σ-algebra in Δ_m, relation (5.5) implies that the joint distribution density for the variables $\tau_1,\ldots,\tau_m$ is equal to

$$p(x_1,\ldots,x_m)=\lambda^m e^{-\lambda x_m}\,\mathbb{I}_{0\leq x_1\leq\cdots\leq x_m}.$$

On the other hand, for independent $\mathrm{Exp}(\lambda)$ random variables $\xi_1,\ldots,\xi_m$, the joint distribution density for the variables $\xi_1,\xi_1+\xi_2,\ldots,\xi_1+\cdots+\xi_m$ is equal to

$$\lambda e^{-\lambda x_1}\,\mathbb{I}_{x_1\geq 0}\prod_{k=2}^m\lambda e^{-\lambda(x_k-x_{k-1})}\,\mathbb{I}_{x_k-x_{k-1}\geq 0}=p(x_1,\ldots,x_m),$$

that is, $(\tau_1,\ldots,\tau_m)\overset{d}{=}(\xi_1,\ldots,\xi_1+\cdots+\xi_m)$. This proves the required statement, because the finite dimensional distributions of a registration process are uniquely defined by the finite-dimensional distributions of the associated sequence $\{\tau_k\}$ (prove the latter statement!).

5.15. X is the Poisson process with intensity $p\lambda$.

5.16. X is the Poisson process with intensity $\lambda_1+\lambda_2$.

5.17. Take $x_1<\cdots<x_n, u_1,\dots,u_n\in\mathbb{R}$ and denote $\Delta X_i=X(x_i)-X(x_{i+1})$, $\Delta F_i=F(x_i)-F(x_{i+1})$. We have

$$\begin{aligned}
&\mathsf{E}\exp[i(u_1X(x_1)+u_2\Delta X_2+\cdots+u_n\Delta X_n)]\\
&=\sum_{k=0}^{\infty}\mathsf{E}(\exp[i(u_1X(x_1)+u_2\Delta X_2+\cdots+u_n\Delta X_n)]/\nu=k)\cdot\frac{\lambda^k e^{-\lambda}}{k!}\\
&=\sum_{k=0}^{\infty}\mathsf{E}\exp\left[i\sum_{j=1}^{k}(u_1\mathbb{I}_{\xi_j\le x_1}+u_2\mathbb{I}_{x_1<\xi_j\le x_2}+\cdots+u_n\mathbb{I}_{x_{n-1}<\xi_j\le x_n})\right]\cdot\frac{\lambda^k e^{-\lambda}}{k!}\\
&=\sum_{k=0}^{\infty}\frac{\lambda^k e^{-\lambda}}{k!}\cdot\left[e^{iu_1}F(x_1)+e^{iu_2}\Delta F_2+\cdots+e^{iu_n}\Delta F_n+(1-F(x_n))\right]^k\\
&=e^{\lambda F(x_1)[e^{iu_1}-1]}e^{\lambda(F(x_2)-F(x_1))[e^{iu_2}-1]}\cdots e^{\lambda(F(x_n)-F(x_{n-1}))[e^{iu_n}-1]}.
\end{aligned}$$

Thus, the variables $X(x_1),\Delta X_2,\dots,\Delta X_n$ are independent and follow the Poisson distribution with parameters $\lambda F(x_1),\lambda\Delta F_2,\dots,\lambda\Delta F_n$ respectively.

5.22. $B=c_0^2, \Pi=\sum_{k=1}^m\lambda_k\delta_{c_k}, a=\sum_{k:|c_k|\le 1}\lambda_k c_k$.

5.23. $B=b^2$; $\Pi(\{u\})=\lambda\alpha^{u-1}(1-\alpha)$, $u\in\mathbb{N}$, $\Pi(\{u\})=\mu\beta^{-u-1}(1-\beta)$, $-u\in\mathbb{N}$; $a=\lambda(1-\alpha)+\mu(1-\beta)$.

$$\mathsf{E}e^{izX(t)}=\exp\left[-\frac{tb^2z^2}{2}+\frac{t\lambda\alpha}{\alpha-iz}+\frac{t\mu\beta}{\beta+iz}\right].$$

$$p_{X(t)}(x)=\frac{1}{2\pi}\int_{\mathbb{R}}\exp\left[-\frac{tb^2z^2}{2}+\frac{t\lambda\alpha^2}{\alpha^2+z^2}+\frac{t\mu\beta^2}{\beta^2+z^2}\right]\cos\left\{zx+\frac{t\lambda z\alpha}{\alpha^2+z^2}-\frac{t\lambda z\beta}{\beta^2+z^2}\right\}dz.$$

5.25. $P(W(t)\in B(0,r))=1-e^{-(r^2/2t)}$.

5.26. $E\int_0^\infty\mathbb{I}_{\{W(t)\in B\}}dt=\int_0^\infty P(W(t)\in B)dt=(2\pi t)^{-(m/2)}\int_0^\infty\int_A e^{-(y^2/2t)}dydt=0$.

5.34. Assume such a process to exist. Denote by ϕ the characteristic function of $X(\frac{1}{2})-X(0)$. One has $\mathsf{E}(X(1)-X(0))^2<+\infty$ and therefore $\mathsf{E}(X(\frac{1}{2})-X(0))^2<+\infty$ (prove this, using that X has independent increments). Thus ϕ should be at least twice differentiable on the whole real line. On the other hand, $\phi^2(z)=\cos z, z\in\mathbb{R}$, hence $|\phi(z)-\phi(\pi/2)|=\sqrt{\cos z}\sim\sqrt{\pi/2-z}$, $z\to(\pi/2)-$. Therefore ϕ is not differentiable at the point $\pi/2$, which contradicts the assumption made above.

5.35. (a) The Poisson process.
(b) The processes from Problems 5.29, 5.30.
(c) The process $X(t)=\sum_{k=1}^\infty(k!)^{-1}N_k(t)$, where $N_k, k\ge 1$ are jointly independent Poisson processes with equal intensities.

5.36. If the process X is stochastically continuous then $\mathsf{E}e^{izX(t)} \to \mathsf{E}e^{izX(s)}$ as $t \to s$ by the dominated convergence theorem. On the other hand, if for every z the function $t \mapsto \mathsf{E}e^{izX(t)}$ is continuous at a point $t = 0$ then the continuity theorem for characteristic functions implies that $X(t) - X(0) \to 0$ in distribution as $t \to 0+$ and thus $X(t) - X(s) \to 0$ in distribution as $t \to s$. Now we can use the fact that the convergence in distribution to a constant implies the convergence in probability.

5.37. Not necessarily. For instance, the process $N(-t)$ is not a process with independent increments because the values $N(1) - N(0)$ and $N(1)$ are not independent.

5.39. The random events $A_n = \{|X((n+1)a) - X(na)| > b\}$, $n \geq 1$ are jointly independent and have the same positive probability (Problem 5.38). By the Borell–Cantelli lemma, an infinite number of the events $A_n, n \geq 1$ occur with probability 1.

5.41. Denote by $\mathcal{C}$ the family of triplets of the sets $A, B, C \in \mathcal{K}_0$ such that $C = A \cup B, A \cap B = \varnothing$. This family is countable since the family of all triplets $A, B, C \in \mathcal{K}_0$ is countable. Therefore, the sets

$$\Omega_0 = \bigcap_{(A,B,C) \in \mathcal{C}} \{\omega |\, \nu(A)(\omega) + \mu(B)(\omega) = \nu(C)(\omega)\},$$

$$\Omega_1 = \bigcap_{(A,B,C) \in \mathcal{C}} \{\omega |\, \nu(A)(\omega), \nu(B)(\omega), \nu(C)(\omega) \in \mathbb{Z}^+\}$$

are random events of probability 1. For a function, defined on sets and taking integer nonnegative values, σ-additivity is equivalent to additivity (prove this!). Thus, for any $\omega \in \Omega_0 \cap \Omega_1$ the function $\mathcal{K}_0 \ni A \mapsto \nu(A)(\omega)$ is σ-additive. By the Carathéodory theorem, this function can be extended to a σ-finite measure on $\sigma(\mathcal{K}_0) = \sigma(\mathcal{K})$. We denote this measure by $\hat{\nu}(\omega, \cdot)$. For $\omega \notin \Omega_0 \cap \Omega_1$ we put $\hat{\nu}(\omega, \cdot) \equiv 0$. Under this construction, for any ω, any $A \subset \sigma(\mathcal{K})$ with $\hat{\nu}(\omega, A) < +\infty$ and $\varepsilon > 0$, there exists $A_\varepsilon \in \mathcal{K}_0$ such that $\hat{\nu}(\omega, A \triangle A_\varepsilon) < \varepsilon$. This implies that $\hat{\nu}$ possesses integer nonnegative values and $\hat{\nu}(A) = \nu(A), A \in \mathcal{K}$ a.s.

Therefore, because every one-point set belongs to $\sigma(\mathcal{K})$, the measure $\hat{\nu}$ is a sum of δ-measures. The properties (2),(3) have been proved. To prove property (1), we use the "principle of the fitting sets". Consider the class $\mathcal{A}$ of the sets $A \in \sigma(\mathcal{K})$ for which $\hat{\nu}(A)$ is an extended random variable. Then $\mathcal{A}$ is a monotone class: $A_1 \subset A_2 \subset \ldots, A = \bigcup_n A_n, A_n \in \mathcal{A}, n \geq 1$ implies that $\hat{\nu}(\omega, A_n) \to \hat{\nu}(\omega, A)$ for every ω and thus $\hat{\nu}(A)$ is an extended random variable; that is, $A \in \mathcal{A}$. Furthermore, $\mathcal{K}_0 \subset \mathcal{A}$ and thus $\mathcal{A} = \sigma(\mathcal{K}_0)$.

5.46. This statement simply follows from the definitions of the Poisson random field and Poisson point measure.

5.47. It doesn't follow. For example, let us put $\mathbb{T} = [-1, 0]$, $X(t) = -N(-t)$, $\nu((s,t]) = X(t) - X(s) = N(-s) - N(-t)$. It follows from the previous problem that ν is a Poisson point measure and Problem 5.37 implies that X is not a process with independent increments.

5.48. The distribution of $\{\xi_k\}$ should be degenerate. That is, $\mathsf{P}(\xi_k = a) = 1$ for some $a \in \mathbb{R}$. The intensity measure in this case is equal to $\lambda(\lambda^1|_{\mathbb{R}^+} \times \delta_a)$ (λ is the intensity of the process N).

5.51. Necessity follows from the estimate $\mathsf{E}|X(t)|^\alpha \geq \mathsf{E}|\xi_1|^\alpha \mathbb{I}_{N(t)=1} = \lambda t e^{-\lambda t}\mathsf{E}|\xi_1|^\alpha$. To prove sufficiency, we use the inequalities

$$|x_1 + \cdots + x_n|^\alpha \leq \begin{cases} n^{\alpha-1}\left(|x_1|^\alpha + \cdots + |x_n|^\alpha\right), & \alpha \geq 1 \\ |x_1|^\alpha + \cdots + |x_n|^\alpha, & \alpha \in (0,1) \end{cases}.$$

The first one is the Jensen inequality; the second one can be verified straightforwardly. Then

$$\mathsf{E}|X(t)|^\alpha \leq \mathsf{E}|\xi_1|^\alpha \sum_{n=1}^{\infty} (n^{\alpha-1} \vee 1)\frac{(\lambda t)^n}{n!}e^{-\lambda t} < +\infty.$$

5.52. It can be verified easily that the characteristic function of the last term in (5.4) is infinitely differentiable. Thus, the norm of this term has a finite moment of an arbitrary order. The norms of the first two terms in (5.4) also have finite moments of an arbitrary order. Thus, $\mathsf{E}\|X(t)\|^\alpha < +\infty$ if and only if $\mathsf{E}\|Y(t)\|^\alpha < +\infty$, where $Y(t)$ is the third term in (5.4). The process Y is the compound Poisson process with the distribution of its jump equal $[\Pi(\|u\|_{\mathbb{R}^d} > 1)]^{-1}\Pi(\cdot \cap \{\|u\|_{\mathbb{R}^d} > 1\})$ (verify this!). Now, we can deduce the required statement using considerations analogous to those used in the previous solution.

5.53. Denote $\mathcal{C}_k = \sigma(\bigcup_{\alpha\in\mathcal{A}\setminus\mathcal{A}_k} \mathcal{G}_\alpha)$. Any set A that belongs to all $\mathcal{B}_k, k \in \mathbb{N}$ is independent of the σ-algebra $\mathcal{C}_k$ for any $k \in \mathbb{N}$. Because $\bigcup_{k=1}^\infty (\mathcal{A}\setminus\mathcal{A}_k) = \mathcal{A}$, this set is independent of $\sigma(\bigcup_{k=1}^\infty \mathcal{C}_k) = \sigma(\bigcup_{\alpha\in\mathcal{A}} \mathcal{G}_\alpha)$. But it is obvious that A belongs to $\sigma(\bigcup_{\alpha\in\mathcal{A}} \mathcal{G}_\alpha)$, and thus A is independent of A. This implies $\mathsf{P}(A) = \mathsf{P}(A \cap A) = \mathsf{P}^2(A)$ and therefore $\mathsf{P}(A) = 0$ or 1.

6

Gaussian processes

Theoretical grounds

Definition 6.1. *Random variables* $\xi_1, \dots, \xi_m$ *are called* jointly Gaussian *if the characteristic function of their joint distribution has the form*

$$\mathsf{E}e^{i\sum_{k=1}^{m} z_k \xi_k} = e^{i(z,a)-(Bz,z)/2}, \quad z = (z_1, \dots, z_m) \in \mathbb{R}^m, \tag{6.1}$$

where $a \in \mathbb{R}^m$, *B is a symmetric nonnegatively defined matrix, and* $(\cdot,\cdot)$ *denotes the scalar product in* $\mathbb{R}^m$.

A random vector $\xi = (\xi_1, \dots, \xi_m)$ *with jointly Gaussian coordinates is said to be* Gaussian *(or follow the* Gaussian distribution*).*

The vector a in the relation (6.1) is the mean vector for the random vector ξ and the matrix B is the covariance matrix for ξ. The distribution of the Gaussian vector with the characteristic function (6.1) is called the *Gaussian measure* with mean a and covariance B, and is denoted $\mathcal{N}(a,B)$. This distribution is uniquely determined by the mean vector a and covariance matrix B, and for any $a \in \mathbb{R}^m$ and symmetric nonnegative matrix B there exists a random vector following the distribution $\mathcal{N}(a,B)$.

Definition 6.1 is equivalent to the following one.

Definition 6.2. *A vector* $\xi = (\xi_1, \dots, \xi_m)$ *is called* Gaussian *if for any nonrandom vector* $z = (z_1, \dots, z_m) \in \mathbb{R}^m$ *the product* $(z, \xi)_{\mathbb{R}^m} = \sum_{j=1}^{m} z_j \xi_j$ *is a Gaussian random variable.*

Proposition 6.1. *(1) Let* $\xi \sim \mathcal{N}(a,B)$ *be an m-dimensional Gaussian vector,* $b \in \mathbb{R}^n$, *and A be an* $n \times m$ *matrix. Then* $\eta = b + A\xi$ *is an n-dimensional Gaussian vector,* $\eta \sim \mathcal{N}(b + Aa, ABA^*)$.

(2) Let $\{\xi^n,\ n \in \mathbb{N}\}$ *be a sequence of Gaussian random vectors weakly convergent to a vector* ξ. *Then* ξ *is a Gaussian vector.*

(3) Let $\xi_1, \dots, \xi_m$, $\eta_1, \dots, \eta_n$ *be jointly Gaussian random variables and* $\mathrm{cov}(\xi_j, \eta_k) = 0$, $j = 1, \dots, m, k = 1, \dots, n$. *Then the random rectors* $\xi = (\xi_1, \dots, \xi_m)$ *and* $\eta = (\eta_1, \dots, \eta_n)$ *are independent.*

D. Gusak et al., *Theory of Stochastic Processes*, Problem Books in Mathematics,
DOI 10.1007/978-0-387-87862-1_6,

For two random vectors $\xi = (\xi_1, \dots, \xi_m)$ and $\eta = (\eta_1, \dots, \eta_n)$ *the joint covariance matrix* is the $m \times n$ matrix $R_{\xi\eta}$ consisting of covariances of the elements of the vectors:

$$\left[R_{\xi\eta}\right]_{jk} = \operatorname{cov}(\xi_j, \eta_k), \quad j = 1, \dots, m,\ k = 1, \dots, n.$$

Remark that, in this notation, $R_{\xi\xi}$ is the covariance matrix R_ξ of the vector ξ.

Theorem 6.1. *(On normal correlation) Let random variables $\xi_1, \dots, \xi_m, \eta_1, \dots, \eta_n$ be jointly Gaussian. Then the conditional distribution of $\xi = (\xi_1, \dots, \xi_m)$ with respect to the σ-algebra, generated by $\eta = (\eta_1, \dots, \eta_n)$, is Gaussian. If the matrix $R_{\eta\eta}$ is nondegenerate, then the mean vector of this conditional distribution equals*

$$a_{\xi|\eta} = a_\xi + R_{\xi\eta} R_{\eta\eta}^{-1} (\eta - a_\eta),$$

and the covariance matrix equals

$$R_{\xi|\eta} = R_{\xi\xi} - R_{\xi\eta} R_{\eta\eta}^{-1} R_{\eta\xi}.$$

Let us emphasize that, in the formulation of the theorem on normal correlation, it is crucial that the vectors ξ and η are parts of one Gaussian vector of the length $m+n$. The statement of this theorem is often interpreted as follows. The conditional distribution of ξ given $\eta = y$ is equal to

$$\mathcal{N}\left(a_\xi + R_{\xi\eta} R_{\eta\eta}^{-1} (y - a_\eta), R_{\xi\xi} - R_{\xi\eta} R_{\eta\eta}^{-1} R_{\eta\xi}\right).$$

Definition 6.3. *The stochastic process $\{X(t),\ t \in \mathbb{T}\}$ is called* Gaussian *if for any $m \geq 1$ and any points $\{t_1, \dots, t_m\} \subset \mathbb{T}$ the vector $(X(t_1), \dots, X(t_m))$ follows the Gaussian distribution.*

Theorem 6.2. *(1) Let $a : \mathbb{T} \to \mathbb{R}$ be an arbitrary function, and $R : \mathbb{T} \times \mathbb{T} \to \mathbb{R}$ be a nonnegatively defined function. Then there exist a probability space $(\Omega, \mathcal{F}, \mathsf{P})$ and a Gaussian stochastic process $\{X(t),\ t \in \mathbb{T}\}$, defined on this space, for which a and R are the mean and covariance functions, respectively.*

(2) The mean and covariance functions uniquely determine the finite-dimensional distributions of the Gaussian process.

The Wiener process is Gaussian (see Problem 6.4). According to Theorem 6.2, the following characterization is available for this process.

Proposition 6.2. *A real-valued stochastic process $\{X(t),\ t \in \mathbb{R}^+\}$ is the Wiener process if and only if it is a Gaussian process with $a_X \equiv 0, R_X(t,s) = t \wedge s,\ t,s \in \mathbb{R}^+$.*

Let us give some examples of the most important Gaussian processes.

Example 6.1. The Brownian bridge is the centered Gaussian process $\{B(t),\ t \in [0,1]\}$ with covariance $R_B(t,s) = t \wedge s - st,\ s,t \in [0,1]$. Sometimes it is called the Brownian bridge of length 1, starting from the point 0 and arriving at the point 0 (see Problem 6.21).

Example 6.2. The Ornstein–Uhlenbeck process is the centered Gaussian process $\{X(t),\, t\in\mathbb{R}\}$ with covariance $R_X(t,s)=e^{-|t-s|},\ s,t\in\mathbb{R}$.

Example 6.3. The fractional Brownian motion with Hurst index $H\in(0,1)$ is the centered Gaussian process $\{B^H(t),\, t\in\mathbb{R}\}$ with covariance $R_{B^H}(t,s)=\frac{1}{2}\left(|t|^{2H}+|s|^{2H}-|t-s|^{2H}\right),\ t,s\in\mathbb{R}$.

Remark 6.1. For $t,s\geq 0$, one has $R_{B^{1/2}}(t,s)=t\wedge s$. Therefore, the process $B^{1/2}$ restricted to $\mathbb{R}^+$ coincides with the Wiener process (see Proposition 6.2). Sometimes, the process $B^{1/2}$ is called the *two-sided Wiener process* (see also Problem 6.8).

Proposition 6.3. *The process*

$$X^H(t)=\int_{\mathbb{R}}\left(\left((t-s)^+\right)^{1/2-H}-\left((-s)^+\right)^{1/2-H}\right)dW(s),\quad t\in\mathbb{R}$$

is the fractional Brownian motion with Hurst index $H\in(0,1)$. Remember that a positive part of a number $b\in\mathbb{R}$ is $b^+=b\mathbb{1}_{b>0}$.

The proposition presented above shows that the function R_{B^H} is nonnegatively defined.

The simple conditions are available, sufficient for existence of the continuous and Hölder modifications of a centered Gaussian process.

Theorem 6.3. *Let $\{X(t),\, t\in[0,T]\}$ be a Gaussian process with $a_X\equiv 0$. Denote*

$$\sigma_X^2(t,h)=\mathsf{E}(X(t+h)-X(t))^2=R_X(t+h,t+h)-2R_X(t,t+h)+R_X(t,t).$$

(1) If $\sigma_X^2(t,h)\leq c|\ln|h||^{-p}, t\in[0,T], h>0$ for some $p>3$ and $c>0$, then the process X has continuous modification $\tilde{X}$.

(2) If there exist $p>0$ and $c>0$ such that

$$\sigma_X^2(t,h)\leq c|h|^p,\quad t\in[0,T], h>,0$$

then for any $\varepsilon>0$ there exist $\Omega'\subset\Omega$, $\mathsf{P}(\Omega')=1$, and a function $c=c(\omega):\Omega'\to\mathbb{R}$ such that for any $t,s\in[0,T]$,

$$\left|\tilde{X}(t,\omega)-\tilde{X}(s,\omega)\right|\leq c(\omega)|t-s|^{p/2}\left|\ln|t-s|\right|^{1+\varepsilon}.$$

Corollary 6.1. *The trajectories of the Wiener process for any $\varepsilon>0$ a.s. satisfy the following inequality*

$$\sup_{t,s\in[0,T],s\neq t}\frac{|W(t)-W(s)|}{\sqrt{|t-s|}\left|\ln|t-s|\right|^{1+\varepsilon}}<+\infty.$$

Bibliography

[9], Chapter II; [24], Volume 1, Chapter III, §1; [25], Chapter I, §2; [15], Chapter II, §3; [36].

Problems

6.1. Prove that Definitions 6.1 and 6.2 are equivalent.

6.2. Prove Proposition 6.1.

6.3. Let $\xi_1,\dots,\xi_m,\eta_1,\dots,\eta_n$ be jointly Gaussian random variables, denote $\zeta_i = \eta_i - \mathsf{E}[\eta_i/\xi_1,\dots,\xi_m], i = 1,\dots,n$. Prove that the random vectors $\xi = (\xi_1,\dots,\xi_m)$ and $\zeta = (\zeta_1,\dots,\zeta_n)$ are independent. Write the covariance matrix for the vector ζ assuming that the matrix R_ξ is nondegenerate.

6.4. Prove that the Wiener process is Gaussian.

6.5. Let $\{W(t),\ t\in\mathbb{R}^+\}$ be a Wiener process. Verify that the following processes are Wiener processes as well ($c,T>0$ are arbitrary constants).
(a) $W^c(t) = \sqrt{c}W(t/c),\ t\in\mathbb{R}^+$.
(b) $W_c(t) = W(t+c) - W(c),\ t\in\mathbb{R}^+$.
(c) $W_T(t) = W(t),\ t\le T,\ W_T(t) = 2W(T) - W(t),\ t>T$.
(d) $\hat{W}(t) = tW(1/t),\ t>0,\ \hat{W}(0) = 0$.
(e) $\overleftarrow{W}_T(t) = W(T) - W(T-t),\ t\in[0,T]$.
Draw a trajectory of the process W and corresponding trajectories of the processes $W^c, W_c, W_T, \overleftarrow{W}_T$.

6.6. Let W be a Wiener process. Does there exist a nonrandom function $c(t)$ such that the process
(a) $X(t) = c(t)W(2t)$
(b) $X(t) = c(t)W(1/t),\ t>0,\ X(0) = 0$
(c) $X(t) = c(t)W(e^t)$
is a Wiener process? If the answer is positive, find the function $c(t)$.

6.7. Give an example of a stochastic process, not being a Wiener process, with its mean and covariance functions equal to $a_X\equiv 0,\ R_X(t,s) = t\wedge s,\ t,s\in\mathbb{R}^+$.

6.8. (1) Let $\{W_1(t), t\in\mathbb{R}^+\}$ and $\{W_2(t), t\in\mathbb{R}^+\}$ be two independent Wiener processes. Prove that

$$X(t) = \begin{cases} W_1(t), & t\ge 0,\\ W_2(-t), & t\le 0\end{cases}$$

is a two-sided Wiener process.
(2) Is the two-sided Wiener process a process with independent increments?

6.9. Prove that the fractional Brownian motion (in particular, the Wiener process) is a *process with stationary increments*, that is, the process $\{\widetilde{B}^H(s) := B^H(t+s) - B(t),\ s\ge 0\}$ has the same distribution for all $t\ge 0$.

6.10. Let $\{X(t),\ t\in\mathbb{R}\}$ be the Ornstein–Uhlenbeck process. Prove that $\{Y(t) = t^{1/2}\, X(\frac{1}{2}\ln t), t>0\}$ is the Wiener process.

6.11. Let $\{X(t),\ t\in[0,1]\}$ be the Brownian bridge. Prove that $\{Y(t)=(1+t)X(1-(t+1)^{-1}),\ t\in\mathbb{R}^+\}$ is the Wiener process.

6.12. Let $\{W(t),\ t\in\mathbb{R}^+\}$ be the Wiener process. Prove that $\{X(t)=e^{-t}W(e^{2t}),\ t\in\mathbb{R}\}$ is the Ornstein–Uhlenbeck process.

6.13. Let $\{W(t),\ t\in[0,1]\}$ be the Wiener process. Prove that the process $\{X(t)=W(t)-tW(1),\ t\in[0,1]\}$ is the Brownian bridge.

6.14. Let $\{W(t),\ t\in[0,1]\}$ be the Wiener process. Prove that all
(a) one-dimensional
(b) m-dimensional
conditional distributions of the process W given that $\{W(1)=0\}$ are equal to the corresponding (unconditional) distributions of the Brownian bridge.

6.15. Let $\{X(t),\ t\in[0,1]\}$ be the Brownian bridge and ξ be a random variable independent of X and following the distribution $\mathcal{N}(0,1)$. Prove that $W(t)=X(t)+t\xi$ is the Wiener process.

6.16. Construct a stochastic process $\{X(t),\ t\in\mathbb{R}^+\}$ such that all its one-dimensional distributions are Gaussian but the process is not a Gaussian one.

6.17. Let X be a centered Gaussian process with covariance R_X. Find mean and covariance functions for the process $Y(t)=X^2(t)$.

6.18. Let $0<s_1<s_2<s_3$. Find the conditional distribution of $X(s_3)$ given $\{X(s_1)=x_1, X(s_2)=x_2\}$ if X is
(a) The Wiener process.
(b) The Brownian bridge (we suppose in this case that $s_3<1$).
(c) The Ornstein–Uhlenbeck process.
(d) The fractional Brownian motion with Hurst index $H\neq\frac{1}{2}$.
What is a difference between items (a) – (c) and item (d)?

6.19. Let $0<s_1<s_2<s_3$. Find the conditional distribution of $W(s_1)$ given $\{W(s_2)=x_2, W(s_3)=x_3\}$.

6.20. Let X be the Gaussian process, and s_1,s_2,s_3 be the points from its domain. Consider the following property. The conditional distribution $X(s_3)$ w.r.t. the σ-algebra $\sigma(X(s_1),X(s_2))$ equals the conditional distribution $X(s_3)$ w.r.t. the σ-algebra $\sigma(X(s_2))$.

(1) Prove that this property takes place if and only if

$$R_X(s_1,s_2)R_X(s_2,s_3)=R_X(s_1,s_3)R_X(s_2,s_2) \tag{6.2}$$

and, additionally, $R_X(s_1,s_3)=0$ as soon as $R_X(s_2,s_2)=0$.

(2) Give an example such that the equality (6.2) holds but the property described above does not hold true.

Compare with Problem 6.18. See also the definition of the Markov process in Chapter 12 and Problems 12.10,12.11.

6.21. *The Brownian bridge of length l, starting from point x and arriving at point y,* is the process $B_{l,x,y}$, defined on $[0,l]$, for which the corresponding finite-dimensional distributions are equal to the conditional distributions of $x+W(t_1),\dots,x+W(t_m)$ given $\{x+W(l)=y\}$ (W is the Wiener process). Prove that $B_{l,a}$ is a Gaussian process and find its mean and covariance functions.

6.22. Let W be the Wiener process. Show that the process

$$X(t)=x+\frac{y-x}{l}t+\frac{l-t}{l}W\left(\frac{lt}{l-t}\right),\quad 0\le t\le l$$

is the Brownian bridge of the length l starting from x and arriving at y.

6.23. *The Ornstein–Uhlenbeck bridge of length l, starting from point x and arriving at point y,* is the process $U_{l,x,y}$, defined on $\mathbb{R}$, for which the corresponding finite-dimensional distributions are equal to the conditional distributions of the values $U(t_1),\dots,U(t_m)$ of the Ornstein–Uhlenbeck process given $\{U(0)=x,U(l)=y\}$. Prove that $U_{l,x,y}$ is a Gaussian process and find its mean and covariance functions.

6.24. *The Ornstein–Uhlenbeck process with the initial value x* is the process U_x, defined on $\mathbb{R}$, for which the corresponding finite-dimensional distributions are equal to the conditional distributions of the values $U(t_1),\dots,U(t_m)$ of the Ornstein–Uhlenbeck process given $\{U(0)=x\}$. Prove that U_x is a Gaussian process and find its mean and covariance functions.

6.25. Verify that the condition on correlation function R_X formulated in Problem 6.20 hold true for any $s_1<s_2<s_3$ if X is

(1) The Ornstein–Uhlenbeck process with the initial value x (see Problem 6.24)

2) The Ornstein–Uhlenbeck bridge of length l, starting from point x and arriving at point y (see Problem 6.23)

6.26. Let $K\in C([a;b]\times[a,b])$ be a nonnegatively defined function, $\{\lambda_k,\ k\ge 1\}$ and $\{\phi_k,\ k\ge 1\}$ be the eigenvalues and corresponding orthonormal eigenfunctions of the integral operator A_K (see Problem 2.28). Let $\{\xi_k\}$ be i.i.d. random variables following the standard normal distribution. Prove that:

(1) The series $X(t)=\sum_{k=1}^{\infty}\sqrt{\lambda_k}\xi_k\cdot\phi_k(t)$ converges in mean square for any $t\in[a,b]$.

(2) The process $\{X(t),\ t\in[a,b]\}$ is a Gaussian one with $a_X\equiv 0, R_X=K$.

6.27. Find the eigenvalues $\{\lambda_k,k\ge 1\}$ and orthonormal eigenfunctions $\{\phi_k,k\ge 1\}$ of the operator A_K in the case when K is the covariance of:

(a) Wiener process on $[0,1]$

(b) Brownian bridge

6.28. Using Problem 6.26 and item (a) of Problem 6.27, prove that the Wiener process on $[0,1]$ has the following representation:

$$W(t)=\sum_{k=1}^{\infty}\xi_k\cdot\frac{\sqrt{2}}{\pi(k-\frac{1}{2})}\cdot\sin\left[\pi\left(k-\frac{1}{2}\right)\cdot t\right],\quad t\in[0,1],$$

where $\{\xi_k,\ k \geq 1\}$ are i.i.d. random variables following the standard normal distribution.

6.29. Prove that the Wiener process on $[0,T]$ can be written as:

$$W(t) = \sum_{k=1}^{\infty} \xi_k \cdot \frac{\sqrt{2T}}{\pi\left(k-\frac{1}{2}\right)} \cdot \sin\left[\left(\frac{\pi}{T}\right)\left(k-\frac{1}{2}\right)\cdot t\right], \quad t \in [0,T],$$

where $\{\xi_k,\ k \geq 1\}$ are i.i.d. random variables following the standard normal distribution. If $T \neq 1$ then, on $[0, 1 \wedge T]$, this representation differs from the one given in the previous problem.

6.30. Using Problem 6.26 and item (b) of Problem 6.27 prove that the Brownian bridge has the following representation,

$$X(t) = \sum_{k=1}^{\infty} \xi_k \cdot \frac{\sqrt{2}}{\pi k} \cdot \sin\left[\pi k \cdot t\right], \quad t \in [0,1],$$

where $\{\xi_k, k \geq 1\}$ are i.i.d. random variables following the standard normal distribution.

6.31. Using the previous problem prove that the Wiener process on $[0,1]$ can be written as

$$W(t) = \xi_0 \cdot t + \sum_{k=1}^{\infty} \xi_k \cdot \frac{\sqrt{2}}{\pi k} \cdot \sin\left[\pi k \cdot t\right], \quad t \in [0,1],$$

where $\{\xi_k, k \geq 0\}$ are i.i.d. random variables following the standard normal distribution.

6.32. (1) Let an operator $B : L_2([0,T]) \to L_2([0,T])$ be defined by the equality $(Bx)(t) = \int_0^t x(r)\,dr, t \in [0,T]$. Find the adjoint operator B^* and prove that $A_K = BB^*$, where $K(t,s) = t \wedge s,\ t,s \in [0,T]$.

(2) Let $\{e_k,\ k \geq 1\}$ be some orthonormal basis (ONB) in $L_2([0,T])$. Take $f_k = Be_k,\ k \geq 1$ and put

$$W(t) = \sum_{k=1}^{\infty} \xi_k \cdot f_k(t), \quad t \in [0,T], \tag{6.3}$$

where $\{\xi_k, k \geq 1\}$ are i.i.d. random variables following the standard normal distribution. Prove that the series converges for any t in the mean square sense and its sum is the Wiener process on $[0,T]$.

(3) Let $T = 1$. Specify the bases $\{e_k,\ k \geq 1\}$ such that the corresponding representations (6.3) coincide with those given in Problems 6.28, 6.31.

6.33. Let a stochastic process $\{X(t),\ t \in \mathbb{R}^+\}$ be defined by the stochastic integral: $X(t) = \int_0^t f(s)dW(s)$, where W is the Wiener process and $f \in L_2([0,T])$ for every $T > 0$. Prove that X is a Gaussian process. Find its mean and covariance functions.

6.34. (1) Prove that the process X from the previous problem has the same finite-dimensional distributions with the process $Y(t) = W\left(\int_0^t f^2(s)\,ds\right),\ t \in \mathbb{R}^+$.

(2) Prove that the process X has a continuous modification.

6.35. Let $K(t,s),\ t,s \in [0,T]$ be a symmetric nonnegatively defined function. Assume there exists a function $Q \in L_2([0,T]^2)$ such that

$$K(t,s) = \int_0^T Q(t,r)Q(s,r)\,dr, \quad t,s \in [0,T]. \tag{6.4}$$

Prove that $X(t) = \int_0^T Q(t,s)dW(s),\ t \in [0,T]$ is a centered Gaussian process and its covariance function is equal to K.

6.36. Let, in the previous problem, $K(t,s) = t \wedge s - ts, t,s \in [0,1]$. Give an example of a function Q satisfying (6.4). Write the corresponding integral representation for the Brownian bridge.

6.37. Let $K(t,s) = t \wedge s,\ t,s \in [0,1]$. Give two *different* functions Q satisfying (6.4). Compare the corresponding representations for the Wiener process with Problem 6.5.

6.38. Let $X(t) = \int_{-\infty}^t e^{s-t}dW(s), t \in \mathbb{R}$. Prove that X is the Ornstein–Uhlenbeck process.

6.39. Let $\{X(t),\ t \in [0,T]\}$ be a Gaussian process and $\mathsf{E}|X(t+h) - X(t)|^2 \le \psi^2(h)$ with $\int_{\mathbb{R}^+} \psi(e^{-x^2})dx < \infty$. Prove that the process X has a continuous modification.

Hints

6.3. Prove that every variable ζ_i is uncorrelated with ξ_j and use Proposition 6.1.
6.5. Prove that every process from items (a)–(e) is a Gaussian one and find the mean and covariance functions.
6.10 — 6.13,6.15. Prove that the process in the formulation of the problem is a Gaussian one, and find its mean and covariance functions.
6.14. The first version of the solution: use the theorem on normal correlation; the second version: let W be the Wiener process,and $X(t) = W(t) - tW(1), t \in [0,1]$ be the Brownian bridge (see Problem 6.13). We have $\text{cov}(X(t),W(1)) = 0$ (check this!), thus the process X and the random variable $W(1)$ are independent (prove this!). This implies the required statement.
6.21, 6.23, 6.24. Use the theorem on normal correlation.
6.27. (1) The equation for eigenvalues and eigenfunctions of A_K has the form $\lambda f(t) = \int_0^t sf(s)\,ds + t\int_t^1 f(s)\,ds, t \in [0,1]$. If $\lambda = 0$ then after differentiation by t we get $\int_t^1 f(s)\,ds = 0, t \in [0,1]$ and thus $f \equiv 0$. Therefore, $\lambda > 0$ (recall that $A_K \ge 0$; see Problems 2.28, 2.29). Then f is differentiable and $\lambda f'(t) = \int_t^1 f(s)\,ds$; furthermore, $f(0) = f'(1) = 0$. Thus the eigenvalues and eigenfunctions of the operator A_K satisfy the boundary problem $\lambda f'' = -f, f(0) = f'(1) = 0$.

(2) Similar arguments lead to the following boundary problem for the eigenvalues and eigenfunctions of the operator A_K: $\lambda f'' = -f$, $f(0) = f(1) = 0$.

6.29. Use similar arguments as in Hint 6.27, item (a) on a segment $[0,T]$. Another opportunity is to use Problem 6.28 and Problem 6.5, item (a).

6.33. Use Proposition 6.1 and the definition of the stochastic integral as the limit of the integrals of stepwise functions.

6.34. (1) Use Theorem 6.2.

(2) Use Theorem 3.2.

6.39. Use Theorem 3.3.

Answers and Solutions

6.1. If ξ satisfies Definition 6.1 then for any $z \in \mathbb{R}^m$ the characteristic function of the variable $(z,\xi)_{\mathbb{R}^m}$ is equal to $\psi_z(t) = \mathsf{E}\exp\{it(z,\xi)_{\mathbb{R}^m}\} = \exp[ita_z - \frac{1}{2}t^2\sigma_z^2]$, $t \in \mathbb{R}$, where $a_z = (a,z)_{\mathbb{R}^m}, \sigma_z^2 = (Bz,z)_{\mathbb{R}^m}$. So, $(z,\xi)_{\mathbb{R}^m} \sim \mathcal{N}(a_z,\sigma_z^2)$. If ξ satisfies Definition 6.2 then $\mathsf{E}\exp\{i(z,\xi)_{\mathbb{R}^m}\} = \exp[i\mathsf{E}(z,\xi)_{\mathbb{R}^m} - \frac{1}{2}\mathsf{D}(z,\xi)_{\mathbb{R}^m}]$, $z \in \mathbb{R}^m$. Taking into account that $\mathsf{E}(z,\xi)_{\mathbb{R}^m} = (z,a_\xi)_{\mathbb{R}^m}, \mathsf{D}(z,\xi)_{\mathbb{R}^m} = (B_\xi z,z)_{\mathbb{R}^m}$ we obtain (6.1).

6.2. (1) If ξ is a Gaussian vector and $z \in \mathbb{R}^n$, then $(z,\eta)_{\mathbb{R}^n} = (z,b)_{\mathbb{R}^n} + (A^*z,\xi)_{\mathbb{R}^m} = c + (z_1,\xi)_{\mathbb{R}^m}$ is also Gaussian ($c = (z,b)_{\mathbb{R}^n} \in \mathbb{R}$, $z_1 = A^*z \in \mathbb{R}^m$, and $(z_1,\xi)_{\mathbb{R}^m}$ is a Gaussian random variable). The expressions for a_η, B_η are obtained from the following relations: $(z,a_\eta)_{\mathbb{R}^n} = \mathsf{E}(z,b+A\xi)_{\mathbb{R}^n} = (z,b)_{\mathbb{R}^n} + \mathsf{E}(A^*z,\xi)_{\mathbb{R}^m} = (z,b)_{\mathbb{R}^n} + (A^*z,a)_{\mathbb{R}^m}$, $(B_\eta z,z)_{\mathbb{R}^n} = \mathsf{D}(z,\eta)_{\mathbb{R}^n} = \mathsf{D}(A^*z,\xi)_{\mathbb{R}^m} = (BA^*z,A^*z)_{\mathbb{R}^m}$, $z \in \mathbb{R}^n$.

(2) If $\xi^n \to \xi$ weakly then $\phi_{\xi^n}(z) \to \phi_\xi(z)$, $z \in \mathbb{R}^m$. Therefore, there exist functions $\psi_{1,2}$ such that $\phi_\xi(z) = \exp[i\psi_1(z) - \psi_2(z)]$, $(a_{\xi^n},z)_{\mathbb{R}^m} \to \psi_1(z)$, $\frac{1}{2}(B_{\xi^n}z,z)_{\mathbb{R}^m} \to \psi_2(z)$ for every $z \in \mathbb{R}^m$. It follows from the last two relations that $a_{\xi^n} \to a, B_{\xi^n} \to B$, where a is a vector and B is a symmetric nonnegatively defined matrix. This implies (6.1) for the characteristic function of ξ.

(3) The characteristic function of the vector $\zeta = (\xi_1,\dots,\xi_m,\eta_1,\dots,\eta_n)$, due to the formula (6.1), can be written in the form

$$\phi_\zeta(z_1,\dots,z_{n+m}) = \phi_\xi(z_1,\dots,z_m)\phi_\eta(z_{m+1},\dots,z_{n+m}),$$

$z = (z_1,\dots,z_{n+m}) \in \mathbb{R}^{n+m}$. The independence of ξ and η follows from this expression.

6.3. Without loss of generality we can assume the variables ξ_i,η_j to be centered. Using item (1) of Proposition 6.1, we obtain that ξ and $\zeta = \eta - R_{\eta\xi}R_{\xi\xi}^{-1}\xi$ are jointly Gaussian, uncorrelated, and $R_\zeta = R_\eta - R_{\eta\xi}R_{\xi\xi}^{-1}R_{\xi\eta}$. Using item (3) of the same proposition we get the independence of ξ and ζ.

6.4. Let $t_1 < \cdots < t_m$. The random variables $\xi_1 = W(t_1), \xi_2 = W(t_2) - W(t_1), \dots, \xi_m = W(t_m) - W(t_{m-1})$ are Gaussian and independent, and thus, the vector $\xi = (\xi_1,\dots,\xi_m)$ is also Gaussian (this can be easily obtained from Definition 6.1). The vector $(W(t_1),\dots,W(t_m))$ is the image of the vector ξ under the linear mapping $A : (x_1,\dots,x_m) \mapsto (x_1,x_1+x_2,\dots,x_1+\cdots+x_m)$ and thus, is Gaussian (Proposition 6.1, item (1)).

6.6. (a) $c(t) = 1/\sqrt{2}$.
(b) $c(t) = t$.
(c) Such a function does not exist.
6.7. The compensated Poisson process with intensity $\lambda = 1$.
6.17. $a_{X^2}(t) = R_X(t,t), R_{X^2}(t,s) = 2R_X^2(t,s)$.
6.18. By the theorem on normal correlation, every required conditional distribution is a Gaussian one with the constant variance σ^2 and the mean a being a linear combination of the values x_1 and x_2. Further on, we give the values a, σ^2 for every item:
(a) $x_2, s_3 - s_2$.
b)

(b)
$$\frac{1-s_3}{1-s_2}x_2, \quad \frac{(1-s_3)(s_2-s_1)}{1-s_2}.$$

(c) $e^{s_2-s_3}x_2, 1-e^{2(s_2-s_3)}$.

d)
$$a = \frac{r_{31}r_{22}-r_{32}r_{12}}{r_{11}r_{22}-r_{12}^2}x_1 + \frac{r_{32}r_{11}-r_{31}r_{12}}{r_{11}r_{22}-r_{12}^2}x_2,$$

$$\sigma^2 = r_{33} - \frac{r_{11}r_{32}^2 + r_{22}r_{31}^2 - 2r_{12}r_{31}r_{32}}{r_{11}r_{22}-r_{12}^2},$$

where $r_{ij} = \frac{1}{2}(s_i^{2H} + s_j^{2H} - |s_i - s_j|^{2H})$.

In items (a) – (c), in contrast to item (d), the conditional distribution does not depend on the value of X at the point s_1.

6.19.
$$\mathcal{N}\left(\frac{s_1}{s_2}x_2, \frac{s_1}{s_2}(s_2-s_1)\right).$$

6.20. (1) Denote $r_{ij} = R_X(s_i, s_j)$. If $r_{11}r_{22} - r_{12}^2 > 0$ (i.e., the covariance matrix of the vector $(X(s_1), X(s_2))$ is nondegenerate), then using the theorem on normal correlation we obtain the explicit expressions for the conditional distributions $X(s_3)$ w.r.t. $\sigma(X(s_2)), \sigma(X(s_1), X(s_2))$; these distributions are Gaussian with the means

$$a_{3|2} = \frac{r_{32}}{r_{22}}X(s_2), \quad a_{3|12} = \frac{r_{31}r_{22}-r_{32}r_{12}}{r_{11}r_{22}-r_{12}^2}X(s_1) + \frac{r_{32}r_{11}-r_{31}r_{12}}{r_{11}r_{22}-r_{12}^2}X(s_2)$$

and the variances

$$\sigma_{3|2}^2 = r_{33} - \frac{r_{32}^2}{r_{22}}, \quad \sigma_{3|12}^2 = r_{33} - \frac{r_{11}r_{32}^2 + r_{22}r_{31}^2 - 2r_{12}r_{31}r_{32}}{r_{11}r_{22}-r_{12}^2}.$$

When the required property holds, the coefficient near $X(t_1)$ in the expression for $a_{3|12}$ has to be zero, and thus the condition $r_{31}r_{22} - r_{32}r_{12} = 0$ is necessary for this property to hold. One can verify straightforwardly that, under this condition, $\sigma_{3|2}^2 = \sigma_{3|12}^2$ and

$$\frac{r_{32}}{r_{22}} = \frac{r_{32}r_{11}-r_{31}r_{12}}{r_{11}r_{22}-r_{12}^2}.$$

Therefore, this condition is also sufficient.

The case $r_{11}r_{22} - r_{12}^2 = 0$ should be considered separately. If, in addition, $r_{22} > 0$ then $X(s_1) = cX(s_2) + d$ for some constants c, d and thus $\sigma(X(s_1), X(s_2)) = \sigma(X(s_2))$ and the needed property holds true automatically. At the same time $r_{31}r_{22} = cr_{32}r_{22} = r_{32}(cr_{22}) = r_{32}r_{12}$. If $r_{22} = 0$ then the σ-algebra $\sigma(X(s_2))$ is degenerate and thus the conditional distribution of $X(s_3)$ w.r.t. it being equal to the ordinary distribution of $X(s_3)$. Consequently, when $r_{22} = 0$, we can reformulate the required property in the following form. The conditional distribution $X(s_3)$ w.r.t. $\sigma(X(s_1))$ equals to the ordinary distribution of $X(s_3)$. The latter claim is known to be equivalent to the independence of the random variables $X(s_1)$ and $X(s_3)$, and consequently (because $X(s_1)$, $X(s_3)$ are jointly Gaussian) to their noncorrelatedness.

(2) $X(s_2) = 0, X(s_1) = X(s_3) \sim \mathcal{N}(0,1)$.

6.21. $a_{B_{l,x,y}}(t) = [(y-x)/l]t, R_{B_{l,x,y}}(t,s) = t \wedge s - (ts/l)$.

6.23. $$a_{U_{l,x,y}}(t) = \begin{cases} xe^t, & t \le 0, \\ \frac{\operatorname{sh}(l-t)}{\operatorname{sh} l}x + \frac{\operatorname{sh} t}{\operatorname{sh} l}y, & t \in [0,l], \\ ye^{l-t}, & t \ge l \end{cases}$$

$$R_{U_{l,x,y}}(t,s) = \begin{cases} e^{-|t-s|} - e^{s+t}, & s,t \le 0, \\ e^{-|t-s|} - e^{2l-s-t}, & s,t \ge l, \\ e^{-|t-s|} - \left(1 - e^{-2l}\right) \frac{\operatorname{sh}(l-s)\operatorname{sh}(l-t)}{(\operatorname{sh} l)^2}, & s,t \in [0,l], \\ 0 & \text{in other cases.} \end{cases}$$

6.24. $a_{U_x}(t) = xe^{-|t|}, R_{U_x}(t,s) = e^{-|t-s|} - e^{-|t|-|s|}$.

6.26. By the *Mercer theorem*, $K(t,s) = \sum_{k=1}^{\infty} \lambda_k \phi_k(t)\phi_k(s)$, $s,t \in [a,b]$ and the series converges uniformly on $[a,b] \times [a,b]$. Thus, for any $t \in [a,b]$ we have that $\sum_{k\ge n} \lambda_k \phi_k^2(t) \to 0$, $n \to \infty$ and we obtain the statement of item (1).

(2) The series expansion for the kernel K yields $K = R_X$. Proposition 6.1 provides that the process X is a Gaussian one.

6.27. (a) $\lambda_k = \left[\pi\left(k - \frac{1}{2}\right)\right]^{-2}, \phi_k(t) = \sqrt{2}\sin\left[\pi\left(k-\frac{1}{2}\right)\cdot t\right]$, $k \ge 1$.

(b) $\lambda_k = [\pi k]^{-2}, \phi_k(t) = \sqrt{2}\sin[\pi k \cdot t]$, $k \ge 1$.

6.32. (1) For any $f, g \in L_2([0,1])$,

$$(Bf,g)_{L_2([0,T])} = \int_0^T \left(\int_0^t f(s)\,ds\right) g(t)\,dt = \int_0^T \left(\int_s^T g(t)\,dt\right) f(s)\,ds,$$

and thus $B^*g(t) = \int_t^T g(s)\,ds$. Then

$$BB^*f(t) = \int_0^t \left(\int_s^T f(r)\,dr\right) ds = \int_0^T \left(\int_0^{t\wedge r} ds\right) f(r)\,dr = A_K f(t)$$

(2) For any $f, g \in L_2([0,T])$ we have $\mathsf{E}(W,f) = \sum_k (f_k, f)\mathsf{E}\xi_k = 0$,

$$\begin{aligned}\operatorname{cov}((W,f),(W,g)) &= \operatorname{cov}\left(\sum_j (f_j,f)\xi_j, \sum_k (f_k,g)\xi_k\right)\\ &= \sum_{k,j}(f_j,f)(f_k,g)\operatorname{cov}(\xi_j,\xi_k) = \sum_k (f_k,f)(f_k,g)\\ &= \sum_k (e_k,B^*f)(e_k,B^*g) = (B^*f,B^*g) = (A_K f,g)\end{aligned}$$

(here $(\cdot,\cdot)$ denotes the inner product in $L_2([0,T])$). This yields $a_W = 0, R_W = K$. Proposition 6.1 implies that the process X is Gaussian.
(3) In Problem 6.28: $e_k(t) = \sqrt{2}\cos\left[\pi(k-\frac{1}{2})\cdot t\right], k \ge 1$. In Problem 6.31: $e_1 \equiv 1$, $e_k(t) = \sqrt{2}\cos[\pi(k-1)\cdot t], k \ge 2$.
6.33. $a_X \equiv 0$, $R_X(t,s) = \int_0^{t\wedge s} f^2(r)\,dr$.
6.35. The process is Gaussian due to Proposition 6.1 and the definition of the stochastic integral as the limit of integrals of stepwise functions. By the properties of the stochastic integral (see Theorem 13.1), $\mathsf{E}X(t) = 0$,

$$\begin{aligned}\operatorname{cov}(X(t),X(s)) &= \mathsf{E}\left[\int_0^T Q(t,r)dW(r)\int_0^T Q(s,r)dW(r)\right]\\ &= \int_0^T Q(t,r)Q(s,r)\,dr = K(t,s), \quad t,s \in [0,T].\end{aligned}$$

6.36. One can take $Q(t,s) = 1\!\!1_{s\le t} - t, s,t \in [0,1]$. The corresponding representation has the form $X(t) = \int_0^1 1\!\!1_{s\le t}dW(s) - t\int_0^1 dW(s) = W(t) - tW(1)$, $t \in [0,1]$ (compare with Problem 6.13).
6.37. One can take $Q_1(t,s) = 1\!\!1_{s\le t}, Q_2(t,s) = 1\!\!1_{s\ge 1-t}, s,t \in [0,1]$. Then the processes obtained as the integral transformations of the given Wiener process W with the kernels Q_1, Q_2 are equal to $X_1 = W$, $X_2 = \overleftarrow{W} = W(1) - W(1-\cdot)$. See Problem 6.5 (e).

7

Martingales and related processes in discrete and continuous time. Stopping times

Theoretical grounds

Let $\mathbb{T}$ be a set with linear order. For instance, $\mathbb{T}=\mathbb{R}^+$ or $\mathbb{T}=\mathbb{Z}^+:=\mathbb{N}\cup\{0\}$ with the usual type of relation $\le$. Let also $\{\Omega,\mathcal{F},\{\mathcal{F}_t\}_{t\in\mathbb{T}},\mathsf{P}\}$ be a probability space with complete right-hand continuous filtration (this kind of space is sometimes called *a stochastic basis*).

Definition 7.1. *A stochastic process $\{X(t),\ t\in\mathbb{T}\}$ is said to be a* martingale *if it satisfies the following three conditions.*

(1) For any $t\in\mathbb{T}$ the random variable $X(t)\in L_1(\mathsf{P})$ (i.e., $\mathsf{E}|X(t)|<\infty$; or sometimes we say that the process X is integrable on $\mathbb{T}$).

(2) For any $t\in\mathbb{T}$ the random variable $X(t)$ is $\mathcal{F}_t$-measurable (sometimes we say that the process $X(t)$ is $\mathcal{F}_t$-adapted).

(3) For any $s\le t$, $s,t\in\mathbb{T}$ it holds that $\mathsf{E}(X(t)/\mathcal{F}_s)=X(s)$ P-a.s.

If we change in condition (3) the sign $=$ for $\ge$ and obtain $\mathsf{E}(X(t)/\mathcal{F}_s)\ge X(s)$ P-a.s, then we have the definition of a submartingale; if $\mathsf{E}(X(t)/\mathcal{F}_s)\le X(s)$ P-a.s. for any $s\le t$, $s,t\in\mathbb{T}$, then we have a supermartingale. Furthermore the property that takes place P-a.s. we denote simply "a.s." A vector process we call a (sub-, super-) martingale if the corresponding property has each of its components.

In what follows we denote the fact that a stochastic process $\{X(t),\ t\in\mathbb{T}\}$ is $\mathcal{F}_t$-adapted as $\{X(t),\mathcal{F}_t,\ t\in\mathbb{T}\}$.

Definition 7.2. *A mapping $\tau:\Omega\to\mathbb{T}\cup\{\infty\}$ is said to be the Markov moment if for any $t\in\mathbb{T}$ an event $A:=\{\omega\in\Omega|\ \tau(\omega)\le t\}\in\mathcal{F}_t$. A stopping time is the Markov moment τ for which $\tau<\infty$ a.s. A sigma-algebra generated by the Markov moment τ is the class of events*

$\mathcal{F}_\tau:=\{A\in\mathcal{F}|\ A\cap\{\tau\le t\}\in\mathcal{F}_t,\ t\in\mathbb{R}^+\}$ *(for continuous time) and*

$\mathcal{F}_\tau:=\{A\in\mathcal{F}|\ A\cap\{\tau\le n\}\in\mathcal{F}_n, n\ge 0\}$ *(for discrete time).*

Definition 7.3. *The Markov moment $\tau(\omega)$ is called predictable if there exists a sequence $\{\tau_n, n\ge 1\}$ of Markov moments such that:*

D. Gusak et al., *Theory of Stochastic Processes*, Problem Books in Mathematics,
DOI 10.1007/978-0-387-87862-1_7, © Springer Science+Business Media, LLC 2010

(1) $\tau_n(\omega)$ is the increasing sequence a.s. and $\lim_{n\to\infty}\tau_n(\omega)=\tau(\omega)$ a.s.
(2) For any $n\ge 1$ it holds that $\tau_n(\omega)<\tau(\omega)$ a.s. on the set $\{\tau(\omega)>0\}$.

Sometimes one says that the sequence τ_n predicts the Markov moment τ.

Definition 7.4. *A σ-algebra is called predictable on $[0,\infty)\times\Omega$ if it is generated by random intervals $[\tau,\sigma):=\{(t,\omega)|\ \tau(\omega)\le t<\sigma(\omega)\}$, where τ and σ are predictable Markov moments.*

Definition 7.5. *A stochastic process $\{A(t),\ t\in\mathbb{T}\}$ defined on $(\Omega,\mathcal{F})$ and with values in a measurable space $(\mathbb{X},\mathfrak{X})$ is said to be predictable if the mapping $A:[0,\infty)\times\Omega\to\mathbb{X}$ is measurable regarding the predictable σ-algebra on $[0,\infty)\times\Omega$.*

Usually we consider $\mathbb{X}=\mathbb{R}$ or $\mathbb{R}^m$.

In the case $\mathbb{T}=\mathbb{Z}^+$ Definition 7.5 is transformed to the following one. A discrete-time process $A(t)$ is predictable if $A(0)$ is a constant and $A(t)$ is $\mathcal{F}_{t-1}$-measurable r.v. for $t\in\mathbb{N}$. If $\mathbb{T}=\mathbb{R}^+$ then left-hand continuous (in particular, continuous) processes are predictable.

Below in this chapter we denote the discrete-time processes as X_n, M_n, and so on (with the lower index as time) and continuous-time processes, as before, are denoted as $X(t), M(t)$, and so on (the time index is inside the parentheses).

Theorem 7.1. *(1) A supermartingale $\{X(t),\mathcal{F}_t,\ t\in\mathbb{R}^+\}$ has càdlàg modification if and only if $\mathsf{E}X(t)$ is a right-hand continuous function of $t\in\mathbb{R}^+$ (you can find the definition of càdlàg modification in Theoretical grounds of Chapter 3).*

(2) Let $\{X(t),\mathcal{F}_t,\ t\in\mathbb{R}^+\}$ be a right-hand continuous supermartingale. Then $X(t)$ has left-hand limits a.s. and almost all trajectories are bounded on every segment $[0,T]$, $T>0$.

Definition 7.6. *A family of random variables $\{\xi_\alpha,\alpha\in\mathfrak{A}\}$ is said to be uniformly integrable if*

$$\lim_{C\to\infty}\sup_{\alpha\in\mathfrak{A}}\int_{\{|\xi_\alpha|>C\}}|\xi_\alpha|d\mathsf{P}=0.$$

A stochastic process $\{X(t),\ t\in\mathbb{T}\}$ is said to be uniformly integrable if the family of random variables $\{X(t),\ t\in\mathbb{T}\}$ is uniformly integrable.

Definition 7.7. *A right-hand continuous uniformly integrable (sub-, super-) martingale $\{X(t),\ t\in\mathbb{R}^+\}$ belongs to a class* D *if a family of random variables $\{X(\tau),\tau$ is a Markov moment$\}$ is uniformly integrable.*

In what follows we use $X(\infty)$ to denote a limit of a stochastic process: $X(\infty)=\lim_{t\to\infty}X(t)$ if the limit exists a.s.

Theorem 7.2. *(Doob–Meyer decomposition for supermartingales from the class* D*) Let $\{X(t),\mathcal{F}_t,\ t\in\mathbb{R}^+\}$ be the right-hand continuous supermartingale from the class* D*. In this case, there exists the unique predictable right-hand continuous nondecreasing process $\{A(t),\mathcal{F}_t,\ t\in\mathbb{R}^+\}$ such that $A(0)=0$, $\mathsf{E}A(\infty)<\infty$ and the process $M(t):=X(t)+A(t)$ is the uniformly integrable martingale.*

Definition 7.8. *A stochastic process $\{M(t), \mathfrak{F}_t,\ t \in \mathbb{R}^+\}$ is said to be a*

(1) Local martingale if there exists a sequence of Markov moments $\{\tau_n, n \geq 1\}$ such that: (a) $0 \leq \tau_n \leq \tau_{n+1}$ a.s., (b) $\tau_n \to \infty$ a.s., (c) for any $n \geq 1$ a stopped process $M^{\tau_n}(t) := M(t \wedge \tau_n)$ is an $\mathfrak{F}_t$-martingale

(2) Square-integrable martingale if it is a martingale and $\mathsf{E}M^2(t) < \infty$ for any $t \in \mathbb{R}^+$

(3) The locally square-integrable martingale if it is a local martingale and every stopped process $M^{\tau_n}(t)$ is a square integrable martingale.

We say that the sequence of Markov moments localizes the corresponding process.

Theorem 7.3. *(Doob–Meyer decomposition for general supermartingales) Let $\{X(t), \mathfrak{F}_t,\ t \in \mathbb{R}^+\}$ be a right-hand continuous supermartingale. Then there exists the unique decomposition of the form*

$$X(t) = M(t) - A(t),\ t \in \mathbb{R}^+,$$

where $\{A(t), \mathfrak{F}_t,\ t \in \mathbb{R}^+\}$ is a nondecreasing predictable process with $A(0) = 0$ a.s.; $\{M(t), \mathfrak{F}_t,\ t \in \mathbb{R}^+\}$ is the local martingale.

Let us introduce the following notations. $\mathscr{M}$ is a class of all martingales determined on some fixed stochastic basis; $\mathscr{M}_{\text{loc}}$ is a class of all local martingales; $\mathscr{M}^2$ is a class of square-integrable martingales; and finally, $\mathscr{M}^2_{\text{loc}}$ is a class of locally square-integrable martingales.

The stopping time or Markov moment τ is called bounded if there exists a constant $C > 0$ such that $\tau \leq C$ a.s.

Theorem 7.4. *(Doob's stopping theorem, or the optional sampling theorem for discrete time) Let $\{X_n, \mathfrak{F}_n,\ n \in \mathbb{Z}^+\}$ be an integrable stochastic process. Then the following conditions are equivalent.*

(1) $\{X_n, \mathfrak{F}_n,\ n \in \mathbb{Z}^+\}$ is a martingale (submartingale).

(2) $\mathsf{E}(X_\tau/\mathfrak{F}_\sigma) = (\geq) X_{\tau\wedge\sigma}$ for any bounded Markov moment τ and any Markov moment σ.

(3) $\mathsf{E}X_\tau = (\geq)\mathsf{E}X_\sigma$ for all bounded Markov moments τ and σ such that $\tau \geq \sigma$ a.s.

Corollary 7.1. *Let $\{X_n, \mathfrak{F}_n,\ n \in \mathbb{Z}^+\}$ be a martingale, σ and τ be stopping times for which $\mathsf{E}|X_\sigma| < \infty$, $\mathsf{E}|X_\tau| < \infty$ and $\liminf_{n\to\infty} \int_{\{\sigma \geq n\}} |X_n| d\mathsf{P} = \liminf_{n\to\infty} \int_{\{\tau \geq n\}} |X_n| d\mathsf{P} = 0$. Then $\mathsf{E}(X_\tau/\mathfrak{F}_\sigma) \cdot \mathbb{I}_{\tau \geq \sigma} = X_\sigma \cdot \mathbb{I}_{\tau \geq \sigma}$ a.s. In particular, if $\sigma \leq \tau$ a.s., then $\mathsf{E}X_\tau = \mathsf{E}X_\sigma$.*

Theorem 7.5. *(A version of Doob's optional sampling theorem for continuous time) Let $\{X(t), \mathfrak{F}_t,\ t \in \mathbb{R}^+\}$ be a submartingale with right-hand continuous trajectories, such that for some random variable η with $\mathsf{E}|\eta| < \infty$ the inequality holds $X(t) \leq \mathsf{E}(\eta/\mathfrak{F}_t)$ a.s., $t \in \mathbb{R}^+$.*

Then for any Markov moments σ and τ it holds that

$$X(\sigma \wedge \tau) \leq \mathsf{E}(X(\tau)/\mathfrak{F}_\sigma).$$

Corollary 7.2. *Let $\{X(t), \mathcal{F}_t,\ t \in \mathbb{R}^+\}$ be a uniformly integrable martingale with right-hand continuous trajectories. Then, for some r.v. η with $\mathsf{E}|\eta| < \infty$, it holds that $X(t) = \mathsf{E}(\eta/\mathcal{F}_t)$ a.s., $t \in \mathbb{R}^+$, and thus, for the Markov moments σ and τ we obtain the equality (see Problem 7.98)*

$$X(\sigma \wedge \tau) = \mathsf{E}(X(\tau)/\mathcal{F}_\sigma).$$

Now, let $\mathbb{T} = \mathbb{Z}^+$.

Definition 7.9. *(1) The stochastic process $[M,M]_n := \sum_{k=1}^n (M_k - M_{k-1})^2$ is called the quadratic variation of a martingale $\{M_n, \mathcal{F}_n,\ n \in \mathbb{Z}^+\}$.*

(2) The stochastic process $[M,N]_n := \sum_{k=1}^n (M_k - M_{k-1})(N_k - N_{k-1})$ is called the joint quadratic variation of two martingales $\{M_n, N_n, \mathcal{F}_n,\ n \in \mathbb{Z}^+\}$.

(3) Let $\{M_n, N_n, \mathcal{F}_n,\ n \in \mathbb{Z}^+\}$ be square integrable martingales; that is, for all $n \in \mathbb{Z}^+$ it holds that $\mathsf{E}|M_n|^2 < \infty$, $\mathsf{E}|N_n|^2 < \infty$. The process $\langle M,M\rangle_n := \sum_{k=1}^n \mathsf{E}((M_k - M_{k-1})^2/\mathcal{F}_{k-1})$, is called a quadratic characteristic of a martingale M and the process $\langle M,N\rangle_n := \sum_{k=1}^n \mathsf{E}((M_k - M_{k-1})(N_k - N_{k-1})/\mathcal{F}_{k-1})$ is called the joint quadratic characteristic of martingales M and N.

In many cases the notations $[M]$ and $\langle M\rangle$ are used instead of $[M,M]$ and $\langle M,M\rangle$, correspondingly.

Suppose that $\mathbb{T} = \mathbb{R}^+$. Denote by $\widetilde{\mathcal{M}}^2$ a class of all square integrable martingales $\{M(t), \mathcal{F}_t,\ t \in \mathbb{R}^+\}$ such that their trajectories belong a.s. to the space $\mathbb{D}(\mathbb{R}^+)$ of functions without discontinuities of the second kind, right-hand continuous with left-hand limits at every point, and such that $\sup_{t\in\mathbb{R}^+} \mathsf{E}M^2(t) < \infty$.

If $M \in \widetilde{\mathcal{M}}^2$ then a process $M^2(t)$ is a submartingale from the class D (see Problem 7.90). According to the Doob–Meyer decomposition for submartingales from the class D (it is an evident modification of Theorem 7.2) there exists a unique integrable nondecreasing predictable process $A(t)$ and the martingale $L(t)$ such that

$$M^2(t) = A(t) + L(t),\ t \in \mathbb{R}^+. \tag{7.1}$$

Definition 7.10. *(1) A predictable process $A(t)$ in decomposition* (7.1) *is said to be a quadratic characteristic of the martingale M and is denoted as $\langle M,M\rangle(t)$ or $\langle M\rangle(t)$.*

(2) The predictable process $\langle M,N\rangle(t) = \frac{1}{2}(\langle M+N\rangle(t) - \langle M\rangle(t) - \langle N\rangle(t))$ is a joint quadratic characteristic of two martingales M and N from $\widetilde{\mathcal{M}}^2$.

Note that the space $\widetilde{\mathcal{M}}^2$ is a Hilbert space with inner product $(M,N) := \mathsf{E}M(\infty) \times N(\infty)$ (see Problem 7.89) and a subset of continuous square integrable martingales. $\mathcal{M}^{2,c}$ is a closed subspace in $\mathcal{M}^2$. The orthogonal complement to $\mathcal{M}^{2,c}$ in $\widetilde{\mathcal{M}}^2$ is called the space of "purely discontinuous" martingales and is denoted by $\mathcal{M}^{2,d}$.

Theorem 7.6. *Any martingale $M \in \widetilde{\mathcal{M}}^2$ admits a unique decomposition of the form*

$$M(t) = M^c(t) + M^d(t),$$

where $M^c \in \mathcal{M}^{2,c}$ is a continuous martingale, and $M^d \in \mathcal{M}^{2,d}$ is a "purely discontinuous" martingale.

Definition 7.11. *A nondecreasing integrable stochastic process*

$$[M](t) := \langle M^c \rangle(t) + \sum_{s \le t} (\Delta_s M^d)^2$$

is called a quadratic variation of a martingale $M \in \widetilde{\mathcal{M}^2}$. *Here* $\Delta_s M = \Delta_s M^d = M^d(s) - M^d(s-)$ *is a jump of the martingale* M^d *at point* $s \in \mathbb{R}^+$. *The series mentioned above converges a.s. for all* $t \in \mathbb{R}^+$. *Moreover,* $\mathsf{E}([M](t) - \langle M \rangle(t)/\mathfrak{F}_s) = [M](s) - \langle M \rangle_s$ *for all* $s \le t$. *In particular,* $\mathsf{E}[M](t) = \mathsf{E}\langle M \rangle(t)$, *and* $M^2(t) - [M]_t$ *is a martingale. Evidently,* $[M](t) = \langle M \rangle(t)$ *for* $M \in \mathcal{M}^{2,c}$.

The definition of the quadratic variation $[M]$ can be extended to processes $M \in \mathcal{M}^2_{\text{loc}}$. The point is that $M \in \mathcal{M}^2_{\text{loc}}$ admits a unique decomposition $M = M^c + M^d$ into the continuous and "purely discontinuous" processes $M^c, M^d \in \mathcal{M}^2_{\text{loc}}$. So, it is possible to define

$$[M](t) = \langle M^c \rangle_t + \sum_{s \le t} (\Delta_s M)^2,$$

and the series converges for all $t \in \mathbb{R}^+$ a.s.

Let us present general inequalities for martingales.

Theorem 7.7. *(Doob's inequalities for discrete-time sub- and supermartingales)*

(1) Let $\{X_n, \mathfrak{F}_n,\ n \in \mathbb{Z}^+\}$ *be a submartingale. Then for any* $C > 0$, $N \in \mathbb{Z}^+$ *it holds that*

$$\mathsf{P}\left(\max_{0 \le n \le N} X_n \ge C\right) \le \frac{\mathsf{E}X_N^+}{C}.$$

(2) Let $\{X_n, \mathfrak{F}_n,\ n \in \mathbb{Z}^+\}$ *be a supermartingale. Then for any* $C > 0$, $N \in \mathbb{Z}^+$ *it holds that*

$$\mathsf{P}(\max_{0 \le n \le N} X_n \ge C) \le \frac{2}{C} \max_{0 \le n \le N} \mathsf{E}|X_n|.$$

(3) Let $\{X_n, \mathfrak{F}_n,\ n \in \mathbb{Z}^+\}$ *be a positive supermartingale. Then for any* $C > 0$, $N \in \mathbb{Z}^+$ *it holds that*

$$\mathsf{P}\left(\max_{0 \le n \le N} X_n \ge C\right) \le \frac{\mathsf{E}X_0}{C}.$$

Corollary 7.3. *Let* $\{X_n, \mathfrak{F}_n,\ n \in \mathbb{Z}^+\}$ *be a martingale,* $p \ge 1$*; then*

$$\mathsf{P}\left(\max_{0 \le n \le N} |X_n| \ge C\right) \le \frac{\mathsf{E}|X_N|^p}{C^p},\ C > 0,\ N \in \mathbb{Z}^+.$$

Recall the notation $\|X\|_p := (\mathsf{E}|X|^p)^{1/p}$ for any $p \ge 1$ and r.v. X.

Theorem 7.8. *(Maximum integral Doob's inequalities for discrete-time submartingales) Let $\{X_n, \mathfrak{F}_n,\ n \in \mathbb{Z}^+\}$ be a submartingale.*

(1) If $p > 1$, then

$$\left\| \sup_{0 \le n \le N} X_n^+ \right\|_p \le \frac{p}{p-1} \|X_N^+\|_p.$$

(2) If $p = 1$, then

$$\left\| \sup_{0 \le n \le N} X_n^+ \right\|_1 := \mathsf{E} \sup_{0 \le n \le N} X_n^+ \le \frac{e}{e-1}(1 + \mathsf{E}X_N^+(\ln^+ X_N^+)).$$

Definition 7.12. *Let $\{X_n, \mathfrak{F}_n,\ n \in \mathbb{Z}^+\}$ be a submartingale, and (a,b) be an interval with $a < b$. We define the Markov moments $\tau_k,\ k \ge 0$ in the following way: $\tau_0 = 0$; $\tau_{2m-1} = \min\{n : n > \tau_{2m-2}, X_n \le a\}$; $\tau_{2m} = \min\{n : n > \tau_{2m-1}, X_n \ge b\}$; $m \in \mathbb{N}$.*

In the case when the set in braces is empty we assume that the corresponding τ_k and all τ_j with $j > k$ are infinite. Let us also define

$$\beta_N(a,b) := \begin{cases} 0, & \text{if } \tau_2 > N, \\ \max\{m : \tau_{2m} \le N\}, & \text{if } \tau_2 \le N. \end{cases}$$

This value is a "number of bottom-up crossings" by a process X of a strip (a,b) during the time period from 0 until N.

Theorem 7.9. *(Doob's theorem on number of crossings) Let $\{X_n, \mathfrak{F}_n,\ n \in \mathbb{Z}^+\}$ be a submartingale, and $a < b$. Then it holds that*

$$\mathsf{E}\beta_N(a,b) \le \frac{\mathsf{E}(X_N - a)^+}{b-a} \le \frac{\mathsf{E}|X_N| + |a|}{b-a}.$$

Corollary 7.4. *(Doob's theorem on convergence of the discrete-time submartingale) Let $\{X_n, \mathfrak{F}_n,\ n \in \mathbb{Z}^+\}$ be a submartingale for which $\sup_n \mathsf{E}|X_n| < \infty$. Then there exists the limit $\lim_{n\to\infty} X_n = X_\infty$ with probability one, and $\mathsf{E}|X_\infty| < \infty$. (This statement evidently holds true for a discrete-time supermartingale as well.)*

Theorem 7.10. *Let $\{X_n, \mathfrak{F}_n,\ n \in \mathbb{Z}^+\}$ be a martingale. The following conditions are equivalent.*

(1) X_n converges in $L_1(\mathsf{P})$ as $n \to \infty$.

(2) $\sup_n \mathsf{E}|X_n| < \infty$ (due to this condition there exists a limit $\lim_{n\to\infty} X_n = X_\infty$ a.s.) and $X_n = \mathsf{E}(X_\infty / \mathfrak{F}_n)$ a.s.

(3) There exists an integrable random variable η such that $X_n = \mathsf{E}(\eta/\mathfrak{F}_n)$, $n \in \mathbb{Z}^+$.

(4) The martingale X_n is uniformly integrable.

Let us present inequalities for martingale transformations (you can find the notation for $\xi \circ M$ in Problem 7.12).

Theorem 7.11. *Let $\{M_n, \mathcal{F}_n,\ n \in \mathbb{Z}^+\}$ be a martingale, $\{\xi_n, \mathcal{F}_n,\ n \in \mathbb{Z}^+\}$ be a predictable sequence, and $|\xi_n| \le 1$ a.s. for any $n \in \mathbb{N}$. Then the following inequalities hold.*

(1) $\mathsf{P}(\sup_n |(\xi \circ M)_n| \ge C) \le 17C^{-1} \sup_n \mathsf{E}|M_n|$, $C > 0$.

(2) $\mathsf{E}|(\xi \circ M)_n|^p \le C_p \mathsf{E}|M_n|^p$, $p > 1$, $n \ge 0$.

Theorem 7.12. *Let $\{X_n, \mathcal{F}_n,\ n \in \mathbb{Z}^+\}$ be either a martingale, or a nonnegative submartingale, $[X]_n$ be its quadratic variation, and $[X] = \lim_{n\to\infty}[X]_n$ (the limit exists a.s. because the stochastic process $[X]_n$ is nondecreasing in n). Then for any $C > 0$ the inequality $\mathsf{P}([X] > C) \le 3C^{-1} \sup_{n\ge 0} \mathsf{E}|X_n|$ holds true.*

Theorem 7.13. *(Burkholder–Davis inequalities for discrete-time martingale) If $\{X_n, \mathcal{F}_n,\ n \in \mathbb{Z}^+\}$ is a martingale and $X_0 = 0$ then for all $p \ge 1$ there exist constants c_p and C_p such that*

$$c_p \mathsf{E}[X]^{p/2} \le \mathsf{E}\sup_{n\ge 0} |X_n|^p \le C_p \mathsf{E}[X]^{p/2}.$$

Theorem 7.13 admits the following generalization. Let a function $\Phi : \mathbb{R}^+ \to \mathbb{R}^+$ be nondecreasing, $\Phi(0) = 0$, and Φ be a function of a bounded growth; that is, there exists $c_0 > 0$ such that $\Phi(2x) \le c_0\Phi(x)$, $x \in \mathbb{R}^+$.

Theorem 7.14. *Let, additionally to a bounded growth, Φ be a convex function. Then there exist positive constants c and C independent of X, such that*

$$c\mathsf{E}\Phi([X]^{1/2}) \le \mathsf{E}\Phi(\sup_{n\ge 0} |X_n|) \le C\mathsf{E}\Phi([X]^{1/2}),$$

where $\{X_n, \mathcal{F}_n,\ n \in \mathbb{Z}^+\}$ is a martingale with $X_0 = 0$.

Theorem 7.15. *Let $\{X_n, \mathcal{F}_n,\ n \in \mathbb{Z}^+\}$ be a martingale with $X_0 = 0$. Then:*

(1) For all $0 < p \le 2$ there exists $C_p > 0$ such that

$$\mathsf{E}\sup_{n\ge 0} |X_n|^p \le C_p \mathsf{E}\langle X\rangle^{p/2},$$

where $\langle X\rangle = \lim_{n\to\infty}\langle X\rangle_n$,

(2) For all $p \ge 2$ there exists $c_p > 0$ such that

$$c_p \mathsf{E}\langle X\rangle^{p/2} \le \mathsf{E}\sup_{n\ge 0} |X_n|^p.$$

Theorem 7.16. *(Doob's inequalities for continuous-time martingales and submartingales)*

(1) Let $\{X(t), \mathcal{F}_t,\ t \in [0,T]\}$ be a continuous submartingale. Then for any $C > 0$,

$$\mathsf{P}(\sup_{0\le t\le T} X(t) \ge C) \le \frac{\mathsf{E}X^+(T)}{C}.$$

(2) Let $\{X(t), \mathcal{F}_t,\ t \in [0,T]\}$ be a continuous square integrable martingale. Then for any $C > 0$,

$$\mathsf{P}(\sup_{0\le t\le T} |X(t)| \ge C) \le \frac{\mathsf{E}X^2(T)}{C^2}.$$

Theorem 7.17. *(Burkholder–Davis inequalities for continuous-time martingales) Let* $\{X(t), \mathcal{F}_t,\ t \in \mathbb{R}^+\}$ *be a martingale with* $X(0) = 0$. *Then for all* $p \geq 1$ *there exist* $c_p > 0$ *and* $C_p > 0$ *such that*

$$c_p \mathsf{E}[X]^{p/2} \leq \mathsf{E} \sup_{t \in \mathbb{R}^+} |X(t)|^p \leq C_p \mathsf{E}[X]^{p/2},$$

where $[X] = \lim_{t \to \infty} [X](t)$, *and the existence of the limit follows from the fact the process* $[X](t)$ *is the quadratic variation of the martingale* X *and is monotonically nondecreasing in* t.

For $p > 1$ the inequalities from Theorem 7.17 were first proved by Burkholder; for $p = 1$ they were proved by Davis.

For discrete time we have the following version of the Burkholder inequalities. If $\{X_n, \mathcal{F}_n,\ n \in \mathbb{Z}^+\}$ is a martingale, $X_0 = 0$, then for all $p > 1$ and all $n \geq 1$

$$c_p \mathsf{E}[X]_n^{p/2} \leq \mathsf{E}|X_n|^p \leq C_p \mathsf{E}[X]_n^{p/2},$$

where $c_p = (18p^{3/2}/(p-1))^{-p}$, $C_p = (18p^{3/2}/(p-1)^{1/2})^p$. It is evident that $\mathsf{E}|X_n|^p$ can be replaced by $\mathsf{E}\sup_{n \geq 1} |X_n|^p$. If $p = 1$, then the left-hand part of this inequality does not hold (see Problem 7.67, item (3)).

The Burkholder–Davis inequalities are generalization of the Khinchin and Marcinkievich–Zygmund inequalities for the sums of independent random variables. *Khinchin inequalities.* Let $\{\xi_i, i \geq 1\}$ be i.i.d. random variables, $\mathsf{P}(\xi_i = 1) = \mathsf{P}(\xi_i = -1) = \frac{1}{2}$, $\{c_i, i \geq 1\}$ be some real sequence. Then for any $p \in (0, +\infty)$ there exist A_p and B_p such that for any $n \geq 1$

$$A_p \Big(\sum_{i=1}^n c_i^2 \Big)^{p/2} \leq \mathsf{E} \Big| \sum_{i=1}^n c_i \xi_i \Big|^p \leq B_p \Big(\sum_{i=1}^n c_i^2 \Big)^{1/2}.$$

Marcinkievich–Zygmund inequalities. If $\{\xi_i, i \geq 1\}$ are independent integrable random variables $\mathsf{E}\xi_i = 0$ then for any $p \geq 1$ there exist A_p and B_p independent of ξ_i and such that

$$A_p \mathsf{E} \Big(\sum_{i=1}^n \xi_i^2 \Big)^{p/2} \leq \mathsf{E} \Big| \sum_{i=1}^n \xi_i \Big|^p \leq B_p \mathsf{E} \Big(\sum_{i=1}^n \xi_i^2 \Big)^{p/2}.$$

Theorem 7.18. *Let* $M \in \mathscr{M}^2_{\mathrm{loc}}$ *and its quadratic characteristic* $\langle M \rangle$ *be continuous. Then for all* $\varepsilon > 0$, $N > 0$, $T > 0$,

$$\mathsf{P}\big(\sup_{0 \leq t \leq T} |M(t)| \geq \varepsilon \big) \leq \frac{N}{\varepsilon^2} + \mathsf{P}(\langle M \rangle_T \geq N).$$

Theorem 7.18 is a particular case of the Lenglart inequality [58].

The following theorem can be useful on numerous occasions.

Theorem 7.19. *(Lévy) Let* $(\Omega, \mathcal{F}, \{\mathcal{F}_n\}_{n \in \mathbb{Z}^+}, \mathsf{P})$ *be a probability space equipped with filtration* $\mathcal{F}_n \subset \mathcal{F}$. *Let also* $\mathcal{F}_\infty = \sigma\{\bigcup_{n \in \mathbb{Z}^+} \mathcal{F}_n\}$ *and* ξ *be a r.v. with* $\mathsf{E}|\xi| < \infty$. *Then the following sequence is uniformly integrable and converges,* $\mathsf{E}(\xi/\mathcal{F}_n) \to \mathsf{E}(\xi/\mathcal{F}_\infty)$ *as* $n \to \infty$, *a.s. and in the space* $L_1(\mathsf{P})$.

Bibliography

[9], Chapter IV; [90], Chapter 7; [24], Volume 1, Chapter II, §2; Volume 3, Chapter I, §1; [25], Chapter III, §1; [51], Chapter 13; [57]; [58], Part I; [79], Chapters 4 and 12; [22], Chapter VI; [23], Chapter 5; [82], Volume 2, Chapter VII; [20], Chapters 3 – 10; [8], Chapters 3 and 4; [46], Chapter 12, §12.3, Chapter 13; [54], Chapter 1, §1.2, Chapter 3, §3.3; [68], Chapters 9 and 10; [85], Chapters 2, 4, and 14.

Problems

7.1. Prove the following. If $\{\xi_n, n \geq 1\}$ is a square integrable sequence of random variables on a probability space $(\Omega, \mathcal{F}, \mathsf{P})$ and $\xi_n \to \xi$ P-a.s., where ξ is also a random variable, then $\mathsf{E}\xi_n \to \mathsf{E}\xi$ and for any σ-algebra $\mathcal{G} \subset \mathcal{F}$ it holds that $\mathsf{E}(\xi_n/\mathcal{G}) \to \mathsf{E}(\xi/\mathcal{G})$. Extend this statement to a uniformly integrable sequence of random variables.

7.2. Prove that every martingale is still a martingale with respect to its natural filtration (see Definition 3.4).

7.3. Prove that a process $\{X(t), \mathcal{F}_t, t \in \mathbb{T}\}$ is a submartingale if and only if $\{-X(t), \mathcal{F}_t, t \in \mathbb{T}\}$ is a supermartingale.

7.4. Prove that a process $\{X(t), \mathcal{F}_t, t \in \mathbb{T}\}$ is a martingale if and only if for all $s, t \in \mathbb{T}$, $s \leq t$ and any event $A \in \mathcal{F}_s$,

$$\int_A X(s) d\mathsf{P} = \int_A X(t) d\mathsf{P}.$$

7.5. Let $\{X_n, \mathcal{F}_n, n \in \mathbb{Z}^+\}$ be an integrable-adapted sequence. Prove that X_n is a martingale if and only if for any $n \in \mathbb{Z}^+$ it holds that $\mathsf{E}(X_{n+1}/\mathcal{F}_n) = X_n$.

7.6. Prove that a linear combination of any finite number of martingales is a martingale as well.

7.7. Prove that for a (sub-, super-) martingale $\{X(t), \mathcal{F}_t, t \in \mathbb{T}\}$ the expectation is a constant (nondecreasing, nonincreasing) function in $t \in \mathbb{T}$.

7.8. Let $\mathbb{T} = \{0, 1, \ldots, N\}$ or $\mathbb{T} = [0, T]$, $\{\mathcal{F}_t\}_{t \in \mathbb{T}}$ be a filtration on a probability space $(\Omega, \mathcal{F}, \mathsf{P})$, and $X \in L_1(\mathsf{P})$ be an integrable r.v. Prove that the stochastic process $\{X(t) := \mathsf{E}(X/\mathcal{F}_t), t \in \mathbb{T}\}$ is a martingale (this is the so-called Lévy martingale).

7.9. Let $\{X(t), \mathcal{F}_t, t \in \mathbb{T}\}$, $\mathbb{T} = \mathbb{Z}^+$ or $\mathbb{R}^+$, be an integrable process with independent increments. Prove that $\{X(t) - \mathsf{E}X(t), \mathcal{F}_t, t \in \mathbb{T}\}$ is a martingale with respect to its natural filtration (see Definition 3.4). In particular a Wiener process is a martingale. If $\{N(t), \mathcal{F}_t, t \in \mathbb{R}^+\}$ is a homogeneous Poisson process with intensity λ, then $\{N(t) - \lambda t, \mathcal{F}_t, t \in \mathbb{R}^+\}$ is a martingale.

7.10. (Martingale generated by a random walk, particular case of Problem 7.9) Let $\{\xi_n,\ n \in \mathbb{Z}^+\}$ be a sequence of integrable independent random variables with $\mathsf{E}\xi_n = 0$, $n \in \mathbb{Z}^+$. Prove that the sequence $\{X_n := \sum_{k=0}^n \xi_k,\ \mathcal{F}_n,\ n \in \mathbb{Z}^+\}$ is a martingale; moreover the equality holds, $\mathcal{F}_n := \sigma\{\xi_0, \dots, \xi_n\} = \sigma\{S_0, \dots, S_n\}$.

7.11. Let $\{\xi_k, k \in \mathbb{Z}^+\}$ be a sequence of independent integrable random variables with $\mathsf{E}\xi_k = 1$. Prove that the process $M_n := \prod_{k=0}^n \xi_k$ is a martingale with respect to both its natural filtration $\mathcal{F}_n = \sigma\{M_k, 0 \le k \le n\}$ and filtration $\sigma\{\xi_k, 0 \le k \le n\}$. Explain why these two filtrations can be different.

7.12. Let $\{\xi_n, \mathcal{F}_n,\ n \in \mathbb{N}\}$ be a predictable bounded sequence, $\{M_n, \mathcal{F}_n,\ n \in \mathbb{Z}^+\}$ be a martingale, and $Y_n := \sum_{k=1}^n \xi_k (M_k - M_{k-1})$. A process $\{Y_n, \mathcal{F}_n,\ n \in \mathbb{N}\}$ is said to be a martingale transformation of the martingale M or a discrete stochastic integral with respect to M, and denoted as $Y_n = (\xi \circ M)_n$.

(1) Prove that $\{Y_n, \mathcal{F}_n,\ n \in \mathbb{N}\}$ is a martingale.

(2) Let $\xi_k \ge 0$ a.s. and M be a submartingale. Prove that Y is a submartingale as well.

7.13. Let Q be a probability measure on $\{\Omega, \mathcal{F}, \{\mathcal{F}_t\}_{t\in\mathbb{T}}\}$. The measure Q is locally absolutely continuous with respect to a measure P (it is denoted as $Q \overset{\text{loc}}{\ll} \mathsf{P}$) if for every $t \in \mathbb{T}$ the restriction $Q_t = Q|_{\mathcal{F}_t}$ of the measure Q on $\mathcal{F}_t$ is absolutely continuous with respect to the restriction $\mathsf{P}_t = \mathsf{P}|_{\mathcal{F}_t}$ of the measure P on $\mathcal{F}_t$. Let $Q \overset{\text{loc}}{\ll} \mathsf{P}$ and $X(t) := dQ_t/d\mathsf{P}_t$, $t \in \mathbb{T}$ is the Radon–Nikodym derivative. Prove that the process $\{X(t), \mathcal{F}_t,\ t \in \mathbb{T}\}$ is a P-martingale, that is, a martingale provided all conditional expectations are computed with respect to the measure P.

7.14. (1) Let $\{X(t), \mathcal{F}_t,\ t \in \mathbb{T}\}$ be a martingale, $h : \mathbb{R} \to \mathbb{R}$ be a convex function, and $\mathsf{E}|h(X(t))| < \infty$, $t \in \mathbb{T}$. Prove that $\{h(X(t)), \mathcal{F}_t,\ t \in \mathbb{T}\}$ is a submartingale.

(2) Let $\{X(t), \mathcal{F}_t,\ t \in \mathbb{T}\}$ be a submartingale, the function h be convex as above, and, in addition, nondecreasing. Moreover, let $\mathsf{E}|h(X(t))| < \infty$, $t \in \mathbb{T}$. Prove that $\{h(X(t)), \mathcal{F}_t,\ t \in \mathbb{T}\}$ is a submartingale.

(3) Let $\{X(t), Y(t), \mathcal{F}_t,\ t \in \mathbb{T}\}$ be martingales with $\mathsf{E}X^2(t) < \infty$. Prove that $X^2(t)$, $|X(t)|$, and $|X(t)| \vee |Y(t)|$ are submartingales.

7.15. (Kendall's example) We suppose that an alarm clock should ring at 6 a.m. but the person wakes up in the middle of the night and cannot sleep any more. He or she doesn't know what time it is and does not want to look at the clock. Let $X(t)$ be the conditional probability that the alarm clock will ring at the latest, 60 minutes after awakening, computed after t minutes after awakening. The σ-algebra $\mathcal{F}_t$, $t \ge 0$ is assumed to be a σ-algebra generated by the information on the ringing of the clock from the moment of awakening till the moment t. Show that $\{X(t), \mathcal{F}_t,\ t \in \mathbb{R}^+\}$ is a martingale.

7.16. Let $\{\xi_n,\ n \in \mathbb{Z}^+\}$ be a sequence of integrable random variables, $\mathcal{F}_n = \sigma\{\xi_0, \dots, \xi_n\}$ and $X_n = \sum_{k=0}^n \xi_k$, $n \in \mathbb{Z}^+$. Prove that $\{X_n, \mathcal{F}_n,\ n \in \mathbb{Z}^+\}$ is a martingale if and only if

$$\mathsf{E}\,(\xi_{n+1} f_n(\xi_0, \dots, \xi_n)) = 0$$

for any bounded Borel function $f_n : \mathbb{R}^n \to \mathbb{R}$ and all $n \ge 0$.

7.17. Let $\{\xi_n, \mathcal{F}_n, n \geq 1\}$ be a sequence of independent, centered, and integrable random variables and $\mathcal{F}_n = \sigma\{\xi_1, \ldots, \xi_n\}, n \geq 1$. Prove that for every $m \geq 1$ the stochastic process $\{X_{n,m}, \mathcal{F}_n, n \geq m\}$ of a form

$$X_{n,m} = \sum_{1 \leq i_1 < \cdots < i_m \leq n} \xi_{i_1} \cdots \xi_{i_m}, \; n \geq m$$

is a martingale with zero mean.

7.18. Let $\{X_n, \mathcal{F}_n, n \in \mathbb{Z}^+\}$ be a supermartingale, $\sup_n \mathsf{E}(X_n^-) < \infty$, and $X_n^- = -X_n + X_n^+ = -X_n \cdot \mathbb{I}_{X_n < 0}$.

(1) Prove that there exists $\lim_{n\to\infty} X_n =: X_\infty$ a.s.

(2) Prove that $\mathsf{E}|X_\infty| < \infty$.

(3) Prove that conditions $\sup_n \mathsf{E}(X_n^-) < \infty$ and $\sup_n \mathsf{E}|X_n| < \infty$ are equivalent.

7.19. (Polya scheme) An urn initially contains b black and r red marbles. One is chosen randomly. Then it is put back in the urn along with another c marbles of the same color. Let $Y_0 = b/b+r$, Y_n be the fraction of the black marbles in the urn after n iterations of the procedure. Prove that $\{Y_n, n \in \mathbb{Z}^+\}$ is a martingale with respect to a natural filtration.

7.20. (Probability ratio as martingale) Let $\{X_n, n \geq 1\}$ be a sequence of random variables and it is known that the joint distribution of the random variables $(X_1, \ldots, X_n)$ has either densities p_n or densities q_n but it is not known of which type exactly. Consider a new sequence of random variables

$$Y_n = \frac{q_n(X_1, \ldots, X_n)}{p_n(X_1, \ldots, X_n)},$$

with the assumption that the real densities are p_n and they are positive and continuous. Prove that $\{Y_n, \mathcal{F}_n, n \geq 1\}$ is a martingale with $\mathcal{F}_n = \sigma\{X_1, \ldots, X_n\}$. The r.v. Y_n is said to be the probability ratio.

7.21. Let $\rho = (\rho_t, t \in \mathbb{T} = \{0, 1, \ldots, T\})$ be a sequence of i.i.d. random variables taking two values b and a with probabilities p and q, respectively. Let $\mathsf{E}\rho_0 = r$ and $-1 < a < r < b$.

(1) Check that $p = (r-a)/(b-a)$.

(2) Specify that the stochastic process $m(t) = \sum_{k=0}^{t} (\rho_k - r)$ is a martingale with respect to the flow of σ-algebras $\mathcal{F}_t = \sigma\{\rho_0, \ldots, \rho_t\}$ (the so-called "basic" martingale).

(3) Prove that every martingale $\{X(t), \mathcal{F}_t, t \in \mathbb{T}\}$ with $\mathsf{E}X(t) = 0$ admits such a decomposition regarding basic martingale: $X(t) = \sum_{k=0}^{t} \alpha_k \Delta m_k$, where $\alpha_0 = 0$, α_k is $\mathcal{F}_{k-1}$-measurable, $k \geq 1$, and $\Delta m_k = \rho_k - r$.

7.22. (Snell envelope) Let $\{X_n, \mathcal{F}_n, 0 \leq n \leq N\}$ be a random sequence of integrable random variables. Denote $Y_N := X_N$, $Y_n = X_n \vee \mathsf{E}(Y_{n+1}/\mathcal{F}_n)$, $1 \leq n < N$. The process $\{Y_n, \mathcal{F}_n, 0 \leq n \leq N\}$ is the Snell envelope of the process X.

(1) Prove that the Snell envelope $\{Y_n, \mathcal{F}_n, 0 \leq n \leq N\}$ is a supermartingale.

(2) Prove that it is the least supermartingale that dominates X. Namely, if $Z_n \geq X_n, 0 \leq n \leq N$ and $\{Z_n, \mathcal{F}_n, 0 \leq n \leq N\}$ is a supermartingale then $Z_n \geq Y_n$.

7.23. Let for a random sequence $\{X_n, \mathcal{F}_n, n \in \mathbb{Z}^+\}$ there exists an integrable r.v. ξ such that $X_n = \mathsf{E}(\xi/\mathcal{F}_n)$, $n \in \mathbb{Z}^+$. Prove that $X_n \to \mathsf{E}(\xi/\mathcal{F}_\infty)$ a.s. and in $L_1(\mathsf{P})$ as $n \to \infty$. Here, as above, $\mathcal{F}_\infty := \sigma\{\bigcup_{n\in\mathbb{Z}^+} \mathcal{F}_n\}$ (compare with Theorem 7.10).

7.24. Let $\{\mathcal{F}_n, n \in \mathbb{Z}^+\}$ be a flow of σ-algebras on $(\Omega, \mathcal{F}, \mathsf{P})$ and $Y \in L_1(\Omega, \mathcal{F}_\infty, \mathsf{P})$, $X_n = \mathsf{E}(Y/\mathcal{F}_n)$. Prove that $\{X_n, \mathcal{F}_n, n \in \mathbb{Z}^+\}$ is a uniformly integrable martingale and $\lim_{n\to\infty} X_n = Y$ a.s.

7.25. (An example of a nonuniformly integrable martingale) Let $\{\xi_k, k \geq 1\}$ be i.i.d. random variables with $\mathsf{P}(\xi_1 = 0) = \mathsf{P}(\xi_1 = 2) = \frac{1}{2}$. Set $X_n = \prod_{k=1}^n \xi_k$, $n \geq 1$. Prove that $\{X_n, \mathcal{F}_n, n \geq 1\}$ is a nonuniformly integrable martingale.

7.26. Let $(\Omega, \mathcal{F}, \mathsf{P}) = ([0,1], \mathcal{B}([0,1]), \lambda^1|_{[0,1]})$ and $\{\pi_n, n \geq 1\}$ be a sequence of partitions of the segment $[0,1]$ such that

$$\pi_n = \{0 = t_0^n < \cdots < t_{k_n}^n = 1\}$$

and $\pi_n \subset \pi_{n+1}$, $n \geq 1$. Let $\mathcal{F}_n$ be a σ-algebra on $[0,1]$ generated by the sets $\Delta_{n,1} = [0, t_1^n]$, $\Delta_{n,2} = (t_1^n, t_2^n], \dots, \Delta_{n,k_n} = (t_{k_n-1}^n, t_{k_n}^n]$. For a function $f : [0,1] \to \mathbb{R}$ we put $X_n(\omega) = \sum_{k=1}^{k_n} f(t_k^n) - f(t_{k-1}^n)(t_k^n - t_{k-1}^n)^{-1} \mathbb{I}_{\omega \in \Delta_{n,k}}$, $n \geq 1$. Prove that $\{X_n, \mathcal{F}_n, n \geq 1\}$ is a martingale.

7.27. Find the values of a and b, under which the stochastic processes $aW^2(t) + bt$ and $\exp\{aW(t) + bt\}$ are (sub-, super-) martingales with respect to the filtration generated by a Wiener process $\{W(t), t \in \mathbb{R}^+\}$.

7.28. It is said that a discrete-time martingale $\{X_n, \mathcal{F}_n, n \geq 1\}$ is bounded in $L_2(\mathsf{P})$ if $\sup_{n\geq 1} \mathsf{E}X_n^2 < \infty$. Assume now that the martingale $\{X_n, \mathcal{F}_n, n \geq 1\}$ is square integrable. Show that it is bounded in $L_2(\mathsf{P})$ if and only if $\sum_{n=1}^\infty \mathsf{E}(X_n - X_{n-1})^2 < \infty$.

7.29. (Random signs) Let $\{X_n, n \geq 1\}$ be a sequence of independent Bernoulli variables with $\mathsf{P}(X_n = 1) = \mathsf{P}(X_n = -1) = \frac{1}{2}$, $n \geq 1$, and $\{\alpha_n, n \geq 1\}$ be a sequence of real numbers. Show that a series $\sum_{n=1}^\infty \alpha_n X_n$ converges a.s. if $\sum_{n=1}^\infty \alpha_n^2 < \infty$.

7.30. (1) Let σ and τ be Markov moments w.r.t. a filtration $\{\mathcal{F}_n, n \in \mathbb{Z}^+\}$. Prove that $\sigma + \tau$, $\sigma \wedge \tau$, $\alpha\tau$ for integer $\alpha \geq 2$, and $\sigma \vee \tau$ are Markov moments as well with respect to this filtration.

(2) Let $\{\tau_k, k \geq 1\}$ be the Markov moments with respect to the filtration $\{\mathcal{F}_n, n \in \mathbb{Z}^+\}$. Prove that $\sup_{k\geq 1} \tau_k$, $\inf_{k\geq 1} \tau_k$, $\limsup_{k\to\infty} \tau_k$, and $\liminf_{k\to\infty} \tau_k$ are Markov moments as well with respect to this filtration.

(3) Generalize the previous statements to the case of Markov moments with respect to the filtration $\{\mathcal{F}_t, t \in \mathbb{R}^+\}$.

(4) Let τ be a Markov moment. Prove that the random variables $\tau - 1$ and $[\tau/2]$ are not Markov moments w.r.t. the same filtration.

7.31. Let $\{X_n, \mathcal{F}_n, 1 \leq n \leq N\}$ be an integrable adapted process. Prove that the following statements are equivalent.

(1) X is a martingale.

(2) For any Markov moment the stopped process $X_n^\tau := X_{n\wedge\tau}$ is a martingale.

(3) For any Markov moment τ it holds that $\mathsf{E}X_{\tau\wedge N} = \mathsf{E}X_0$.

7.32. Let $\{X_n, \mathcal{F}_n, n \in \mathbb{Z}^+\}$ be a martingale and τ be a stopping time such that $|X_{\tau\wedge n}| \le Y$ a.s. for some r.v. $Y \in L_1(\mathsf{P})$ and all $n \ge 0$. Prove that $\mathsf{E}X_\tau = \mathsf{E}X_0$.

7.33. Let $\{X_n, \mathcal{F}_n, n \in \mathbb{Z}^+\}$ be a martingale and τ be a stopping time such that $\mathsf{E}|X_\tau| < \infty$ and $\liminf_{n\to\infty} \mathsf{E}(|X_n| 1\!\!1_{\tau>n}) = 0$.

(1) Prove that for any Markov moment σ $\mathsf{E}(X_\tau/\mathcal{F}_\sigma) = X_{\tau\wedge\sigma}$.

(2) Prove: if σ is a stopping time such that $\sigma \le \tau$ a.s. and $\mathsf{E}|X_\sigma| < \infty$, then $\mathsf{E}X_\tau = \mathsf{E}X_\sigma$.

(3) State and prove the corresponding statements in the case where X is a submartingale.

7.34. Let $\{X_n, \mathcal{F}_n, n \in \mathbb{Z}^+\}$ be a martingale with $\mathcal{F}_n = \sigma\{X_0, \dots, X_n\}$ and τ be a stopping time with $\mathsf{E}\tau < \infty$. Moreover, let there exist $C > 0$ such that for any $n \ge 0$,

$$\mathsf{E}\Big(|X_{n+1} - X_n|/\mathcal{F}_n\Big) \cdot 1\!\!1_{\tau>n} \le C \text{ a.s.}$$

Prove: (a) $\mathsf{E}|X_\tau| < \infty$; (b) $\mathsf{E}X_\tau = \mathsf{E}X_0$.

7.35. Prove that for any submartingale $\{X_n, \mathcal{F}_n, n \in \mathbb{Z}^+\}$ and any stopping time τ

$$\mathsf{E}|X_\tau| \le \liminf_{n\to\infty} \mathsf{E}|X_n|.$$

7.36. Let $\{X_n, \mathcal{F}_n, n \ge 0\}$ be a supermartingale and $X_n \ge \mathsf{E}(\xi/\mathcal{F}_n), n \ge 0$ a.s. where $\mathsf{E}|\xi| < \infty$. Let also σ and τ be stopping times with $\sigma \le \tau$ a.s. Prove that $X_\sigma \ge \mathsf{E}(X_\tau/\mathcal{F}_\sigma)$ a.s. State and prove the corresponding statements for the submartingale.

7.37. Let $\{X_n, \mathcal{F}_n, n \in \mathbb{Z}^+\}$ be a real random sequence and a set $B \in \mathcal{B}(\mathbb{R})$. Prove that a moment of the first visit of the set B:

$$\sigma_B := \inf\{n \ge 0 |\, X_n \in B\},\ \sigma_B = \infty, \quad \text{if } X_n \notin B \text{ for any } n \ge 0,$$

and the moment of the first departure from the set B:

$$\tau_B := \inf\{n \ge 0 |\, X_n \notin B\},\ \tau_B = \infty, \quad \text{if } X_n \in B \text{ for any } n \ge 0,$$

are Markov moments (it is evident that $\sigma_B = \tau_{B^c}$, where $B^c = \mathbb{R}\backslash B$).

7.38. Let $\{X_n, \mathcal{F}_n, n \in \mathbb{Z}^+\}$ be a martingale (submartingale) and τ be a Markov moment with respect to a filtration $\{\mathcal{F}_n, n \in \mathbb{Z}^+\}$. Prove that the stopped process $X_n^\tau := X_{\tau\wedge n}$ is an $\mathcal{F}_n$-martingale (submartingale) as well.

7.39. Let $\{X_n, \mathcal{F}_n, n \in \mathbb{Z}^+\}$ be a martingale, constants $N \ge 1$ and $C > 0$ be fixed, and $\nu = \min\{k \ge 0 \mid X_k > C\} \wedge N$ with $\nu = N$, if $X_k \le C$, $k \le N$. Prove that ν is a stopping time, and a random sequence $\{Y_n = X_n \cdot 1\!\!1_{n\le\nu} + (2X_\nu - X_n) \cdot 1\!\!1_{n>\nu}, n \in \mathbb{Z}^+\}$ is an $\mathcal{F}_n$-martingale as well.

7.40. Let $\{M_n, \mathcal{F}_n, n \in \mathbb{Z}^+\}$ be a process such that $\mathsf{E}|M_n| < \infty$ and $\mathsf{E}M_\tau = 0$ for any stopping moment τ. Prove that $\{M_n, \mathcal{F}_n, n \in \mathbb{Z}^+\}$ is a martingale.

7.41. Let τ be a Markov moment and $\mathcal{F}_\tau$ be a σ-algebra generated by τ (see Definition 7.2). In this problem as well as in the next one we assume that time is discrete.

(1) Prove that $\mathcal{F}_\tau$ is indeed a σ-algebra.

(2) Prove that $\mathcal{F}_\tau$ can be defined in the following way: $\mathcal{F}_\tau := \{A \in \mathcal{F} | A \cap \{\tau = n\} \in \mathcal{F}_n, n \geq 0\}$.

(3) Prove that τ is an $\mathcal{F}_\tau$-measurable r.v.

(4) Prove that if $\{X_n, \mathcal{F}_n, n \in \mathbb{Z}^+\}$ is a real-valued random sequence, then

$$X_\tau, \max_{1\leq k\leq \tau} X_k \text{ and } X_1 + X_2 + \cdots + X_k,$$

are $\mathcal{F}_\tau$-measurable random variables.

(5) Let σ and τ be Markov moments. Prove that events $\{\sigma = \tau\}$, $\{\sigma \leq \tau\}$, and $\{\sigma < \tau\}$ belong to $\mathcal{F}_{\sigma\wedge\tau}$.

(6) Prove that $\mathcal{F}_\sigma \subset \mathcal{F}_\tau$ if $\sigma \leq \tau$ a.s., and σ and τ are Markov moments.

(7) Let $X \in L_1(\mathsf{P})$ and τ be a stopping time on a set $\{0,\ldots,N\}$. Prove that $\mathsf{E}(X/\mathcal{F}_\tau) = \sum_{j=0}^N \mathsf{E}(X/\mathcal{F}_j)\mathbb{I}_{\tau=j}$.

(8) Show that $\mathsf{E}(\mathsf{E}(Y/\mathcal{F}_\tau)/\mathcal{F}_\sigma) = \mathsf{E}(\mathsf{E}(Y/\mathcal{F}_\sigma)/\mathcal{F}_\tau) = \mathsf{E}(Y/\mathcal{F}_{\tau\wedge\sigma})$.

7.42. Let σ and τ be Markov moments. Prove that events $\{\sigma = \tau\}$ and $\{\sigma \leq \tau\}$ belong to $\mathcal{F}_\sigma \bigcap \mathcal{F}_\tau$.

7.43. Let τ be a Markov moment on a probability space $(\Omega, \mathcal{F}, \mathsf{P})$, and a set $A \in \mathcal{F}$. The restriction of the moment τ to the set A we denote by τ_A and define as follows.

$$\tau_A(\omega) := \begin{cases} \tau(\omega), & \omega \in A, \\ \infty, & \omega \notin A. \end{cases}$$

Prove that the restriction τ_A is a Markov moment if and only if $A \in \mathcal{F}_\tau$.

7.44. Let $\{X(t), \mathcal{F}_t, t \geq 0\}$ be a continuous real-valued stochastic process, $A \subset \mathbb{R}$ be a closed set, and $\tau = \inf\{t \geq 0 | X(t) \in A\} \cup \{+\infty\}$ be a hitting moment of the set A. Prove that τ is a Markov moment.

7.45. (1) Prove that for any stopping time τ and any constant $r > 0$ a r.v. $\tau + r$ is a predictable stopping time and it is predicted by the sequence $\tau_n := \tau + r(1 - 1/n)$.

(2) Prove that any stopping time is a limit of a decreasing sequence of predictable stopping times.

7.46. The σ-algebra $\mathcal{F}_{\tau-}$, generated by all elements from $\mathcal{F}_0$ and all sets of a form $A \bigcap \{t < \tau\}$, where $t \in \mathbb{R}^+, A \in \mathcal{F}_t$, is called the σ-algebra of events that are strictly prior to Markov moment τ. Prove the following statements.

(1) $\mathcal{F}_{\tau-} \subset \mathcal{F}_\tau$.

(2) τ is an $\mathcal{F}_{\tau-}$-measurable r.v.

(3) If $\tau \leq \sigma$, τ, σ are Markov moments then $\mathcal{F}_{\tau-} \subset \mathcal{F}_{\sigma-}$.

(4) For any $A \in \mathcal{F}_\sigma$ the following inclusion holds true $A \bigcap \{\sigma < \tau\} \in \mathcal{F}_{\tau-}$.

7.47. Let $\mathscr{Y}_1 \supset \mathscr{Y}_2 \supset \cdots$ be a nonincreasing family of σ-algebras and ξ be an integrable r.v. Prove that a sequence $\{X_n, \mathscr{Y}_n, n \geq 1\}$ with $X_n = \mathsf{E}(\xi/\mathscr{Y}_n)$ produces an inverse martingale; that is, $\mathsf{E}(X_n/\mathscr{Y}_{n+1}) = X_{n+1}$ a.s. for any $n \geq 1$.

7.48. Let $\{X_k, k \geq 1\}$ be i.i.d. random variables, $Z_n = \sum_{k=1}^n X_k$, and $\mathscr{G}_n = \sigma\{Z_n, Z_{n+1}, \ldots\}$. Prove that the random sequence $\{Z_n/n, \mathscr{G}_n, n \geq 1\}$ produces an inverse martingale.

7.49. (The first Wald identity) Let $\{\xi_k, k \geq 1\}$ be i.i.d. random variables, $S_n = \sum_{k=1}^n \xi_k$, τ be a Markov moment with respect to the filtration $\mathfrak{F}_n = \sigma\{\xi_1, \ldots, \xi_n\}$, $\mathsf{E}|\xi_1| < \infty$, and $\mathsf{E}\tau < \infty$. Prove that $\mathsf{E}S_\tau = \mathsf{E}\tau\mathsf{E}\xi_1$, where $S_{\tau(\omega)}(\omega) := \sum_{k=1}^{\tau(\omega)} \xi_k(\omega)$ as $\omega \in \{\tau < \infty\}$ and $S_{\tau(\omega)}(\omega) := 0$ as $\omega \in \{\tau = \infty\}$.

7.50. (Corollary to the first Wald identity) Let $\{\xi_i, i \geq 1\}$ be i.i.d. random variables with $\mathsf{P}(\xi_i = 1) = \mathsf{P}(\xi_i = -1) = \frac{1}{2}$. Consider the random walk $S_0 = 0$, $S_n = \sum_{i=1}^n \xi_i$, $n \geq 1$, $\tau = \inf\{n \geq 1 : S_n = 1\}$.

(1) Prove that $\mathsf{P}(\tau < \infty) = 1$; that is, τ is a stopping time.

(2) Prove that $\mathsf{P}(S_\tau = 1) = 1$ and $\mathsf{E}S_\tau = 1$.

(3) Derive from the first Wald identity that $\mathsf{E}\tau = \infty$.

7.51. (The second Wald identity) Let the conditions of Problem 7.49 hold true and additionally $\mathsf{D}\xi_1 < \infty$. Prove that $\mathsf{E}(S_\tau - \tau\mathsf{E}\xi_1)^2 = \mathsf{E}\tau\mathsf{D}\xi_1$.

7.52. (The fundamental Wald identity) Let $\{\xi_i, i \geq 1\}$ be a sequence of i.i.d. random variables, $S_n = \sum_{i=1}^n \xi_i$, $n \geq 1$, a function $\varphi(t) = \mathsf{E}e^{t\xi_1}$, $t \in \mathbb{R}$, and for some $t_0 \neq 0$ the following relation holds, $1 \leq \varphi(t_0) < \infty$. Let also $\mathfrak{F}_n = \sigma\{\xi_1, .., \xi_n\}, n \geq 1$, τ be a stopping time with respect to $\{\mathfrak{F}_n, n \geq 1\}$ and such that $\mathsf{E}\tau < \infty$, and $|S_n| \cdot \mathbb{1}_{\tau > n} \leq C$ a.s. for a constant $C > 0$. Prove that

$$\mathsf{E}\left(\frac{e^{t_0 S_\tau}}{(\varphi(t_0))^\tau}\right) = 1.$$

7.53. (The generalized Wald identity) Let $\{\xi_i, i \geq 1\}$ be independent random variables (not necessarily identically distributed), $\mathsf{E}\xi_i = 0$, $\mathsf{E}\xi_i^2 = \sigma_i^2, S_n = \sum_{i=1}^n \xi_i$, $\mathfrak{F}_n = \sigma\{\xi_1, \ldots, \xi_n\}$, and τ be a stopping time. Prove the following.

(1) If $\mathsf{E}\sum_{i=1}^\tau |\xi_i| < \infty$, then $\mathsf{E}S_\tau = 0$.

(2) If $\mathsf{E}\sum_{i=1}^\tau \xi_i^2 < \infty$, then $\mathsf{E}S_\tau^2 = \mathsf{E}\sum_{i=1}^\tau \sigma_i^2$.

7.54. (Galton—Watson process) Let $\{\xi_k^{(n)}, k, n \geq 1\}$ be i.i.d. random variables taking their values in $\mathbb{Z}^+$ and $\mathsf{E}\xi_1^{(1)} = \mu > 0$. We set $S_0 = 1$, $S_1 = \xi_1^{(1)}$ and $S_n = \sum_{k=1}^{S_{n-1}} \xi_k^{(n)}$ as $n \geq 2$ (here $\sum_{k=1}^0 \xi_k^{(n)} := 0, n \geq 1$). A sequence $\{S_n,\ n \in \mathbb{Z}^+\}$ is said to be the branching process describing the population which is developing in the following manner. At starting moment $n = 0$ we have one population representative that gives birth to a random number $\xi_1^{(1)}$ of successors. Every successor gives birth independently of others to the random number of successors and so on. A degeneration of population is possible at some step.

(1) Prove that a random sequence $M_n := \mu^{-n}S_n, n \geq 0$ produces a martingale with respect to a natural filtration.

(2) Prove that $\sup_n \mathsf{E}|M_n| = 1$. Derive that $M_n \to M_\infty$ a.s. as $n \to \infty$ and $\mathsf{E}M_\infty < \infty$.

(3) Prove that for $\mu < 1$ $S_n \to 0$ a.s.; that is, the population degenerates asymptotically with probability one.

7.55. Let $\{\xi_k, k \geq 1\}$ be i.i.d. random variables with $\mathsf{P}(\xi_k = 1) = p$, $\mathsf{P}(\xi_k = -1) = q = 1-p$, $0 < p < 1$. We put $X_0 = x$, $x \in \mathbb{Z}$, $X_n = x + \sum_{k=1}^n \xi_k, n \geq 1$. Let a and b be integers with $a < x < b$, $\tau_a = \inf\{n \geq 1 | X_n = a\}$, $\tau_b = \inf\{n \geq 1 | X_n = b\}$, and $\tau = \inf\{n \geq 1 | X_n \notin (a,b)\}$.

(1) Prove that τ_a, τ_b, and τ are Markov moments with respect to the filtration $\mathfrak{F}_n = \sigma\{\xi_1, \ldots, \xi_n\}$.

(2) Using the properties of random walks (see Problem 7.49) prove that τ is a stopping time.

(3) Show that the process $M_n := (q/p)^{X_n}$ is a martingale with respect to $\{\mathfrak{F}_n, n \geq 0\}$, where $\mathfrak{F}_0 = \{\varnothing, \Omega\}$.

(4) Using Problem 7.32 or item (2), prove that for $p \neq q$,

$$\mathsf{P}(X_\tau = a) = \mathsf{P}(\tau_a < \tau_b) = \frac{(q/p)^x - (q/p)^b}{(q/p)^a - (q/p)^b}.$$

(5) Prove that for $p = q = \frac{1}{2}$,

$$\mathsf{P}(\tau_a < \tau_b) = \frac{b-x}{b-a}.$$

(6) Calculate $\mathsf{E}\tau$ separately for $p \neq q$ and for $p = q = \frac{1}{2}$. (Achieving the level a can be interpreted as the bankruptcy of a gambler with initial capital x, and achieving the level b as the winning of a gambler, respectively).

7.56. Let $\{X_n, \mathfrak{F}_n, n \in \mathbb{Z}^+\}$ be a supermartingale, and σ and τ be bounded stopping times with $\sigma \leq \tau$ a.s. Prove that $\mathsf{E}(X_\tau / \mathfrak{F}_\sigma) \leq X_\sigma$ a.s.

7.57. Let $\{\xi_i, i \geq 1\}$ be i.i.d. random variables, $\mathsf{P}(\xi_i = 1) = \mathsf{P}(\xi_i = -1) = \frac{1}{2}, 0 < a < b$, $X_n = a\sum_{i=1}^n \mathbb{I}_{\xi_i = 1} - b\sum_{i=1}^n \mathbb{I}_{\xi_i = -1}$, and $\tau_r = \inf\{n \geq 1 | X_n \leq -r\}, r > 0$. Prove that $\mathsf{E}e^{\lambda\tau_r} < \infty$ if $\lambda \leq \alpha$ and $\mathsf{E}e^{\lambda\tau_r} = \infty$ if $\lambda > \alpha$, where

$$\alpha = \frac{b}{a+b} \ln \frac{2b}{a+b} + \frac{a}{a+b} \ln \frac{2a}{a+b}.$$

7.58. Let $\{X_n, \mathfrak{F}_n, n \in \mathbb{Z}^+\}$ be a square integrable martingale, and τ and σ be bounded stopping times with $\sigma \leq \tau$. Show that $\mathsf{E}X_\tau^2$ and $\mathsf{E}X_\sigma^2$ are finite and $\mathsf{E}((X_\tau - X_\sigma)^2 / \mathfrak{F}_\sigma) = \mathsf{E}(X_\tau^2 - X_\sigma^2 / \mathfrak{F}_\sigma)$.

7.59. Let $\{X_n, \mathfrak{F}_n, n \in \mathbb{Z}^+\}$ be a square integrable martingale with $\sup_{n \geq 1} \mathsf{E}X_n^2 < \infty$ and $X_0 = 0$. Let τ be a stopping time and $\liminf_{n \to \infty} \mathsf{E}X_n^2 \mathbb{I}_{\tau > n} = 0$. Prove that in this case $\mathsf{E}X_\tau^2 = \mathsf{E}\langle X \rangle_\tau$. Here $\langle X \rangle_\tau = \sum_{i=1}^\tau \mathsf{E}((X_i - X_{i-1})^2 / \mathfrak{F}_{i-1})$.

7.60. Let $\{X_n, Y_n, \mathfrak{F}_n,\ n \in \mathbb{Z}^+\}$ be square integrable martingales. Prove that $\{X_nY_n - \langle X,Y\rangle_n, \mathfrak{F}_n, n \in \mathbb{Z}^+\}$ is a martingale.

7.61. Consider martingales $X_n = \sum_{i=1}^n \xi_i$, $Y_n = \sum_{i=1}^n \eta_i$, where $\{(\xi_i, \eta_i), i \geq 1\}$ is a sequence of independent random vectors with $\mathsf{E}\xi_i = 0, \mathsf{E}\eta_i = 0, \mathsf{E}\xi_i^2 < \infty$ and $\mathsf{E}\eta_i^2 < \infty$. Prove that $\langle X,Y\rangle_n = \sum_{i=1}^n \operatorname{cov}(\xi_i, \eta_i)$.

7.62. (Doob's decomposition for discrete-time stochastic processes) Let $\{X_n, \mathfrak{F}_n, n \in \mathbb{Z}^+\}$ be an integrable random sequence.

(1) Prove that X admits the unique decomposition of the form

$$X_n = M_n + A_n,$$

where $\{M_n, \mathfrak{F}_n, n \in \mathbb{Z}^+\}$ is a martingale, $A_0 = 0$, and $\{A_n, \mathfrak{F}_n, n \in \mathbb{Z}^+\}$ is a predictable process.

(2) Prove that process A is nondecreasing (nonincreasing) if and only if X is a discrete-time submartingale (supermartingale).

7.63. Let $\{X_n, \mathfrak{F}_n, n \in \mathbb{Z}^+\}$ be a nonnegative submartingale with $X_0 = 0$ and $X_n = M_n + A_n$ be its Doob's decomposition. Prove that for any $C > 0$ and $N \in \mathbb{Z}^+$

$$\mathsf{P}(\sup_{0\leq n\leq N} X_n \geq C) \leq \frac{\mathsf{E}A_N}{C}.$$

7.64. Prove the correctness of the next Krickeberg decomposition: every martingale with $\sup_{n\geq 1} \mathsf{E}|X_n| < \infty$ can be presented as a difference of two nonnegative martingales.

7.65. (Riesz decomposition) Let $X = \{X_n, \mathfrak{F}_n,\ n \in \mathbb{Z}^+\}$ be a uniformly integrable supermartingale. Prove that one can present X in the form $X_n = M_n + P_n$, where $\{M_n, \mathfrak{F}_n, n \in \mathbb{Z}^+\}$ is a uniformly integrable martingale and $\{P_n, \mathfrak{F}_n, n \in \mathbb{Z}^+\}$ is a potential. A potential is defined as a uniformly integrable nonnegative supermartingale such that $P_n \to 0$ a.s. as $n \to \infty$.

7.66. Using the Khinchin inequalities, prove the Burkholder inequalities.

7.67. Let $\{\xi_i,\ i \geq 1\}$ be i.i.d. random variables with $\mathsf{P}(\xi_i = 1) = \mathsf{P}(\xi_i = -1) = \frac{1}{2}$, and $X_n = \sum_{i=1}^{\tau\wedge n} \xi_i$, where $\tau = \inf\{n \geq 1 |\ \sum_{i=1}^n \xi_i = 1\}$.

(1) Prove that $\mathsf{E}|X_n| \to 2$ as $n \to \infty$.

(2) Prove that $\mathsf{E}\sqrt{[X]_n} = \mathsf{E}\sqrt{\tau \wedge n} \to \infty$ as $n \to \infty$.

(3) Derive from this that the Davis inequality in the form $c\mathsf{E}\sqrt{[X]_n} \leq \mathsf{E}|X_n|, n \geq 1$, with some fixed positive constant c, does not hold true.

7.68. Prove Theorem 7.7, statement (2).

7.69. Let $\{X_n, \mathfrak{F}_n, n \in \mathbb{Z}^+\}$ be a submartingale. Prove that for any $C > 0$ and for any $N \geq 0$

$$C \cdot \mathsf{P}(\min_{0\leq n\leq N} X_n \leq -C) \leq \mathsf{E}(X_N - X_0) - \int_{\{\min_{0\leq n\leq N} X_n \leq -C\}} X_N d\mathsf{P} \leq \mathsf{E}X_N^+ - \mathsf{E}X_0.$$

7.70. Let $h : \mathbb{R} \to \mathbb{R}^+$ be a nondecreasing convex function and $\{X_n, \mathcal{F}_n, 1 \le n \le N\}$ be a submartingale. Prove that for any $u \in \mathbb{R}$ and $r > 0$,

$$\mathsf{P}(\max_{1\le n\le N} X_n \ge u) \le \mathsf{E}h(rX_N)/h(ru).$$

7.71. Let $\{M(t), \mathcal{F}_t, t \in \mathbb{R}^+\}$ be a uniformly integrable supermartingale (martingale) with right-hand continuous trajectories. Prove that $\{M(t \wedge \tau), \mathcal{F}_{t\wedge\tau}, t \in \mathbb{R}^+\}$ is a supermartingale (martingale) as well for any Markov moment τ.

7.72. Let $\{W(t), \mathcal{F}_t, t \in \mathbb{R}^+\}$ be a Wiener process, and τ be a stopping time with respect to the filtration $\{\mathcal{F}_t, t \in \mathbb{R}^+\}$. Prove that $\{W(t \wedge \tau), \mathcal{F}_t, t \in \mathbb{R}^+\}$ is a square integrable martingale. (We mention that a Wiener process is not a uniformly integrable martingale; i.e., one cannot use the previous problem to prove the present statement.)

7.73. Let $\{W(t), \mathcal{F}_t, t \in \mathbb{R}^+\}$ be a Wiener process. For $\alpha, x \in \mathbb{R}$ and $t \in \mathbb{R}^+$ define a function $h(\alpha,x,t) := \exp\{\alpha x - \alpha^2 t/2\}$ and put

$$h_k(x,t) = \frac{\partial^k h(\alpha,x,t)}{\partial \alpha^k}\Big|_{\alpha=0}, \quad k \ge 1.$$

Prove that $\{h_k(W(t),t), \mathcal{F}_t, t \in \mathbb{R}^+\}$ is a martingale for every $k \ge 1$. (According to Problem 7.27, $\{h(\alpha, W(t),t), \mathcal{F}_t, t \in \mathbb{R}^+\}$ is a martingale for any $\alpha \in \mathbb{R}$.)

7.74. Let $\{W(t), \mathcal{F}_t, t \in \mathbb{R}^+\}$ be a Wiener process. Consider a Markov moment $\tau_a := \inf\{t \in \mathbb{R}^+ \mid |W(t)| = a\}$. Prove that $\mathsf{E}\tau_a = a^2$, $\mathsf{E}\tau_a^2 = 5a^4/3$.

7.75. Let $X(t) = \gamma t + \sigma W(t)$, $t \in \mathbb{R}^+$, and $\sigma_a := \inf\{t \in \mathbb{R}^+ \mid X(t) = a\}$. Prove that for $\gamma > 0, \sigma \ne 0$, and $a < 0$ the equality $\mathsf{P}(\sigma_a < \infty) = e^{2\gamma a/\sigma^2}$ holds.

7.76. A function $f : \mathbb{R}^m \to \mathbb{R}$ is said to be superharmonic if for any $x \in \mathbb{R}^m$ and $r > 0$ it holds that $f(x) \ge \lambda_{m-1}^{-1}(S(x,r)) \int_{S(x,r)} f(y)dy$, where $S(x,r)$ is a sphere of radius r and with the center at x, $\lambda_{m-1}(S(x,r))$ is its surface Lebesgue measure in $\mathbb{R}^{m-1}$. Let $\{W(t), \mathcal{F}_t^W, t \in \mathbb{R}^+\}$ be m-dimensional Brownian motion, and f be a continuous superharmonic function with $\mathsf{E}|f(W(t))| < \infty$ for any $t \in \mathbb{R}^+$. Prove that $\{f(W(t)), \mathcal{F}_t^W, t \in \mathbb{R}^+\}$ is a supermartingale.

7.77. Let $\{W(t), \mathcal{F}_t^W, t \in \mathbb{R}^+\}$ be m-dimensional Brownian motion, and $\|W(t)\|$ be its Euclidean norm in $\mathbb{R}^m$. (This is an m-dimensional Bessel process.) Prove that $\{\|W(t)\|, \mathcal{F}_t^W, t \in \mathbb{R}^+\}$ is a submartingale. Prove also that for any $t > 0$ and $x > \sqrt{mt}$,

$$\mathsf{P}(\sup_{s\in[0,t]} \|W(s)\| \ge x) \le \left(\frac{tm}{x^2}\right)^{-m/2} \exp\left(-\frac{x^2}{2t}\right).$$

7.78. Let $\{W(t), \mathcal{F}_t^W, t \in \mathbb{R}^+\}$ be m-dimensional Brownian motion, $m \ge 3$.

(1) Prove that $\|W(t)\| \to \infty$ a.s. as $t \to \infty$.

(2) Prove that the process W does not return to 0 with probability 1.

(3) Prove that for $m = 1$ the statements (1) and (2) are wrong.

(4) Prove that for $m = 2$ statement (1) is wrong, whereas statement (2) is still correct.

7.79. Prove that every continuous-time martingale is a local martingale.

7.80. Prove that any linear combination of a finite number of local martingales is still a local martingale.

7.81. Prove that any nonnegative local martingale is a supermartingale.

7.82. Let $X \in \mathscr{M}_{\text{loc}}$ with $X(t) \geq 0,\ X(0) = 0$. Prove that for every $t \in \mathbb{R}^+$ it holds that $X(t) = 0$ a.s.

7.83. Prove that the sequence $\{X_n, \mathcal{F}_n, n \in \mathbb{Z}^+\}$ is a local martingale provided that $\mathsf{E}(|X_{n+1}|/\mathcal{F}_n) < \infty$ a.s. and $\mathsf{E}(X_{n+1}/\mathcal{F}_n) = X_n$ a.s. for every $n \geq 0$.

7.84. Let $X \in \mathscr{M}_{\text{loc}}$ and $\mathsf{E}\sup_{t \in \mathbb{R}^+} |X(t)| < \infty$. Prove that for any Markov moment τ it holds that $\mathsf{E}X(\tau) = \mathsf{E}X(0)$.

7.85. Let $p \in [1, \infty)$, M be a local martingale with a localizing sequence $\{\tau_n, n \geq 1\}$, and for every $t \in \mathbb{R}^+$ the sequence of the random variables $\{|M(t \wedge \tau_n)|^p, n \geq 1\}$ is uniformly integrable. Prove that $\{M(t), \mathcal{F}_t, t \in \mathbb{R}^+\}$ is a martingale and $\mathsf{E}|M(t)|^p < \infty$ for all $t \in \mathbb{R}^+$.

7.86. Let $\{M(t), \mathcal{F}_t, t \geq 0\}$ be a local martingale, and for every $t > 0$ it holds that $\mathsf{E}M^*(t) < \infty$ where $M^*(t) = \sup_{s \leq t} |M(s)|$. Prove that $\{M(t), \mathcal{F}_t, t \geq 0\}$ is a martingale.

7.87. Let $\{M(t), \mathcal{F}_t, t \geq 0\}$ be a locally square integrable martingale and there exist $C > 0$ and a localizing sequence of stopping times $\{\tau_n, n \geq 1\}$, such that $\tau_{n+1} \geq \tau_n \to \infty$ as $n \to \infty$ a.s. and $\mathsf{E}M^2(t \wedge \tau_n) \leq C$. Prove that $\{M(t), \mathcal{F}_t, t \geq 0\}$ is a square integrable martingale.

7.88. Prove that every continuous martingale is a locally square integrable martingale.

7.89. Assume that $\widetilde{\mathscr{M}}$ is a set of right-hand continuous, uniformly integrable martingales determined on a stochastic basis $\{\Omega, \mathcal{F}, \{\mathcal{F}_t\}_{t \in \mathbb{R}^+}, \mathsf{P}\}$. For martingale $M \in \widetilde{\mathscr{M}}$ and $p \in [1, \infty]$ we put

$$\|M\|_{\mathscr{H}_p} := \|M^*(\infty)\|_p,$$

where $M^*(t) = \sup_{0 \leq s \leq t} |M(t)|$, $\|\cdot\|_p$ is a norm in $L_p(\Omega, \mathcal{F}, \mathsf{P})$. Denote by $\mathscr{H}_p$ the space of $M \in \widetilde{\mathscr{M}}$ with $\|M\|_{\mathscr{H}_p} < \infty$. Prove the following statements.

(1) If we identify martingales that are equal a.s., that is,

$$\mathsf{P}(\omega \in \Omega | \sup_{0 \leq t \leq \infty} |M(t) - N(t)| > 0) = 0,$$

then $\mathscr{H}_p$ becomes Banach space equipped with the norm $\|\cdot\|_{\mathscr{H}_p}$.

(2) If $1 \leq p' \leq p$ then $\mathscr{H}_p \subset \mathscr{H}_{p'}$.

(3) If $1 < p < \infty$ and $M \in \widetilde{\mathscr{M}}$, then $\|M(\infty)\|_p \leq \|M^*(\infty)\|_p \leq q\|M(\infty)\|_p$, where $1/p + 1/q = 1$. Deduce from here that $\mathscr{H}_2 = \widetilde{\mathscr{M}^2}$.

(4) If $p = \infty$ and $M \in \widetilde{\mathscr{M}}$, then $\|M(\infty)\|_\infty = \|M^*(\infty)\|_\infty$.

(5) If $1 < p \le \infty$ and $\{M^n(t), n \ge 1\}$ is a sequence of martingales converging to a martingale M in the norm of the space $\mathscr{H}_p$, then there exists a subsequence $\{M^{n_k}(t), k \ge 1\}$, that uniformly converges to $M(t)$ on $[0,\infty)$ a.s.

(6) Prove that the limit in $\mathscr{H}_p$, $1 < p \le \infty$, of a sequence of continuous martingales is again a continuous martingale.

7.90. Let martingale $M \in \widetilde{\mathscr{M}^2}$ (or $\mathscr{H}_2$, see Problem 7.89). Prove that M^2 is a right-hand continuous uniformly integrable submartingale that belongs to the class D (see Definition 7.7).

7.91. Two martingales $M, N \in \widetilde{\mathscr{M}}$ are said to be orthogonal if their product $MN \in \widetilde{\mathscr{M}}$ and $M(0)N(0) = 0$.

(1) Let M and N belong to $\mathscr{H}_2$ and be orthogonal martingales. Prove that for any stopping time τ it holds that $\mathsf{E}M(\tau)N(\tau) = 0$ and $MN \in \mathscr{H}_1$.

(2) Prove the inverse statement: if M and N belong to $\mathscr{H}_2$, $M(0)N(0) = 0$ a.s., and $\mathsf{E}M(\tau)N(\tau) = 0$ for any stopping time τ, then M and N are orthogonal martingales.

7.92. Define a conditional covariance of two square integrable random variables X and Y by $\mathrm{cov}(X,Y/\mathfrak{F}) := \mathsf{E}(XY/\mathfrak{F}) - \mathsf{E}(X/\mathfrak{F})\mathsf{E}(Y/\mathfrak{F})$. Now, let $\{M_n, N_n, \mathfrak{F}_n, n \in \mathbb{Z}^+\}$ be two square integrable discrete-time martingales. We say that they are orthogonal if their product MN is a martingale and $M_0N_0 = 0$ a.s. Prove the equivalence of the following statements.

(1) M and N are orthogonal.

(2) $M_0N_0 = 0$ and $\mathrm{cov}(M_{n+1} - M_n, N_{n+1} - N_n/\mathfrak{F}_n) = 0$ a.s., for any $n \in \mathbb{Z}^+$.

7.93. Consider discrete time $\mathbb{T} = \{0, 1, \ldots, T\}$. Denote by $\mathscr{H}_{2,\mathbb{T}}$ a space of square integrable martingales $\{M_n, \mathfrak{F}_n, n \in \mathbb{T}\}$. A subspace $S \subset \mathscr{H}_{2,\mathbb{T}}$ is said to be stable if $N^\tau \in S$ for all $N \in S$ and for all stopping times τ on $\mathbb{T}$. For a stable subspace S and martingale $M \in \mathscr{H}_{2,T}$ with $M_0 = 0$ prove the equivalence of the following statements.

(1) $\mathsf{E}M_T N_T = 0$ for any $N \in S$.

(2) For any $N \in S$ the equality $\mathrm{cov}(M_{n+1} - M_n, N_{n+1} - N_n/\mathfrak{F}_n) = 0$ holds true a.s. for any $n \in \mathbb{T}$.

(3) The product MN is a martingale for any $N \in S$.

7.94. Prove the following discrete version of the Kunita–Watanabe decomposition. If $\{X_n, \mathfrak{F}_n, n \in \mathbb{T}\}$ is a square integrable martingale, then any martingale $M \in \mathscr{H}_{2,\mathbb{T}}$ can be expanded in the following way.

$$M_n = M_0 + \sum_{k=1}^{n} \xi_k (X_k - X_{k-1}) + L_n = M_0 + (\xi \circ X)_n + L_n,$$

where $\{\xi_k, \mathfrak{F}_k, 1 \le k \le T\}$ is a predictable stochastic process, $\xi_k(X_k - X_{k-1}) \in L_2(\Omega, \mathfrak{F}, \mathsf{P})$ for any $1 \le k \le T$, and L is a square integrable martingale, which is orthogonal to X and such that $L_0 = 0$.

7.95. Prove the following statements.

(1) Let P and Q be probability measures on a measurable space $(\Omega,\mathcal{F})$, and $\mathsf{Q} \ll \mathsf{P}$ on $\mathcal{F}$ with density ρ. If $\mathcal{F}_0$ is a σ-algebra contained in $\mathcal{F}$, then $\mathsf{Q} \ll \mathsf{P}$ on $\mathcal{F}_0$ and the corresponding density can be defined by equality $(d\mathsf{Q}/d\mathsf{P})\,|_{\mathcal{F}_0} = \mathsf{E}(\rho/\mathcal{F}_0)$ P-a.s.

(2) Let $\mathsf{Q} \ll \mathsf{P}$ on $\mathcal{F}$ with density ρ and $\mathcal{F}_0 \subset \mathcal{F}$. Then for any $\mathcal{F}$-measurable nonnegative r.v. ξ the equality holds true P-a.s.,

$$\mathsf{E}_{\mathsf{Q}}(\xi/\mathcal{F}_0) = \frac{1}{\mathsf{E}(\rho/\mathcal{F}_0)}\mathsf{E}(\xi\rho/\mathcal{F}_0).$$

(3) Let $\mathbb{T} = \{0,1,\dots,T\}$, $\{\Omega,\mathcal{F},\{\mathcal{F}_t\}_{t\in\mathbb{T}},\mathsf{P}\}$ be a probability space with filtration, and Q be a probability measure with $\mathsf{Q} \sim \mathsf{P}$. Prove that an adapted process M is a Q-martingale if and only if the process $M_t \cdot \mathsf{E}((d\mathsf{Q}/d\mathsf{P}) \mid \mathcal{F}_t)$, $t \in \mathbb{T}$ is a P-martingale.

7.96. (Discrete version of stochastic Doleans–Dade exponent) Let, as in the previous problem, $\mathbb{T} = \{0,1,\dots,T\}$, $\{\Omega,\mathcal{F},\{\mathcal{F}_t\}_{t\in\mathbb{T}},\mathsf{P}\}$ be a probability space with filtration, and Q be a probability measure with $\mathsf{Q} \sim \mathsf{P}$.

(1) Prove that there exists such a P-martingale L that: $L(0) = 1$, and $L(t+1) - L(t) > -1$ P-a.s. for all $t \in \{0,1,\dots,T-1\}$, and the martingale $Z(t) := \mathsf{E}((d\mathsf{Q}/d\mathsf{P})/\mathcal{F}_t)$, $t \in \mathbb{T}$ can be expanded as

$$Z(t) = \prod_{k=1}^{t}(1+L(s)-L(s-1)),\ t \in \mathbb{T}. \tag{7.2}$$

(2) Prove the inverse statement: if L is a P-martingale, $L(0) = 1$, and $L(t+1) - L(t) > -1$ P-a.s., and moreover the equality (7.2) determines the P-martingale Z, then the equality $d\mathsf{Q} := Z(T)d\mathsf{P}$ determines the probability measure $\mathsf{Q} \sim \mathsf{P}$.

7.97. (Discrete version of the Girsanov formula) Let the conditions of item (1) of Problem 7.96 hold and L be a martingale from the relation (7.2) for a sequence of densities $Z(t) := \mathsf{E}((d\mathsf{Q}/d\mathsf{P})/\mathcal{F}_t)$, $t \in \mathbb{T}$. If $\widehat{M}$ is such a Q-martingale that $\widehat{M}(t) \in L_1(\mathsf{P})$ for all $t \in \mathbb{T}$, then the process

$$M(t) := \widehat{M}(t) + \sum_{k=1}^{t}\mathsf{E}\Big((L(k)-L(k-1))(\widehat{M}(k)-\widehat{M}(k-1))/\mathcal{F}_{k-1}\Big) \tag{7.3}$$

is a P-martingale.

7.98. Prove: if $\{M(t),\mathcal{F}_t,\ t \in \mathbb{R}^+\}$ is a uniformly integrable martingale with right-hand continuous trajectories, whereas σ and τ are Markov moments with $\sigma \le \tau$, then $M(\sigma) = \mathsf{E}(M(\tau)/\mathcal{F}_\sigma)$ and $\mathsf{E}|M(\tau)| < \infty$.

7.99. Let $\{X(t),\mathcal{F}_t,\ t \in [0,T]\}$ be a supermartingale and $\mathsf{E}X(0) = \mathsf{E}X(T)$. Prove that X is a martingale.

7.100. Prove: if η is an integrable r.v., $X(t) = \mathsf{E}(\eta/\mathcal{F}_t)$ and τ is a Markov moment, then $X(\tau) = \mathsf{E}(\eta/\mathcal{F}_\tau)$.

7.101. (1) Let $\{X_n, \mathcal{F}_n, 0 \le n \le N\}$ be an integrable stochastic process with $X_0 = 0$ and $\mathsf{E}X_\tau = 0$ for any stopping time τ. Prove that X is a martingale.

(2) Let $\{X(t), \mathcal{F}_t, t \in \mathbb{R}^+\}$ be an integrable càdlàg stochastic process with $X(0) = 0$, $\mathsf{E}X(\tau) = 0$ for any stopping time τ, and for any $T > 0$ the family of random variables $\{X(s), 0 \le s \le T\}$ is uniformly integrable. Prove that X is a martingale.

7.102. Let $\{X(t), \mathcal{F}_t, t \in \mathbb{R}^+\}$ be a nonnegative right-hand continuous supermartingale. We put $\tau = \inf\{t \in \mathbb{R}^+ | X(t) = 0 \text{ or } X_{t-} = 0\}$ if the corresponding set is empty and $\tau = \infty$ otherwise. Prove that $X(t) = 0$ a.s. for all $t \ge \tau$ and for such ω that $\tau(\omega) < \infty$.

7.103. Let $\{M_n, \mathcal{F}_n, n \in \mathbb{Z}^+\}$ be a square integrable martingale. Prove that for any $\varepsilon > 0$, $C > 0$, and $N \ge 1$,

$$\mathsf{P}(\max_{0 \le n \le N} |M_n| \ge \varepsilon) \le \frac{C + \mathsf{E}\max_{1 \le n \le N}(\langle M \rangle_n - \langle M \rangle_{n-1})}{\varepsilon^2} + \mathsf{P}(\langle M \rangle_N \ge C).$$

7.104. Prove Theorem 7.18.

7.105. Let $\{N(t), \mathcal{F}_t, t \in \mathbb{R}^+\}$ be a Poisson process with intensity λ. Prove that a square integrable martingale $M(t) := N(t) - \lambda t$ has a quadratic characteristic $\langle M \rangle(t) = \lambda t$.

7.106. Let $M \in \mathscr{H}^2$, and $M(0) = 0$. Prove that $\mathsf{E}M^2(\tau) = \mathsf{E}[M](\tau) = \mathsf{E}\langle M \rangle(\tau)$ for any Markov moment τ.

7.107. Let $\{M(t), \mathcal{F}_t, t \in [0, T]\}$ be a continuous square integrable martingale.

(1) Prove that for any $u > 0$ it holds that $u \cdot \mathsf{P}(M^*(T) \ge u) \le \mathsf{E}(|M(T)| \mathbb{I}_{M^*(T) > u})$, where $M^*(T) = \sup_{0 \le t \le T} |M(t)|$.

(2) Prove that for any $A > 0$ it holds that

$$\mathsf{E}\left((M^*(T) \wedge A)^2\right) \le 2\mathsf{E}\left((M^*(T) \wedge A)|M(T)|\right).$$

(3) Prove that $\mathsf{E}M^*(T) < \infty$ and $\mathsf{E}(M^*(T))^2 \le 4\mathsf{E}M^2(T)$.

7.108. (Joint distribution for $W(t)$ and $\sup_{s \le t} W(s)$) Let $\{W(t), \mathcal{F}_t, t \in \mathbb{R}^+\}$ be a Wiener process.

(1) Assume that τ is a bounded stopping time. Prove that for all $u < v$ we have the equality $\mathsf{E}(\exp\{iz(W(v+\tau) - W(u+\tau))\}/\mathcal{F}_{u+\tau}) = \exp\{-z^2(v-u)/2\}$.

(2) Prove that $W^\tau(u) := W(u+\tau) - W(u)$ is an $\mathcal{F}_{u+\tau}$-Wiener process independent of the σ-algebra $\mathcal{F}_\tau$.

(3) Let $\{Y(t), t \in \mathbb{R}^+\}$ be a continuous stochastic process, independent of σ-algebra $\mathscr{A}$ and such that $\mathsf{E}\sup_{0 \le s \le T} |Y_s| < \infty$. Let T_1 be a nonnegative $\mathscr{A}$-measurable r.v. that is bounded from above. Prove that $\mathsf{E}(Y_{T_1}/\mathscr{A}) = \mathsf{E}Y(t)\big|_{T_1 = t}$.

(4) Let $\tau^\lambda = \inf\{s \ge 0 | W(s) \ge \lambda\}$.

(a) Prove that for a bounded Borel function f,

$$\mathsf{E}f(W(t))\mathbb{I}_{\tau^\lambda \le t} = \mathsf{E}\phi(t - \tau^\lambda)\mathbb{I}_{\tau^\lambda \le t},$$

where $\phi(u) = \mathsf{E}f(W_u + \lambda)$.

(b) Show that $\mathsf{E}f(W(u)+\lambda) = \mathsf{E}f(-W(u)+\lambda)$ and derive that

$$\mathsf{E}f(W(t))\mathbb{I}_{\tau^\lambda \le t} = \mathsf{E}f(2\lambda - W(t))\mathbb{I}_{\tau^\lambda \le t}.$$

(5) Let $W^*(t) = \sup_{s\le t} W(s)$. Show that for any $\lambda \ge 0$,

$$\mathsf{P}(W(t) \le \lambda, W^*(t) \ge \lambda) = \mathsf{P}(W(t) \ge \lambda, W^*(t) \ge \lambda) = \mathsf{P}(W(t) \ge \lambda).$$

Derive that $W^*(t)$ and $|W(t)|$ are equally distributed.

(6) Prove that for $\lambda \ge (\mu \vee 0)$,

$$\begin{aligned}\mathsf{P}\left(W(t) \le \mu, W^*(t) \ge \lambda\right) &= \mathsf{P}\left(W(t) \ge 2\lambda - \mu, W^*(t) \ge \lambda\right)\\ &= \mathsf{P}\left(W(t) \ge 2\lambda - \mu\right),\end{aligned}$$

and for $0 \le \lambda \le \mu$,

$$\mathsf{P}\left(W(t) \le \mu, W^*(t) \ge \lambda\right) = 2\mathsf{P}(W(t) \ge \lambda) - \mathsf{P}(W(t) \ge \mu).$$

(7) Prove that the joint distribution of $(W(t), W^*(t))$ is given by the expression

$$\mathbb{I}_{0\vee x \le y} \frac{2(2y-x)}{\sqrt{2\pi t^3}} \exp\left(-\frac{(2y-x)^2}{2t}\right).$$

7.109. (Reflection principle for Wiener process) Let $\{W(t), \mathcal{F}_t, t \in \mathbb{R}^+\}$ be a Wiener process and τ be a stopping time w.r.t. $\{\mathcal{F}_t\}$. Also, let $W_\tau(t) = W(t)\mathbb{I}_{t\le\tau} + (2W(\tau) - W(t))\mathbb{I}_{t>\tau}, t \in \mathbb{R}^+$. Prove that $\{W_\tau(t), \mathcal{F}_t, t \in \mathbb{R}^+\}$ is a Wiener process. Represent in diagram form a trajectory of the process W and the corresponding trajectory of the process W_τ.

Hints

7.3–7.13. Use the definition of a (sub-, super-) martingale.
7.14. Use the definition of a (sub-) martingale and Jensen's inequality.
7.15. Use Problem 7.8.
7.16. Prove that $\mathsf{E}(\xi_{n+1}/\mathcal{F}_n) = 0$ if and only if $\mathsf{E}(\xi_{n+1} \cdot f_n(\xi_0, \ldots, \xi_n)) = 0$ for any bounded Borel function $f_n : \mathbb{R}^n \to \mathbb{R}$.
7.17. Rewrite the r.v. $X_{n,m}$ for $n > m$ as

$$\sum_{1\le i_1<\ldots<i_m\le n-1} \xi_{i_1}\cdots\xi_{i_m} + \sum_{1\le i_1<\ldots<i_{m-1}\le n-1, i_m=n} \xi_{i_1}\cdots\xi_{i_m},$$

where the first term is equal to $X_{n-1,m}$. Prove that the conditional expectation of the second term with respect to the σ-algebra $\mathcal{F}_{n-1}$ is zero.
7.18. Use Corollary 7.4.
7.19. Let X_n be a number of black marbles inside the urn after the nth step, and $Y_n = X_n/(b+r+cn)$ be a fraction for which we are searching. Write $X_n = (X_{n-1} +$

c) $\mathrm{II}_A + X_{n-1}\mathrm{II}_{\overline{A}}$, where the event A consists in the choice of the black marble on the nth step, and calculate the conditional probability of the event A with respect to the σ-algebra $\mathfrak{F}_{n-1} = \sigma\{X_1,\dots,X_{n-1}\} = \sigma\{Y_1,\dots,Y_{n-1}\}$.

7.22. (1) Use the "inverse" induction. It means to start with $\mathsf{E}(Y_N/\mathfrak{F}_{N-1})$.

(2) It is evident that $Z_N \ge Y_N$. Prove that $Z_n \ge \mathsf{E}(Y_{n+1}/\mathfrak{F}_n)$ using the "inverse" induction.

7.23. Due to Theorem 7.10, there exists $\lim_{n\to\infty} X_n =: X_\infty$. Furthermore, you should use the Lévy theorem: if $\mathsf{E}|\xi| < \infty$, then $\mathsf{E}(\xi/\mathfrak{F}_n) \to \mathsf{E}(\xi/\mathfrak{F}_\infty), n \to \infty$ a.s. Convergence in $L_1(\mathsf{P})$ follows now from the fact that $X_\infty = \mathsf{E}(\xi/\mathfrak{F}_\infty)$ a.s. and from Theorem 7.10.

7.24. Use Problem 7.23 and Theorem 7.10.

7.25. Define events $A_k = \{\xi_k = 2\}$ and use the Borel–Cantelli lemma: because $\sum_{k\ge1}\mathsf{P}(A_k) = \infty$ and events A_k are mutually independent, then $\mathsf{P}(\lim_{n\to\infty} X_n = \infty) = 1$. And finally, use Theorem 7.10.

7.27. Use the facts that increments of a Wiener process are independent and that for the r.v. $\xi \sim N(0,\sigma^2)$ $\mathsf{E}e^{\alpha\xi} = e^{(\alpha^2\sigma^2)/2}$. For example, $aW^2(t) + bt$ is a martingale if and only if $b = -a$.

7.30. (1) Use the definition of Markov moment.

(2) For example, an event $\{\limsup_{k\to\infty}\tau_k \le n\} = \{\lim_{k\to\infty}\sup_{m\ge k}\tau_m \le n\} = \bigcup_{r=1}^{\infty}\bigcap_{k=r}^{\infty}\bigcap_{m=k}^{\infty}\{\tau_m \le n\} \in \mathfrak{F}_n$.

7.35. Write $|X_\tau| = \liminf_{n\to\infty}|X_{\tau\wedge n}|$ and then use the Fatou lemma and Theorem 7.4.

7.37. It follows directly from the definition of the Markov moment.

7.38. See Problem 7.31.

7.39. Use the fact that $|X_\nu| \le C$, which implies that $\mathsf{E}|X_\nu| < \infty$. Now, the conditional expectations can be calculated.

7.40 Put $\tau := n\mathrm{II}_A + N\mathrm{II}_{A^c}$, where $A \in \mathfrak{F}_n$ and $0 \le n \le N$, and prove that $\{M_n, \mathfrak{F}_n,\ n \in \mathbb{Z}^+\}$ is a martingale.

7.41. (1) Prove that for $\mathfrak{F}_\tau$ all the properties of a σ-algebra are fulfilled.

(2) Consider the event $\{\tau = k\}$ and prove that it belongs to $\mathfrak{F}_\tau$. For this purpose consider an event $\{\tau = k\} \cap \{\tau = n\}$ for different k and n.

3) Let $B \in \mathfrak{B}(\mathbb{R})$. Represent the event $\{X_\tau \in B\}$ as $\bigcup_{n=0}^{\infty}\{X_n \in B, \tau = n\}$, and prove that this event belongs to $\mathfrak{F}_\tau$.

7.42. Use Problem 7.41, item (4).

7.43. Write the event $\{\tau_A \le t\}$ as $\{\tau \le t\} \cap A, 0 \le t < \infty$.

7.44. $\{\tau \le t\} = \inf_{s\in\mathbb{Q}\cap[0,t]} \operatorname{dist}(X(s),A) = 0$, where $\operatorname{dist}(x,A) = \inf_{y\in A}|x-y|$.

7.45. (1) Follows directly from the corresponding definitions; (1) implies (2).

7.46. (3) Consider the event $A \cap \{t < \tau\} \in \mathfrak{F}_t \cap \mathfrak{F}_{\tau-}$ for $A \in \mathfrak{F}_t, t \in \mathbb{R}^+$, and represent it as $A \cap \{t < \tau\} \cap \{\sigma < \tau\}$.

(4) Represent $A \cap \{\sigma < \tau\}$ as a union $\bigcup_{r\in\mathbb{Q}^+} A \cap \{\sigma < r\} \cap \{r < \tau\}$, where $\mathbb{Q}^+$ is a set of nonnegative rational numbers.

7.47. Follows directly from the definition of the inverse martingale and from the fact that σ-algebras do not increase.

7.49. Represent S_τ as

$$S_\tau = \sum_{n=1}^{\infty}\sum_{k=1}^{n}\xi_k(\omega)\mathrm{II}_{\tau=n} = \sum_{k=1}^{\infty}\xi_k(\omega)\sum_{n=k}^{\infty}\mathrm{II}_{\tau=n} = \sum_{k=1}^{\infty}\xi_k(\omega)\mathrm{II}_{\tau\ge k},$$

and take into account the fact that the event $\{\tau \ge k\} = \Omega \setminus \{\tau < k\} \in \mathcal{F}_{k-1}$, that is, does not depend on ξ_k.

7.51. Write the identity $S_\tau - \tau\mathsf{E}\xi_1 = \sum_{k=1}^{\infty}\sum_{i=1}^{k}(\xi_i - \mathsf{E}\xi_i)\,\mathbb{I}_{\tau=k}$ and use a transformation similar to the one used for proving the first Wald identity.

7.59. Prove that there exists a r.v. $\xi \in L_2(\mathsf{P})$ such that $X_n = \mathsf{E}(\xi/\mathcal{F}_n)$. Using the maximum Doob's inequality prove that $\mathsf{E}\sup_{n\ge 1} X_n^2 < \infty$. Derive that $\mathsf{E}X_\tau^2 < \infty$. Prove that the process $Y_n := X_n^2 - \langle X\rangle_n$ is a martingale. Use the problem situation and the statement of Problem 7.36, transformed for a submartingale, and prove that for any $n \ge 1$ it holds that $\mathsf{E}X_{\tau\wedge n}^2 \le \mathsf{E}X_\tau^2$. Derive that $\mathsf{E}\langle X\rangle_\tau \le \mathsf{E}X_\tau^2$, and the opposite inequality.

7.60. Transform the increment $X_nY_n - X_{n-1}Y_{n-1} - \mathsf{E}\big((X_n - X_{n-1})(Y_n - Y_{n-1})/\mathcal{F}_{n-1}\big)$.

7.61. Use the definition of $\langle X, Y\rangle_n$.

7.62. (1) Put $A_0 = 0$, $\Delta A_n := \mathsf{E}(X_n - X_{n-1}/\mathcal{F}_{n-1})$.

(2) Use the shape and uniqueness of the decomposition obtained in the item (1).

7.63. Consider a stopping time $\tau = \inf\{n \le N \mid X_n \ge C\} \wedge N$ and use the fact that $\mathsf{E}X_\tau = \mathsf{E}M_\tau + \mathsf{E}A_\tau = \mathsf{E}A_\tau$.

7.64. Use Theorem 7.10 and write X_n as $\mathsf{E}(\xi/\mathcal{F}_n)$; then, decompose the r.v. ξ as $\xi^+ - \xi^-$.

7.65. Prove that there exists $\lim_{n\to\infty} X_n =: X_\infty$ a.s. Put $M_n = \mathsf{E}(X_\infty/\mathcal{F}_n)$ and $P_n = X_n - \mathsf{E}(X_\infty/\mathcal{F}_n)$.

7.67. (1) The symmetry implies that $\mathsf{E}|X_n| = 2\mathsf{E}X_n^+$; τ is the stopping time (see Problem 7.55). Hence, $X_n^+ \to X_\tau^+ = 1$ a.s., and it is possible to use the Lebesgue dominated convergence theorem. So, $\mathsf{E}|X_n| \to 2$, $n \to \infty$.

(2) Check that $[X_n]^{1/2} = \sqrt{\tau \wedge n}$. Furthermore, it follows from the theory of random walks that

$$\mathsf{P}(\tau = 2k-1) = \frac{(-1)^{k-1}1/2(1/2-1)\dots(1/2-k+1)}{k!},$$

and it is evident that $\mathsf{P}(\tau = 2k) = 0$. Derive the equality $\mathsf{E}\sqrt{\tau} = \infty$ ($\mathsf{E}\tau = \infty$ was already proved in a different way in Problem 7.55).

7.68. Consider the stopping time $\tau = \inf\{0 \le n \le N \mid X_n \ge C\} \wedge N$ and use the fact that $\mathsf{E}X_0 \ge \mathsf{E}X_\tau \ge C \cdot \mathsf{P}(\max_{0\le n\le N} X_n \ge C) - \mathsf{E}X_N^-$.

7.69. Consider the stopping time $\tau = \inf\{0 \le n \le N \mid X_n \le -C\} \wedge N$ and use the inequality $\mathsf{E}X_\tau \ge \mathsf{E}X_0$.

7.70. Use the facts that $\{h(tX_n), 1 \le n \le N\}$ is a submartingale and the function h is nonnegative. Also use the statement (1) of Theorem 7.7.

7.71. Perform a discretization of both time and Markov moment (note that there exists a limit $X(\infty) = \lim_{t\to\infty} X(t)$ and you can put $X(\tau) = X(\infty)$ if the Markov moment $\tau = \infty$). Then, you can pass to the limit assuming uniform integrability and right-hand continuity.

7.73. Prove by induction that $(\partial^k h(\alpha, W(t), t))/\partial\alpha^k$ is a martingale for any $k \ge 1$ and $\alpha \in \mathbb{R}$. For this purpose you should write a derivative as a limit, taking into account that the prelimit expression is a martingale and check whether the prelimit values are uniformly integrable. Fix some $\alpha_0 \in \mathbb{R}$, consider its neighborhood (e.g., $|\alpha - \alpha_0| < 1$), and prove that inside the neighborhood

$$\sup_{|\alpha-\alpha_0|<1} \mathsf{E}\left|\frac{\partial^k h(\alpha,W(t),t)}{\partial\alpha^k}\right|^2 < \infty \text{ for all } k \geq 1 \text{ and } t \in \mathbb{R}^+.$$

7.74. Derive from the statement of Problem 7.73 that the processes $\{W(t), W^2(t) - t, W^3(t) - 3tW(t), W^4(t) - 6tW^2(t) + t^2,\ t \geq 0\}$ are martingales. Prove that τ_a is a stopping time, $|W(\tau_a)| = a$ a.s. Prove that $\mathsf{E}W(\tau_a) = 0$ with $\mathsf{E}W^2(\tau_a) = \mathsf{E}\tau_a$, and $\mathsf{E}\left(W^4(\tau_a) - 6\mathsf{E}\tau_a W^2(\tau_a) + 3\mathsf{E}\tau_a^2\right) = 0$.
7.75. First, find such a $\lambda \in \mathbb{R}$ that a process $\exp\{\lambda X(t)\}$ is a martingale. Then, it holds that $\mathsf{E}\exp\{\lambda X(t\wedge\sigma_a)\} = 1$. Furthermore, pass to a limit as $t \to \infty$ and obtain the inequality $\mathsf{E}\exp\{\lambda X(\sigma_a)\}\mathbb{I}_{\sigma_a<\infty} \leq 1$ which implies the statement.
7.76. Transform $\mathsf{E}\left(f(W(t))/\mathcal{F}_s^W\right)$ as $\mathsf{E}\left(f(W(t))/W(s)\right)$ (which Wiener process property should you use?). Furthermore,

$$\mathsf{E}(f(W(t))/W(s)) = \mathsf{E}(f(W(t) - W(s) + W(s))/W(s)).$$

Use the following equality. If ξ and η are independent random vectors and a Borel function g is such that $\mathsf{E}|g(\xi,\eta)| < \infty$, then $\mathsf{E}(g(\xi,\eta)/\eta) = \mathsf{E}f(\xi,y)\big|_{y=\eta}$. And finally, prove that $\mathsf{E}f(W(t) - W(s) + y) \leq f(y)$ using the superharmonic property of f. For this you should transform $\mathsf{E}f(W(t) - W(s) + y)$ as $\int_{\mathbb{R}^m} f(x+y)p_{t-s,m}(x)dx$, where $p_{t-s,m}(x)$ is the density of the distribution of m-measurable Gaussian vector $W(t) - W(s)$, and rewrite $\int_{\mathbb{R}_m} = \int_0^\infty \int_{S(y,r)}$.
7.77. The fact that $\{\|W(t)\|, \mathcal{F}_t^W, t \in \mathbb{R}^+\}$ is a submartingale follows from the convexity of the function $f : \mathbb{R}^m \to \mathbb{R}^+$, $f(x) = \|x\|$ (check whether this function is indeed convex). Then use the statement of Problem 7.70 that is valid as well for continuous-time submartingales. For this purpose put $h(rx) = e^{r^2x^2}$ and choose

$$r^2 = \frac{x^2}{2s^2m} + \frac{1}{2s}.$$

7.79. Consider any sequence of stopping times $\tau_1 \leq \tau_2 < \cdots$ a.s., such that $\tau_n \to \infty$ a.s., and use the fact that $\{X(t\wedge\tau), t \geq 0\}$ is a martingale if $\{X(t), \geq 0\}$ is a martingale (the proof of this is similar to Theorem 7.4).
7.81. Use Fatou's lemma directly for the equality that defines the local martingale.
7.82. Direct corollary of Problem 7.81.
7.83. First prove that $\mathsf{E}(|X_{\tau\wedge(n+1)}|/\mathcal{F}_n) < \infty$ for any Markov moment τ, and then use similar reasoning as in the proof of the implication $(1) \Longrightarrow (2)$ in Problem 7.31.
7.84. Let $\{\tau_n, n \geq 1\}$ be a localizing sequence. First prove that a martingale $X(t\wedge\tau_n)$ is uniformly integrable for every $n \geq 1$. And then apply Corollary 7.2 for it. Finally, tend $n \to \infty$ using the Lebesgue dominated convergence.
7.85. Because $\lim_{n\to\infty} M(t\wedge\tau_n) = M(t)$, derive from the uniform integrability that the convergence holds in $L_p(\Omega,\mathcal{F},P)$ too. You can derive the inequality $\mathsf{E}|M_t|^p < \infty$ for all $t \in \mathbb{R}^+$. Check whether it is possible to tend $n \to \infty$ in equation $\mathsf{E}(M(t\wedge\tau_n)/\mathcal{F}_s) = M(s\wedge\tau_n)$.
7.86. As in Problems 7.87 and 7.85, the equality $\mathsf{E}(M(t\wedge\tau_n)/\mathcal{F}_s) = M(s\wedge\tau_n)$ holds a.s., and you need to prove the uniform integrability of the sequence $\{M(t\wedge\tau_n),$

$n \geq 1\}$. For this purpose you can use the evident inequality $|M(t \wedge \tau_n)| \leq M^*(t)$ and prove that $\mathsf{P}(|M(t \wedge \tau_n)| \geq C) \to 0$ as $C \to \infty$.
7.87. Write the relation $\mathsf{E}(M(t \wedge \tau_n)/\mathcal{F}_s) = M(s \wedge \tau_n)$ and prove that $\{M(t \wedge \tau_n), n \geq 1\}$ is a uniformly integrable sequence and tend $n \to \infty$. You can use Problem 7.1. Use Fatou's lemma to prove the square integrability.
7.88. Consider a sequence of stopping times $\tau_n = \inf\{t \in \mathbb{R}^+ | \; |M_t| \geq n\} \wedge n$.
7.89. (1) Check directly that $\mathscr{H}_p$ is Banach space.

(2) Use the Hölder inequality.

(3) The first inequality is evident. The second one can be derived from the maximal integral Doob's inequality that holds in the continuous time case as well.

(4) You can derive the inequality $\|M(\infty)\|_\infty \leq \|M^*(\infty)\|_\infty$ from the following: $|M(t)| \leq \|M^*(\infty)\|_\infty$ for all $t \in \mathbb{R}^+$. For the inverse inequality consider for any $\varepsilon > 0$ a Markov moment $\tau_\varepsilon = \inf\{t \in \mathbb{R}^+ | \; |M(t)| \geq \|M^*(\infty)\|_\infty - \varepsilon\}$ and use the fact that $\mathsf{P}(|M^*(\infty)| > \|M^*(\infty)\|_\infty - \varepsilon) > 0$ (check this), and equality $M(\tau_\varepsilon) = \mathsf{E}(M(\infty)/\mathcal{F}_{\tau_\varepsilon})$ which implies that $\|M(\tau_\varepsilon)\|_\infty \leq \|M(\infty)\|_\infty$.

(5) It is sufficient to ensure that $\lim_{n\to\infty} \|(M^n - M)^*(\infty)\|_p = 0$, and derive the existence of the sequence for which we are searching.

(6) Derive from item (5).
7.90. First prove that $\mathsf{E}\sup_{t\geq 0} M^2(t) < \infty$; that is, $M \in \mathscr{H}_2$. Next, $\sup_\tau \mathsf{E}M^2(\tau) \times \mathbb{I}_{M^2(\tau)\geq C} \leq \sup_\tau \mathsf{E}M^2(\infty)\mathbb{I}_{M^2(\tau)\geq C}$, where τ runs through the family of all Markov moments. Prove that the latter expression tends to zero as $C \to \infty$.
7.92. It can be checked via direct computation.
7.94. Consider a family G of all processes of the form $\{Y(t) := \sum_{k=1}^t \xi_k \cdot (X_k - X_{k-1}), t \in \mathbb{T}\}$ where ξ_k are predictable random variables and

$$\xi_k(X_k - X_{k-1}) \in L_2(\Omega, \mathcal{F}, \mathsf{P}).$$

Prove that it is a subspace in $\mathscr{H}_{2,\mathbb{T}}$. Use the projection theorem in Hilbert space $\mathscr{H}_{2,\mathbb{T}}$.
7.95. (1) The fact that $\mathsf{Q} \ll \mathsf{P}$ on $\mathcal{F}_0$ can be derived directly from the definition of absolute continuity. Now, let $A \in \mathcal{F}_0$. Check the equalities $\mathsf{Q}(A) = \int_A \rho d\mathsf{P} = \int_A \mathsf{E}(\rho/\mathcal{F}_0) d\mathsf{P}$ and derive that $\mathsf{E}(\rho/\mathcal{F}_0)$ is a required density.

(2) Let a r.v. ξ_0 be nonnegative and $\mathcal{F}_0$-measurable. Prove equalities $\mathsf{E}_\mathsf{Q}(\xi_0\xi) = \mathsf{E}(\xi_0\xi\rho) = \mathsf{E}(\xi_0\mathsf{E}(\xi\rho/\mathcal{F}_0))$. Denote $\rho_0 = \mathsf{E}(\rho/\mathcal{F}_0)$. Derive from item (1) that $\rho_0 > 0$ Q-a.s. So, we can assume that P-a.s. $\xi_0 = 0$ on the set $\{\rho_0 = 0\}$ and obtain the equality $\mathsf{E}(\xi_0\mathsf{E}(\xi\rho/\mathcal{F}_0)) = \mathsf{E}_\mathsf{Q}(\xi_0 \cdot (1/(\mathsf{E}(\rho/\mathcal{F}_0)))\mathsf{E}(\xi\rho/\mathcal{F}_0))$. Derive from this the required statement.

(3) Denote $Z_t = \mathsf{E}((d\mathsf{Q}/d\mathsf{P}) \,|_{\mathcal{F}_t})$. Check that $M_t \in L_1(\mathsf{Q})$ if and only if $M_t Z_t \in L_1(\mathsf{P})$ and also that the process Z is positive P-a.s. Then, derive from item (2) the equality $Z_t \cdot \mathsf{E}_\mathsf{Q}(M_{t+1}/\mathcal{F}_t) = \mathsf{E}(M_{t+1}Z_{t+1}/\mathcal{F}_t)$. And finally, obtain that $\mathsf{E}_\mathsf{Q}(M_{t+1}/\mathcal{F}_t) = M_t$ if and only if the equality $\mathsf{E}(M_{t+1}Z_{t+1}/\mathcal{F}_t) = M_t Z_t$ holds true.
7.98. Use the corresponding result for discrete-time martingales, make a discretization of stopping times and use the uniform integrability.
7.99. You can assume that for some $0 \leq s < t \leq T$ the following inequalities hold: $\mathsf{E}(X(t)/\mathcal{F}_s) \geq X(s)$ and $\mathsf{P}(\mathsf{E}(X(t)/\mathcal{F}_s) > X(s)) > 0$. Consider an event $A = \{\omega \in \Omega | \mathsf{E}(X(t)/\mathcal{F}_s) > X(s)\} \in \mathcal{F}_s$ and produce a chain of the inequalities $\mathsf{E}X(T) \geq \mathsf{E}X(t) = \mathsf{E}(\mathsf{E}(X(t)/\mathcal{F}_s)) > \mathsf{E}X(s)\mathbb{I}_{\overline{A}} + \mathsf{E}X(s)\mathbb{I}_A = \mathsf{E}X(s) \geq \mathsf{E}X(0)$.

7.100. Check whether the process X is uniformly integrable. Next, make a discretization of the stopping time and pass to the limit using the uniform integrability.
7.101. (1) Write Doob's decomposition for the process X and consider stopping times $\tau := \inf\{n|\ A_{n+1} > 0\}$ and $\sigma := \inf\{n|\ A_{n+1} < 0\}$, where A is a predictable process in the Doob's decomposition. Prove that $A = 0$.

(2) Use item (1) and the uniform integrability.
7.102. Assume that $X(\infty) = 0$ a.s. First prove that this supermartingale is uniformly integrable and satisfies the conditions of Theorem 7.5. Then, use this theorem for Markov moments $\nu := \tau + t$ with rational $t > 0$ and $\sigma := \tau_n$ where $\tau_n = \inf\{s \in \mathbb{R}^+|\ X(s) \le 1/n\}$.
7.105. Find a decomposition of the process $\{(N(t) - \lambda t)^2, t \ge 0\}$ into a martingale and a nonrandom continuous nondecreasing function.
7.106. To prove the identity $\mathsf{E}M^2(\tau) = \mathsf{E}\langle M\rangle(\tau)$ we note that $\{M^2(t) - \langle M\rangle(t), t \ge 0\}$ is a uniformly integrable martingale with initial value zero. In order to prove the equality $\mathsf{E}M^2(\tau) = \mathsf{E}[M](\tau)$ consider separately continuous and discontinuous components and take into account their orthogonality.
7.107. (1) Consider $\tau = \inf\{t \in \mathbb{R}^+|\ |M_t| > u\} \wedge T$.

(2) Use the fact that

$$(M^* \wedge A)^p = \int_0^{M^* \wedge A} p x^{p-1} dx,\ p = 1, 2.$$

(3) The proof is similar to the one of the maximal integral Doob's inequality in case of discrete time (see, e.g., the proof of Theorem 5, Chapter IV [9]).
7.108. (1) Use Doob's theorem of optional sampling for the martingale $M_t = \exp\{izW(t) + (z^2 t)/2\}$, where $z \in \mathbb{R}$.

(2) Follows from item (1).

(3) First consider $T_1 = \sum_{i=1}^n t_i \mathbb{I}_{A_i}$ where $A_i \in \mathscr{A}$, A_i are disjoint, and $0 < t_1 < \cdots < t_n = T$.
7.109. Use items (2) and (3) of Problem 7.108. The trajectory of W_τ can be created from the trajectory of W using the reflection of that part of the latter that corresponds to the values of $t > \tau$ with respect to the line $y = W(\tau)$.

Answers and Solutions

7.18. (1) Because $X_n = X_n^+ - X_n^-$, $|X_n| = X_n^+ + X_n^-$, and $\mathsf{E}X_n \le \mathsf{E}X_1 < \infty$ for any supermartingale $\{X_n, n \in \mathbb{Z}^+\}$, then $\sup_n \mathsf{E}X_n^+ < \infty$; moreover $\sup_n \mathsf{E}|X_n| < \infty$. It follows now from Corollary 7.4 that there exists a limit $\lim_{n\to\infty} X_n = X_\infty$ a.s.

(2) According to Fatou's lemma $\mathsf{E}|X_\infty| = \mathsf{E}\liminf_{n\to\infty}|X_n| \le \liminf_{n\to\infty}\mathsf{E}|X_n| \le \sup_n \mathsf{E}|X_n| < \infty$.

(3) One-way implication is proved (see item (1)). Because $X_n^- \le |X_n|$, the opposite implication is evident.
7.20. It is evident that if the true densities are p_n, then the conditional density of the distribution X_n given $X_1, .., X_{n-1}$ is equal to $(p_{n+1})/p_n$, and then

$$\mathsf{E}(Y_n/X_1=x_1,..,X_{n-1}=x_{n-1})=\int_{\mathbb{R}}\frac{q_n(x_1,..,x_{n-1},x)}{p_n(x_1,..,x_{n-1},x)}\frac{p_n(x_1,..,x_{n-1},x)}{p_{n-1}(x_1,..,x_{n-1})}dx=$$
$$=\int_{\mathbb{R}}q_n(x_1,..,x_{n-1},x)dx\cdot(p_{n-1}(x_1,..,x_{n-1}))^{-1}=\frac{q_{n-1}(x_1,..,x_{n-1})}{p_{n-1}(x_1,..,x_{n-1})}.$$

7.26. Let us calculate $\xi_{n-1,k} := \mathsf{E}(\mathrm{I\!I}_{\omega\in\Delta_{n,k}}/\mathfrak{F}_{n-1})$. For any event $\Delta_{n-1,k}$ the following equality holds.

$$\int_A \xi_{n-1,k}d\mathsf{P}=\int_A \mathrm{I\!I}_{\omega\in\Delta_{n,k}}d\lambda^1=\begin{cases}\lambda^1(\Delta_{n,k}), & \text{if } \Delta_{n,k}\subset\Delta_{n-1,k},\\ 0, & \text{if } \Delta_{n,k}\cap\Delta_{n-1,k}=\emptyset.\end{cases}$$

That is why

$$\xi_{n-1,k}=\frac{\lambda^1(\Delta_{n,k})}{\lambda^1(\Delta_{n-1,k})}\mathrm{I\!I}_{\omega\in\Delta_{n-1,k}}.$$

This immediately implies the proof, if to decompose $\sum_{k=1}^{k_n}$ in the definition of X_n into sums with respect to the partition π_{n-1}.

7.31. Implication (1) $\Rightarrow$ (2) follows from the next chain of equalities: $\mathsf{E}(X_n^\tau - X_{n-1}^\tau/\mathfrak{F}_{n-1}) = \mathsf{E}(X_{\tau\wedge n} - X_{\tau\wedge(n-1)}/\mathfrak{F}_{n-1}) = \mathsf{E}(X_n - X_{n-1}/\mathfrak{F}_{n-1})\mathrm{I\!I}_{\tau>n-1} = 0$. Implication (2) $\Rightarrow$ (3) follows from Theorem 7.12, item (3). Implication (3) $\Rightarrow$ (1): first we put $\tau = N$; then $\mathsf{E}X_N = \mathsf{E}X_0$. Let an event $A\in\mathfrak{F}_n$, and $\tau = n\mathrm{I\!I}_A + N\mathrm{I\!I}_{\overline{A}}$ (check that τ is a stopping time). Then item 2) implies that $\mathsf{E}X_\tau = \mathsf{E}X_N$; that is, $\mathsf{E}X_n\mathrm{I\!I}_A + \mathsf{E}X_N\mathrm{I\!I}_{\overline{A}} = \mathsf{E}X_N$, thus, $\mathsf{E}X_n\mathrm{I\!I}_A = \mathsf{E}X_N\mathrm{I\!I}_A$ or $\int_A X_n dP = \int_A \mathsf{E}(X_N/\mathfrak{F}_n)d\mathsf{P}$, which is equivalent to the martingale property of the process X.

7.32. According to Theorem 7.4, $\mathsf{E}X_{\tau\wedge n} = \mathsf{E}X_0$. The Lebesgue dominated convergence theorem allows us to go to the limit in the integral as $n\to\infty$.

7.33. (1) Theorem 7.4 implies that for any $n\in\mathbb{Z}^+$ it holds that $\mathsf{E}(X_{\tau\wedge n}/\mathfrak{F}_\sigma) = X_{\tau\wedge n\wedge\sigma}$. Note that $X_{\tau\wedge n\wedge\sigma}\to X_{\tau\wedge\sigma}$ a.s. as $n\to\infty$. Further more, we choose such a sequence $n_k\to\infty$ that $\lim_{n_k\to\infty}\mathsf{E}|X_{n_k}|\mathrm{I\!I}_{\tau>n_k}=0$. Then $\mathsf{E}|\mathsf{E}(X_{\tau\wedge n_k}/\mathfrak{F}_\sigma) - \mathsf{E}(X_\tau/\mathfrak{F}_\sigma)| \le \mathsf{E}|\mathsf{E}((X_{n_k}-X_\tau)\mathrm{I\!I}_{\tau>n_k}/\mathfrak{F}_\sigma)| \le \mathsf{E}|X_{n_k}|\mathrm{I\!I}_{\tau>n_k} + \mathsf{E}|X_\tau|\mathrm{I\!I}_{\tau>n_k}\to 0$ as $n_k\to\infty$, because $|X_\tau|\mathrm{I\!I}_{\tau>n_k}\to 0$ a.s. and X_τ is an integrable r.v. As $\mathsf{E}(X_{\tau\wedge n_k}/\mathfrak{F}_\sigma)\to\mathsf{E}(X_\tau/\mathfrak{F}_\sigma)$ in $L_1(\mathsf{P})$, then there exists $n_{k_j}\to\infty$ such that $\mathsf{E}(X_{\tau\wedge n_{k_j}}/\mathfrak{F}_\sigma)\to\mathsf{E}(X_\tau/\mathfrak{F}_\sigma)$ a.s., and the first statement follows. Item (2) follows immediately from item (1), and item (3) we leave for you to do on your own.

7.34. First we prove that $\mathsf{E}|X_\tau|<\infty$. Indeed, $|X_\tau|\le|X_0|+\sum_{k=1}^{\tau}|X_k-X_{k-1}|\mathrm{I\!I}_{\tau>0}$, and

$$\sum_{n=1}^{\infty}\sum_{k=1}^{n}\mathsf{E}(|X_k-X_{k-1}|\mathrm{I\!I}_{\tau=n})=\sum_{k=1}^{\infty}\sum_{n=k}^{\infty}\mathsf{E}(|X_k-X_{k-1}|1_{\tau=n})$$
$$=\sum_{k=1}^{\infty}\mathsf{E}(|X_k-X_{k-1}|1_{\tau\ge k}).$$

Because $\mathrm{I\!I}_{\{\tau\ge k\}}=1-\mathrm{I\!I}_{\{\tau\le k-1\}}\in\mathfrak{F}_{k-1}$, then

$$\mathsf{E}(|X_k-X_{k-1}|\mathrm{I\!I}_{\tau\ge k})=\mathsf{E}(\mathrm{I\!I}_{\tau>k-1}\mathsf{E}(|X_k-X_{k-1}|/\mathfrak{F}_{k-1}))\le C\mathsf{P}(\tau\ge k)$$

and therefore $\mathsf{E}\sum_{k=1}^{\tau}|X_k-X_{k-1}|\mathrm{I\!I}_{\tau>0}\le C\sum_{k=1}^{\infty}\mathsf{P}(\tau\ge k)=C\mathsf{E}\tau<\infty$, and thus $\mathsf{E}|X_\tau|<\infty$. Now $\mathsf{E}|X_n|\mathrm{I\!I}_{\tau>n}\le\mathsf{E}(|X_0|+\sum_{k=1}^{\tau}|X_k-X_{k-1}|\mathrm{I\!I}_{\tau>0})\mathrm{I\!I}_{\tau>n}$ and we just proved that a r.v. $|X_0|+\sum_{k=1}^{\tau}|X_k-X_{k-1}|\mathrm{I\!I}_{\tau>0}$ is integrable and $\mathrm{I\!I}_{\tau>n}\to 0$, $n\to\infty$ a.s. So, $\mathsf{E}|X_n|\mathrm{I\!I}_{\tau>n}\to 0$; that is, we obtain the conditions of Problem 7.33. According to item (2) of that problem $\mathsf{E}X_\tau=\mathsf{E}X_0$ as $\sigma=0$.

7.36. Let $Y_n = -X_n$. Then Y_n is a submartingale, $Y_n \le \mathsf{E}(-\xi/\mathfrak{F}_n)$, and $\mathsf{E}|\xi| < \infty$. According to Theorem 7.4, for any $n \ge 0$ it holds that $Y_{\sigma\wedge n} \le \mathsf{E}(Y_{\tau\wedge n}/\mathfrak{F}_\sigma)$. Note that $Y_{\sigma\wedge n} \to Y_\sigma$, $Y_{\tau\wedge n} \to Y_\tau$, as $n \to \infty$ a.s. The only thing we need to ground is the limit change for the conditional mathematical expectation. In order to do this, we need to prove that the family of random variables $\{Y_{\tau\wedge n}, n \ge 1\}$ is uniformly integrable. But

$$\begin{aligned}\mathsf{E}|Y_{\tau\wedge n}|\mathbb{I}_{Y_{\tau\wedge n}\ge C} \le \textstyle\sum_{k=1}^n \mathsf{E}(\mathsf{E}(|\xi|/\mathfrak{F}_k)\mathbb{I}_{\tau=k}\mathbb{I}_{\mathsf{E}(|\xi|/\mathfrak{F}_k)\ge C})\\ +\mathsf{E}(\mathsf{E}(|\xi|/\mathfrak{F}_n)\mathbb{I}_{\tau>n}\mathbb{I}_{\mathsf{E}(|\xi|/\mathfrak{F}_n)\ge C}).\end{aligned}$$

Note that for any set $A \in \mathfrak{F}_k$

$$\begin{aligned}\textstyle\int_A \mathsf{E}(|\xi|/\mathfrak{F}_k)\mathbb{I}_{\tau=k}d\mathsf{P} &= \textstyle\int_{A\cap\{\tau=k\}} |\xi| d\mathsf{P}\\ &= \textstyle\int_{A\cap\{\tau=k\}} \mathsf{E}(|\xi|/\mathfrak{F}_\tau)d\mathsf{P} = \int_A \mathsf{E}(|\xi|/\mathfrak{F}_\tau)\mathbb{I}_{\tau=k}d\mathsf{P},\end{aligned}$$

because $A\cap\{\tau=k\} \in \mathfrak{F}_\tau$. So, $\mathsf{E}(|\xi|/\mathfrak{F}_k)\mathbb{I}_{\tau=k} = \mathsf{E}(|\xi|/\mathfrak{F}_\tau)\mathbb{I}_{\tau=k}$ and

$$\begin{aligned}\sup_{n\ge 1}\textstyle\sum_{k=1}^n \mathsf{E}(\mathsf{E}(|\xi|/\mathfrak{F}_k)\mathbb{I}_{\tau=k}\mathbb{I}_{\mathsf{E}(|\xi|/\mathfrak{F}_k)\ge C}) &= \mathsf{E}(|\xi|\mathbb{I}_{\mathsf{E}(|\xi|/\mathfrak{F}_\tau)\ge C}\mathbb{I}_{\tau\le n})\\ &\le \mathsf{E}(|\xi|\mathbb{I}_{\mathsf{E}(|\xi|/\mathfrak{F}_\tau)\ge C}) \to 0,\ C\to\infty.\end{aligned}$$

We now put $Z_n = \mathsf{E}(|\xi|/\mathfrak{F}_n)$. This is a uniformly integrable martingale according to Theorem 7.10, thus $\lim_{C\to\infty}\sup_n \mathsf{E}(Z_n\mathbb{I}_{Z_n\ge C}) = 0$. The obtained relations mean that $\{Y_{\tau\wedge n}, n\ge 1\}$ is a uniformly integrable sequence.

7.39. It is evident that $\mathbb{I}_{n\le\nu} = 1 - \mathbb{I}_{\nu\le n-1} \in \mathfrak{F}_{n-1}$. Thus,

$$\begin{aligned}\mathsf{E}(Y_n/\mathfrak{F}_{n-1}) &= \mathsf{E}(X_n/\mathfrak{F}_{n-1})\mathbb{I}_{n\le\nu} + \mathsf{E}(2X_\nu - X_n/\mathfrak{F}_{n-1})\mathbb{I}_{n>\nu}\\ &= X_{n-1}\mathbb{I}_{n-1\le\nu} - X_{n-1}\mathbb{I}_{\nu=n-1} + 2\mathsf{E}(X_\nu\mathbb{I}_{\nu<n-1}/\mathfrak{F}_{n-1}) - X_{n-1}\mathbb{I}_{\nu<n-1}\\ &\quad -X_{n-1}\mathbb{I}_{\nu=n-1} + 2X_{n-1}\mathbb{I}_{\nu=n-1} = X_{n-1}\mathbb{I}_{n-1\le\nu} + (2X_\nu - X_{n-1})\mathbb{I}_{\nu<n-1}\\ &= Y_{n-1}.\end{aligned}$$

7.41. (4) Consider an event $\{\sigma\le\tau\} = \{\sigma\wedge\tau=\sigma\} = \cup_{k=0}^\infty\{\sigma\wedge\tau=k, \sigma=k\}$. Further more, for any $n\ge 0$

$$\{\sigma\wedge\tau=k,\sigma=k\}\cap\{\sigma\wedge\tau=n\} = \begin{cases}\emptyset, & k\ne n,\\ \{\sigma\wedge\tau=n,\sigma=n\}\in\mathfrak{F}_n, & k=n.\end{cases}$$

It is easy to check in a similar way that the event $\{\sigma=\tau\}\in\mathfrak{F}_{\sigma\wedge\tau}$, and then $\{\sigma<\tau\} = \{\sigma\le\tau\}\backslash\{\sigma=\tau\}\in\mathfrak{F}_{\sigma\wedge\tau}$.

(5) Let $B\in\mathfrak{F}_\sigma$. Then for any $i\ge 0$ the event $B\cap\{\sigma=i\}\in\mathfrak{F}_i$. Now, $B\cap\{\tau=n\} = \bigcup_{i=0}^n B\cap\{\tau=n\}\cap\{\sigma=i\}$ and all these events belong to $\mathfrak{F}_n$. It means that $B\in\mathfrak{F}_\tau$.

7.46. (1) Let $B\in\mathfrak{F}_{\tau-}$. If $B\in\mathfrak{F}_0$, then we can assume that $\sigma\equiv 0$ is a stopping time, and $\sigma\le\tau$. It means that $\mathfrak{F}_0 = \mathfrak{F}_\sigma\subset\mathfrak{F}_\tau$ and $B\in\mathfrak{F}_\tau$. Let now $B = A\cap\{t<\tau\}$, where $t\in\mathbb{R}^+$, $A\in\mathfrak{F}_t$. We need to prove that for any $s\in\mathbb{R}^+$ an event $C := B\cap\{\tau\le s\}\in\mathfrak{F}_s$. If $s\le t$, then $C=\emptyset\in\mathfrak{F}_s$. If $s>t$ then $C = A\cap\{t<\tau\le s\} = A\cap(\{\tau\le s\}\backslash\{\tau\le t\})\in\mathfrak{F}_s$.

(2) We need to prove that for any $s\in\mathbb{R}^+$ an event $A := \{\tau\le s\}\in\mathfrak{F}_{\tau-}$. It is evident that this event belongs to $\mathfrak{F}_s$. Its complement $A^c = \{\tau>s\}$ belongs to $\mathfrak{F}_s$ as well. So, $A^c = A^c\cap\{s<\tau\}\in\mathfrak{F}_{\tau-}$ according to the definition of $\mathfrak{F}_{\tau-}$. Thus, $A\in\mathfrak{F}_{\tau-}$.

7.48. It follows directly from the symmetry that

$$\mathsf{E}(Z_n/\mathscr{G}_{n+1}) = (1/n)\sum_{k=1}^n \mathsf{E}(X_k/\mathscr{G}_{n+1}) = \mathsf{E}(X_1/\mathscr{G}_{n+1}) = \mathsf{E}(1/(n+1)\sum_{k=1}^{n+1} X_k/\mathscr{G}_{n+1}) \\ = Z_{n+1}.$$

7.50. (1) Denote $p_k := \mathsf{P}(S_n \le k, n \ge 1/S_0 = 0)$. Then $p_0 = \frac{1}{2}\mathsf{P}(S_n \le 0, n \ge 1/\xi_1 = -1) = \frac{1}{2}p_1$. Furthermore, it is easy to check that $p_1 = \frac{1}{2}p_0 + \frac{1}{2}p_2$ and, in general, for $k > 1$ the equality $p_k = \frac{1}{2}p_{k-1} + \frac{1}{2}p_{k+1}$ implies $p_0 = \frac{1}{k}p_k$. Because $p_k \le 1$ then $p_0 = p_1 = \cdots = 0$. It means that $\mathsf{P}(\tau < \infty) = 1$.

Item (2) is an immediate corollary of item (1).

(3) Because $\mathsf{E}\xi_1 = 0$ and $\mathsf{E}S_\tau = 1$, then the first Wald identity implies that $\mathsf{E}\tau = \infty$, otherwise we would obtain that $1 = c \cdot 0$ with certain $c \in \mathbb{R}^+$.

7.52. Put $X_n = e^{t_0 S_n}/((\varphi(t_0))^n)$. Then

$$\mathsf{E}(X_n/\mathfrak{F}_{n-1}) = \frac{e^{t_0 S_{n-1}}}{(\varphi(t_0))^{n-1}} \text{ and } \mathsf{E}\left(\frac{e^{t_0\xi_n}}{\varphi(t_0)}\right) = X_{n-1},$$

that is, $\{X_n, \mathfrak{F}_n, n \in \mathbb{Z}^+\}$ is a martingale with $\mathsf{E}X_n = 1$, and

$$\mathsf{E}\left(|X_n - X_{n-1}|/\mathfrak{F}_{n-1}\right) \mathbb{I}_{\tau > n-1} = X_{n-1}\mathbb{I}_{\tau>n-1}\mathsf{E}\left|\frac{e^{t_0\xi_n}}{\varphi(t_0)} - 1\right| \\ \le \frac{e^{t_0 C}}{(\varphi(t_0))^n} \cdot \mathsf{E}\left|\frac{e^{t_0\xi_n}}{\varphi(t_0)} - 1\right| \le 2e^{t_0 C}.$$

It follows from Problem 7.48 that $\mathsf{E}X_\tau = \mathsf{E}X_0 = 1$. Therefore, $\mathsf{E}e^{t_0 S_\tau}/((\varphi(t_0))^\tau) = 1$.

7.53. We prove only item (1) (item (2) can be proved in a similar way). As above, $S_\tau = \sum_{i=1}^\infty \xi_i \mathbb{I}_{\tau \ge i}$. At that time ξ_i and $\mathbb{I}_{\tau\ge i}$ are independent and $\mathsf{E}\xi_i\mathbb{I}_{\tau\ge i} = \mathsf{E}\xi_i \mathsf{P}(\tau \ge i) = 0$. In addition, $\mathsf{E}|S_\tau| \le \mathsf{E}\sum_{i=1}^\tau |\xi_i| < \infty$. So, we only need to justify changing the order for expectations. It is sufficient to prove that $\mathsf{E}\sum_{i=1}^\infty |\xi_i|\mathbb{I}_{\tau\ge i} < \infty$. But $\sum_{i=1}^\infty |\xi_i|\mathbb{I}_{\tau \ge i} = \sum_{i=1}^\tau |\xi_i|$ and, according to the condition, $\mathsf{E}\sum_{i=1}^\tau |\xi_i| < \infty$.

7.54. We prove only item (1). It is evident that

$$\mathsf{E}\left(M_n/\mathfrak{F}_{n-1}\right) = \mu^{-n}\mathsf{E}\left(\sum_{k=1}^{S_{n-1}} \xi_k^{(n)}/\mathfrak{F}_{n-1}\right)$$

$$= \mu^{-n}\mathsf{E}\left(\sum_{N=1}^\infty \sum_{k=1}^N \xi_k^{(n)}\mathbb{I}_{S_{n-1}=N}/\mathfrak{F}_{n-1}\right) = \mu^{-n}\mathsf{E}\left(\sum_{k=1}^\infty \xi_k^{(n)}\mathbb{I}_{S_{n-1}\ge k}/\mathfrak{F}_{n-1}\right).$$

(We change the order of summation and conditional expectation and take into account that S_{n-1} is $\mathfrak{F}_{n-1}$-measurable and $\xi_k^{(n)}$ is independent of $\mathfrak{F}_{n-1}$.) So,

$$\mathsf{E}\left(M_n/\mathfrak{F}_{n-1}\right) = \mu^{-n}\sum_{k=1}^\infty \mathsf{E}\xi_k^{(n)}\mathbb{I}_{S_{n-1}\ge k} = \mu^{-n+1}\cdot S_{n-1} = M_{n-1}.$$

The order of summation and conditional expectation can be changed because $\xi_k^{(n)}$ are nonnegative.

7.55. (2) We need to prove that $\mathsf{P}(\tau < \infty / X_0 = x) = 1$. Consider the events $A_1 = \{\tau < \infty, X_\tau = b\}$ and $A_2 = \{\tau < \infty, X_\tau = a\}$. It is evident that the event $\{\tau < \infty\} = A_1 \cup A_2$. Now, denote $\alpha(x) = \mathsf{P}(A_1/X_0 = x)$ and $\beta(x) = \mathsf{P}(A_2/X_0 = x)$. In this case $\alpha(x)$ and $\beta(x)$ satisfy the following difference equations with corresponding boundary conditions.

$$\begin{aligned}\alpha(x) &= p\alpha(x+1) + q\alpha(x-1), \;\; \alpha(b) = 1, \;\; \alpha(a) = 0.\\ \beta(x) &= p\beta(x+1) + q\beta(x-1), \;\; \beta(b) = 0, \;\; \beta(a) = 1.\end{aligned}$$

Let $p \neq q$. The equation for $\alpha(x)$ has two obvious solutions $\alpha_1(x) = c_1$, and $\alpha_2(x) = c_2 (q/p)^x$, where c_1 and c_2 are some nonnegative constants. If we find the solution of the form $\alpha(x) = c_1 + c_2 (q/p)^x$, then we obtain, taking into account boundary conditions, that

$$\alpha(x) = \frac{\left(\frac{q}{p}\right)^x - \left(\frac{q}{p}\right)^a}{\left(\frac{q}{p}\right)^b - \left(\frac{q}{p}\right)^a}.$$

Similarly,

$$\beta(x) = \frac{\left(\frac{q}{p}\right)^b - \left(\frac{q}{p}\right)^x}{\left(\frac{q}{p}\right)^b - \left(\frac{q}{p}\right)^a}.$$

In particular, this implies $\mathsf{P}(\tau < \infty / X_0 = x) = 1$. The only thing we need is to prove that the solution of every difference equation with corresponding boundary conditions is unique. Indeed, let $\widehat{\alpha}(x)$ be some solution with $\widehat{\alpha}(b) = 1$, $\widehat{\alpha}(a) = 0$. Then we can find two constants c_1 and c_2 such that $c_1 + c_2 (q/p)^a = \widehat{\alpha}(a)$ and $c_1 + c_2 (q/p)^{a+1} = \widehat{\alpha}(a+1)$. So, we can obtain from the difference equation for $\widehat{\alpha}(x)$ that $\widehat{\alpha}(a+2) = c_1 + c_2 (q/p)^{a+2}$ and, in general, $\widehat{\alpha}(x) = c_1 + c_2 (q/p)^x$, $a \le x \le b$. Consequently, the solution $\alpha(x)$ is unique. $\beta(x)$ can be treated in a similar way. If $p = q = \frac{1}{2}$, then the solutions $\alpha_1(x)$ and $\alpha_2(x)$ are equal, but there exists one more obvious solution $\alpha(x) = c_2 x$. Taking into account boundary conditions we obtain that

$$\alpha(x) = \frac{x-a}{b-a}, \quad \beta(x) = \frac{b-x}{b-a}$$

and again the solution is unique. Thus, $\mathsf{P}(\tau < \infty / X_0 = x) = 1$ for any $a \le x \le b$.

Statements (4) and (5) have been proved in fact within proving statement (2). Another way to prove our statement in the case when $p \neq q$: a martingale $|X^M_{n\wedge\tau}| \le |a| + |b|$ and is bounded; so, according to Problem 7.32, $\mathsf{E}M_\tau = \mathsf{E}M_0 = (q/p)^x$, but $\mathsf{E}M_\tau = (q/p)^a \mathsf{P}(\tau = \tau_a / X_0 = x) + (q/p)^b \mathsf{P}(\tau = \tau_b / X_0 = x)$, and $\mathsf{P}(\tau = \tau_a / X_0 = x) + \mathsf{P}(\tau = \tau_b / X_0 = x) = 1$, hence

$$\mathsf{P}(\tau = \tau_a / X_0 = x) = \frac{\left(\frac{q}{p}\right)^x - \left(\frac{q}{p}\right)^b}{\left(\frac{q}{p}\right)^a - \left(\frac{q}{p}\right)^b}.$$

(6) Because the r.v. X_τ is bounded and $\mathsf{E}\xi_1 = 0$ as $p = q = 1/2$, then it follows from the first Wald identity that $\mathsf{E}\tau = \infty$ in this case. If $p \neq q$, then $\mathsf{E}\xi_1 = p - q \neq 0$, and

$$\mathsf{E}X_\tau = \frac{a\left(\left(\frac{q}{p}\right)^x - \left(\frac{q}{p}\right)^b\right) + b\left(\left(\frac{q}{p}\right)^a - \left(\frac{q}{p}\right)^x\right)}{\left(\frac{q}{p}\right)^b - \left(\frac{q}{p}\right)^a}.$$

7.66. Let $\{r_n(t), n \geq 1, t \in [0,1]\}$ be a sequence of Rademaher functions. It means that $r_n(t) = \pm 1$ and $r_n(t)$ are mutually independent with respect to Lebesgue measure on $[0,1]$ (we consider the probability space $\Omega = [0,1]$, $\mathcal{F}$ consisting of Lebesgue measurable sets from $[0,1]$ and $\mathsf{P} = \lambda^1|_{[0,1]}$). This sequence satisfies the Khinchin inequality where expectation should be treated as an integral on Lebesgue measure. Consider the martingale transformation $(r(t) \circ X)_n$ and put $r_0 = 0$. For this transformation we have $r(t) \circ (r(t) \circ X)_n = (r^2(t) \circ X)_n = X_n$. According to the second statement of Theorem 7.11, it holds that $\mathsf{E}|X_n|^p \leq C_p \mathsf{E}|(r(t) \circ X)_n|^p \leq C_p^2 \mathsf{E}|X_n|^p$ as $p > 1$ and $t \in [0,1]$. (In this formula and below we deal with ordinary expectation). The following estimates are valid due to the Khinchin inequality: $c_p [X]_n^{p/2} \leq \int_0^1 |(r(t) \circ X)_n|^p dt \leq C_p [X]_n^{p/2}$ for $p > 0$. It follows from these inequalities that for $p > 1$ it holds $c_p \mathsf{E}[X]_n^{p/2} \leq \mathsf{E}|X_n|^p \leq C_p \mathsf{E}[X]_n^{p/2}$.

7.72. It is obvious that $\{W(t), \mathcal{F}_t, t \in \mathbb{R}^+\}$ is a martingale. The integral Doob's inequality holds true for continuous-time martingales: for any $t > 0$ and $p > 1$ it holds that $\mathsf{E}\sup_{0 \leq s \leq t} |X(s)|^p \leq (p/(1-p))^p \mathsf{E}|X(t)|^p$ (see Theorem 7.8 for discrete time and,e.g., [57] for continuous time). Thus, $\mathsf{E}|W(t \wedge \tau)|^2 \leq \mathsf{E}\sup_{0 \leq s \leq t} |W(s)|^2 \leq 4\mathsf{E}|W(t)|^2 = 4t < \infty$, and then $W(t \wedge \tau) \in L_2(\mathsf{P}) \subset L_1(\mathsf{P})$. Now, we need to use a version of Theorem 7.5 but instead of ∞ we consider any $T > 0$ and prove the martingale property on $[0,T]$.

7.91. (1) According to the problem situation, $M^*(\infty)$ and $N^*(\infty)$ belong to $L_2(\Omega, \mathcal{F}, \mathsf{P})$. That is why the product $M^*(\infty)N^*(\infty)$ belongs to $L_1(\Omega, \mathcal{F}, \mathsf{P})$. Furthermore,

$$(MN)^*(\infty) = \sup_{t \in \mathbb{R}^+} |M(t)N(t)| \leq M^*(\infty)N^*(\infty)$$

and then the product MN belongs to $\widetilde{\mathcal{M}}_{\text{loc}}$ and $\mathsf{E}(MN)^*(\infty) < \infty$. We show that in this case MN is a uniformly integrable martingale. Indeed, there exists a sequence $\tau_n \uparrow \infty$ of stopping times such that $\mathsf{E}((MN)(\tau_n \wedge t)/\mathcal{F}_s) = (MN)(\tau_n \wedge s)$, and we are able to tend $n \to \infty$, because there exists an integrable dominant $(MN)^*(\infty)$. Thus, $\mathsf{E}((MN)(t)/\mathcal{F}_s) = (MN)(s)$ and MN is a martingale. We can derive its integrability from the fact that $\mathsf{E}(MN)^*(\infty) < \infty$, because in the case $\mathsf{E}\sup_n |\xi_n| < \infty$, it holds that $\sup_n \mathsf{E}|\xi_n| \mathbb{I}_{|\xi_n| > C} \leq \mathsf{E}\sup_n |\xi_n| \mathbb{I}_{\sup_n |\xi_n| > C} \to 0$ as $n \to \infty$, and it can be easily generalized for any set of parameters. Now, to prove that $\mathsf{E}M(\tau)N(\tau) = \mathsf{E}M(0)N(0) = 0$, we write $\mathsf{E}(MN)(\tau \wedge n) = \mathsf{E}(MN)(0) = 0$ and tend $t \to \infty$. It is possible due to the uniform integrability.

(2) In this case for any stopping time τ we have that $M(\tau)N(\tau) \in L_1(\Omega, \mathcal{F}, \mathsf{P})$. Thus, $\mathsf{E}|M(\tau)N(\tau)| < \infty$ and $\mathsf{E}M(\tau)N(\tau) = 0$. It follows from Theorem 7.1 and from the generalization of Problem 7.40 to continuous-time processes (check whether it

indeed holds true), that MN has the right-hand continuous modification with finite limits from the left at every point. Now, we put $\tau(\omega) := t\,\mathbb{I}_A + T\,\mathbb{I}_{\overline{A}}$ for any $t \in [0,T)$ and any $A \in \mathfrak{F}_t$ and obtain that $\mathsf{E}M(\tau)N(\tau) = \mathsf{E}M(t)N(t)\,\mathbb{I}_A + \mathsf{E}M(T)N(T)\,\mathbb{I}_{\overline{A}} = 0 = \mathsf{E}M(T)N(T)$. So, $\mathsf{E}M(t)N(t)\,\mathbb{I}_A = \mathsf{E}M(T)N(T)\,\mathbb{I}_A$, thus, MN is a martingale. The uniform integrability follows if we note that under our conditions $(MN)^*(\infty) \in \mathscr{H}_1$ (see also the proof of the first statement of this problem).

7.93. The equivalence of statements (2) and (3) has been proved in Problem 7.92. It is evident that (3) implies (2) if we remember that $M_0 = 0$. Let us assume that statement (1) holds. Because $N^\tau \in S$ for any τ on $\mathbb{T}$ then $\mathsf{E}M_T N_T^\tau = \mathsf{E}M_T N_\tau = \mathsf{E}M_\tau N_\tau = 0$. Now, we can use a discrete-time version of Problem 7.91.

7.96. (1) We define a process L for a given measure $\mathsf{Q} \sim \mathsf{P}$ by the equations $L(0) = 1$ and $L(t+1) = L(t) + (Z(t+1) - Z(t))/Z(t)$, $t \in \{0,1,\dots,T-1\}$. It is obvious that equality (7.2) holds within this choice of L. Moreover, L satisfies the condition $L(t+1) - L(t) > -1$ because the equivalence of the measures P and Q implies the positivity P-a.s. of random variables $Z(t)$ for all $t \in \mathbb{T}$. Now, by induction in t we show that $L(t) \in L_1(\mathsf{P})$. It is evident for $t = 0$. Assume that $L(t) \in L_1(\mathsf{P})$. As the process Z is nonnegative, the conditional expectation of the r.v. $(Z(t+1))/Z(t)$ is correctly defined and P-a.s. satisfies the equality

$$\mathsf{E}\left(\frac{Z(t+1)}{Z(t)}/\mathfrak{F}_t\right) = \frac{1}{Z(t)}\mathsf{E}(Z(t+1)/\mathfrak{F}_t) = 1.$$

It means that $(Z(t+1))/Z(t) \in L_1(\mathsf{P})$ and hence,

$$L(t+1) = L(t) - 1 + \frac{Z(t+1)}{Z(t)} \in L_1(\mathsf{P}).$$

Now we can derive the martingale property of the process L: because the r.v. Z is positive, we can divide both parts of the equation $\mathsf{E}(Z(t+1)/\mathfrak{F}_t) = Z(t)$ by $Z(t)$ and obtain the equality $\mathsf{E}(L(t+1) - L(t)/\mathfrak{F}_t) = 0$.

(2) If L satisfies the indicated conditions and the equality (7.2) determines a positive P-martingale Z, then the equalities $\mathsf{E}Z(t) = Z(0) = 1$ are obvious.

7.97. First note that

$$\begin{aligned}&(L(t) - L(t-1))(\widehat{M}(t) - \widehat{M}(t-1)) \\ &= \frac{1}{Z(t-1)} \cdot Z(t)(\widehat{M}(t) - \widehat{M}(t-1)) - (\widehat{M}(t) - \widehat{M}(t-1)).\end{aligned} \tag{7.4}$$

Furthermore, note that $\mathsf{E}Z(t)|\widehat{M}(t)| = \mathsf{E}_{\mathsf{Q}}|\widehat{M}(t)| < \infty$ and the same is true for $Z(t)|\widehat{M}(t-1)|$. Hence, $Z(t)(\widehat{M}(t) - \widehat{M}(t-1)) \in L_1(\mathsf{P})$. We put $\tau_n = \inf\{t \in \mathbb{T} |\, Z(t) < 1/n\} \wedge T$, $n \in \mathbb{N}$. Then $(L(t) - L(t-1))(\widehat{M}(t) - \widehat{M}(t-1))\,\mathbb{I}_{\tau_n \geq t} \in L_1(\mathsf{P})$. In particular, the conditional expectations from (7.3) are P-a.s. correctly defined. Furthermore, it follows from (7.4) that the following equalities hold true P-a.s. on the set $\{\tau_n \geq t\}$.

$$\begin{aligned}&\mathsf{E}\left(\widehat{M}(t) - \widehat{M}(t-1)/\mathfrak{F}_{t-1}\right) \\ &\quad = \tfrac{1}{Z(t-1)}\mathsf{E}\left(Z(t)(\widehat{M}(t) - \widehat{M}(t-1))/\mathfrak{F}_{t-1}\right) \\ &\quad - \mathsf{E}\left((L(t) - L(t-1))(\widehat{M}(t) - \widehat{M}(t-1))/\mathfrak{F}_{t-1}\right).\end{aligned}$$

The equality means that Doob's decomposition with respect to the measure P of the process $\widehat{M}$ is

$$\widehat{M}(t) = M(t) - \sum_{k=1}^{t} \mathsf{E}\Big(\big(L(k)-L(k-1)\big)\big(\widehat{M}(k)-\widehat{M}(k-1)\big)/\mathfrak{F}_{k-1}\Big),\ t \in \mathbb{T}.$$

7.103. Consider a stopping time $\tau = \inf\{n \geq 1 : \langle M\rangle_n \geq C\} \wedge N$. Thus, according to the maximal Doob's inequality,

$$\mathsf{P}(\max_{0\leq n\leq N} |M_n| \geq \varepsilon) \leq \mathsf{P}(\tau < N) + \mathsf{P}(\max_{0\leq n\leq N} |M_n| \geq \varepsilon) \leq \mathsf{P}(\langle M\rangle_N \geq C)$$

$$+\mathsf{P}(\max_{0\leq n\leq N} |M_{\tau\wedge n}| \geq \varepsilon) \leq \mathsf{P}(\langle M\rangle_N \geq C) + \frac{\mathsf{E}|M_{\tau\wedge N}|^2}{\varepsilon^2} = \mathsf{P}(\langle M\rangle_N \geq C) + \frac{\mathsf{E}|M_\tau|^2}{\varepsilon^2}.$$

Now,

$$\mathsf{E}|M_\tau|^2 = \mathsf{E}\langle M\rangle_\tau \leq C + \mathsf{E}\max_{1\leq n\leq N}\big(\langle M\rangle_n - \langle M\rangle_{n-1}\big).$$

7.104. It is sufficient to prove the inequality we are searching only for martingales from the class $\mathscr{M}^2$ and then generalize it for $\mathscr{M}^2_{\mathrm{loc}}$ via passing to the limit. Put $\tau_N = \inf\{t |\ \langle M\rangle(t) \geq N\} \wedge T$. Then, due to the continuity $\langle M\rangle(\tau) \leq N$ a.s. On the other hand, $\mathsf{P}(\sup_{0\leq t\leq T} |M(t)| \geq \varepsilon) = \mathsf{P}((\sup_{0\leq t\leq T} |M(t)| \geq \varepsilon) \cap (\tau < T)) + \mathsf{P}((\sup_{0\leq t\leq T} |M(t)| \geq \varepsilon) \cap (\tau = T)) \leq \mathsf{P}(\tau < T) + \mathsf{P}(\sup_{0\leq \tau\leq T} |M(t)| \geq \varepsilon)$. Taking into account that $\mathsf{P}(\tau < T) \leq \mathsf{P}(\langle M\rangle(T) \geq N)$, $\mathsf{P}(\sup_{0\leq t\leq \tau} |M(t)| \geq \varepsilon) \leq N\varepsilon^{-2}$, we obtain the required inequality.

8

Stationary discrete- and continuous-time processes. Stochastic integral over measure with orthogonal values

Theoretical grounds

In this chapter we consider complex-valued stochastic processes. Let us recall the definition of covariance for complex-valued random variables X,Y:

$$\operatorname{cov}(X,Y) = \mathsf{E}(X - \mathsf{E}X)\overline{(Y - \mathsf{E}Y)}.$$

Definition 8.1. *A stochastic process* $\{X(t), t \in \mathbb{R}\}$, $\mathsf{E}|X(t)|^2 < \infty$ *is called a wide-sense stationary process if* $\mathsf{E}X(t) = \mathsf{E}X(0)$ *for all* $t \in \mathbb{R}$, *and* $\operatorname{cov}(X(t+s), X(s)) = \operatorname{cov}(X(t), X(0))$ *for all* $s,t \in \mathbb{R}$. *The function* $R_X(t) := \operatorname{cov}(X(t), X(0))$ *is called the covariance function for the process* $\{X(t), t \in \mathbb{R}\}$.

A definition of a wide-sense stationary random sequence $\{X_n, n \in \mathbb{Z}\}$ can be given in a similar way.

The following two theorems state that the covariance function of a wide-sense stationary process or sequence is a Fourier transform of a finite measure.

Theorem 8.1. *(Bochner–Khinchin theorem) Assume that the covariance function* R_X *of a wide-sense stationary stochastic process* $\{X(t), t \in \mathbb{R}\}$ *is continuous. Then there exists a finite measure* F_X *on* $(\mathbb{R}, \mathcal{B}(\mathbb{R}))$ *such that*

$$R_X(t) = \int_{-\infty}^{\infty} e^{itu} F_X(du), \quad t \in \mathbb{R}.$$

Theorem 8.2. *(Herglotz theorem) Assume that* $\{X_n, n \in \mathbb{Z}\}$ *is the wide-sense stationary sequence. Then there exists a finite measure* F_X *on* $((-\pi,\pi], \mathcal{B}((-\pi,\pi]))$ *such that*

$$R_X(n) = \int_{-\pi}^{\pi} e^{inu} F_X(du), \quad n \in \mathbb{Z}.$$

The measure F_X from the Bochner–Khinchin (Herglotz) theorem is called the *spectral measure* of the process $\{X(t)\}$ (sequence $\{X_n\}$). Its distribution function, which we also denote by F_X, is called the *spectral function*. If the spectral function F_X is absolutely continuous then its derivative $p_X(x) = F_X'(x)$ is called the *spectral density* of the process.

D. Gusak et al., *Theory of Stochastic Processes*, Problem Books in Mathematics, DOI 10.1007/978-0-387-87862-1_8,

It is turns out that wide-sense stationary random sequences and processes can be represented as a Fourier transform of processes with orthogonal increments. In order to formulate the corresponding result, we present a construction of the stochastic integral.

Definition 8.2. *Let $(\Omega,\mathcal{F},\mathsf{P})$ be a probability space, $(E,\mathcal{E})$ be a space with σ-finite measure μ, and $\mathcal{E}_\mu=\{A\in\mathcal{E}:\mu(A)<+\infty\}$ be a ring of sets of finite measure μ. A function $Z:\ \mathcal{E}_\mu\to L_2(\Omega,\mathcal{F},\mathsf{P})$ is said to be the orthogonal stochastic measure with the structural measure μ if:*

(1) $\forall\,\Delta_1,\Delta_2\in\mathcal{E}_\mu,\ \Delta_1\cap\Delta_2=\varnothing\ :\ Z(\Delta_1\cup\Delta_2)=Z(\Delta_1)+Z(\Delta_2)$ *a.s.*

(2) $\forall\,\Delta\in\mathcal{E}_\mu\ :\ \mathsf{E}|Z(\Delta)|^2=\mu(\Delta)$.

(3) $\forall\,\Delta_1,\Delta_2\in\mathcal{E}_\mu,\ \Delta_1\cap\Delta_2=\varnothing\ :\ \mathsf{E}Z(\Delta_1)\overline{Z(\Delta_2)}=0$.

Define a stochastic integral for a simple function of the type $f=\sum_{k=1}^n c_k 1\!\!\mathrm{I}_{\Delta_k}$, where $c_k\in\mathbb{C}$, $\Delta_k\in\mathcal{E}_\mu$, as

$$\int_E f(\zeta)Z(d\zeta)=\sum_{k=1}^n c_k Z(\Delta_k). \tag{8.1}$$

Theorem 8.3. *The stochastic integral defined by* (8.1) *can be uniquely extended to a continuous linear operator which acts from $L_2(E,\mathcal{E},\mu)$ to $L_2(\Omega,\mathcal{F},\mathsf{P})$. Moreover, this extension is an isometry; that is, for all $f,g\in L_2(E,\mathcal{E},\mu)$,*

$$\mathsf{E}\int_E f(\zeta)Z(d\zeta)\overline{\int_E g(\zeta)Z(d\zeta)}=\int_E f(\zeta)\overline{g(\zeta)}\mu(d\zeta).$$

If the space $(E,\mathcal{E})$ is a real line with the Borel σ-algebra, then there exists a one-to-one correspondence between a process with orthogonal increments which is right-continuous in the mean square and an orthogonal stochastic measure with locally finite structural measures.

Namely, consider a process $\{X(t),t\in\mathbb{R}\}$, such that $\mathsf{E}|X(t)|^2<\infty$ and:

(1) $\forall\,t\in\mathbb{R}:\ \lim_{s\to t+}\mathsf{E}|X(s)-X(t)|^2=0$.

(2) $\forall\,t_1\le t_2\le t_3\le t_4:\ \mathsf{E}(X(t_2)-X(t_1))\overline{(X(t_4)-X(t_3))}=0$.

It can be checked easily that the set function

$$F_X((a,b]):=\mathsf{E}|X(b)-X(a)|^2,\quad a<b,$$

can be uniquely extended to a locally finite measure on the Borel σ-algebra on $\mathbb{R}$. This measure (and, correspondingly, its distribution function) is said to be the structural measure (structural function) of the stochastic process with orthogonal increments $\{X(t),t\in\mathbb{R}\}$.

Theorem 8.4. *A mapping*

$$Z_X((a,b]):=X(b)-X(a),$$

defined on intervals can be uniquely extended to the orthogonal stochastic measure on $\mathbb{R}$ with structural function F_X.

Conversely, if an orthogonal stochastic measure Z on $\mathbb{R}$ and its structural measure are locally finite, then the stochastic process

$$X(t) := \begin{cases} Z((0,t]), & t \geq 0, \\ -Z((t,0]), & t < 0, \end{cases}$$

is a process with orthogonal increments and X is right-continuous in the mean square.

Because of Theorem 8.4, an orthogonal stochastic measure sometimes is identified with the corresponding orthogonal process, that is a process with orthogonal increments.

The following result is one of the most important in the theory of stationary processes.

Theorem 8.5. *(Spectral representation) Let $\{X(t), t \in \mathbb{R}\}$ be a wide-sense stationary continuous in mean square process, $\mathsf{E}X(t) = 0$. Then there exists an orthogonal stochastic measure Z_X on $\mathbb{R}$ such that*

$$X(t) = \int_{\mathbb{R}} e^{i\zeta t} Z_X(d\zeta), \quad t \in \mathbb{R}.$$

If $\{X_n, n \in \mathbb{Z}\}$ is a wide-sense stationary sequence, with $\mathsf{E}X_n = 0$, then it holds

$$X_n = \int_{-\pi}^{\pi} e^{i\zeta n} Z_X(d\zeta), \quad n \in \mathbb{Z},$$

for some orthogonal stochastic measure Z_X on $(-\pi, \pi]$. Moreover, the spectral measure of the process (or sequence) X coincides with the structural measure of the corresponding orthogonal stochastic measure.

In what follows, only wide-sense stationary stochastic processes and sequences, which satisfy conditions of Theorem 8.5, are considered.

Definition 8.3. *A stationary sequence $\{\varepsilon_n, n \in \mathbb{Z}\}$ is called a white noise if $\mathsf{E}\varepsilon_n = 0$, $n \in \mathbb{Z}$ and $R_\varepsilon(n) = 0$, $n \neq 0$, $R_\varepsilon(0) = 1$.*

Let us introduce one more type of stationarity of random sequences.

Definition 8.4. *A random sequence $\{X_n, n \geq 0\}$ is said to be strictly stationary if for any m the distribution of sequences $\{X_n, n \geq 0\}$ and $\{X_{n+m}, n \geq 0\}$ is the same. The definition of a strictly stationary stochastic process can be given in a similar way.*

Remark 8.1. Unless otherwise specified, by the term "stationary process" we mean a wide-sense stationary process.

The following construction provides one of the most important examples of a strictly stationary sequence.

Example 8.1. Let $(\Omega, \mathfrak{F}, \mathsf{P})$ be a probability space. Assume that T is a measurable transformation of Ω preserving a measure P. That is, an image of the measure P under the transformation T equals P :

$$\forall A \in \mathfrak{F}: \ \mathsf{P}(T^{-1}(A)) = \mathsf{P}(A).$$

Let $\xi = \xi(\omega)$ be a random variable. Then a random sequence $\{\xi_n(\omega) = \xi(T^n(\omega)), n \geq 0\}$ is strictly stationary.

Let's introduce a σ-algebra $\mathfrak{J}$ consisting of sets $A \in \mathfrak{F}$ that are invariant under T; that is, $A = T^{-1}(A)$.

Definition 8.5. *A measure-preserving transformation T is ergodic if every invariant set A has probability either 0 or 1.*

Theorem 8.6. *(Birkhoff–Khintchin theorem) Consider a measure-preserving transformation T. Then for any integrable random variable ξ it holds*

$$\lim_{n\to\infty} \frac{\sum_{k=0}^{n-1} \xi\left(T^k(\omega)\right)}{n} = \mathsf{E}(\xi/\mathfrak{J})$$

for almost all ω and in the mean as well.

If T is ergodic then the corresponding limit almost everywhere is equal to $\mathsf{E}\xi$.

If $\Omega = \mathbb{R}^\infty$ and a stationary random sequence is the coordinate sequence (i.e., $\xi_n(\overline{x}) = x_n$ where $\overline{x} = (x_0, x_1, \ldots) \in \mathbb{R}^\infty$), then ξ_n can be expressed as a composition of an initial random variable ξ_0 and a measure-preserving mapping $T : (x_0, x_1, \ldots) \mapsto (x_1, x_2, \ldots)$ (compare with Example 8.1). In a case of a general strictly stationary random sequence $\{\xi_n, n \geq 0\}$, there is no measure-preserving mapping T such that $\xi_n(\omega) = \xi_0(T^n(\omega))$ for all ω. Therefore, the following construction which is close to Example 8.1 is proposed. Define *shift operator* on the set of functions $\{f(\xi_0, \ldots, \xi_n) |\, f \text{ is bounded and measurable}, n \geq 0\}$ by $Uf(\xi_0, \ldots, \xi_n) = f(\xi_1, \ldots, \xi_{n+1})$. Let η be a random variable measurable with respect to $\sigma(\xi_k, k \geq 0)$, the σ-algebra generated by the random sequence $\{\xi_k, k \geq 0\}$. Then there exists a sequence of random variables $\{\eta_n := f_n(\xi_0, \ldots, \xi_n),\ n \geq 1\}$ converging to η in probability. It can be checked that the sequence of shifts $U\eta_n = f_n(\xi_1, \ldots, \xi_{n+1})$ is also convergent in probability, and its limit depends on η but does not depend on the approximating sequence $\{\eta_n, n \geq 1\}$. Denote this limit by $U\eta$.

Definition 8.6. *Let $\mathfrak{J}_\xi$ be the σ-algebra of sets $A \in \sigma(\xi_k, k \geq 0)$ for which $\mathbb{I}_A = U\mathbb{I}_A$ a.s. A stationary random sequence $\xi = \{\xi_n, n \geq 0\}$ is called ergodic if every event from $\mathfrak{J}_\xi$ has probability either 0 or 1.*

Theorem 8.7. *Let η be a $\sigma(\xi_k, k \geq 0)$-measurable and integrable random variable. Then*

$$\lim_{n\to\infty} \frac{\sum_{k=0}^{n-1} U^k \eta}{n} = \mathsf{E}\left(\eta/\mathfrak{J}_\xi\right) \quad \textit{for almost all } \omega.$$

If $\{\xi_k, k \geq 0\}$ is ergodic then the corresponding limit a.s. equals $\mathsf{E}\eta$. In particular, if $\mathsf{E}|\xi_0| < \infty$ then $\lim_{n\to\infty} n^{-1} \sum_{k=0}^{n-1} \xi_k = \mathsf{E}\xi_0$.

Bibliography

[82] Chapters V,VI; [24], volume 1, Chapter II §8 and Chapter IV §1–7; [79] Chapters VI, VII, XV, XVI; [72]; [69] Chapter III §6,7; [90] Chapter V; [9] Chapters X and XI; [15] Chapters X and XI, [49], Chapters 15 and 16.

Problems

8.1. Let $\{\varepsilon_n, n \ge 0\}$ be i.i.d. random variables with $\mathsf{E}\varepsilon_n = 0, \mathsf{D}\varepsilon_k = 1$. Define the stochastic process

$$X(t) = \sum_n c_n \varepsilon_n e^{i\lambda_n t}, \quad t \in \mathbb{R},$$

where $\{c_n, n \ge 0\} \subset \mathbb{C}$, $\sum_n |c_n|^2 < \infty$, $\{\lambda_n, n \ge 0\} \subset \mathbb{R}$. Prove that $\{X(t), t \in \mathbb{R}\}$ is a wide-sense stationary stochastic process. Find the covariance function R_X.

8.2. Let $\{W(t), t \in \mathbb{R}\}$ be a Wiener process. Prove that $X(t) = W(t+1) - W(t), t \in \mathbb{R}$ is a wide-sense stationary process. Find R_X and F_X.

8.3. Let $\{X(t), t \in \mathbb{R}\}$ be a wide-sense stationary process with the spectral function F_X. Denote

$$Y_1(t) = \sum_{k=1}^{n} c_k X(t + \lambda_k), \quad Y_2(t) = \sum_{j=1}^{m} d_j X(t + \mu_j),$$

where $c_k, d_j \in \mathbb{C}$ and $\lambda_k, \mu_j \in \mathbb{R}$. Prove that the processes $\{Y_1(t), t \in \mathbb{R}\}$ and $\{Y_2(t), t \in \mathbb{R}\}$ are wide-sense stationary. Find their covariance and spectral functions. Find also the joint covariance function $R_{Y_1,Y_2}(s,t) = \operatorname{cov}(Y_1(s), Y_2(t))$.

8.4. Let $\{X_n, n \in \mathbb{Z}\}$ be a wide-sense stationary sequence with zero mean and covariance function $R_X(n) = 2^{-|n|}$. Find $\operatorname{cov}(X(3), X(5))$, $\mathsf{E}|X(3)|^2$, $\operatorname{cov}(2X(1) + 3X(2), 3X(1) - 2iX(3))$, and $\mathsf{E}\left|\sum_{k=0}^{\infty} 3^{-k} X_k\right|^2$.

8.5. Let $\{X(t), t \in \mathbb{R}\}$ be a wide-sense stationary measurable stochastic process and $f : \mathbb{R} \to \mathbb{C}$ be an integrable function.

(1) Prove that the stochastic process $Y(t) := \int_{\mathbb{R}} f(t-s) X(s) ds$ is correctly defined and stationary.

(2) Express F_Y in terms of F_X and an orthogonal random measure Z_Y in terms of Z_X.

8.6. Assume that $\{X_n, n \in \mathbb{Z}\}$ is a wide-sense stationary random sequence with zero mean and the spectral function F_X. Denote $Y_n = \sum_{k \in \mathbb{Z}} c_{n-k} X_k$, where $\{c_n\} \subset \mathbb{C}$ with $\sum_n |c_n| < \infty$.

(1) Prove that the sequence $\{Y_n, n \in \mathbb{Z}\}$ is wide-sense stationary. Express the spectral function F_Y in terms of F_X.

(2) Prove that if $\{X_n, n \in \mathbb{Z}\}$ is a white noise and $\sum_n |c_n|^2 < \infty$ then the series $\sum_k c_{n-k} X_k$ is convergent in the mean square and stationary.

8.7. Assume that the covariance function R_X of a wide-sense stationary random sequence satisfies $\sum_{n \in \mathbb{Z}} |R_X(n)|^2 < \infty$. Prove that the spectral measure has a density

$$p_X(\zeta) = \frac{1}{2\pi} \sum_n R_X(n) e^{-in\zeta}, \quad \zeta \in (-\pi, \pi],$$

where the series converges in $L_2((-\pi, \pi])$.

8.8. Let $X(t) = \alpha(-1)^{N(t)}$, where N is a Poisson process, and the random variable α is independent of N. Which conditions should α satisfy in order for the process X to be: (a) wide sense stationary; (b) strictly stationary?

Find R_X and F_X in case (a).

8.9. Assume that $\{X_n, n \in \mathbb{Z}\}$ is a wide-sense stationary random sequence with zero mean and the covariance function:

(a) $R_X(n) = \begin{cases} 1, & n = 0, \\ 0, & n \neq 0. \end{cases}$

(b) $R_X(n) = \begin{cases} 4, & n = 0, \\ 1, & |n| = 1, \\ 0, & |n| > 1. \end{cases}$

(c) $R_X(n) = a^n$, where $a \in \mathbb{C}, |a| < 1$.

(d) $R_X(n) = a^n$, where $a \in \mathbb{C}, |a| = 1$.

(e) $R_X(n) = \begin{cases} 1, & n \text{ is even}, \\ 0, & n \text{ is odd}. \end{cases}$

(f) $R_X(n) = \begin{cases} 3, & n = 3k, \\ 1, & n \neq 3k. \end{cases}$

(g) $R_X(n) = \frac{1}{1+|n|}$.

(h) $R_X(n) = \begin{cases} 10, & n = 0, \\ \frac{1}{(|n|)!}, & n \neq 0. \end{cases}$

Find the spectral measure F_X. Describe the structure of $\{X_n, n \in \mathbb{Z}\}$ in items (d), (e), and (f).

8.10. Prove that $R(n) = 1/(|n|)!$, $n \in \mathbb{Z}$ cannot be a covariance function of either wide-sense stationary sequence.

8.11. A covariance function of a wide-sense stationary stochastic process $\{X(t), t \in \mathbb{R}\}$ is equal to

(a) $e^{-|t|}$.
(b) $e^{-t^2/2}$.
(c) $\frac{1}{1+t^2}$.
(d) $e^{i\lambda t}$.
(e) $\cos t$.
(f) $\cos^2 2t + 1$.
(g) $e^{\lambda(e^{it}-1)}$.
(h) $R(t) = 1 - |t|$, $|t| \leq 1$ and R has a period $T = 2$.
(i) $\frac{\sin at}{at}$.

(j) $\frac{1-\cos t}{t^2}$.
(k) $\frac{e^{iat}-1}{iat}$.
Find the spectral function F_X.

8.12. Let $\{X(t), t \in \mathbb{R}\}$ be a process with orthogonal increments and its structural measure be Lebesgue measure. Find
(a) $\mathsf{E}\left|\int_0^\pi \sin t dX(t)\right|^2$.
(b) $\mathsf{E}\int_0^1 t dX(t) \overline{\int_0^1 (2+t^2) dX(t)}$.
(c) $\mathsf{E}\int_0^2 (3+t) dX(t) \overline{\int_1^3 t^2 dX(t)}$.

8.13. Prove that $\{X(t), t \geq 0\}$ is a process with orthogonal increments and find its structural measure if:
(a) $X(t) = W(t)$ is a Wiener process.
(b) $X(t) = N(t) - \lambda t$, where N is a Poisson process with intensity $\lambda > 0$.
(c) $X(t) = W^2(t) - t$, where W is a Wiener process.

8.14. Let $\{W(t), t \in \mathbb{R}\}$ be a Wiener process and $h_1, \ldots, h_n \in L_2(\mathbb{R})$. Prove that $\{\int_{\mathbb{R}} h_i(s) dW(s), i = \overline{1,n}\}$ is a Gaussian vector with zero mean and the covariance matrix $\| \int_{\mathbb{R}} h_i(s) h_j(s) ds \|_{i,j=1}^n$.

8.15. (a) Prove that a stochastic process $\{X(t), t \in \mathbb{R}\}$, where $X(t) = \int_{-\infty}^t e^{\alpha(s-t)} dW(s)$, $\alpha > 0$, is stationary.
(b) Find its covariance and spectral measure.
(c) Prove that the increments of the corresponding process with orthogonal increments are Gaussian.

8.16. Find
(a) $\mathsf{E}\left(\int_0^1 t dW(t)\right)^4$.
(b) $\mathsf{E}W(3) \int_0^{\pi/2} \sin s dW(s)$.
(c) $\mathsf{E}\int_0^1 s dW(s) \int_0^1 W(s) ds$.

8.17. Assume that $\{X(t),\ t \in \mathbb{R}\}$ is a process with orthogonal increments and the structural measure F_X. Let $\varphi : \mathbb{R} \to \mathbb{R}$ be a nondecreasing function. Put $Y(t) = X(\varphi(t))$. Prove that $\{Y(t)\}$ is the process with orthogonal increments too. Find its structural measure.

8.18. Let $\{X(t), t \geq 0\}$, $X(0) = 0$ be a Gaussian process with zero mean and orthogonal increments, and $\{W(t), t \geq 0\}$ be a Wiener process.

(1) Find a function $\varphi : [0;\infty) \to [0;\infty)$ such that both stochastic processes $\{X(t), t \ge 0\}$ and $\{W(\varphi(t)), t \ge 0\}$ have the same distributions.

(2) Assume that $\{X(t), t \ge 0\}$ is stochastically continuous and $\lim_{t\to+\infty} \mathsf{E}X^2(t) = \infty$. Prove that there exists a nonrandom function φ and a Wiener process $\widetilde{W}$ on the initial probability space such that $X(t) = \widetilde{W}(\varphi(t)), t \ge 0$.

(3) Prove that, in the general case, there exist a nonrandom function $\psi : [0;\infty) \to [0;\infty)$ and a Wiener process $\widetilde{W}(t)$ on some extension of the probability space, for which $X(t) = \widetilde{W}(\psi(t)),\ t \ge 0$.

(4) Solve the problem for a Gaussian process $X(t)$ defined for all $t \in \mathbb{R}$.

8.19. Let $\{X(t), t \in \mathbb{R}\}$ be a stochastic process with orthogonal increments and the structural measure F_X. Suppose that $f \in L_2(\mathbb{R}, F_X)$. Prove that $Y(t) := \int_{-\infty}^{t} f(s)dX(s)$ has orthogonal increments as well. Find its structural function.

8.20. Let $f \in L_2(\mathbb{R})$. Prove that $X(t) := \int_{-\infty}^{\infty} f(t-s)dW(s)$ is a stationary process. Find its spectral function.

8.21. Let $\{c_j, \delta_j, j = \overline{1,n}\}$ be real numbers. Find a constant $c \in \mathbb{R}$ for which the process

$$X(t) = cW(t) + \sum_{j=1}^{m} c_j W(t+\delta_j), \quad t \in \mathbb{R}$$

is stationary. Find its spectral function.

8.22. Let $\{X_n, n \ge 0\}$ be a white noise. A sequence $\{Y_n, n \ge 0\}$ is defined by the recurrence relation

$$Y_{n+1} = \frac{1}{2}Y_n + X_n, n \ge 0,$$

where the random variable Y_0 doesn't depend on $\{X_n, n \ge 0\}$. What should the expectation and variance of Y_0 be to ensure the wide-sense stationarity of the sequence $\{Y_n, n \ge 0\}$? Find $\mathsf{E}(Y_5\overline{Y}_3 + 2Y_1\overline{Y}_2 + |Y_3|^2)$ in this case.

8.23. Let $\{X_n, n \in \mathbb{Z}\}$ be a white noise.

(1) Prove that the equation $Y_{n+1} = \alpha Y_n + X_n,\ n \in \mathbb{Z}$ has only one stationary solution if $|\alpha| \ne 1$.

(2) Express Y_n as the series $\sum_k c_k X_{n-k}$. Consider the cases when: (a) $|\alpha| < 1$; (b) $|\alpha| > 1$.

8.24. The covariance function of a stationary sequence $\{X_n, n \in \mathbb{Z}\}$ is

$$R_X(n) = \begin{cases} 5, & n = 0, \\ 2, & |n| = 1, \\ 0, & |n| > 1. \end{cases}$$

(1) Represent X_n in the form $\sum_k c_k \varepsilon_{n-k}$, where $\{\varepsilon_n, n \in \mathbb{Z}\}$ is a white noise.

(2) Express ε_n in terms of Z_X and also represent it in the form $\sum_k a_k X_{n-k}$, where the series converge in the mean-square.

8.25. Prove that

$$X(t) = \sin t \int_{t-\pi}^{t+\pi} \cos s dW(s) - \cos t \int_{t-\pi}^{t+\pi} \sin s dW(s)$$

is a stationary stochastic process.

8.26. Solve Problem 8.23 for the equations:
(a) $6Y_{n+2} - 5Y_{n+1} + Y_n = X_n$.
(b) $2Y_{n+1} - 5Y_n + 2Y_{n-1} = X_n$.
(c) $4Y_n + 4Y_{n-1} + Y_n = 3X_n + 2X_{n-1}$.

8.27. Let $\{X_n, n \in \mathbb{Z}\}$ be a white noise. Prove that the equation

$$\sum_{k=0}^{m} a_k Y_{n-k} = \sum_{j=0}^{r} b_j X_{n-j}$$

has unique solution for any collection $b_0, \ldots, b_r$ if and only if the absolute value of every root of the polynomial $P(z) = \sum_{k=0}^{m} a_k z^k$ is not equal to 1. And in this case, the solution can be expressed in a form of moving average $Y_n = \sum_k c_{n-k} X_k$, where $\{c_n, n \in \mathbb{Z}\} \subset \mathbb{C}$ is a summable sequence.

8.28. Prove that for any wide-sense stationary random sequence $\{X_n, n \in \mathbb{Z}\}$ the condition $|P(z)| \neq 1$ as $|z| = 1$ ensures the existence of a stationary solution to the equation from Problem 8.27.

8.29. Prove that the process $X(t) = W(t+1) - \int_t^{t+1} W(s) ds$ is stationary.

8.30. Let $\{X(t), t \in \mathbb{R}\}$ be a wide-sense stationary stochastic process. Does the stationary solution exist for the following equations? If it does, is it unique?
(a) $Y''(t) + Y(t) = X(t)$, $dF_X(\zeta) = \mathbb{I}_{\zeta \in [-5,5]} d\zeta$.
(b) $Y''(t) + 4Y(t) = X(t) - X'(t)$, $dF_X(\zeta) = (\zeta - 2)^2 \, \mathbb{I}_{\zeta \in [0,5]} d\zeta$.

8.31. Let $\{X_n, n \in \mathbb{Z}\}$ be a wide-sense stationary random sequence. Does the stationary solution exist for the following equations? If it does, is it unique?
(a) $Y_{n+2} - 2iY_{n+1} \sin \alpha - Y_n = X_n$, $dF_X(\zeta) = \mathbb{I}_{\zeta \in [-\pi/2, \pi/2]} d\zeta$.
(b) $\sum_{k=1}^{\infty} (Y_{n-k})/k! = X_n$, $dF_X(\zeta) = \zeta^2 \mathbb{I}_{\zeta \in (-\pi,\pi]} d\zeta$.

8.32. Represent stationary solutions of the equations in Problem 8.26 in the form

$$Y_n = \sum_{k=0}^{\infty} c_k \widetilde{X}_{n-k},$$

where $\{\widetilde{X}_n, n \in \mathbb{Z}\}$ is a white noise. Evaluate $\widetilde{X}_n$ in terms of $\{Y_k, k \in \mathbb{Z}\}$.

8.33. Let $\{X(t), t \in \mathbb{R}\}$ be a wide-sense stationary stochastic process. Prove that the equation

$$Y'(t) + \alpha Y(t) = X(t), t \in \mathbb{R},$$

has a stationary solution if $\mathrm{Re}\,\alpha \neq 0$, and solutions to this equation are $\int_{-\infty}^{t} e^{-\alpha(t-s)} dX(s)$, $-\int_t^{\infty} e^{-\alpha(t-s)} dX(s)$ when $\mathrm{Re}\,\alpha > 0$ and $\mathrm{Re}\,\alpha < 0$, respectively.

8.34. Prove that the equation

$$\frac{d^nY(t)}{dt^n}+a_1\frac{d^{n-1}Y(t)}{dt^{n-1}}+\cdots+a_nY(t)=X(t)$$

has a stationary solution for any stationary process $\{X(t), t \in \mathbb{R}\}$ if and only if the polynomial $P(z)=z^n+a_1z^{n-1}+\cdots+a_n$ does not have roots on the imaginary axis. Represent the solution in the form $\int_{-\infty}^{\infty} f(t-s)X(s)ds$.

8.35. Let $\{X(t),\ Y(t), t \in (-\pi,\pi]\}$ be processes with orthogonal increments. Assume that a function $f \in L_2((-\pi,\pi],F_Y)$ is not vanishing at any point and it holds

$$\int_{(-\pi,\pi]} e^{int} f(t)Z_Y(dt) = \int_{(-\pi,\pi]} e^{int} Z_X(dt), \quad n \in \mathbb{Z}.$$

Prove that $1/f \in L_2((-\pi,\pi],F_X)$ and for all $a,b \in (-\pi,\pi]$:

$$Z_Y((a,b]) = \int_{(a,b]} \frac{1}{f(t)} Z_X(dt).$$

8.36. Let $\{X(t),\ Y(t),\ t \in \mathbb{R}\}$ be processes with orthogonal increments. Assume that the function $f \in L_2(\mathbb{R},F_Y)$ is not vanishing at any point and it holds

$$\int_{\mathbb{R}} e^{iut} f(t)Z_Y(dt) = \int_{\mathbb{R}} e^{iut} Z_X(dt), \quad u \in \mathbb{R}.$$

Prove that for all $a,b \in \mathbb{R}$:

$$Z_Y((a,b]) = \int_{(a,b]} \frac{1}{f(t)} Z_X(dt).$$

8.37. Let Z be an orthogonal measure with the structural measure μ. Prove that:
(a) $Z(\emptyset)=0$ a.s.
(b) If $\{\Delta_n,\ n\geq 1\}$ is a sequence of measurable disjoint sets with $\mu(\cup_n\Delta_n)<\infty$, then the series $\sum_n Z(\Delta_n)$ converges to $Z(\cup_n\Delta_n)$ in the mean square.

8.38. Let Z be an orthogonal stochastic measure on $[0,1]$ generated by a Wiener process $\{W(t), t \in [0,1]\}$. Prove that for almost all ω the mapping $Z(\omega,\cdot):\mathcal{B}([0,1])\to\mathbb{R}$ is not a signed measure. (Compare with Problem 8.37.)

8.39. Prove that a wide-sense stationary centered stochastic process $X(t)$ is continuously differentiable in the mean square if and only if $\int_{\mathbb{R}} s^2F_X(ds)<\infty$. And in this case $X'(t)=\int_{\mathbb{R}} i\zeta e^{i\zeta t}Z_X(d\zeta)$.

8.40. (1) Let $\{X_n, Y_n, n \in \mathbb{Z}\}$ be wide-sense stationary random sequences. Prove that $\mathrm{cov}(X_n,Y_m)=0$ for all n,m if, and only if, for any Borel sets Δ_1 and Δ_2 it holds $\mathsf{E}Z_X(\Delta_1)\overline{Z_Y(\Delta_2)}=0$.

(2) Let $\{X(t), Y(t), t \in \mathbb{R}\}$ be continuous in the mean square wide-sense stationary stochastic processes. Prove that $\mathrm{cov}(X(t),Y(s))=0$ for all t,s if, and only if, for any Borel sets Δ_1 and Δ_2 it holds $\mathsf{E}Z_X(\Delta_1)\overline{Z_Y(\Delta_2)}=0$.

8.41. (a) Can a wide-sense stationary sequence not be strictly stationary?
(b) Can a strictly stationary sequence not be wide-sense stationary?

8.42. (1) Let $\{X_n,\ n \geq 0\}$ be a homogeneous Markov chain. Assume that X_0 has a stationary distribution. Prove that $\{X_n,\ n \geq 0\}$ is a strictly stationary random sequence.

(2) Generalize the statement to the case of a continuous-time Markov process.

8.43. Let $\{X_n,\ n \geq 0\}$ be a strictly stationary random sequence. Prove that there exists (possibly, at another probability space) a strictly stationary sequence $\{\tilde{X}_n,\ n \in \mathbb{Z}\}$ such that the distributions of $\{X_n,\ n \geq 0\}$ and $\{\tilde{X}_n,\ n \geq 0\}$ coincide.

8.44. Prove that processes (a) $X(t) = W(t+1) - W(t), t \geq 0$, (b) $X(t) = N(t+1) - N(t), t \geq 0$ are strictly stationary.

8.45. Prove that the notions of the strict and wide-sense stationarity coincide for Gaussian random sequences or processes.

8.46. Let $\Omega = \mathbb{R}^2$ and P be a normal distribution in $\mathbb{R}^2$ with zero mean and identity matrix of covariances. Assume that a transformation $T : \Omega \to \Omega$ acts in polar coordinates as $T((r,\phi)) = (r,2\phi),\ r \geq 0,\ 0 \leq \phi < 2\pi$.
(a) Prove that T preserves the measure P.
(b) Find the limit

$$\lim_{n\to\infty} \frac{1}{n}\left[f(x) + f(T(x)) + \cdots + f(T^{n-1}(x))\right], \quad x \in \mathbb{R}^2$$

for

$$f(x) = x_1^2,\quad f(x) = x_1x_2 \text{ and } f(x) = \arccos\left(\frac{x_1}{\sqrt{x_1^2+x_2^2}}\right), \quad x = (x_1,x_2) \in \mathbb{R}^2.$$

8.47. Let $\{X_n,\ n \geq 0\}$ be a strictly stationary random sequence. Prove that there exists a probability space $(\tilde{\Omega}, \tilde{\mathfrak{F}}, \tilde{\mathsf{P}})$, a measure $\tilde{\mathsf{P}}$ preserving mapping $T : \tilde{\Omega} \to \tilde{\Omega}$, and a random variable $\tilde{\xi}$ (in $\tilde{\Omega}$) such that $\{X_n,\ n \geq 0\}$ and $\{\tilde{\xi}(T^{n-1}),\ n \geq 0\}$ are stochastically equivalent in the general sense.

8.48. Let $\Omega = \mathbb{R}^\infty$, $\mathfrak{F}$ be a σ-algebra generated by the cylindric sets, and T be a function transforming the sequence $(x_1, x_2, \ldots)$ into $(x_2, x_3, \ldots)$. Assume that P is a measure corresponding to a sequence of independent identically distributed random variables. Prove that T is the measure P preserving transformation and T is ergodic.

8.49. Assume that a strictly stationary sequence $\{\xi_n, n \geq 0\}$ is m-dependent; that is, families of random variables $\{\xi_k, k \leq n\}$ and $\{\xi_j, j \geq n+m\}$ are independent for any n. Prove that the sequence $\{\xi_n, n \geq 0\}$ is ergodic.

8.50. Assume that a mapping T is measure-preserving. Prove that T is ergodic if and only if for any random variable ξ with $\mathsf{E}|\xi| < \infty$ it holds $\mathsf{E}(\xi/\mathfrak{J}) = \mathsf{E}\xi$, a.s.

Prove a similar statement about the ergodicity of a strictly stationary random sequence.

8.51. Let $\{X_n,\ n \geq 0\}$ be a homogeneous Markov chain with a finite number of states. Assume that all states of $\{X_n,\ n \geq 0\}$ are connected and have unity period, and X_0 has a stationary distribution. Prove that $\{X_n,\ n \geq 0\}$ is an ergodic sequence.

8.52. Let $\{\xi_n, n \ge 0\}$ be a strictly stationary sequence, and $f : \mathbb{R}^\infty \to \mathbb{R}$ be a measurable function. Prove that the random sequence

$$\eta_n := f(\xi_{n+1}, \xi_{n+2}, \ldots), \quad n \ge 0$$

is strictly stationary as well. Prove that if $\{\xi_n, n \ge 0\}$ is ergodic then the sequence $\{\eta_n, n \ge 0\}$ is ergodic too. Is the opposite statement correct?

8.53. Let $\{X_n, n \in \mathbb{Z}\}$ be a wide-sense stationary random sequence with zero mean. Prove that $n^{-1}\sum_{k=0}^{n-1} X_k \to Z_X(\{0\})$ as $n \to \infty$ in the mean square.

8.54. Assume that a sequence $\{X_n, n \in \mathbb{Z}\}$ is both strictly stationary and wide-sense stationary. Suppose additionally that the spectral measure has a singular component. Prove that $\{X_n, n \in \mathbb{Z}\}$ is not ergodic.

8.55. (1) Assume that T is a measure-preserving transformation. Prove that T is ergodic if and only if for some $p \ge 1$ there exists a total set S in $L_p(\Omega, \mathcal{F}, \mathsf{P})$ (i.e., the completion of the linear hull is the whole space L_p) such that for any random variable $\xi \in S$ the sequence $n^{-1}\sum_{k=0}^{n-1} \xi(T^k)$ converges a.s. to a constant.

(2) Prove a similar statement about the ergodicity of a strictly stationary random sequence.

8.56. Let $\Omega = [0,1)$, P be a Lebesgue measure in Ω, and $T(x) = x + \alpha \pmod 1$, where α is fixed.

(1) Prove that T is a measure-preserving transformation.
(2) Prove that T is ergodic if and only if α is an irrational number.
(3) Describe the invariant sets if α is rational.

8.57. Let $\Omega = [0,1)$, and P be Lebesgue measure in Ω. Assume that the mapping T transforms the number $x = 0.x_1x_2x_3\ldots$ into $T(x) = 0.x_2x_3x_4\ldots$. Prove that T is a measure-preserving transformation. Is T ergodic?

8.58. Let $\{\xi(t), t \in \mathbb{R}\}$ be a continuous, strictly stationary stochastic process, and $a : \mathbb{R}^2 \to \mathbb{R}$ be a continuous and bounded function. Assume that a satisfies the Lipschitz condition with respect to the first coordinate with a constant L, and $\alpha > L$ is some constant. Prove that there exists a unique strictly stationary solution to the differential equation

$$\frac{dX(t)}{dt} = -\alpha X(t) + a(X(t), \xi(t)), \quad t \in \mathbb{R}.$$

8.59. Let $\{\xi_n, n \in \mathbb{Z}\}$ be a strictly stationary random sequence, and $a : \mathbb{R}^2 \to \mathbb{R}$ be a continuous and bounded function. Assume that a satisfies the Lipschitz condition with respect to the first coordinate with constant L, and $\alpha \in (-1,1)$ is a constant with $|\alpha| + L < 1$. Prove that there exists a unique strictly stationary sequence $\{X_n, n \in \mathbb{Z}\}$ satisfying the equation

$$X_{n+1} = \alpha X_n + a(X_n, \xi_n), \quad n \in \mathbb{Z}.$$

8.60. (Poincaré theorem on returns) Let $(\Omega, \mathcal{F}, \mathsf{P})$ be a probability space, T be a measure-preserving mapping, and $A \in \mathcal{F}$. Prove that for almost all points $\omega \in A$ the number of members of the sequence $\{T^n(\omega), n \geq 1\}$ belonging to A, is infinite.

8.61. Let $\{X(t), t \geq 0\}$ be a continuous strictly stationary process.
Assume that $\mathsf{E}\max_{t\in[0,1]} |X(t)| < \infty$.
(a) Prove that $\lim_{t\to+\infty} e^{-\alpha t} X(t) = 0$ a.s. for any $\alpha > 0$.
(b) Is it correct that $\lim_{t\to+\infty} (X(t))/t^3 = 0$ a.s.?

8.62. Give an example of a continuous and strictly stationary process $\{X(t), t \geq 0\}$ such that for any $\alpha > 0$ it holds $\limsup_{t\to+\infty} X(t)e^{-\alpha t} = +\infty$ a.s.

8.63. Let a function $a : \mathbb{R}^m \to \mathbb{R}^m$ satisfy the Lipschitz condition. Assume that there exists a point x_0 such that solution to the equation

$$\frac{dy(t)}{dt} = a(y(t)), \quad t \geq 0, \tag{8.2}$$

with initial condition $y(0) = x_0$ is bounded. Prove that there exists a strictly stationary stochastic process $\{X(t), t \in \mathbb{R}\}$, satisfying (8.2) for all ω.

8.64. Let functions $a, b : \mathbb{R}^m \to \mathbb{R}^m$ satisfy the Lipschitz condition. Assume that there exists a point x_0 such that the solution of the stochastic differential equation

$$dy(t) = a(y(t))dt + b(y(t))dW(t), \quad t \geq 0, \tag{8.3}$$

with initial condition $y(0) = x_0$ is bounded in the mean square. Prove that there exists a strictly stationary stochastic process $\{X(t), t \geq 0\}$ satisfying (8.3).

Hints

8.14. Prove the corresponding statement for simple functions and then pass to a limit.
8.25. Observe that

$$X(t) = \int_{-\infty}^{\infty} \mathbb{I}_{t-s\in[-\pi,\pi]} \sin(t-s)dW(s)$$

and use Problem 8.20.
8.26. See the solution to Problem 8.27.
8.27. Similarly to the solution of Problem 8.24 it can be proved that

$$\sum_{k=0}^{m} a_k e^{-ik\zeta} Z_Y(d\zeta) = \sum_{j=0}^{r} b_j e^{-ij\zeta} Z_X(d\zeta). \tag{8.4}$$

Denote $P(z) = \sum_{k=0}^{m} a_k z^k$, and $Q(z) = \sum_{j=0}^{m} b_j z^j$. The equality for the structural measures follows from (8.4):

$$\left|P\left(e^{-i\zeta}\right)\right|^2 F_Y(d\zeta) = \frac{1}{2\pi}\left|Q\left(e^{-i\zeta}\right)\right|^2 d\zeta.$$

The measure F_Y must be finite, and the assumptions of the problem imply that the solution $\{Y_n\}$ exists for any coefficients $\{b_j\}$. Thus the polynomial $P(z)$ does not vanish when $|z| = 1$. It follows from (8.4) that

$$Z_Y(d\zeta) = \frac{Q\left(e^{-i\zeta}\right)}{P\left(e^{-i\zeta}\right)} Z_X(d\zeta).$$

Let us represent $Q(z)/P(z)$ as a sum:

$$\frac{Q(z)}{P(z)} = \sum_{k=0}^{l} \alpha_k z^k + \sum_{j=1}^{m_1} \sum_{k=1}^{m_2} \frac{\beta_{jk}}{(\gamma_{jk} - z)^j}, \tag{8.5}$$

where $\alpha_k, \beta_{jk}, \gamma_{jk} \in \mathbb{C}$.

Because P doesn't have roots on the unit circumference, then $|\gamma_{jk}| \neq 1$ for all j,k. Let us expand every term of the second sum (8.5) into a uniformly convergent power series about z in the neighborhood of the unit circumference. Use spectral representation to complete the proof.

8.28. See the hint to Problem 8.27.

8.29. Use the result of Problem 13.2 and verify that

$$X(t) = \int_t^{t+1} (t-s)dW(s) = \int_{-\infty}^{\infty} f(t-s)dW(s),$$

where $f(t) = t\mathbb{I}_{t\in[0,1]}$.

8.32. The following can be obtained similarly to the solution of Problem 8.27:

$$Y_n = \int_{-\pi}^{\pi} \alpha e^{i(n+m)\zeta} \frac{\prod_k \left(\beta_k - e^{-i\zeta}\right)}{\prod_j \left(\gamma_k - e^{-i\zeta}\right)} Z_X(d\zeta),$$

where $|\gamma_j| \neq 1$, $m \in \mathbb{Z}$, and $\alpha, \beta_k, \gamma_j \in \mathbb{C}$. Note the following two facts.

(1) For any $\gamma \in \mathbb{C}$ it holds $|\gamma - e^{i\zeta}| = |\overline{\gamma} - e^{-i\zeta}|$.

(2) If $\{X_n\}$ is a white noise then for any measurable function f with $|f(\zeta)| = 1$ as $\zeta \in (-\pi, \pi]$, the sequence

$$\widetilde{X}_n = \int_{-\pi}^{\pi} e^{in\zeta} f(\zeta) Z_X(d\zeta), \quad n \in \mathbb{Z}$$

is a white noise as well.

Let Λ be a set of indices for which $|\gamma_j| < 1$. Check that the function

$$g(z) = \alpha \frac{\prod_k (\beta_k - z)}{\prod_{j\notin\Lambda} (\gamma_j - z) \prod_{j\in\Lambda} \left(\overline{\gamma}_j - z^{-1}\right)}$$

can be expanded into the Taylor series in a neighborhood of the unit circle, and observe that

$$\alpha e^{i(n+m)\zeta} \frac{\prod_k \left(\beta_k - e^{-i\zeta}\right)}{\prod_j \left(\gamma_k - e^{-i\zeta}\right)}$$

can be represented as $g(e^{-i\zeta})f(\zeta)$, where $|f(\zeta)| = 1$ as $\zeta \in (-\pi, \pi]$.

8.34. If the solution exists then $P(e^{i\zeta})Z_Y(d\zeta) = Z_X(d\zeta)$. The stochastic measure $Z_Y(d\zeta) = P(e^{i\zeta})^{-1} Z_X(d\zeta)$ has the finite structural measure for any wide-sense stationary process X if and only if $P(z) \neq 0$ as $|z| = 1$. Rewrite $P(z)^{-1}$ as

$$\sum_{j=1}^{n_1} \sum_{k=1}^{n_2} \frac{\alpha_{kj}}{(\beta_{kj} - z)^j}$$

and use Problem 8.5 (see also the solution to Problem 8.33).
8.35. At first check that it holds

$$\int_{(-\pi,\pi]} g(t)f(t)Z_Y(dt) = \int_{(-\pi,\pi]} g(t)Z_X(dt),$$

for any $g \in C([-\pi,\pi])$. Then prove this formula for every g, such that $fg \in L_2([-\pi;\pi],F_Y)$.
8.43. Use the Kolmogorov theorem on finite-dimensional distributions (Theorem 1.1).
8.45. Finite-dimensional distributions of a Gaussian process are uniquely determined by the mean and covariance functions.
8.46. The corresponding limits of averages are equal to the conditional expectation of initial functions with respect to the σ-algebra generated by the random variable φ. At first, prove this fact for the functions of the form $f(r,\varphi) = \sum_{k=0}^{m} c_k \mathbb{I}_{\varphi\in[\alpha_k,\beta_k]} \mathbb{I}_{r\in[x_k,y_k]}$, and then pass to a limit.
8.47. Take $\widetilde{\Omega} = \mathbb{R}^\infty$, and let the finite-dimensional distributions of $\widetilde{\mathsf{P}}$ coincide with the finite-dimensional distributions of $\{X_n,\ n \geq 0\}$. In this case the shift-transformation $T : (x_1,x_2,\dots) \mapsto (x_2,x_3,\dots)$ satisfies the condition of the problem.
8.48. To prove ergodicity use the result of Problem 8.55 and the law of large numbers.
8.49. See the hint to Problem 8.55.
8.51. Use Theorem 10.3 and the result of Problem 8.55.
8.54. The sequence $\{X_n, n \in \mathbb{Z}\}$ is not regular (see Theorem 9.3), therefore it cannot be ergodic.
8.56. (2) Use Problem 8.55. Take exponential functions $\{e^{inx}, n \in \mathbb{Z}\}$ as a total set.
8.57. To prove the ergodicity of T use Problem 8.48 (the coordinates are independent random variables).
8.58. *Uniqueness.* Let $X(t)$ and $Y(t)$ be solutions to the initial equation, $t > s$. Check that

$$(X(t)-Y(t))^2 e^{\alpha t} \leq (X(s)-Y(s))^2 e^{\alpha s} + \int_s^t L(X(z)-Y(z))^2 e^{\alpha z} dz.$$

The Gronwall–Bellman lemma (see Problem 14.17) implies that $(X(t)-Y(t))^2 \leq (X(s)-Y(s))^2 \exp\{-(\alpha-L)(t-s)\}$. Tending $s \to -\infty$ prove that $X(t) = Y(t)$ a.s.

Existence. Denote the solution to the following equation by $X_s(t), t \geq s$.

$$\begin{cases} \frac{dX_s(t)}{dt} = a(X_s(t),\xi(t)), & t \geq s, \\ X_s(s) = 0. \end{cases}$$

Use the same bounds as in the proof of uniqueness and check that for every ω the processes $X_s(t)$ uniformly converge on any interval $[a,b]$ as $s \to -\infty$, and the limiting process is stationary.
8.59. Check that for any ξ the function $x \mapsto \alpha x + a(x,\xi)$ is a contractive mapping. Furthermore, use reasoning similar to the hint of Problem 8.58.
8.60. Let $N = A \cap \bigcap_{n\geq 1} T^{-n}(\Omega \setminus A)$. Check that for any numbers $m \neq n$ it holds $T^{-n}(N) \cap T^{-m}(N) = \emptyset$ and $\mathsf{P}(N) = \mathsf{P}(T^{-n}(N))$. Conclude that $\mathsf{P}(N) = 0$, that is, almost all points return to the set A at least once.
8.61. Let f be a nonnegative and decreasing function. Then

$$\mathsf{E}\sup_{t\geq 0}|X(t)|f(t) \leq \sum_{n=0}^{\infty}\mathsf{E}\max_{t\in[n,n+1]}|X(t)|f(n) = \mathsf{E}\max_{t\in[0,1]}|X(t)|\sum_{n=0}^{\infty}f(n).$$

8.62. Denote $X(t) = f(W(t+1) - W(t))$, where f is a continuous function with $\mathsf{P}(f(W(1)) > e^{n^2}) > 1/n$, $n > 2$. To prove the statement use the Borel–Cantelli lemma.

8.63. Denote by $y(t,x)$ the solution to (8.2) with initial condition $y(0,x) = x$. Let μ_T be the image of Lebesgue measure on [0;T] under the mapping $y(\cdot,x_0)$. Because the function $y(t,x_0), t \geq 0$ is bounded, the family of measures $(1/T)\mu_T$ is relatively compact. Let ν be a limit point of this family, and ξ be a random variable with distribution ν. Check that $X(t) = y(t,\xi)$ is the desired process.

8.64. See the hint to Problem 8.63.

Answers and Solutions

8.1. $R_X(t) = \sum_n |c_n|^2 e^{i\lambda_n t}$.

8.2. $R_X(t) = \begin{cases} 1-|t|, & |t|<1, \\ 0, & |t|\geq 1. \end{cases}$

$dF_X(\zeta) = p(\zeta)d\zeta$, where $p(\zeta) = (1/2\pi)\int_{-1}^{1}(1-|t|)e^{-i\zeta t}dt = (1-\cos\zeta)/\pi\zeta^2$.

8.4. 0.25; 1; $6R_X(0) + 4iR_X(-2) + 9R_X(1) + 6iR_X(-1) = 10,5 + 4i$; 63/40.

8.5. Justify the transformations:

$$\int_{\mathbb{R}} f(t-s)X(s)ds = \int_{\mathbb{R}} f(s)X(t-s)ds = \int_{\mathbb{R}} f(s)\left(\int_{\mathbb{R}} e^{i\zeta(t-s)}Z_X(d\zeta)\right)ds$$
$$= \int_{\mathbb{R}} e^{i\zeta t}\left(\int_{\mathbb{R}} f(s)e^{-i\zeta s}ds\right)Z_X(d\zeta) = \int_{\mathbb{R}} e^{i\zeta t}\check{f}(\zeta)Z_X(d\zeta),$$

where $\check{f}$ is inverse Fourier transform of f. Therefore,

$$F_Y(d\zeta) = \left|\check{f}(\zeta)\right|^2 Z_X(d\zeta).$$

8.6. $F_Y(d\zeta) = \left|\sum_k c_k e^{-ik\zeta}\right|^2 F_X(d\zeta)$.

8.7. Verify that $\int_{-\pi}^{\pi} e^{in\zeta}p_X(\zeta)d\zeta = R_X(n)$. Because p_X is a function from $L_2((-\pi,\pi])$, then $p_X(\zeta)d\zeta$ is a finite signed-measure on $(-\pi,\pi]$. It follows from

$$\int_{-\pi}^{\pi} e^{in\zeta}p_X(\zeta)d\zeta = \int_{-\pi}^{\pi} e^{in\zeta}F_X(d\zeta), \quad n\in\mathbb{Z},$$

that $p_X(\zeta)d\zeta = F_X(d\zeta)$, thus, p_X is a spectral density.

8.8. (a) $\mathsf{E}\alpha(-1)^{N(t)} = \mathsf{E}\alpha\mathsf{E}(-1)^{N(t)} = e^{-2\lambda t}\mathsf{E}\alpha$. Thus, $\mathsf{E}\alpha = 0$.

Let $t \geq s$ and $\sigma^2 = \mathsf{E}\alpha^2 < \infty$. Then

$$\mathsf{E}X(t)X(s) = \sigma^2\mathsf{E}(-1)^{N(t-s)} = \sigma^2 e^{-2\lambda(t-s)}.$$

So, the process X is stationary and $R_X(t) = \sigma^2 e^{-\lambda|t|}$. The spectral density is

$$p_X(\zeta) = \frac{\sigma^2}{2\pi}\int_{-\infty}^{\infty} e^{-2\lambda|t|}e^{-it\zeta}dt = \frac{2\sigma^2\lambda}{\pi(4\lambda^2+\zeta^2)}.$$

(b) α must have a symmetric distribution.

8.9. (a) $\frac{1}{2\pi}d\zeta$.

(b) $\frac{2+\cos\zeta}{\pi}d\zeta$.

(c) $\frac{1}{2\pi}\frac{1-|a|^2}{|1-ae^{-i\zeta}|^2}d\zeta$.

(d) The unit mass concentrated at the point φ with $e^{i\varphi}=a$. In this case $X_n=a^nX_0$.

(e) Check that $X_{n+2}=X_n$. Thus, for any $n\in\mathbb{Z}$:

$$\int_{-\pi}^{\pi} e^{i\zeta(n+2)}Z_X(d\zeta)=\int_{-\pi}^{\pi} e^{i\zeta n}Z_X(d\zeta).$$

Therefore, $(e^{2i\zeta}-1)Z_X(d\zeta)=0$. So, the measure $F_X(d\zeta)$ is concentrated at points ζ with $e^{2i\zeta}=1$; that is, $F_X(d\zeta)=c_1\delta_{\{0\}}+c_2\delta_{\{\pi\}}$.

Substitute $n=0$ and $n=1$ into the equality $R_X(n)=\int_{-\pi}^{\pi}e^{i\zeta n}F_X(d\zeta)$ and obtain the system of linear equations

$$\begin{cases} c_1+c_2=1, \\ c_1-c_2=0. \end{cases}$$

So, $c_1=c_2=1/2$.

(f) Check that $X_{n+3}=X_n$ and use reasoning of the previous item.

(g) $p_X(\zeta)=\frac{1}{2\pi}\sum\limits_n\frac{1}{1+|n|}e^{-i\zeta n}=\frac{1}{2\pi}\left(-e^{-i\zeta}\ln(1-e^{i\zeta})-e^{i\zeta}\ln(1-e^{-i\zeta})-1\right)$;

(h) $p_X(\zeta)=\frac{1}{2\pi}\left(e^{e^{i\zeta}}+e^{e^{-i\zeta}}+9\right)$.

8.10. Assume the contrary. In this case the spectral density is

$$p_X(\zeta)=\frac{1}{2\pi}\left(e^{e^{i\zeta}}+e^{e^{-i\zeta}}-1\right).$$

This function is negative in some neighborhood of the point π. We get the contradiction.

8.11. (a) $\frac{1}{\pi}\frac{1}{1+\zeta^2}d\zeta$.

(b) $\frac{1}{\sqrt{2\pi}}e^{-\frac{\zeta^2}{2}}d\zeta$.

(c) $\frac{e^{-|\zeta|}}{2}d\zeta$ (see item (a)).

(d) $\delta_{\{\lambda\}}$.

(e) $\cos t=\frac{e^{it}+e^{-it}}{2}$, thus, $F_X(d\zeta)=\frac{\delta_{\{1\}}+\delta_{\{-1\}}}{2}$.

(f) $\cos^2 2t+1=\frac{\cos 4t+3}{2}=\frac{e^{4it}+e^{-4it}+3}{2}$, thus, $F_X(d\zeta)=\frac{\delta_{\{4\}}+\delta_{\{-4\}}+3\delta_{\{0\}}}{2}$.

(g) $e^{\lambda(e^{it}-1)}=e^{-\lambda}\sum_{n=0}^{\infty}\frac{e^{int}\lambda^n}{n!}$.

So, F_X has a Poisson distribution with parameter λ.

(h) Express $R_X(t)$ as the series $R_X(t)=\sum\limits_n c_ne^{\pi int}$ with

$$c_n=\frac{1}{\sqrt{2}}\int_{-1}^{1}R_X(t)e^{-\pi int}dt=\begin{cases} 1/\sqrt{2}, & n=0, \\ 0, & n \text{ even}, n\neq 0, \\ \frac{2\sqrt{2}}{\pi|n|}, & n \text{ odd}. \end{cases}$$

So,

$$F_X(d\zeta) = \sum_n c_n \delta_{(\pi n)}.$$

(i) Uniform distribution on $[-a,a]$.
(j) $\frac{1-|\zeta|}{2} \mathbb{I}_{\zeta\in[-1,1]} d\zeta$.
(k) Uniform distribution on $[0,a]$.
8.12. (a) $\int_0^\pi \sin^2 t dt = \pi/2$.
(b) $\int_0^1 t(2+t^2)dt = 1.25$.
(c) $\mathsf{E}\int_0^2 (3+t)dX(t)\overline{\int_1^3 t^2 dX(t)} =$

$$\mathsf{E}\int_0^3 (3+t)\mathbb{I}_{t\in[0,2]} dX(t)\overline{\int_0^3 t^2 \mathbb{I}_{t\in[1,3]} dX(t)} = \int_1^2 (3+t)t^2 dt = 11.75.$$

8.13. (a) $F_X(dt) = dt, t \geq 0$.
(b) $F_X(dt) = \lambda dt, t \geq 0$.
(c) $F_X(dt) = 4t dt,\ t \geq 0$.
8.15. (a) $R_X(t) = (1/2\alpha)e^{-\alpha|t|}$, $F_X(d\zeta) = \alpha/(\pi(\alpha^2+\zeta^2))d\zeta$.
(b) Use the solution to Problem 8.5 and check that for any $a,b \in \mathbb{R}$ a random variable $Z_X(b) - Z_X(a) = \int_{-\infty}^{\infty} \mathbb{I}_{\zeta\in[a,b]} Z_X(d\zeta)$ can be expressed as a limit of a Gaussian random variables of the form

$$\lim_{n\to\infty} \int_{-\infty}^{\infty} f_n(t-s)X(s)ds$$

where $\{f_n, n \geq 1\}$ is a sequence of integrable functions.
8.16. (a) It follows from Problem 8.14 that $\int_0^1 t dW(t) \sim N(0,1/3)$. So,

$$\mathsf{E}\left(\int_0^1 t dW(t)\right)^4 = 3\sigma^4 = 1/3.$$

(b) Because $W(3) = \int_0^3 1 dW(s)$, it holds

$$\mathsf{E}\int_0^3 1 dW(s)\int_0^{\pi/2} \sin s dW(s) = \int_0^{\pi/2} 1 \sin s ds = 1.$$

(c) $$\mathsf{E}\int_0^1 s dW(s)\int_0^1 W(s)ds = \int_0^1 \left(\mathsf{E}\int_0^1 z dW(z)\int_0^1 \mathbb{I}_{z\in[0,s]} dW(z)\right) ds$$
$$= \int_0^1 \int_0^s z dz\, ds = \frac{1}{6}.$$

8.17. $F_Y(t) = F_X(\varphi(t))$.
8.18. (a) Let $\sigma^2(t) = EX^2(t)$. Verify that the function σ^2 is monotonically decreasing. Put $\varphi(t) = \sigma^2(t)$.
(b) $\widetilde{W}(t) = X(\varphi^{(-1)}(t))$ where $\varphi^{(-1)}(t) = \inf\{s: \ \varphi(s) = t\}$.
(c) Define a monotone function $\psi(t) = \inf\{s: \ \varphi(t) \geq s\}$. There exists an enumerable set of disjoint intervals (a_n,b_n) on which the function ψ is constant. Let $\{Y_n(t),\ t\in[a_n,b_n]\}$ be a sequence of independent Brownian bridges on $[a_n,b_n]$, and also independent on the process X. Then

$$\widetilde{W}(t) := \begin{cases} X(\psi(t)), & \forall n: \ t \notin (a_n,b_n), \\ X(a_n) + Y_n(t) + \frac{t-a_n}{b_n-a_n}(X(b_n)-X(a_n)), & \exists n: t\in(a_n,b_n) \end{cases}$$

is the desired Winer process.
8.19. $F_Y(dt) = |f(t)|^2 F_X(dt)$.
8.20. $R_X(t) = \int_{\mathbb{R}} f(t-s)\overline{f(-s)}ds$, $F_X(d\zeta) = |\widehat{f}(\zeta)|^2 d\zeta$.
8.21. Let $0 < \delta_1 < \cdots < \delta_n$. The variance of the process X is constant only for $c = -(c_1 + \cdots + c_n)$. In order to check that this implies the stationarity, observe that

$$X(t) = \sum_{k=1}^{n} (c_k + \cdots + c_n)\left(W(t+\delta_k) - W(t+\delta_{k-1})\right) = \int_{\mathbb{R}} f(t-s)dW(s),$$

where

$$\delta_0 = 0, \quad f(t) = \sum_{k=1}^{n} (c_k + \cdots + c_n) 1\!\!\mathrm{I}_{t \in [\delta_{k-1}, \delta_k]}.$$

To complete the proof, use the result of Problem 8.20.
8.22. If $\{Y_n, n \geq 0\}$ is stationary and $a = EY_n, \sigma^2 = DY_n$, then

$$a = EY_{n+1} = \frac{1}{2}EY_n + EX_n = \frac{1}{2}EY_n = \frac{1}{2}a.$$

Thus, $a = 0$. Moreover

$$\mathsf{D}Y_1 = \frac{1}{2}\mathsf{D}Y_0 + \mathsf{D}X_0 = \frac{1}{2}\sigma^2 + 1.$$

Therefore, $\sigma^2 = 2$. In this case $R_Y(n) = 2^{-|n|+1}, n \in \mathbb{Z}$, and

$$\mathsf{E}\left(Y_5\overline{Y}_3 + 2Y_1\overline{Y}_2 + |Y_3|^2\right) = R_Y(2) + 2R_Y(-1) + R_Y(0) = 5.$$

8.23. *Solution 1.* Let $|\alpha| < 1$.

$$\begin{aligned} Y_{n+1} &= \alpha Y_n + X_n = \alpha^2 Y_{n-1} + \alpha X_{n-1} + X_n = \cdots \\ &= \alpha^{m+1} Y_{n-m} + \sum_{k=0}^{m} \alpha^k X_{n-k}, \quad m \in \mathbb{N}. \end{aligned} \tag{8.6}$$

Because $\mathsf{E}|Y_{n-m}|^2 = \mathsf{E}|Y_0|^2 = \text{const}$, the first term in (8.6) tends to zero in the mean square, as $m \to \infty$. This implies the equality

$$Y_n = \sum_{k=0}^{\infty} \alpha^k X_{n-k-1}.$$

Solution 2. It follows from the spectral representation that for any $n \in \mathbb{Z}$,

$$\int_{-\pi}^{\pi} e^{i(n+1)\zeta} Z_Y(d\zeta) = \int_{-\pi}^{\pi} \alpha e^{in\zeta} Z_Y(d\zeta) + \int_{-\pi}^{\pi} e^{in\zeta} Z_X(d\zeta),$$

and thus,

$$\int_{-\pi}^{\pi} e^{in\zeta}\left(e^{i\zeta} - \alpha\right) Z_Y(d\zeta) = \int_{-\pi}^{\pi} e^{in\zeta} Z_X(d\zeta), \quad n \in \mathbb{Z}.$$

Problem 8.35 implies that $Z_Y(d\zeta) = (e^{i\zeta} - \alpha)^{-1} Z_X(d\zeta)$. Therefore,

$$Y_n = \int_{-\pi}^{\pi} \frac{e^{in\zeta}}{e^{i\zeta} - \alpha} Z_X(d\zeta).$$

If $|\alpha| < 1$, then

$$\frac{e^{in\zeta}}{e^{i\zeta}-\alpha} = e^{i(n-1)\zeta} \sum_{m=0}^{\infty} \alpha^m e^{-im\zeta}.$$

Because the series converges in $L_2((-\pi,\pi], F_X) = L_2((-\pi,\pi], (2\pi)^{-1}d\zeta)$, then

$$Y_n = \sum_{k=0}^{\infty} \alpha^k X_{n-k-1}.$$

If $|\alpha| > 1$, then

$$\frac{e^{in\zeta}}{e^{i\zeta}-\alpha} = \sum_{k=0}^{\infty} \alpha^{-k-1} e^{i\zeta(n+k)}$$

and

$$Y_n = \sum_{k=0}^{\infty} \alpha^{-k-1} X_{n+k}.$$

8.24. Observe that (see Problem 8.7):

$$F_X(d\zeta) = \left|2+e^{i\zeta}\right|^2 \frac{d\zeta}{2\pi}, \quad \zeta \in (-\pi,\pi].$$

So, (Problem 8.19) the orthogonal random measure $Z_\varepsilon(d\zeta) := (2+e^{i\zeta})^{-1}Z_X(d\zeta)$ has the structural density $1/(2\pi)$. It means that

$$\varepsilon_n := \int_{-\pi}^{\pi} e^{in\zeta}\left(2+e^{i\zeta}\right)^{-1} Z_X(d\zeta) = \sum_{k=0}^{\infty} 2^{-k-1} X_{n+k}$$

is a white noise. Therefore,

$$X_n = \int_{-\pi}^{\pi} \left(2+e^{i\zeta}\right) e^{in\zeta} Z_\varepsilon(d\zeta) = 2\varepsilon_n + \varepsilon_{n+1}.$$

8.30. (a) If a solution exists then (see Problem 8.39) the following equalities hold:

$$(1-\zeta^2)\, Z_Y(d\zeta) = Z_X(d\zeta),$$

$$\left|1-\zeta^2\right|^2 F_Y(d\zeta) = F_X(d\zeta) = \mathbb{I}_{\zeta\in[-5,5]} d\zeta. \tag{8.7}$$

Neither finite measure F_Y satisfies the relation (8.7). Therefore, there is no stationary solution.

(b) $(4-\zeta^2)Z_Y(d\zeta) = (1-i\zeta)Z_X(d\zeta)$.

It is easy to check that the orthogonal random measure $(1-i\zeta)(4-\zeta^2)^{-1}Z_X(d\zeta)$ is correctly defined and the process

$$\widetilde{Y}(t) = \int_{-\infty}^{\infty} e^{it\zeta} \frac{1-i\zeta}{4-\zeta^2} Z_X(d\zeta)$$

is a stationary solution to the initial equation. This solution is not unique. A general stationary solution is of the form

$$Y(t) = \alpha_1 e^{2it} + \alpha_2 e^{-2it} + \widetilde{Y}(t)$$

where the random variables α_1 and α_2 satisfy $\mathsf{E}|\alpha_i|^2 < \infty$, $\mathsf{E}\alpha_i = 0$, and α_1 and α_2 are orthogonal in L_2 and orthogonal to $\{X(t), t \in \mathbb{R}\}$.

8.31. (a) It follows from the spectral representation that:

$$\left(e^{2i\zeta} - 2ie^{i\zeta}\sin\alpha - 1\right) Z_Y(d\zeta) = Z_X(d\zeta),$$

or

$$\left(e^{i\zeta}-e^{i\alpha}\right)\left(e^{i\zeta}+e^{i\alpha}\right)Z_Y(d\zeta)=Z_X(d\zeta).$$

So,

$$\left|e^{i\zeta}-e^{i\alpha}\right|^2\left|e^{i\zeta}-e^{-i\alpha}\right|^2 F_Y(d\zeta)=\mathbb{I}_{[-(\pi/2),(\pi/2)]}(\zeta)\,d\zeta.$$

There is no finite measure F_Y satisfying the last equality.

Answer: there is no stationary solution.

(b) $\sum_{k=1}^{\infty}(e^{-i\zeta}/k!)=(e^{e^{-i\zeta}}-1)$.

Because $e^{e^{-i\zeta}}-1\neq 0$ as $\zeta\in(-\pi,\pi]$, there exists a unique stationary solution to the equation

$$Y_n=\int_{-\pi}^{\pi}e^{in\zeta}\left(e^{e^{-i\zeta}}-1\right)^{-1}Z_X(d\zeta).$$

8.33. Consider only the case $\mathrm{Re}\,\alpha>0$. $\int_{-\infty}^{t}e^{-\alpha(t-s)}X(s)ds=\int_{-\infty}^{\infty}f(t-s)X(s)ds$, where $f(s)=e^{-\alpha s}\mathbb{I}_{s\in[0,\infty)}$. It follows from the solution to Problem 8.5 that

$$Y(t)=\int_{-\infty}^{\infty}e^{i\zeta t}\check{f}(\zeta)Z_X(d\zeta),$$

where $\check{f}(\zeta)=\int_{-\infty}^{\infty}e^{-i\zeta t}f(t)dt=1/(\alpha+i\zeta)$. To complete the proof, use the result of Problem 8.39

8.38. Use the fact that trajectories of a Wiener process have infinite variation a.s. on any interval (Problem 3.19).

8.41. (a) Yes.

(b) Yes, if the sequence doesn't have finite second moment.

8.55. The transformation $U:L_p(\Omega,\mathfrak{F},\mathsf{P})\ni\xi\mapsto\xi(T)\in L_p(\Omega,\mathfrak{F},\mathsf{P})$ is an isometry, therefore, the norm in $L_p(\Omega,\mathfrak{F},\mathsf{P})$ of the operator $P_n:\xi\to(1/n)\sum_{k=0}^{n-1}\xi(T^n)$ is bounded by unity. The Birkhoff–Khinchin theorem implies that the sequence P_n strongly converges to the conditional expectation. So, to prove that the limit is a constant a.s., it is sufficient to check this on a total set.

9

Prediction and interpolation

Theoretical grounds

Let $\{X_n, n \in \mathbb{Z}\}$ be a mean zero wide-sense stationary random sequence, $\Lambda \subset \mathbb{Z}$ be a set of indices, and $H_\Lambda := \mathcal{L}^{cl}(X_n, n \in \Lambda)$ be a closure in $L_2(\Omega, \mathfrak{F}, \mathsf{P})$ of a linear hull of the system of random variables $\{X_n, n \in \Lambda\}$.

In this chapter we study methods to find the projection of some element X_k on the space H_Λ in $L_2(\Omega, \mathfrak{F}, \mathsf{P})$, that is, to find the best approximation of X_k by a linear combination of variables $\{X_n, n \in \Lambda\}$. Denote this projection by $\pi_{H_\Lambda}(X_k)$. The most important cases for technical applications are $\Lambda = \{-n, n \ge 0\}$ and $\Lambda = \mathbb{Z} \setminus \{n_1, \dots, n_m\}$. The first case is called the prediction problem, when we need to approximate in the best way the "future" X_n via "past", that is, via observations $X_0, X_{-1}, X_{-2}, \dots$. The second case is called the interpolation problem and the task is to reconstruct in the best way some "lost" elements of the sequence $\{X_n, n \in \mathbb{Z}\}$.

Consider the prediction problem first. Introduce the spaces:

$$H_k(X) = \mathcal{L}^{cl}(X_n, n \le k); \qquad H(X) = \mathcal{L}^{cl}(X_n, n \in \mathbb{Z});$$
$$S(X) = \cap_{k \in \mathbb{Z}} H_k(X); \qquad R(X) = H(X) \ominus S(X).$$

Let $\pi_k = \pi_{H_k(X)}$ be a projection in $L_2(\mathsf{P})$ on $H_k(X)$, and $\pi_{-\infty} = \pi_{S(X)}$.

Definition 9.1. *A sequence $\{X_n, n \in \mathbb{Z}\}$ is called regular if $H(X) = R(X)$ and singular if $H(X) = S(X)$.*

Write X_n as a sum $X_n^r + X_n^s$, where $X_n^s = \pi_{-\infty}(X_n)$ and $X_n^r = X_n - X_n^s$. It can be proved that the sequences $\{X_n^s\}$ and $\{X_n^r\}$ are mutually orthogonal and wide-sense stationary. Furthermore, the sequence $\{X_n^r\}$ is regular and $\{X_n^s\}$ is singular. Because $\pi_0(X_n^s) = X_n^s$, it makes sense to consider the prediction problem for the regular sequences only. The following result is a key tool for solving the prediction problem.

Theorem 9.1. *(Wald decomposition) Let $X = \{X_n, n \in \mathbb{Z}\}$ be a nondegenerate wide-sense stationary sequence. It is regular if and only if there exist a white noise $\varepsilon = \{\varepsilon_n, n \in \mathbb{Z}\}$ and a sequence $\{a_n, n \in \mathbb{Z}^+\} \subset \mathbb{C}$ with $\sum_{n \in \mathbb{Z}^+} |a_n|^2 < \infty$, such that:*

(a) $X_n = \sum_{k=0}^\infty a_k \varepsilon_{n-k}$ and the series converges in the mean square.

(b) $H_n(X) = H_n(\varepsilon)$, $n \in \mathbb{Z}$.

DOI 10.1007/978-0-387-87862-1_9,

Theorem 9.2. *Let* $X_n = \sum_{k=0}^{\infty} a_k \varepsilon_{n-k}$ *be the Wald decomposition of a regular wide-sense stationary sequence* $\{X_n, n \in \mathbb{Z}\}$. *The solution to the prediction problem is given by the formula:*

$$\pi_0(X_n) = \sum_{k=n}^{\infty} a_k \varepsilon_{n-k}.$$

The following statement provides a criterion for the regularity in the terms of spectral function.

Theorem 9.3. *(Kolmogorov theorem) A nondegenerate wide-sense stationary random sequence is regular if, and only if, it has the spectral density p with*

$$\int_{-\pi}^{\pi} \log p(\zeta) d\zeta > -\infty.$$

In particular cases it is possible to express $\pi_0(X_n)$ in terms of the spectral representation.

Theorem 9.4. *Assume that the spectral density of a wide-sense stationary sequence* $\{X_n, n \in \mathbb{Z}\}$ *is of the form* $|\Phi(e^{-i\zeta})|^2$, *where the function* $\Phi(z) = \sum_{k=0}^{\infty} a_k z^k$ *is analytic in the circle* $\{|z| \le 1\}$ *and* $\Phi(z) \ne 0$ *as* $|z| \le 1$. *Denote* $\Phi_n(z) = \sum_{k=n}^{\infty} a_k z^k$. *Then*

$$\pi_0(X_n) = \int_{-\pi}^{\pi} e^{in\zeta} \frac{\Phi_n(e^{-i\zeta})}{\Phi(e^{-i\zeta})} Z_X(d\zeta). \tag{9.1}$$

In the case where Λ consists of a single point, the solution to the interpolation problem is provided by the following theorem.

Theorem 9.5. *Assume that the spectral density p of a regular sequence* $\{X_n, n \in \mathbb{Z}\}$ *satisfies the condition* $\int_{-\pi}^{\pi} (1/p(\lambda)) d\lambda < +\infty$. *Then the projection* $\pi_{\mathcal{L}^{cl}(X_n, n \ne 0)}(X_0)$ *can be found by the formula* $\int_{-\pi}^{\pi} \varphi(\zeta) Z_X(d\zeta)$ *where*

$$\varphi(\zeta) = 1 - \frac{2\pi}{p(\zeta) \int_{-\pi}^{\pi} \frac{d\lambda}{p(\lambda)}}.$$

Bibliography

[82] Chapter VI; [24], Volume 1, Chapter V §8-9; [79] Chapters XVII and XVIII; [72]; [69] Chapter V; [90] Chapter V; [15] Chapter XII.

Problems

9.1. Consider a random vector (X_1,X_2,X_3) with zero mean and the covariance matrix

$$\|\operatorname{cov}(X_i,X_j)\| = \begin{Vmatrix} 1 & 1 & -1 \\ 1 & 4 & -1 \\ -1 & -1 & 10 \end{Vmatrix}.$$

(1) Find the best in mean square linear approximation $\widehat{X}_3$ of a random variable X_3 via X_1 and X_2.

(2) Find $\mathsf{E}(X_3-\widehat{X}_3)^2$.

(3) Find the best in mean square linear approximation of a random variable X_3-2X_1 via X_1+X_2.

9.2. Prove Theorem 9.4.

9.3. Let $\{X_n, n\in\mathbb{Z}\}$ be a white noise, and $\widehat{Y}_n := \pi_0(Y_n), n\ge 1$. Express $\widehat{Y}_n$ via $\{Y_k, k\le 0\}$, if $\{Y_k, k\in\mathbb{Z}\}$ is a stationary solution to the following equations.
(a) $Y_{n+1} = \frac{1}{2}Y_n + X_n, n\in\mathbb{Z}$.
(b) $Y_{n+1} = 2Y_n + X_n$.
(c) $Y_{n+1} = 3X_{n+1} + X_n$.
(d) $2Y_{n+2} - 3Y_{n+1} - 2Y_n = X_n$.
(e) $Y_{n+3} - 3Y_{n+2} - 4Y_{n+1} + 12Y_n = X_n$.
(f) $4Y_{n+2} + 4Y_{n+1} + Y_n = X_n + X_{n+1}$.
(g) $Y_n = 2X_{n+2} - 3X_{n+1} - 2X_n$.
(h) $Y_{n+2} + 2Y_{n+1} + 10Y_n = X_n$.
(i) $Y_n = 4X_n + 4X_{n-1} + X_{n-2}$.
(j) $Y_n = 4X_{n+1} + 4X_n + X_{n-1}$.
(k) $Y_{n+2} + 4Y_{n+1} + Y_n = X_n$.
(l) $Y_n = 10X_n + 3X_{n-1} - X_{n-2}$.
(m) $Y_n = 3X_n + 11X_{n-1} - 4X_{n-2}$.
(n) $Y_n = 2X_{n+1} + 13X_n - 7X_{n-1}$.
(o) $4Y_{n+2} - 7Y_{n+1} - 2Y_n = -2X_n + X_{n-1}$.
(p) $Y_{n+1} + Y_n = X_{n+1} + X_n$.
(q) $Y_{n+1} + 2Y_n = X_{n+1} + X_n$.
(r) $3Y_{n+1} + 2Y_n = X_n$.
Find $\mathsf{E}(Y_n - \widehat{Y}_n)^2$ for items (a) – (c).

9.4. Express $\widehat{X}_n := \pi_0(X_n), n\ge 1$ via $\{X_k, k\le 0\}$ if:
(a) $R_X(n) = a^{|n|},\ |a|<1$.
(b) $R_X(n) = \begin{cases} 5, & n=0, \\ -2, & |n|=1, \\ 0, & |n|>1. \end{cases}$
(c) $R_X(n) = 1, n\in\mathbb{Z}$.
(d) $R_X(n) = e^{i\lambda n}, \lambda\in(-\pi,\pi], n\in\mathbb{Z}$.
(e) $R_X(n) = 2 + e^{in\lambda}, \lambda\in(-\pi,\pi], n\in\mathbb{Z}$.

Is the sequence $\{X_n, n\in\mathbb{Z}\}$ regular? Singular?

9.5. Assume that $\{X_n, n \in \mathbb{Z}\}$ is a stationary sequence with $F_X(d\zeta) = \delta_{\{1\}} + \mathbb{I}_{\zeta\in[0,\pi]}d\zeta$, where $\delta_{\{1\}}$ is a unity mass concentrated at point 1. Find the regular and singular components of the sequence $\{X_n, n \in \mathbb{Z}\}$.

9.6. Assume that $F_X(d\zeta) = \delta_{\{0\}} + (1/2\pi)|2 - e^{-i\zeta}|^2 d\zeta$.

(1) Prove that

$$\frac{X_0 + \cdots + X_{-n+1}}{n} \xrightarrow{L_2(\mathsf{P})} Z_X(\{0\}) \text{ as } n \to \infty.$$

(2) Find the covariance function for the sequence $X_n - Z_X(\{0\})$, $n \in \mathbb{Z}$.

(3) Describe the decomposition $X_n = X_n^s + X_n^r$.

9.7. Assume that $F_X(d\zeta) = \delta_{\{1\}} + (1/2\pi)|3 - e^{i\zeta}|^2 d\zeta$.

(1) Find a sequence of numbers $\{c_{n,k}, k = \overline{0,n}, n \in \mathbb{N}\}$, such that

$$\sum_{k=0}^{n} c_{n,k} X_{-k} \xrightarrow{L_2} Z_X(\{1\}) \text{ as } n \to \infty.$$

(2) Specify the decomposition $X_n = X_n^s + X_n^r$.

9.8. Let $\{X_n, n \in \mathbb{Z}\}$ be a wide-sense stationary random sequence with spectral function F_X. Denote by L_2^I the closure of a linear hull of $\{e^{in\zeta}, n \in I\}$ in $L_2((-\pi,\pi], F_X)$ and H_I the closure of a linear hull of $\{X_n, n \in I\}$ in $L_2(\Omega, \mathfrak{F}, \mathsf{P})$, where I is a subset of $\mathbb{Z}$.

(1) Prove that $\eta \in H_I$ if, and only if, there exists a function $f \in L_2^I$ such that

$$\eta = \int_{-\pi}^{\pi} f(\zeta) Z_X(d\zeta).$$

(2) Prove that $\eta = \pi_{H_I}(X)$ (projection of X on H_I) if and only if $\eta \in H_I$ and $\mathsf{E}(\eta - X)\overline{X}_n = 0, n \in I$.

9.9. Assume that polynomials $P(z) = \sum_{k=0}^{r} a_k z^k$ and $Q(z) = \sum_{j=0}^{m} b_j z^j$ do not have roots on the unit circumference. Prove that the equation

$$\sum_{k=0}^{r} a_k Y_{n-k} = \sum_{j=0}^{m} b_j X_{n-j}, \quad n \in \mathbb{Z}$$

has a stationary solution, where $\{X_n, n \in \mathbb{Z}\}$ is a white noise.

Prove that

(a) $\widehat{Y}_n := \pi_0(Y_n)$ can be expressed as the series $\sum_{k=0}^{\infty} c_k Y_{-k}$ convergent in the mean square with $\limsup_{k\to\infty} \sqrt[k]{|c_k|} < 1$.

(b) If $Q(z) \equiv 1$ then this series has only a finite number of nonvoid terms.

9.10. Prove Theorem 9.5 in the case where $c_1 \le p(\zeta) \le c_2$, $\zeta \in (-\pi,\pi]$, for some positive numbers c_1, c_2.

9.11. Let a wide-sense stationary random sequence $\{Y_n, n \in \mathbb{Z}\}$ satisfy the relations:

(a) $Y_{n+1} = 2Y_n + X_n$,

(b) $Y_n = 2X_n + X_{n-1}$, $n \in \mathbb{Z}$,

where $\{X_n, n \in \mathbb{Z}\}$ is a white noise. Find $\pi_H(Y_0)$ with $H = \mathscr{L}^{cl}(Y_n, n \neq 0)$.

9.12. Suppose that the assumptions of Problem 9.10 are satisfied. Construct an algorithm for $\pi_H(X_0)$ calculation with $H = \mathscr{L}^{cl}(X_k, k \notin I)$, where $I \subset \mathbb{Z}$ is a finite set, containing 0.

9.13. Find $\pi_H(Y_0)$ with $H = \mathscr{L}^{cl}(Y_n, n \notin \{0;1\})$ for the sequence $\{Y_n, n \in \mathbb{Z}\}$ from Problem 9.11 (a).

9.14. Let $\{X_n,\ n \in \mathbb{Z}\}$ be a nondegenerate stationary sequence, and its spectral density be of the form

$$p_X(\zeta) = \sum_{k=-m}^{m} c_k e^{i\zeta k}, \quad \zeta \in (-\pi, \pi],$$

where $m \in \mathbb{N}$. Prove that:
(a) $\{X_n, n \in \mathbb{Z}\}$ is a regular sequence.
(b) $X_n = \sum_{k=0}^{m} a_k \varepsilon_{n-k}$, where $\{\varepsilon_n, n \in \mathbb{Z}\}$ is a white noise.

9.15. Let $\{X_n,\ n \in \mathbb{Z}\}$ be a nondegenerate stationary sequence. Prove the following. If $R_X(n) = 0$ as $|n| \geq m$, then the sequence $\{X_n, n \in \mathbb{Z}\}$ is regular. Is this statement correct if $|R_X(n)| \leq ca^{-|n|}, n \in \mathbb{Z}$, where $a \in (0,1)$?

9.16. Can a sum of orthogonal regular and singular sequences be: (a) regular; (b) singular; (c) neither a regular nor singular sequence?

9.17. Can a sum of orthogonal singular sequences be a regular sequence?

9.18. Is it possible that there exist different sequences $\{X_n^1\}$, $\{X_n^2\}$, $\{Y_n^1\}$, and $\{Y_n^2\}$ such that $X_n^1 + X_n^2 = Y_n^1 + Y_n^2$ and the sequences $\{X_n^1\}, \{Y_n^1\}$ are regular, and the sequences $\{X_n^2\}$, $\{Y_n^2\}$ are singular?

9.19. Assume that $\{X_n,\ n \in \mathbb{Z}\}$ is a white noise, and Y is a random variable, with $\mathsf{E}Y = 0$, $\mathsf{E}|Y|^2 < \infty, \mathsf{E}Y\bar{X}_n = 0$, $n \in \mathbb{Z}$.

(1) Find the decomposition $Z_n = Z_n^s + Z_n^r$ for the sequence $Z_n := X_n + Y$.

(2) May another decomposition as a sum of regular and singular sequences exist for the sequence $\{Z_n,\ n \in \mathbb{Z}\}$?

9.20. Let $\{X_n,\ n \in \mathbb{Z}\}$ be a sequence of square integrable random variables (not necessarily stationary). Assume that $\mathsf{E}X_m\bar{X}_k = 0$, $k \leq 0$ for some fixed $m \in \mathbb{N}$. Is it always the case that the projection in $L_2(\Omega, \mathfrak{F}, \mathsf{P})$ of a random variable X_m on $\mathscr{L}^{cl}(X_n,\ n < m)$ coincides with the projection on $\mathscr{L}(X_n,\ n = 1, \ldots, m-1)$?

9.21. Prove Theorem 9.2.

Hints

9.2. Check that a function $(z^n \Phi_n(z))/(\Phi(z))$ is analytic inside the circle $|z| \leq 1$ and, therefore, it can be expanded into the uniformly convergent power series

$\sum_{k=0}^{\infty} \alpha_k z^k$, $|z| \le 1$. Conclude that the right-hand side of (9.1) belongs to the space $H_0(X)$. Verify that

$$\operatorname{cov}\left(\xi_k, \xi_n - \int_{-\pi}^{\pi} e^{in\zeta} \frac{\Phi_n(e^{-i\zeta})}{\Phi(e^{-i\zeta})} Z_X(d\zeta)\right) = 0,\ k \le 0.$$

9.3. Obtain the following representation for the spectral measure of the solution:

$$Z_Y(d\zeta) = \frac{P(e^{-i\zeta})}{Q(e^{-i\zeta})} Z_X(d\zeta),$$

where P, Q are polynomials. Assume that $Q(z) \ne 0$ as $|z| = 1$. The spectral density is

$$p_Y(\zeta) = \frac{1}{2\pi} \left| \frac{P(e^{-i\zeta})}{Q(e^{-i\zeta})} \right|^2.$$

Write the function p_Y as $|\Phi(e^{-i\zeta})|^2$, where $\Phi(z) \ne 0$ as $|z| \le 1$ and Φ is analytic inside a circle $|z| \le 1$. In order to do this observe that, if

$$\frac{P(z)}{Q(z)} = \gamma \frac{\prod_{k=1}^{n} (z - \alpha_k)}{\prod_{j=1}^{m} (z - \beta_k)}, \quad \gamma, \alpha_k, \beta_k \in \mathbb{C},$$

then

$$\left| \frac{P(z)}{Q(z)} \right| = \left| \frac{\widetilde{P}(z)}{\widetilde{Q}(z)} \right|, \quad |z| = 1,$$

where

$$\widetilde{P}(z) = \prod_{k \in \Lambda_1} (z - \alpha_k) \prod_{k \notin \Lambda_1} (1 - \overline{\alpha}_k z),$$

$$\widetilde{Q}(z) = \prod_{j \in \Lambda_2} (z - \beta_j) \prod_{j \notin \Lambda_2} (1 - \overline{\beta}_j z).$$

Here Λ_1 is the index set for which $|\alpha_k| > 1$, and Λ_2 is the index set for which $|\beta_j| > 1$. Consider

$$\Phi(z) = \frac{1}{\sqrt{2\pi}} \frac{\widetilde{P}(z)}{\widetilde{Q}(z)}$$

and use Theorem 9.4.

9.10. Check that the following two statements are true.

(1) For any $n \ne 0$ it holds $(Y - X_0, X_n) = 0$, where $Y = \int_{-\pi}^{\pi} \varphi(\zeta) Z_X(d\zeta)$.

(2) The function φ can be presented as the convergent in the $L_2((-\pi,\pi], F_X)$ series $\varphi(\zeta) = \sum_{n \ne 0} c_n e^{in\zeta}$. To prove this use the equivalence of the norms of the spaces $L_2((-\pi,\pi], F_X)$ and $L_2((-\pi,\pi], \lambda^1)$.

9.12. Search the answer in the form $\int_{-\pi}^{\pi} \varphi(\zeta) Z_X(d\zeta)$, where

$$\varphi(\zeta) = 1 + (p(\zeta))^{-1} \sum_{k \in I} \alpha_k e^{ik\zeta},$$

with

$$\int_{-\pi}^{\pi} \varphi(\zeta) e^{-in\zeta} d\zeta = 0, \quad n \in I.$$

9.15. See the solution to Problem 9.14 (a).

9.21. Because $H_k(X) = H_k(\varepsilon)$ then $\sum_{k=n}^{\infty} a_k \varepsilon_{n-k} \in H_0(X)$.
Check that $\sum_{k=0}^{n-1} a_k \varepsilon_{n-k} \perp H_0(\varepsilon) = H_0(X)$.

Answers and Solutions

9.1. (1) If $\widehat{X}_3 = \alpha X_1 + \beta X_2$ then

$$(X_3 - \widehat{X}_3, X_1) = 0, \quad (X_3 - \widehat{X}_3, X_2) = 0.$$

We obtain the system of linear equations

$$\begin{cases} -1 - \alpha - \beta = 0, \\ -1 - \alpha - 4\beta = 0. \end{cases}$$

Thus, $\alpha = -1, \beta = 0, \widehat{X}_3 = -X_1$. The Gramm–Schmidt orthogonalization algorithm can also be used for obtaining $\widehat{X}_3$.

(2) $\mathsf{E}(X_3 - \widehat{X}_3)^2 = \mathsf{E}(X_3 + X_1)^2 = 10 - 2 + 1 = 9$.

(3) The covariance matrix of the vector $(Y_1, Y_2) := (X_1 + X_2, X_2 + 2X_3)$ equals

$$R_Y = \left\| \begin{matrix} 7 & -6 \\ -6 & 16 \end{matrix} \right\|.$$

If $\widehat{Y}_2 = \alpha Y_1$ then $(Y_2 - \widehat{Y}_2, Y_1) = 0$; that is, $-6 - 7\alpha = 0$. So, $\widehat{Y}_2 = -(6/7)Y_1$.

9.4. (a) The spectral density is equal to $p(\zeta) = (1 - |a|^2)/(2\pi\, |1 - ae^{-i\zeta}|^2)$. It follows from Theorem 9.4 that $\Phi(z) = \sqrt{(1 - |a|^2)/(2\pi)}(1 - ae^{-i\zeta})^{-1}$ and $\widehat{X}_n = a^n X_0$. The sequence is regular.

(b)
$$p(\zeta) = \frac{1}{2\pi}|2 - e^{-i\zeta}|^2, \quad \widehat{X}_n = \begin{cases} 0, & n \geq 2, \\ \sum_{k=0}^{\infty} \frac{X_{-k}}{2^k}, & n = 1. \end{cases}$$

The sequence is regular.

(c) $\widehat{X}_n = X_n = X_0$. The sequence is singular.

(d) $\widehat{X}_n = X_n = X_0 e^{in\lambda}$. The sequence is singular.

(e) The sequence X_n is of the type $X_n = 2Z_X(\{0\}) + e^{in\lambda} Z_X(\{\lambda\})$. Express $Z_X(\{0\})$, $Z_X(\{\lambda\})$ via X_0 and X_{-1}, and find the representation of X_n in terms of X_0, X_{-1}. The sequence $\{X_n, n \in \mathbb{Z}\}$ is singular.

9.5. Because the regular and singular components are mutually orthogonal, their structural measures are also orthogonal (see Problem 8.40). That is why the measure F_X has to be equal to the sum of structural measures $F_{X^r} + F_{X^s}$. Theorem 9.3 implies that $F_{X^r} = 0$; that is, $\{X_n, n \in \mathbb{Z}\}$ is the singular sequence.

9.6. (2) $1/(2\pi)|2 - e^{-i\zeta}|^2 d\zeta$.

(3) $X_n^s = Z_X(\{0\})$, $X_n^r = X_n - Z_X(\{0\})$.

9.7. (1) $(1/n)\sum_{k=0}^{n} e^{ik} X_{-k} \to Z_X(\{1\})$ as $n \to \infty$.

(2) $X_n^s = Z_X(\{1\}) e^{in}$.

9.9. (a) See Problem 9.3 hint.

(b) If $Q(z) = 1$ then the function $\Phi(z)$ has a form (see Problem 9.3 hint)

$$\Phi(z) = \frac{\alpha}{\prod_{k=1}^{r}(\beta_k - z)},$$

where $\alpha, \beta_k \in \mathbb{C}$ with $|\beta_k| > 1$. Let us observe that

$$\frac{\Phi_n(z)}{\Phi(z)} = 1 + \frac{\Phi_n(z) - \Phi(z)}{\Phi(z)}.$$

Because $\Phi_n(z) - \Phi(z)$ and $1/\Phi(z)$ are polynomials, a function $\Phi_n(z)/\Phi(z)$ is a polynomial as well. Therefore, Theorem 9.4 implies that

$$\widehat{Y}_n = \int_{-\pi}^{\pi} e^{in\zeta} \frac{\Phi_n(e^{-i\zeta})}{\Phi(e^{-i\zeta})} Z_Y(d\zeta)$$

can be expressed as a sum $\sum_{k=0}^{2r} c_k Y_{-k}$, where c_k are complex numbers.

9.11. (a) $4(Y_{-1} + Y_1)$.

(b) $\sum\limits_{n=1}^{\infty} (-1/2)^n (Y_n - Y_{-n})$.

9.13. The desired random variable Y can be expressed as (see Problem 9.12 hint): $\int_{-\pi}^{\pi} \varphi(\zeta) Z_X(d\zeta)$, where

$$\varphi(\zeta) = 1 + \frac{\alpha_0}{p(\zeta)} + \frac{\alpha_1 e^{i\zeta}}{p(\zeta)}$$

and coefficients α_0, α_1 satisfy linear equations:

$$\begin{cases} 2\pi + \alpha_0 \int_{-\pi}^{\pi} \frac{d\zeta}{p(\zeta)} + \alpha_1 \int_{-\pi}^{\pi} \frac{e^{i\zeta}}{p(\zeta)} d\zeta = 0, \\ \alpha_0 \int_{-\pi}^{\pi} \frac{e^{i\zeta}}{p(\zeta)} d\zeta + \alpha_1 \int_{-\pi}^{\pi} \frac{d\zeta}{p(\zeta)} = 0. \end{cases}$$

Answer: $-(20/9)Y_{-1} + (16/9)Y_2$.

9.14. (a) Because the spectral density $p_X(\zeta)$ is a nondegenerate analytic function in a neighborhood of the interval $[-\pi, \pi]$, then it can have only a finite number of zeros on $[-\pi, \pi]$, and every zero root has finite multiplicity. Use Theorem 9.3.

(b) It follows from the assumptions that $R_X(k) = 0$ as $|k| > m$. Thus, $X_n \perp H_{n-k}(X)$ as $k > m$. Let $X_n = \sum_{k=0}^{\infty} a_k \varepsilon_{n-k}$, where $\{\varepsilon_k, k \in \mathbb{Z}\}$ is the white noise from the Wald decomposition, and $\sum_k |a_k|^2 < \infty$. Because $H_{n-k}(X) = H_{n-k}(\varepsilon)$, we have $X_n \perp \varepsilon_{n-k}$ as $k > m$. Therefore, $a_k = \mathsf{E} X_n \overline{\varepsilon}_{n-k} = 0,\ k > m$, which was to be demonstrated.

9.16. (a) Yes. Assume that the spectral density of the sequence $\{X_n\}$ equals $\mathbb{I}_{x \in (-\pi,\pi]}$, and the spectral density of the sequence $\{Y_n\}$ equals $\mathbb{I}_{x \in [0,\pi]}$. Therefore (see Theorem 9.3) the sequence $\{X_n\}$ is regular, and $\{Y_n\}$ is singular. The spectral density $\{X_n + Y_n\}$ is equal to $\mathbb{I}_{x \in (-\pi,\pi]} + \mathbb{I}_{x \in [0,\pi]}$, and Theorem 9.3 implies that this sequence is regular.

(b) No. (c) Yes.

9.17. Yes. For instance, if $\{X_n\}$ and $\{Y_n\}$ are independent with spectral densities $\mathbb{I}_{x \in [0,\pi]}$ and $\mathbb{I}_{x \in [-\pi,0]}$, respectively.

See the solution of Problem 9.19(2).

9.18. Yes.

9.19. (1) $Z_n^r = X_n$, $Z_n^s = Y$.

(2) Yes. Let $X_n = X_n^1 + X_n^2$, where the sequences $\{X_n^1\}$ and $\{X_n^2\}$ are stationary and orthogonal, $F_{X^1}(d\zeta) = (1/4\pi)\mathbb{I}_{\zeta \in [0,\pi]} d\zeta$, and $F_{X^2}(d\zeta) = (1/2\pi)d\zeta - F_{X^1}(d\zeta)$. Thus, $\{X_n^2\}$ is regular, and $\{X_n^1 + Y\}$ is singular.

9.20. No. Consider $m = 2$ and the next sequence: $X_2 = Y_2$, $X_1 = Y_1 + Y_2$, $X_k = Y_k + Y_1$, $k \le 0$, where $\{Y_n, n \in \mathbb{Z}\}$ is a white noise.

10

Markov chains: Discrete and continuous time

Theoretical grounds

Let phase space $\mathbb{X}$ of a random sequence $\{X_n, n \in \mathbb{Z}^+\}$ be enumerable. The sequence $\{X_n, n \in \mathbb{Z}^+\}$ is called a *Markov chain* if

$\forall n \in \mathbb{N}\ \forall i_1, \dots, i_n, i_{n+1} \in \mathbb{X}\ \forall t_1 \le \dots \le t_n \le t_{n+1} \in \mathbb{Z}_+ :$

$$\mathsf{P}(X_{t_{n+1}} = i_{n+1}/X_{t_1} = i_1, \dots, X_{t_n} = i_n) = \mathsf{P}(X_{t_{n+1}} = i_{n+1}/X_{t_n} = i_n).$$

The system $\mathsf{P}(X_{n+1} = j/X_n = i)$, $i, j \in \mathbb{X}, n \in \mathbb{Z}^+$ is called the system of transition probabilities. If these conditional probabilities are independent of n then the Markov chain is said to be homogeneous. Further on we consider only homogeneous Markov chains.

A matrix $P = \|p_{ij}\| := \|\mathsf{P}(X_1 = j/X_0 = i)\|$ (or $P^n = \|p_{ij}^{(n)}\| := \|\mathsf{P}(X_n = j/X_0 = i)\|$, respectively) is called the *transition matrix* (n-step transition matrix). Transition probabilities satisfy the Kolmogorov–Chapman equations

$$\forall i, j \in \mathbb{X}\ \forall n, m \in \mathbb{Z}_+ :\ p_{ij}^{(n+m)} = \sum_{k \in \mathbb{X}} p_{ik}^{(n)} p_{kj}^{(m)}.$$

The Kolmogorov–Chapman equations can be reformulated as follows.

Proposition 10.1. *The n-step transition matrix is equal to the matrix P raised to the nth power.*

Definition 10.1. *A state j is said to be* accessible *from a state i ($i \to j$) if there is positive probability that in a finite number of steps a Markov chain moves from i to j; that is, $p_{ij}^{(n)} > 0$ for some $n \ge 0$.*

A state i is said to communicate *with state j ($i \leftrightarrow j$) if $i \to j$ and $j \to i$.*

A state i is inessential *if there exists a state j such that $i \to j$ but $j \nrightarrow i$. Otherwise, a state i is said to be* essential.

Let $\tau_i = \inf\{n \ge 1 | X_n = i\}$ be the time moment of the first visit to i. If $\mathsf{P}(\tau_i < \infty/X_0 = i) \ne 0$ then period $d(i)$ *of a state i is the greatest common divisor of numbers n such that $p_{ii}^{(n)} > 0$. If $d(i) = 1$ then the state i is called aperiodic.*

D. Gusak et al., *Theory of Stochastic Processes*, Problem Books in Mathematics,
DOI 10.1007/978-0-387-87862-1_10,

Denote by $f_{ij}^{(n)} = \mathsf{P}(\tau_j = n/X_0 = i)$ *the probability for the chain to make the first visit to a state* j *on the nth step given the chain started from the state i. A state i is said to be* recurrent *if* $\mathsf{P}(\tau_i < \infty/X_0 = i) = 1$ *or equivalently if* $\sum_{n=1}^{\infty} f_{ii}^{(n)} = 1$. *If a state is not recurrent, then it is said to be* transient.

Theorem 10.1. *(Recurrence criterion) A state i is recurrent if and only if* $\sum_n p_{ii}^{(n)} = +\infty$.

The important class of Markov chains is random walks on a lattice $\mathbb{Z}^d$, that is, random sequences of the type $X_n = x + \varepsilon_1 + \cdots + \varepsilon_n$, where $\{\varepsilon_k,\ k \geq 1\}$ are i.i.d. random variables with values in $\mathbb{Z}^d$.

The following criterion provides the necessary and sufficient recurrence condition for a random walk.

Theorem 10.2. *Assume that all states* $\mathbb{Z}^d$ *of a random walk communicate. Let* $\varphi(u) = \mathsf{E}\{\exp i(u,\varepsilon)\}$, $u \in \mathbb{R}^d$ *be a characteristic function of a random walk* $\{X_n, n \geq 0\}$ *jump. The sequence* $\{X_n, n \geq 0\}$ *is recurrent if, and only if,*

$$\int_{(-\pi,\pi)^d} \mathrm{Re}(1 - \varphi(u))^{-1} du = \infty.$$

A system of nonnegative numbers $\pi_i, i \in \mathbb{X}$ is called a distribution if $\sum_i \pi_i = 1$.

Definition 10.2. *A distribution* $\pi_i = \mathsf{P}(X_0 = i), i \in \mathbb{X}$ *is said to be a stationary (or invariant) distribution of a Markov chain if* $\pi_i = \mathsf{P}(X_n = i)$ *for all* $n \in \mathbb{N}$, $i \in \mathbb{X}$.

Theorem 10.3. *(Ergodic theorem) Assume that all states of a Markov chain communicate and have period 1. Then for all* $i, j \in \mathbb{X}$

$$p_{ij}^{(n)} \to 1/\mu_j, \quad n \to \infty,$$

where $\mu_j = \mathsf{E}(\tau_j/X_0 = j) \in (0,\infty]$ *is an average recurrence time to the state j.*

The ergodic theorem can be proved by use of renewing theory methods (see Chapter 11, Problem 11.29).

Theorem 10.4. *Suppose that assumptions of Theorem 10.3 are satisfied. Denote* $\pi_j = 1/\mu_j$. *Then, there are two possibilities. Either all* $\pi_j = 0$ *and the stationary distribution does not exist or* $(\pi_1, \pi_2, \ldots)$ *is the only stationary distribution.*

Theorem 10.5. *(Strong Markov property) Let* τ *be a stopping time. Then* $\forall m \geq 1$, $\forall n_1 \leq \cdots \leq n_m \in \mathbb{N}$, $\forall j_1, \ldots, j_m \in \mathbb{X}$

$$\mathsf{P}(X_{\tau+n_1} = j_1, \ldots, X_{\tau+n_m} = j_m / \mathcal{F}_\tau) = p_{X_\tau j_1}^{(n_1)} p_{j_1 j_2}^{(n_2-n_1)} \cdots p_{j_{m-1} j_m}^{(n_m-n_{m-1})}. \quad (10.1)$$

Continuous-time Markov chains

Assume that a phase space $\mathbb{X}$ of a stochastic process $\{X(t), t \geq 0\}$ is an enumerable set. The process $\{X(t), t \geq 0\}$ is called a continuous-time Markov chain if

$$\forall n \in \mathbb{N}\, \forall i_1, \dots, i_n, i_{n+1} \in \mathbb{X}\, \forall t_1 \le \dots \le t_n \le t_{n+1} \in \mathbb{R}_+ :$$
$$\mathsf{P}(X(t_{n+1}) = i_{n+1}/X(t_1) = i_1, \dots, X(t_n) = i_n) = \mathsf{P}(X(t_{n+1}) = i_{n+1}/X(t_n) = i_n).$$

Further on we consider only continuous-time Markov chains that are time-homogeneous and stochastic continuous. That is, the transition probability $p_{ij}(t) = \mathsf{P}(X(t+s) = j/X(s) = i)$ does not depend on s and is continuous on t.

Theorem 10.6. *For any i, j there exist limits*

$$\alpha_{ij} := \begin{cases} \lim\limits_{t \to 0+} \frac{p_{ij}(t)}{t}, & i \ne j, \\ \lim\limits_{t \to 0+} \frac{p_{ii}(t)-1}{t}, & i = j. \end{cases}$$

The limits α_{ij} are finite as $i \ne j$, and α_{ii} can take value $-\infty$.

The limits α_{ij}, $i \ne j$ are called the *transition intensities* from a state i to a state j and a matrix $A = \|\alpha_{ij}\|$ is called the *generator* of $\{X(t), t \ge 0\}$.

Proposition 10.2. *If $\alpha_{ii} \ne -\infty$ for all i then $\{X(t), t \ge 0\}$ has a right-continuous modification.*

We assume that the right-continuous modification has already been chosen for such processes.

Definition 10.3. *A state i is regular if $\alpha_{ii} > -\infty$ and $\sum_j \alpha_{ij} = 0$.*

Theorem 10.7. *(First, or backward, Kolmogorov system of equation) Assume that i is a regular state. Then the transition probabilities $p_{ij}(t)$ are differentiable in t and satisfy the system of differentiable equations*

$$p'_{ij}(t) = \sum_k \alpha_{ik} p_{kj}(t), \quad t \ge 0. \tag{10.2}$$

In addition, the transition probabilities satisfy the relations

$$p_{ij}(0) = \mathbb{I}_{i=j}, \tag{10.3}$$

$$p_{ij}(t) \ge 0, \quad \sum_j p_{ij}(t) = 1, \quad t > 0. \tag{10.4}$$

Definition 10.4. *A continuous-time Markov chain $\{X(t), t \ge 0\}$ is said to be regular if*

(a) X is a right-continuous process.

(b) For any initial distribution the probability that an infinite number of jumps occurs in finite time is equal to zero.

Theorem 10.8. *A continuous-time Markov chain is regular if and only if one of the following conditions holds true.*

(1) There exists a unique solution to the system of equations (10.2) *with initial condition* (10.3) *satisfying* (10.4).

(2) For any $\lambda > 0$ the system of equations $\lambda g_i = \sum_j \alpha_{ik} g_k$, $i \in E$ does not have any bounded solution $\{g_i\}$ except a trivial zero one.

Theorem 10.9. *(Second, or forward, Kolmogorov system of equations) If X is a regular continuous-time Markov chain then the transition probabilities satisfy the system of differential equations*

$$\begin{cases} p'_{ij}(t) = \sum_k p_{ik}(t)\alpha_{kj}, \quad t \geq 0, \\ p_{ij}(0) = \mathbb{I}_{i=j}. \end{cases}$$

Theorem 10.10. *(Strong Markov property) Let X be a regular continuous-time Markov chain. Then for any stopping time τ:*
$\forall m \geq 1,\ \forall t_1,\dots t_m,\ 0 < t_1 < \cdots < t_m,\ \forall j_1,\dots,j_m \in \mathbb{X}:$

$$\begin{aligned} &\mathsf{P}(X(\tau+t_1) = j_1,\dots,X(\tau+t_m) = j_m / \mathcal{F}_\tau) \\ &\quad = p_{X(\tau)j_1}(t_1)p_{j_1j_2}(t_2-t_1)\dots p_{j_{m-1}j_m}(t_m-t_{m-1}). \end{aligned} \tag{10.5}$$

Bibliography

[82] Chapter I §12, Chapter VIII; [69] Chapter III §2,3; [24], Volume 1, Chapter II §4–7; [79] Chapters XXI, XXIII; [22] Volume 1, Chapters 14–17; [9] Chapters 7–9; [15] Chapters 5,6; [89] Chapters 3-5; [80]; [12].

Problems

10.1. Prove that if $i \to j,\ j \to k$, then $i \to k$.

10.2. Prove that a "communicate" relation is an equivalence relation. That is, it is reflective, symmetric, and transitive.

10.3. Prove that if a state i is recurrent then it is essential. Is the inverse statement true when the phase space is (a) finite; (b) countable?

10.4. Assume that $i \leftrightarrow j$. Prove that the state i is recurrent if and only if the state j is recurrent.

10.5. Prove that a state i is recurrent if and only if

$$\mathsf{P}(X_n = i \text{ infinitely often}/X_0 = i) = 1.$$

10.6. Suppose that $i \to j$. Is it correct that: (a) $d(i) \geq d(j)$; (b) $d(i) \leq d(j)$?

10.7. The transition matrix of a Markov chain $\{X_n, n \geq 0\}$ is of the following type

$$P = \begin{Vmatrix} 0.2 & 0.8 \\ 0.4 & 0.6 \end{Vmatrix}.$$

The initial distribution is: $\mathsf{P}(X_0 = 1) = 0,3;\ \mathsf{P}(X_0 = 2) = 0,7$.
(a) Find the distribution of X_1.
(b) Find the probabilities $\mathsf{P}(X_0 = 1, X_1 = 2)$; $\mathsf{P}(X_0 = 1, X_1 = 2, X_2 = 2)$; $\mathsf{P}(X_1 = 2, X_2 = 2/X_0 = 1)$; $\mathsf{P}(X_1 = 1, X_2 = 2)$; $\mathsf{P}(X_2 = 1)$; $\mathsf{P}(X_2 = 2/X_0 = 1)$; $\mathsf{P}(X_1 = 1, X_3 = 2)$; $\mathsf{P}(X_1 = 1, X_2 = 1, X_4 = 2, X_6 = 1, X_5 = 1/X_0 = 1)$.

10.8. The transition matrix of the a Markov chain $\{X_n, n \geq 0\}$ is

$$P = \begin{Vmatrix} 0.1 & 0.2 & 0.7 \\ 0.3 & 0.4 & 0.3 \\ 0.5 & 0.4 & 0.1 \end{Vmatrix}.$$

The initial distribution is:

$$\mathsf{P}(X_0 = 1) = 0.6; \quad \mathsf{P}(X_0 = 2) = 0.3; \quad \mathsf{P}(X_0 = 3) = 0.1.$$

(a) Find the distribution of X_1.
(b) Find the probabilities $\mathsf{P}(X_2 = 1)$; $\mathsf{P}(X_1 = 1, X_2 = 2, X_3 = 3/X_0 = 1)$; $\mathsf{P}(X_1 = 1, X_3 = 1, X_4 = 3/X_0 = 2)$; $\mathsf{P}(X_1 = 1, X_3 = 2, X_4 = 1, X_6 = 1, X_8 = 1/X_0 = 1)$; $\mathsf{P}(X_0 = 1, X_1 = 2)$; $\mathsf{P}(X_1 = 1, X_2 = 2)$; $\mathsf{P}(X_4 = 1, X_3 = 2, X_1 = 1, X_2 = 4, X_5 = 1/X_0 = 2)$.
(c) Find the distribution of the random variable $\tau_1 = \inf\{n \geq 0 | \ X_n \neq 1\}$.

10.9. Let $\{X_n, n \geq 0\}$ be a Markov chain with a finite phase space, and A be a set of its essential states. Let τ be a time of the first visit of the set A by the Markov chain. Prove that
(a) There exist $M > 0$ and $c \in [0, 1)$ such that $\mathsf{P}(\tau > n) \leq Mc^n$.
(b) $\mathsf{E}\tau < \infty$.

Are these statements correct for a Markov chain with countable phase space?

10.10. Let a transition matrix be

$$P = \begin{Vmatrix} 0.1 & 0.2 & 0.3 & 0.2 & 0.1 & 0.1 \\ 0.3 & 0.1 & 0.1 & 0.2 & 0.1 & 0.2 \\ 0 & 0 & 0.4 & 0.2 & 0.3 & 0.1 \\ 0 & 0 & 0.5 & 0.1 & 0.1 & 0.3 \\ 0 & 0 & 0 & 0 & 0.6 & 0.4 \\ 0 & 0 & 0 & 0 & 0.2 & 0.8 \end{Vmatrix}.$$

(a) Classify the states.
(b) Find the expectation of the time when the chain reaches some essential state given $X_0 = 1$.
(c) Let $\tau = \inf\{n \geq 1 | \ X_n = 3\}$. Find the probability $\mathsf{P}(\tau < \infty / \ X_0 = 1)$.

10.11. John and Peter play a "pennies matching" game till one is ruined. At the beginning John has \$3 and Peter has \$1. One bet costs \$1 per game. The coin is symmetric.
(a) Find the probability that John loses all his money.
(b) Find the expectation of the number of games played till the one's ruin.
(c) How it affects the answer if the bet cost changes to 50 cents?

10.12. Solve the previous problem if at the beginning of the game John has \$$n$, Peter has \$$m$, and probability that John wins in each game is equal to $p \in (0; 1)$.

10.13. Assume that a sequence $\{p_n, \ n \geq 1\}$ of nonnegative numbers satisfies the relation $p_1 + p_2 + \cdots = 1$. Describe the dynamics of the Markov chain and classify the states if the transition matrix is equal to

$$(a)\ \left\|\begin{matrix} p_1 & p_2 & p_3 & \cdots \\ 1 & 0 & 0 & \cdots \\ 1 & 0 & 0 & \\ \vdots & \vdots & & \ddots \end{matrix}\right\|; \quad (b)\ \left\|\begin{matrix} p_1 & p_2 & p_3 & & \cdots \\ 0 & 1 & 0 & 0 & \cdots \\ 0 & 0 & 1 & 0 & \cdots \\ 0 & 0 & 0 & 1 & \cdots \\ \cdots & \cdots & \cdots & \cdots & \cdots \end{matrix}\right\|;$$

$$(c)\ \left\|\begin{matrix} p_1 & p_2 & p_3 & \cdots \\ 1 & 0 & 0 & \cdots \\ 0 & 1 & 0 & \cdots \\ 0 & 0 & 1 & \cdots \\ \cdots & \cdots & \cdots & \cdots \end{matrix}\right\|; \quad (d)\ \left\|\begin{matrix} (1-p_1) & p_1 & 0 & 0 & \cdots \\ (1-p_2) & 0 & p_2 & 0 & \cdots \\ (1-p_3) & 0 & 0 & p_3 & \cdots \\ \cdots & \cdots & \cdots & \cdots & \cdots \end{matrix}\right\|;$$

$$(e)\ \left\|\begin{matrix} (1-p_1) & p_1 & 0 & 0 & 0 & \cdots \\ 0 & (1-p_2) & p_2 & 0 & 0 & \cdots \\ 0 & 0 & (1-p_3) & p_3 & 0 & \cdots \\ \cdots & \cdots & \cdots & \cdots & \cdots & \cdots \end{matrix}\right\|.$$

Calculate $f_{11}^{(n)}$. Find the stationary distribution if it exists.

10.14. Consider transition matrices from the previous problem. Assume additionally that $p_1 = 0$ in item (a). Find $p_{1k}^{(n)}$ in items (a), (b) and $p_{34}^{(5)}$ in items (c)–(e).

10.15. Let $\{X_n, n \geq 0\}$ be a Markov chain, and $\tau = \inf\{n \geq 0 | X_n \in A\}$ be a time of the first visit of a set A by the chain. Denote $r_i = \mathsf{P}(\tau < \infty / X_0 = i)$, $\rho_i = \mathsf{E}(\tau / X_0 = i) \in (0, \infty]$. Prove that:

(a) Probabilities $\{r_i\}$ satisfy the system of linear equations

$$\begin{cases} r_i = \sum_{j \notin A} p_{ij} r_j, & i \notin A, \\ r_i = 1, & i \in A, \\ r_i = 0, & i \nrightarrow A \text{ and } i \notin A. \end{cases}$$

(b) Expectations $\{\rho_i\}$ satisfy the following system of equations

$$\begin{cases} \rho_k = 1 + \sum_j p_{kj} \rho_j, & k \notin A, \\ \rho_k = 0, & k \in A. \end{cases}$$

(c) If the phase space is finite then the expectation ρ_i is finite if and only if for any state j for which $i \to j$ there exists a state $k \in A$ accessible from j $(j \to k)$.

10.16. Find $\lim_{n\to\infty} \mathsf{P}(X_n = i)$ in Problems 10.8 and 10.10 if $X_0 = 1$.

10.17. The lifetime of a detail is equal to 1 day with probability $p \in (0,1)$ and 2 days with probability $q = 1 - p$. At the moment of failure the detail is replaced immediately. Denote by $X_n \in \{1; 2\}$ a total lifetime of the detail that is processing at the nth day. Let $p_n = \mathsf{P}(X_n = 1)$.

(a) Find p, if $\lim_{n\to\infty} p_n = \frac{1}{2}$.

(b) Is $\{X_n, n \geq 0\}$ a Markov chain?

10.18. Let $\{\xi_n, n \geq 1\}$ be i.i.d. random variables, and $f : \mathbb{X} \times \mathbb{R} \to \mathbb{X}$ be a measurable function. Suppose that a random variable X_0 does not depend on $\{\xi_n, n \geq 1\}$. Prove that a sequence $X_n = f(X_{n-1}, \xi_n)$, $n \geq 1$ is a Markov chain. Find a function f for a Markov chain from Problem 10.8 if $\{\xi_n, n \geq 1\}$ has a uniform distribution on $[0,1]$.

10.19. Let $\{X_n, Y_n, n \geq 1\}$ be Markov chains with values in $\mathbb{Z}$ and for any $x \in \mathbb{Z}$,

$$\forall i, j: \ \mathsf{P}(X_{n+1} - X_n \geq x / X_n = i) \geq \mathsf{P}(Y_{n+1} - Y_n \geq x / Y_n = j),$$
$$\mathsf{P}(X_0 \geq x) \geq \mathsf{P}(Y_0 \geq x).$$

Prove that for all $n \in \mathbb{N}, x \in \mathbb{Z}$

$$\mathsf{P}(X_n \geq x) \geq \mathsf{P}(Y_n \geq x).$$

10.20. Let $\{X_n, n \in \mathbb{Z}\}$ be a homogeneous Markov chain. Prove that $\{X_{-n},\ n \in \mathbb{Z}\}$ is also a Markov chain. Is it always homogeneous? Find the transition matrix of the chain $\{X_{-n},\ n \in \mathbb{Z}\}$ if the chain $\{X_n, n \in \mathbb{Z}\}$ is stationary and has a transition matrix $P = \|p_{ij}\|$.

10.21. Let $\{X_n, n \geq 0\}$ be a Markov chain. Is it necessary that for any moments of time $k < m < n$ and for any $A, B, C \subset \mathbb{X}$ the following equality holds true, $\mathsf{P}(X_n \in A |\ X_m \in B, X_k \in C) = \mathsf{P}(X_n \in A |\ X_m \in B)$?

10.22. Let $\{\varepsilon_n, n \geq 0\}$ be a sequence of independent Bernoulli random variables, $\mathsf{P}(\varepsilon_n = 1) = p, \mathsf{P}(\varepsilon_n = -1) = 1 - p$. For which p is a sequence $X_n := \varepsilon_{n+1} \cdot \varepsilon_n, n \geq 0$ a Markov chain?

10.23. Give an example of a Markov chain $\{X_n\}$ and a subset $A \subset \mathbb{X}$ for which the sequence $\{Y_n = \mathbb{I}_{X_n \in A}\}$
(a) Is a Markov chain
(b) Is not a Markov chain.

10.24. Can all states of a Markov chain be inessential if the phase space is (a) finite; (b) infinite?

10.25. Is it possible that all states of a Markov chain are transient if the phase space is (a) finite; (b) infinite?

10.26. Let $\{\varepsilon_n, n \geq 1\}$ be a sequence of i.i.d. random variables, $\mathsf{P}(\varepsilon_n = 0) = \mathsf{P}(\varepsilon_n = 1) = \frac{1}{2}$. A sequence $X_n \in \mathbb{X} \equiv \{0,1\}^m,\ n \geq 1$ is built in a recurrent way as follows: $X_0 = (0, \ldots, 0)$, kth coordinate of X_n is equal to the $(k-1)$th coordinate of $X_{n-1} (k > 1)$ and the first coordinate X_n is equal to ε_n. Is the sequence $\{X_n\}$ a Markov chain? If so, indicate the essential and inessential states, number of essential state classes, and period of every class.

10.27. Let $\{X_n, n \geq 0\}$ be i.i.d. random variables, $\mathsf{P}(X_n = 1) = \frac{1}{3}, \mathsf{P}(X_n = 0) = \frac{2}{3}$. Define Y_n as a number of units among numbers X_n, X_{n+1}, X_{n+2}. Is the sequence $\{Y_n\}$ a Markov chain? If so, indicate essential and unessential states, number of essential state classes, and period of every class.

10.28. Let $\{X_n\}$ be a random walk in $\mathbb{Z}$, $\mathsf{P}(X_{n+1} = i + 1 | X_n = i) = p$, $\mathsf{P}(X_{n+1} = i - 1 | X_n = i) = q = 1 - p, i \in \mathbb{Z}$, where $p \in (0, 1)$. Find the n-step transition matrix. For which p is the walk recurrent?

10.29. A particle jumps 1 step up or down with probability $\frac{1}{2}$ at every even moment of time independently of other steps, and 1 step to the right or to the left with probability $\frac{1}{2}$ at every odd moment of time. Let X_n be a location of the particle at moment n. Find $\mathsf{P}(X_n = (k,l)/X_0 = (i,j))$.

10.30. Let $\{Y_n\}$ be a random walk in $\mathbb{Z}^2$, $\mathsf{P}(Y_{n+1} - Y_n = e) = \frac{1}{4}$, where $e \in \{(0,1), (1,0), (-1,0), (0,-1)\}$. Find the transition probabilities $\mathsf{P}(Y_n = (k,l)/Y_0 = (i,j))$. Is this random walk recurrent?

10.31. Let $\{X_n\}$ be a symmetric random walk on $\mathbb{Z}$, $\mathsf{P}(X_{n+1} = i \pm 1 | X_n = i) = \frac{1}{2}$; $\{Y_n\}$ be a symmetric random walk on $\mathbb{Z}^+$ with reflection at zero, $\mathsf{P}(Y_{n+1} = i \pm 1 | Y_n = i) = \frac{1}{2}$, $i \in \mathbb{N}$, $\mathsf{P}(Y_{n+1} = 1 | Y_n = 0) = 1$. Prove that the distributions of the sequences $\{|X_n|\}$ and $\{Y_n\}$ coincide assuming that the distributions of random variables $|X_0|$ and Y_0 coincide.

10.32. Let $\{X_n\}$ be the symmetric random walk from Problem 10.31, $X_0 = j \in \mathbb{N}$.
(a) Prove that the probability of the event $X_m = -k, k \in \mathbb{N}$ is equal to the probability of the event that on time-interval $0, \dots, m$ the sequence $\{X_n\}$ visits zero and $X_m = k$. That is,

$$\mathsf{P}(X_m = -k | X_0 = j) = \mathsf{P}(\{\exists\, l \in \overline{0, m-1} :\ X_l = 0\} \cap \{X_m = k\} | X_0 = j).$$

(b) Find the probability that $X_m = k, k \in \mathbb{N}$ and the sequence $\{X_n\}$ has never been to zero till the moment m.
(c) Find the probability that the sequence $\{X_n\}$ visits zero the first time at the moment m.
(d) Find $\mathsf{P}(X_1 \geq 0, \dots, X_{m-1} \geq 0, X_m = k | X_0 = j)$, $j, k \in \mathbb{Z}_+$.
(e) Find $\mathsf{P}(X_1 > 0, \dots, X_{m-1} > 0, X_m = k | X_0 = 0, X_m = k)$, $k \in \mathbb{N}$.

10.33. Let $\{Y_n\}$ be a symmetric random walk in $\mathbb{Z}_+$ with capture in zero. That is, $\mathsf{P}(Y_{n+1} = i \pm 1 | Y_n = i) = \frac{1}{2}, i \in \mathbb{N}$ and $\mathsf{P}(Y_{n+1} = 0 | Y_n = 0) = 1$. Find $p_{ij}^{(n)}$.

10.34. Let $\{Y_n\}$ be a symmetric random walk in $\mathbb{Z}_+$ reflecting at zero. That is,

$$\mathsf{P}(Y_{n+1} = i \pm 1 | Y_n = i) = \frac{1}{2}, \quad i \geq 1, \quad \mathsf{P}(Y_{n+1} = 1 | Y_n = 0) = 1.$$

Find $p_{ij}^{(n)}$.

10.35. Find $p_{ij}^{(n)}$ for a symmetric random walk $\{Y_n\}$ on $\mathbb{Z}$ with the elastic barrier at 0: $\mathsf{P}(Y_{n+1} = i \pm 1 | Y_n = i) = \frac{1}{2}$, $i \neq 0$, $\mathsf{P}(Y_{n+1} = 1 | Y_n = 0) = p \in (0,1)$, $\mathsf{P}(Y_{n+1} = -1 | Y_n = 0) = q = 1 - p$.

10.36. Let $\{X_n, n \geq 0\}$ be a nonsymmetric random walk on $\mathbb{Z}^+$ reflecting at 0. That is, $\mathsf{P}(X_{n+1} = i + 1 / X_n = i) = p$ and $\mathsf{P}(X_{n+1} = \max(i-1, 0) |\ X_n = i) = (1-p), i \in \mathbb{Z}^+$.
(a) Find $f_{00}^{(n)}$.
(b) Give the values of p for which the chain X is recurrent.
(c) Find the probability for $\{X_n\}$ to visit zero assuming $X_0 = m$, $m \in \mathbb{N}$.

10.37. A device can break down at moments $n = 1, 2, \ldots$; a broken device is replaced immediately by a new one. A failure probability for the device of age k is equal to $p_k, k \geq 1$. Consider a Markov chain $X_n =$ (the age of the device functioning at the moment n), $n \geq 1$. Find:

(1) $f_{00}^{(n)}, f_{11}^{(n)}, n \geq 1$ and condition on $\{p_k\}$ under which the chain is recurrent.

(2) Condition on $\{p_k\}$ under which the chain is positively recurrent, and find the invariant distribution of the chain.

10.38. A package of requests arrives at the device buffer. The device processes each request in one second. At the moment when the last request has been served the new package arrives and so on. The sizes of incoming packages are i.i.d. random variables $\{\xi_m, m \geq 1\}$, $\mathsf{P}(\xi_m = k) = p_k, k \geq 1$. Answer the questions formulated in the previous problem for the Markov chain X_n =(number of requests being in the buffer at the moment n), $n \geq 1$.

10.39. Transition matrix P of a Markov chain is equal to

$$\text{(a)}\ \left\|\begin{matrix} 0.2 & 0.8 & 0 & 0 & 0 \\ 0.4 & 0.6 & 0 & 0 & 0 \\ 0.1 & 0.2 & 0.3 & 0.2 & 0.2 \\ 0 & 0 & 0 & 0.1 & 0.9 \\ 0 & 0 & 0 & 0.6 & 0.4 \end{matrix}\right\|; \quad \text{(b)}\ \left\|\begin{matrix} 0.1 & 0.2 & 0.3 & 0.1 & 0.1 & 0.2 \\ 0.1 & 0.1 & 0.1 & 0.1 & 0.2 & 0.4 \\ 0 & 0 & 0.4 & 0.6 & 0 & 0 \\ 0 & 0 & 0.3 & 0.7 & 0 & 0 \\ 0 & 0 & 0 & 0 & 0.5 & 0.5 \\ 0 & 0 & 0 & 0 & 0.5 & 0.5 \end{matrix}\right\|;$$

$$\text{(c)}\ \left\|\begin{matrix} 0.3 & 0.7 & 0 & 0 \\ 0.5 & 0.5 & 0 & 0 \\ 0.2 & 0.3 & 0.3 & 0.2 \\ 0 & 0 & 0 & 1 \end{matrix}\right\|; \quad \text{(d)}\ \left\|\begin{matrix} 1 & 0 & 0 & 0 \\ 0.3 & 0.3 & 0.3 & 0.1 \\ 0 & 0 & 0.6 & 0.4 \\ 0 & 0 & 0.4 & 0.6 \end{matrix}\right\|;$$

$$\text{(e)}\ \left\|\begin{matrix} 1 & 0 & 0 & 0 & 0 & 0 \\ 0 & 0.5 & 0.5 & 0 & 0 & 0 \\ 0 & 0.4 & 0.6 & 0 & 0 & 0 \\ 0 & 0 & 0 & 1 & 0 & 0 \\ 0 & 0 & 0 & 0 & 0.2 & 0.8 \\ 0.1 & 0.2 & 0.3 & 0.1 & 0.1 & 0.2 \end{matrix}\right\|.$$

Classify the states. Find $P^{\infty} = \lim_{n\to\infty} P^n$. For any unessential state i, find the mean time before the chain visits an essential state given $X(0) = i$.

Find the mean time before the chain reaches any state with even number if $X(0) = 1$ for the matrix from item (b).

10.40. Let $\{X_n, n \geq 1\}$ be a random walk with the step $\xi_1 = \eta - \zeta$, where: (a) $\eta \sim \zeta \sim \text{Geom}(p)$; (b) $\eta \sim \text{Pois}(1)$, $\zeta = 1$. Prove that this random walk is recurrent.

10.41. N white and N black balls are put into two boxes containing N balls each. At the moments $n = 1, 2, \ldots$ one ball from each box is chosen at random and placed in the other box (*Laplace diffusion model*). Define a Markov chain $X_n =$ (number of balls inside the first box at the moment n), $n \geq 1$.

(a) Find the transition probabilities matrix P and classify the states.

(b) Find $P^{\infty} = \lim_{n\to\infty} P^n$.

10.42. N balls are distributed between two boxes. At the time moments $n = 1, 2, \ldots$ a ball is chosen at random and is shifted into another box (*P. and T. Ehrenfest diffusion model*). Define a Markov chain $X_n =$ (number of balls inside the first box at the moment n), $n \geq 1$.
(a) Find the transition probabilities matrix P and classify the states.
(b) Find $P^\infty = \lim_{n\to\infty} P^n$.

10.43. (Coding by pile of books method) Suppose that every second a router accepts letters from some alphabet $A = \{a_1, \ldots, a_m\}$ with probabilities $p_1, \ldots, p_m$ respectively. A letter a_k at the moment n is coded by a number $x_k(n) \in \{1, \ldots, m\}$ (different letters are coded by different codes). The coding algorithm is built in a recurrent way. Assume that at the moment n the letter a_k arrives. Then $x_k(n+1) = 1$. Letters encoded at the moment n by numbers $1, \ldots, (x_k(n) - 1)$, are re-encoded by numbers $2, \ldots, x_k(n)$, respectively. The codes of the other letters are left unchanged.

Is a letter's a_1 code at the moment n a Markov chain? Find the limits for probabilities of the following event as $n \to \infty$.
(a) A letter a_1 at the moment n is encoded by 1.
(b) Letters a_1, a_2 at the moment n are encoded by 1 and 2, respectively.
Assume that letters come independently.

10.44. (A birth-and-death process). Let a phase space of a Markov chain $\{X_n\}$ be $\mathbb{Z}_+$. Assume that the transition matrix is equal to

$$P = \left\| \begin{matrix} r_0 & p_0 & 0 & 0 & 0\ldots \\ q_1 & r_1 & p_1 & 0 & 0\ldots \\ 0 & q_2 & r_2 & p_2 & 0\ldots \\ 0 & 0 & q_3 & r_3 & p_3\ldots \\ \ldots & \ldots & \ldots & \ldots & \ldots \end{matrix} \right\|,$$

where $q_i + r_i + p_i = 1$, $q_i > 0$, $p_i > 0$.

(1) Classify the states of the Markov chain.

(2) Prove that either the chain is recurrent or $\lim_{n\to\infty} X_n = \infty$ with probability one. Find the recurrence conditions in terms of $\{q_i, r_i, p_i\}$.

(3) Find the invariant distribution if it exists.

10.45. (A cyclic birth-and-death process). Let a Markov chain have the phase space $\{0, \ldots, N\}$ and the transition matrix

$$P = \left\| \begin{matrix} r_0 & p_0 & 0 & \ldots & q_0 \\ q_1 & r_1 & p_1 & \ldots & 0 \\ \ldots & \ldots & \ldots & \ldots & \ldots \\ 0 & \ldots & q_{N-1} & r_{N-1} & p_{N-1} \\ p_N & \ldots & 0 & q_N & r_N \end{matrix} \right\|,$$

where $q_i + r_i + p_i = 1$, $q_i > 0$, $p_i > 0$. Find the invariant distribution of this chain if
(a) $p_0 \cdots p_N = q_0 \cdots q_N$.
(b) There exists $\mu \neq 1$ for which $(p_i/q_i) = \mu, i = 0, \ldots, N$.
(c) $p_0 \cdots p_N \neq q_0 \cdots q_N$.

10.46. (A continuous-time birth-and-death process). Let a phase space of a continuous-time Markov chain $\{X(t), t \geq 0\}$ be $\mathbb{Z}^+$. Suppose that a generator is

$$A = \left\| \begin{matrix} -\lambda_0 & \lambda_0 & 0 & 0 & 0\dots \\ \mu_1 & -(\mu_1+\lambda_1) & \lambda_1 & 0 & 0\dots \\ 0 & \mu_2 & -(\mu_2+\lambda_2) & \lambda_2 & 0\dots \\ 0 & 0 & \mu_3 & -(\mu_3+\lambda_3) & \lambda_3\dots \\ \dots & \dots & \dots & \dots & \dots \end{matrix} \right\|,$$

where $\mu_i > 0,\ \lambda_i > 0$.

Prove that the invariant distribution exists if, and only if,

$$\Lambda = 1 + \sum_{n\geq 0} \frac{\lambda_0 \cdots \lambda_{n-1}}{\mu_1 \cdots \mu_n} < \infty.$$

In this case

$$\pi_k = \Lambda^{-1} \frac{\lambda_0 \cdots \lambda_{k-1}}{\mu_1 \cdots \mu_k}, \quad k \geq 1 \text{ and } \pi_0 = \Lambda^{-1}.$$

10.47. A student tosses a die and makes the corresponding number of steps towards the finish line. The distance left to the finish equals six steps. Find the expected duration of the walk until reaching the finish in the following cases.

(1) The step is not made if the number on the die is greater than the distance to the finish.

(2) The student stops at the finish if the number on the die is more than or equal to the distance to the finish.

3) The random walk reflects from the finish if the number on the die is greater than the distance to the finish.

10.48. A package of requests arrives at a device buffer every second. The device processes 1 request per second. The sizes of incoming packages are i.i.d. random variables $\{\xi_m, m \geq 1\}$, $\mathsf{P}(\xi_m = k) = p_k, k \geq 0$. Describe the transition probabilities of the Markov chain $\{X_n = (\text{number of requests in the buffer at time } n), n \geq 1\}$. Study the chain for recurrence.

10.49. Let $\{X_n, n \geq 0\}$ be a random walk in $\mathbb{Z}^d$. Assume that the step of the walk is uniformly distributed on $\{x \in \mathbb{Z}^d,\ \|x\| = 1\}$. For which d is this random walk recurrent?

10.50. Let the expectation of the step of a random walk be finite and nonzero. Prove that the random walk is not recurrent.

10.51. John and Peter play a game: they toss a coin till either three heads or two tails appear. In the first case John wins \$3 and in the second case he loses \$2. Find the expected amount of John's winnings.

10.52. Suppose that a Markov chain $\{X_n\}$ visits the set $A \subset \mathbb{X}$ infinitely many times with probability 1. Let τ_m be the mth moment. That is, $\tau_m = \inf\{k > \tau_{m-1} | X_k \in A\}$, where $\tau_{-1} \equiv 0$. Prove that $Y_n = X_{\tau_n}, n \geq 0$ is a Markov chain.

10.53. Assume that $\{X_n\}$ is an aperiodic Markov chain, and all states communicate. Prove that for any $d \in \mathbb{N}$ a sequence $Y_n = X_{nd}, n \geq 0$ is an aperiodic Markov chain, and all states of $\{Y_n\}$ communicate.

10.54. Let the transition matrix of a Markov chain $\{X_n\}$ be

$$P = \begin{Vmatrix} 0 & 1 & 0 \\ \frac{1}{2} & 0 & \frac{1}{2} \\ 0 & \frac{1}{2} & \frac{1}{2} \end{Vmatrix}.$$

Find all possible distributions of X_0 such that the distributions of X_0 and X_{24} are equal.

10.55. The transition matrix of a Markov chain $\{X_n, n \geq 0\}$ equals

$$P = \begin{Vmatrix} 0 & 0.5 & 0 & 0.5 \\ 0.3 & 0 & 0.7 & 0 \\ 0 & 0.4 & 0 & 0.6 \\ 0.1 & 0 & 0.9 & 0 \end{Vmatrix}.$$

Classify the states and find the invariant distribution. Prove that $Y_n = X_{2n}, n \geq 0$ is a Markov chain. Find all invariant distributions of $\{Y_n\}$.

10.56. Suppose that $\{X_n, Y_n, n \geq 1\}$ are independent homogeneous Markov chains with the same transition probabilities. Let $\tau = \inf\{n \mid X_n = Y_n\}$.

(1) Prove that sequences

$$V_n := (X_n, Y_n); \quad W_n := \begin{cases} (X_n, Y_n), & n < \tau, \\ (X_n, X_n), & n \geq \tau; \end{cases} \qquad U_n := \begin{cases} Y_n, & n < \tau, \\ X_n, & n \geq \tau \end{cases}$$

are homogeneous Markov chains.

(2) Assume that a phase space is finite, all states communicate, and states are aperiodic. Prove that $\mathsf{P}(\tau < \infty) = 1$.

(3) Generalize previous results for regular continuous-time Markov chains $\{X(t), Y(t), t \geq 0\}$.

10.57. Assume that for every couple of states i, j of the regular Markov chain $\{X(t), t \geq 0\}$ with finite phase space there exists a moment t_{ij} such that $p_{ij}(t_{ij}) > 0$. Prove that there exists a unique stationary distribution, and the distribution of $X(t)$ converges as $t \to \infty$ to the stationary distribution whatever the initial distribution $X(0)$ is.

10.58. Let the phase space $\mathbb{X}$ of the stochastically continuous Markov chain $\{X(t), t \geq 0\}$ be finite. Prove that X has càdlàg modification, and this modification is a regular process.

10.59. Let all states of a Markov chain $\{X(t), t \geq 0\}$ be regular and its intensities satisfy the condition $\sup_i \alpha_{ii} > -\infty$. Prove that the chain X is regular.

10.60. Let a Markov chain $\{X(t), t \geq 0\}$ have a finite phase space and be stochastically continuous. Prove that for any $t > 0$ the determinant of the matrix $P(t)$ is positive.

10.61. Let a stochastically continuous Markov chain $\{X(t), t \geq 0\}$ have a finite phase space and for some i, j there exists a time moment t such that $p_{ij}(t) > 0$. Prove that $p_{ij}(t) > 0$ for all $t > 0$.

10.62. Let $\{X(t), t \geq 0\}$ be a regular Markov chain, $X_0 = i$, τ_1 be a moment of the first jump. Prove that τ_1 has exponential distribution with parameter $(-\alpha_{ii})$ and

$$\mathsf{P}(X(\tau_1) = j / X(0) = i) = -\frac{\alpha_{ij}}{\alpha_{ii}}, \quad i \neq j.$$

10.63. Suppose that a homogeneous continuous-time Markov chain $\{X(t), t \geq 0\}$ is regular and $\alpha_{ii} \neq 0$ for any $i \in \mathbb{X}$. Let $\{\tau_n, n \geq 0\}$ be a moment of the nth jump of $\{X(t), t \geq 0\}$; that is, $\tau_{n+1} = \inf\{t > \tau_n | X(t) \neq X(\tau_n)\}$. Prove that $\mathsf{P}(\tau_n = +\infty) = 0, n \in \mathbb{N}$ and $Y_n = X(\tau_n), n \geq 0$ is a homogeneous Markov chain.

10.64. Prove that the Poisson process is a homogeneous continuous-time Markov chain. Find its transition matrices $P(t), t \geq 0$ and the generator A. Is this chain regular?

10.65. Let a phase space of a Markov chain $\{X(t), t \geq 0\}$ be $\{1,2\}$. A jump $1 \mapsto 2$ intensity is a. No other jumps occur. Find $P(t), t \geq 0$.

10.66. Let a phase space of a Markov chain $\{X(t), t \geq 0\}$ be $\{1,2,3\}$. The jumps $1 \mapsto 2, 2 \mapsto 3$ intensities are a, b, respectively. No other jumps occur. Find $P(t), t \geq 0$. Consider the cases $a = b$ and $a \neq b$.

10.67. Let a phase space of a Markov chain $\{X(t), t \geq 0\}$ be $\{1,2,3,4\}$. The jumps $1 \mapsto 2, 2 \mapsto 3, 3 \mapsto 4$ intensities are equal to a. No other jumps occur. Find $P(t), t \geq 0$. Generalize this problem for the case when the phase space consists of n elements.

10.68. Let a phase space of a Markov chain $\{X(t), t \geq 0\}$ be $\{1,2,3\}$. The jumps $1 \mapsto 2, 1 \mapsto 3$ intensities equal a, b, respectively. No other jumps occur. Find $P(t)$, $t \geq 0$.

10.69. Let a phase space of a Markov chain $\{X(t), t \geq 0\}$ be $\{1,2,3,4\}$. The jumps $1 \mapsto 2, 1 \mapsto 3, 2 \mapsto 4, 3 \mapsto 4$ intensities equal a, b, c, d, respectively. No other jumps occur. Find $P(t), t \geq 0$.

10.70. Let a phase space of a Markov chain $\{X(t), t \geq 0\}$ be $\{1,2,3,4\}$. The jumps $1 \mapsto 2, 1 \mapsto 3, 2 \mapsto 4$ intensities equal a, b, c, respectively. No other jumps occur. Find $P(t), t \geq 0$.

10.71. Let a phase space of a Markov chain $\{X(t), t \geq 0\}$ be $\{1,2,3,4\}$. The jumps $1 \mapsto 2, 2 \mapsto 3, 2 \mapsto 4$ intensities equal a, b, c, respectively. No other jumps occur. Find $P(t), t \geq 0$.

10.72. Let a Markov chain, and $\{X(t), t \geq 0\}$ be a birth-and-death process with $\mu_i = i\mu, \lambda_i = i\lambda, i \in \mathbb{Z}^+$. Find $\mathsf{E}X(t), t \geq 0$, if $X(0) = n$.

10.73. Let $\{X(t), t \geq 0\}$ be a Markov process on $\mathbb{Z}$ with intensities $\alpha_{k,k+1} = \lambda_k > 0, k \in \mathbb{Z}^+$ and $\alpha_{k,0} = \mu_k > 0, k \in \mathbb{N}$ (all other intensities are equal to zero). Find the conditions on $\{\lambda_k\}$ and $\{\mu_k\}$ under which the process is recurrent. When does $\{X(t), t \geq 0\}$ possess the invariant distribution? Find this distribution.

10.74. Let $\{X(t), t \geq 0\}$ be a Markov process on $\mathbb{Z}^+$ with intensities $\alpha_{k,k-1} = \mu_k > 0, k \in \mathbb{Z}^+$ and $\alpha_{0,k} = \lambda_k > 0, k \in \mathbb{N}$. Assume that $\sum_k \lambda_k < +\infty$ (all other intensities are equal to zero). Find the conditions on $\{\lambda_k\}$ and $\{\mu_k\}$ under which the process is recurrent. When does $\{X(t), t \geq 0\}$ possess the invariant distribution? Find this distribution.

10.75. Let $\{X(n), n \geq 0\}$ be a Markov chain, and $\{N(t), t \geq 0\}$ be a Poisson process independent on it. Prove that a stochastic process $\{X(N(t)), t \geq 0\}$ is a Markov process. Express its transition function via transition probabilities of X. Find the generator.

10.76. Let $\{X(t), t \geq 0\}$ be a continuous-time Markov chain with a finite phase space. Assuming that X is stochastic continuous, prove that there exist an independent discrete-time Markov chain $\{Y(n), n \geq 0\}$ and the Poisson process $N(t), t \geq 0$ with intensity λ such that a process $\{Y(N(t)), t \geq 0\}$ has the same distribution as $\{X(t), t \geq 0\}$. What is the least value λ can have?

10.77. Find representations of the form $\{Y(N(t)), t \geq 0\}$ from Problem 10.76 if the generator of the process $\{X(t), t \geq 0\}$ is

$$\left\|\begin{matrix} -4 & 1 & 1 & 2 \\ 1 & -3 & 2 & 0 \\ 0 & 5 & -6 & 1 \\ 1 & 1 & 1 & -3 \end{matrix}\right\|.$$

10.78. Let $\{X(t), t \geq 0\}$ be a Markov process with the generator from the previous problem. Denote by τ_n the moment of the nth jump of X. Find the probability $\mathsf{P}(X(\tau_1) = 1, X(\tau_2) = 4, X(\tau_3) = 3 / X(0) = 2)$.

10.79. Packages arrive at a device at random with intensity λ and are processed with intensity μ, $\lambda < \mu$. If the device is busy then a new package is added at the end of the queue. Denote by $X(t)$ the total number of packages waiting in a queue or processing at the moment t.

(1) Write the Kolmogorov equations for transition probabilities of the Markov process $X(t)$.

(2) Find the expectation of the queue's length and time of processing a package that has just come if the system is in the stationary state. Assume that the number of places in the queue is infinite.

10.80. A device accepts requests with intensity λ and processes requests with intensity μ, $\lambda \leq \mu$. If the device is busy then a new package is added at the end of the queue. The capacity of the queue is n. If the queue is full then any other pending request at the moment is discarded. Let $X(t)$ be defined as in Problem 10.79.

(1) Write the Kolmogorov equations for the transition probabilities for the Markov process $X(t)$.

(2) Find the expectation of the queue's length and the probability that the arrived request is discarded, if the system is in the stationary state.

10.81. A switch unites n channels transferring signals. Signals arrive with intensity λ, and each signal can be transferred by some channel with intensity μ. If all channels are busy then any other pending signal at the moment is discarded.

(1)Write the Kolmogorov equations for the transition probabilities of the Markov process $X(t) =$ (number of busy channels at the moment t).

(2) Find the expectation of the number of busy channels if the system is in the stationary state.

(3) Find the probability that the arrived signal is discarded if the system is in the stationary state.

10.82. Consider the queueing system from the previous problem. Assume that at the initial moment of time all channels were free. Find the probability that the signal that arrived at moment t is discarded if the number of channels is equal to: (a) $n = 1$; (b) $n = 2$. Find the expectation of the number of lost signals until moment t.

10.83. A router accepts packets with intensity λ, and it processes packets with intensity μ. If the router is busy then a new packet is added at the end of the queue. The queue capacity is n. Let $p_n(t)$ be the probability that the packet arrived at the moment t is discarded. Find $\lim_{n\to\infty} \lim_{t\to\infty} p_n(t)$.

10.84. Requests of two types A and B arrive at the device at random with intensities λ_1, λ_2 and are processed with intensities μ_1, μ_2, respectively. If the device is busy then any other pending request at the moment is discarded.

(1) Define the Markov process describing such a queueing system. Write the Kolmogorov equations for the transition probabilities.

(2) Find the probability that an arriving request is discarded if the system is in the stationary state.

10.85. Consider the queueing system from the previous problem, but assume that the type A request has higher priority than the type B request. That is, if the system is busy with processing the type B request, and the type A request arrives, then request B is discarded and request A starts the process immediately. Suppose that the system is in the stationary state.

(1) Find the probability that the arrived request is discarded if it is of: (a) type A; (b) type B.

(2) Find the probability that the request B will be accepted and processed.

10.86. The queueing system consists of the main and standby devices. If the main device is free then the request is processed by it. Otherwise, if the standby device is free then the request is processed to the end by the standby device. The intensities of processing by the main and standby devices are equal to μ_1 and μ_2, respectively. The request arriving intensity equals λ.

(1) Characterize the Markov process describing this queueing system. Write the Kolmogorov equations for the transition probabilities.

(2) Solve this problem under the following two arrangements: if both devices are busy then a new request: (a) is added at the end of the (infinite capacity) queue; (b) is discarded.

10.87. A device accepts requests with intensity λ, and it processes requests with intensity μ. If the device is busy then the new request is added at the end of the queue. The capacity of the queue is n. The intensity of discarding a request during the processing or waiting in a queue is equal to ν. Find the probability that the request will be accepted and processed if the system is in the stationary state.

10.88. The queueing system consists of n devices. Requests arrive at random with intensity λ. Every device processes a request with intensity μ. If all devices are busy, then the new request is added at the end of the common queue. The capacity of the queue is m. Find the stationary distribution.

10.89. The queueing system consists of two devices. The requests of types A and B arrive at the device with intensities λ_1, λ_2, respectively. Every device processes the request A (B) with intensity μ_1 (μ_2). If both devices are busy then a new request is discarded. Characterize the Markov process describing such a queueing system. Write the Kolmogorov equations for the transition probabilities.

10.90. The requests of types A and B arrive at the device with intensities λ_1, λ_2 and are processed with intensities μ_1, μ_2, respectively. The number of the queue capacity is equal to 1. Request A has higher priority than B. That is, if the device is busy with processing of the type B request, another request B is waiting in the queue, and the type A request arrives, then the first request B moves to the queue, the second request B discards, and request A starts process.

(1) Characterize a Markov process describing such a queueing system. Write the Kolmogorov equations for transition probabilities.

(2) Assume that at the initial moment of time the request B is processing. Find the probability that it will be processed sooner or later.

(3) Find the probability that the request A starts the process on arrival. Suppose that the system is in the stationary state.

Hints

10.4. States i and j communicate. So, there exist n and m such that $p_{ij}^{(n)} > 0$ and $p_{ji}^{(m)} > 0$. Prove that $p_{ii}^{(n+k+m)} \geq p_{ij}^{(n)} p_{jj}^{(k)} p_{ji}^{(m)}$ for any $k \in \mathbb{N}$. Therefore, if a series $\sum_k p_{jj}^{(k)}$ is divergent, then a series $\sum_k p_{ii}^{(k)}$ is divergent as well. Apply Theorem 10.1.

10.5. Let v be a probability of return to the initial state. Use Theorem 10.5 and prove that the probability to return to the initial state at least n times is equal to v^n. For the transient state use the Borel–Cantelli lemma.

10.9. Let n_0 be such number that $\alpha := \min_j \max_{1\le n\le n_0} \mathsf{P}(X_n \in A/X_0 = j) > 0$. Then $\mathsf{P}(X_k \notin A,\ k = 1,\ldots,nd) \le \mathsf{P}(X_{ld} \notin A,\ l = 1,\ldots,n) \le \alpha^n$.

For the Markov chain with a countable set of states the corresponding statements, in general, are not correct.

10.19. Construct the chains $\{\bar{X}_n,\ n\ge 0\}$, $\{\bar{Y}_n,\ n\ge 0\}$ that are stochastically equivalent in the wide sense to $\{X_n,\ n\ge 0\}$, $\{Y_n,\ n\ge 0\}$, respectively, and are of the following type $\bar{X}_{n+1} = f(\bar{X}_n,\varepsilon_n)$, $\bar{Y}_{n+1} = g(\bar{Y}_n,\varepsilon_n)$, where $\bar{X}_0 \ge \bar{Y}_0$, $\{\varepsilon_n,\ n\ge 0\}$ are the i.i.d. random variables and $f(x,\varepsilon) \ge g(y,\varepsilon)$ for all $\varepsilon > 0, x, y, x \ge y$.

10.27. Consider the conditional probabilities $\mathsf{P}(Y_2 = 3/Y_1 = 2, Y_0 = 3)$ and $\mathsf{P}(Y_2 = 3/Y_1 = 2)$.

10.29. $\mathsf{P}(X_n = (k,l)/X_0 = (i,j))$ equals the product $p_{ik}^{[n/2]} \cdot p_{jl}^{n-[n/2]}$ where $p_{ik}^{(r)}$ is the transition probability for the symmetric random walk on $\mathbb{Z}$ (see Problem 10.28 with $p = \frac{1}{2}$).

10.30. Let $\{X_n = (X_n^1, X_n^2)\}$ be the sequence from Problem 10.29. Prove that the sequences $\{Y_n\}$ and

$$\left\{\widetilde{Y}_n = \left(\frac{X_{2n}^1 + X_{2n}^2}{2}, \frac{X_{2n}^1 - X_{2n}^2}{2}\right)\right\}$$

have the same distributions. Thus,

$$\begin{aligned}&\mathsf{P}(Y_n = (k,l)/Y_0 = (i,j))\\ &\quad = \mathsf{P}(X_{2n}^1 = k+l-i-j, X_{2n}^2 = k+j-i-l/X_0 = (0,0)).\end{aligned}$$

In particular,

$$\mathsf{P}(Y_{2m} = (0,0)/Y_0 = (0,0)) = \frac{C_{2m}^m}{2^{2m}},$$

$$\mathsf{P}(Y_{2m-1} = (0,0)/Y_0 = (0,0)) = 0.$$

Because the series $\sum_{m=1}^{\infty} 2^{-2m} C_{2m}^m$ diverges then $(0,0)$ is recurrent. By the same reasoning show that any other state is also recurrent.

10.32. (a) Determine the one-to-one correspondence between the route from point j to $(-k)$ and the route from j to k that visits zero.

10.35. The desired probability equals the sum of the transition probabilities from i to j in n steps without visitation zero plus the corresponding probability with visitation zero. Apply the results of Problems 10.32, 10.34.

10.37. See Problem 10.13 (d).

10.38. See Problem 10.13 (c).

10.40. Use Theorem 10.2.

10.45. Prove that for the invariant distribution $(\pi_0,\ldots,\pi_N)$ the following equality holds, $q_i\pi_i - p_{i-1}\pi_{i-1} = c, i = 0,\ldots,N$, where c is some constant (the notion $0-1 = N$ is used). Check that $c = 0 \Leftrightarrow p_0\cdots p_N = q_0\cdots q_N$. Every π_i can be expressed either via π_0 in the case $p_0\cdots p_N = q_0\cdots q_N$ or via c in the case $p_0\cdots p_N \ne q_0\cdots q_N$. The unknown values π_0 or c can be found from the condition $\pi_0 + \cdots + \pi_N = 1$.

10.48. Let ε be the number of requests in one package. Use the law of large numbers and prove that the chain is recurrent if $\mathsf{E}\varepsilon < 1$ and the chain is not recurrent if $\mathsf{E}\varepsilon > 1$.

When $\mathsf{E}\varepsilon = 1$ then the chain is recurrent if and only if the random walk on $\mathbb{Z}$ with the step equal to $\varepsilon - 1$ is recurrent. Use Theorem 10.2 and prove the recurrence.
10.49. Use Theorem 10.2.
10.50. Use the law of large numbers.
10.52. Prove that τ_n is a Markov moment and use the strong Markov property.
10.53. To prove the Markov property of $\{Y_n\}$ use the definition. Check that for any $i \in \mathbb{X}$ there exists a number n_0 such that $p_{ii}^{(n)} > 0,\ n \geq n_0$. It follows from this that all states of the chain $\{Y_n\}$ communicate and have a period 1.
10.54. $\{X_n\}$ is an aperiodic Markov chain and all states of $\{X_n\}$ communicate. Thus, $Y_n = X_{24n}, n \geq 0$ is also an aperiodic Markov chain all states of which communicate (see 10.53). Thus, $\{Y_n\}$ has a unique invariant distribution. The invariant distribution of $\{X_n\}$ is also the invariant distribution of $\{Y_n\}$, so this distribution is the unique solution of the problem.
10.59. Check that the generator A is a bounded linear operator in the space l_∞. That is why the first Kolmogorov system of equations has the unique solution. Use Theorem 10.8.
10.63. To prove that $\mathsf{P}(\tau_n = +\infty) = 0$ use Problem 10.62. To prove the Markov property for $\{Y_n\}$ verify that $\{\tau_n\}$ are Markov moments and use the strong Markov property.
10.69. $p_{11}(t) = e^{-(a+b)t}$, $p_{22}(t) = e^{-ct}$, $p_{24}(t) = 1 - e^{-ct}$, $p_{33}(t) = e^{-dt}$, $p_{34}(t) = 1 - e^{-dt}$, $p_{44}(t) = 1$. Use the Kolmogorov equations for $p_{12}(t)$ and $p_{13}(t)$, and obtain the differential equations with one variable. Then, find $p_{14}(t)$, $p_{13}(t)$. Other probabilities are equal to zero.
10.70. $p_{11}(t) = e^{-(a+b)t}$, $p_{22}(t) = e^{-ct}$, $p_{24}(t) = 1 - e^{-ct}$, $p_{13}(t) = 1 - e^{-bt}$, $p_{33}(t) = p_{44}(t) = 1$, $p_{14}(t) = 1 - p_{13}(t) - p_{11}(t) - p_{12}(t)$. Use the Kolmogorov equations for $p_{12}(t)$ and obtain the differential equations with one variable.
10.71. $p_{11}(t) = e^{-at}$, $p_{22}(t) = e^{-(b+c)t}$, $p_{24}(t) = (c/(b+c))(1 - e^{-(b+c)t})$, $p_{23}(t) = (b/(b+c))(1 - e^{-(b+c)t})$, $p_{33}(t) = p_{44}(t) = 1$, $p_{14}(t) = 1 - p_{13}(t) - p_{11}(t) - p_{12}(t)$. Use the Kolmogorov equations for $p_{13}(t)$ and obtain the differential equations with one variable.

All other probabilities are equal to zero.

Answers and Solutions

10.3. (a) Yes. (b) No.
10.6. If $j \nrightarrow i$, then $d(i)$, $d(j)$ can be arbitrary. If $i \leftrightarrow j$ then $d(i) = d(j)$.
10.11. (a) 0.75; (b) 3; (c) 0.75, 12.
10.12. (a) If $p = \frac{1}{2}$ then the ruin probability for John is equal to $m/(m+n)$. Otherwise it is equal to

$$\frac{(q/p)^{n+m} - (q/p)^n}{(q/p)^{n+m} - 1}.$$

(b) If $p = \frac{1}{2}$ then the expectation of the game duration is equal to nm, and

$$\frac{n}{q-p} - \frac{n+m}{q-p}\frac{1-(q/p)^n}{1-(q/p)^{n+m}}$$

otherwise.

10.13. (a) Let all $p_i > 0$. Then $\pi_1 = 1/(2-\pi_1)$, $\pi_k = p_k/(2-\pi_1)$, $k \geq 2$.
(b) Any distribution for which $p_1 = 0$ is stationary.
(c) Assume that all $p_i > 0$. Denote $a_{n+1} = (1 - p_1 - \cdots - p_n), n \geq 1, a_1 = 1$. The stationary distribution exists if and only if $\sum_n a_n < \infty$. At the same time $\pi_k = a_k / \sum_n a_n$.
(d) Let all $p_i > 0$. Denote $a_{n+1} = p_1 \cdot \cdots \cdot p_n, n \geq 1, a_0 = 1$. The chain is recurrent if and only if $\lim_{n\to\infty} a_n = 0$. The stationary distribution exists if and only if $\sum_n a_n < \infty$, and $\pi_k = a_k / \sum_n a_n$.
10.17. (a) $p = \frac{2}{3}$.
(b) No. However, a sequence $Y_n = (X_{n-1}, X_n)$ is a Markov chain.
10.21. No.
10.22. $p = 0.5$; $p = 0$; $p = 1$.
10.24. (a) No. (b) Yes.
10.25. (a) No. (b) Yes.
10.26. The aperiodic Markov chain with one class of communicating states.
10.27. No.
10.28.

$$p_{ij}^{(n)} = C_n^{\frac{n+j-i}{2}} p^{\frac{n+j-i}{2}} q^{\frac{n+i-j}{2}},$$

if $n + j - i$ is the even number and $|j - i| \leq n$.
The Stirling's formula implies that

$$p_{ii}^{(2m)} \sim \frac{(4p(1-p))^m}{\sqrt{\pi m}}, \quad m \to \infty.$$

Thus, the random walk is recurrent if, and only if, the series

$$\sum_m \frac{(4p(1-p))^m}{\sqrt{\pi m}}$$

diverges, that is, $p = \frac{1}{2}$.
10.32. (b) $\mathsf{P}(X_1 \neq 0, \ldots, X_{m-1} \neq 0, X_m = k | X_0 = j)$

$$= \mathsf{P}(X_m = k|X_0 = j) - \mathsf{P}(\exists\, l \in \{1,\ldots,m\} : X_l = 0, X_m = k|X_0 = j)$$
$$= C_m^{(m+k-j)/2} 2^{-m} - \mathsf{P}(X_m = -k|X_0 = j) = 2^{-m}(C^{(m+k-j)/2} - C^{(m-k-j)/2}),$$

if $|k - j| \leq m, m + k - j$ is an even number.
(c) $\mathsf{P}(X_1 \neq 0, \ldots, X_{m-1} \neq 0, X_m = 0|X_0 = j)$

$$= \mathsf{P}(X_1 \neq 0, \ldots, X_{m-2} \neq 0, X_{m-1} = 1, X_m = 0|X_0 = j)$$
$$= \frac{1}{2}\mathsf{P}(X_1 \neq 0, \ldots, X_{m-2} \neq 0, X_{m-1} = 1|X_0 = j).$$

(d) $\mathsf{P}(X_1 \geq 0, \ldots, X_{m-1} \geq 0, X_m = k|X_0 = j) = \mathsf{P}(X_1 \neq 0, \ldots, X_{m-1} \neq 0, X_m = k + 1|X_0 = j + 1)$.
(e) k/m.
10.33. The required probability coincides with the answer to Problem 10.32 (b) if $i \neq 0, j \neq 0$.

10.34. Due to Problem 10.31:

$$p_{ij}^{(n)} = \frac{C_n^{(n+j-i)/2} + C_n^{(n-i-j)/2}}{2^n},$$

where $i, j \in \mathbb{N}, n+j-i$ is an even number, $|j-i| \le n$.

10.35. Let $i \ge 0$. Then:

$$p_{ij}^{(n)} = \begin{cases} 2^{-n}(C_n^{(n+j-i)/2} - C_n^{(n-i-j)/2}) + 2^{-n+1} p C_n^{(n-i-j)/2}, & j \ge 0; \\ 2^{-n+1} q C_n^{\frac{n-i-j}{2}}, & j < 0. \end{cases}$$

10.36. The chain is recurrent if $p \le 0.5$.

10.41. $p_{k,k+1} = (N-k)^2/N^2$; $p_{k,k-1} = k^2/N^2$; $p_{k,k} = 2k(N-k)/N^2, k = 1, \dots, N-1$; $p_{0,1} = p_{N,N-1} = 1$.

The stationary distribution is $\pi_k = C_N^k/2^k$, $k = 0, \dots, N$.

All states communicate and have period 1.

10.43. The code of the letter a_1 at the moment n is not a Markov chain generally, but the vector $X(n) = (x_1(n), \dots, x_m(n))$ is an aperiodic Markov chain all states of which communicate. That is why there exists only one stationary distribution. It is also assumed that $X(n)$ is already stationary distributed.

(a) Check that the probability that a_1 is encoded by 1 at the moment $n+1$ is equal to p_1.

(b) Let r be a sought probability. Then $r = \mathsf{P}(x_1(n+1) = 1, x_2(n+1) = 2) = \mathsf{P}(x_1(n) = 1, x_2(n) = 2,$ and a letter a_1 has arrived at the moment $n) + \mathsf{P}(x_2(n) = 1,$ and a letter a_1 has arrived at the moment $n) = rp_1 + p_2 p_1$. Therefore, $r = p_1 p_2/(1-p_1)$.

10.44. Denote by $y^{n,k}, k = 1, \dots, n$ the probability that the chain visits the point 0 earlier than $n+1$, $y^{n,0} = 1$, $y^{n,n+1} = 0$. Then

$$y^{n,k} = q_k y^{n,k-1} + r_k y^{n,k} + p_k y^{n,k+1}. \tag{10.6}$$

Taking into account that $r_k = 1 - q_k - p_k$ we obtain:

$$y^{n,k} - y^{n,k-1} = \frac{q_k}{p_k}(y^{n,k-1} - y^{n,k-2}) = \dots = \frac{q_1}{p_1} \times \dots \times \frac{q_k}{p_k}(y^{n,1} - 1).$$

We add these identities and obtain $y^{n,n} - y^{n,1} = (y^{n,1} - 1)\sum_{k=1}^{n} \prod_{i=1}^{k} p_i q_i^{-1}$. Due to (10.6) we have that $-y^{n,n} = \prod_{i=1}^{n+1} p_i q_i^{-1} (y^{n,1} - 1)$ for $k = n+1$.

Thus,

$$y^{n,1} = \frac{\sum_{k=1}^{n+1} \prod_{i=1}^{k} p_i q_i^{-1}}{1 + \sum_{k=1}^{n+1} \prod_{i=1}^{k} p_i q_i^{-1}}.$$

The chain is recurrent if and only if $\lim_{n\to\infty} y^{n,1} = 1$ or $\sum_{k=1}^{\infty} \prod_{i=1}^{k} p_i q_i^{-1} = +\infty$. The stationary distribution exists if $A := 1 + \sum_{k=1}^{\infty} \prod_{i=1}^{k} p_{i-1} q_i^{-1} < +\infty$. In this case

$$\pi_0 = A^{-1}, \quad \pi_k = A^{-1} \prod_{i=1}^{k} p_{i-1} q_i^{-1}, \quad k \ge 1.$$

10.45. (a) $\pi_i = \sigma_i/(\sum_{j=0}^{N} \sigma_j)$, where $\sigma_0 = 1, \sigma_j = (p_{j-1} \cdots p_0)/(q_j \cdots q_1), j = 1, \dots, N$.

(b) $\pi_i = p_i^{-1}(\sum_{j=0}^{N} p_j^{-1})^{-1}, i = 0, \dots, N$.

(c) Denote

$\rho_i = p_i/q_i$, $\theta_i = \Big[q_i(1-\rho_0\cdots\rho_N)\Big]^{-1}\Big(1+\rho_{i-1}+\rho_{i-1}\rho_{i-2}+\cdots+\rho_{i-1}\cdots\rho_0\rho_N\cdots$ $\rho_{i+1}\Big)$; then $\pi_i = \theta_i/(\sum_{j=0}^N \theta_j)$, $i = 0,\ldots,N$.

10.47. (a), (c) 6.

(b) Let r_k, $k = 1,\ldots,6$ be the expected time to finish if the distance to finish equals k. Then r_k can be found from the recurrent formula: $r_1 = 1$, $r_{k+1} = 1+\frac{1}{6}\sum_{j=1}^k r_j$.

10.49. The chain is recurrent if and only if $d \le 2$.

10.54. $\mathsf{P}(X_0 = 1) = 0.2$, $\mathsf{P}(X_0 = 2) = \mathsf{P}(X_0 = 3) = 0.4$.

10.60. $\det P(t) = \det\exp\{At\} = (\det\exp\{At/n\})^n$. If n is large, then the matrix $\exp\{At/n\}$ is close to the identity matrix, and therefore its determinant is positive.

Furthermore, it is known from the theory of differential equations that $\det P(t) = \exp\{(\mathrm{tr}A)t\} > 0$.

10.64. The Poisson process is a regular Markov chain with

$$p_{ij}(t) = \frac{(\lambda t)^{j-i}}{(j-i)!}e^{-\lambda t}\mathbb{I}_{j\ge i}, \quad A = \|\alpha_{ij}\|,$$

where $\alpha_{ii} = -\lambda, \alpha_{i(i+1)} = \lambda$ and $\alpha_{ij} = 0$ for all other i, j.

10.72. $\mathsf{E}X(t) = ne^{(\lambda-\mu)t}$.

10.75. $P(t) = \sum_{n\ge 0} e^{-\lambda t}((\lambda t)^n/n!)P^n, \quad A = \lambda(P - \mathbb{I})$.

10.76. $P_\lambda = \lambda^{-1}A + \mathbb{I}$, where $\lambda \ge \max_{1\le i\le n}(-\lambda_{ii})$.

10.78. See Problem 10.62: $\frac{1}{3}\cdot\frac{2}{4}\cdot\frac{1}{3} = \frac{1}{18}$.

10.79. $X(t)$ is the number of packages in the system; transition intensities $\alpha_{n,n+1} = \lambda, a_{n,n-1} = \mu$. The stationary distribution is

$$\pi_k = \frac{\left(\frac{\lambda}{\mu}\right)^k}{1-\frac{\lambda}{\mu}}, \; k \in \mathbb{Z}_+.$$

10.80. The stationary distribution is

$$\pi_k = \left(\frac{\lambda}{\mu}\right)^k\left(\sum_{j=0}^{n+1}\left(\frac{\lambda}{\mu}\right)^j\right)^{-1}, \qquad k = 0,\ldots,n+1.$$

10.81. Stationary distribution is

$$\pi_k = \frac{1}{k!}\left(\frac{\lambda}{\mu}\right)^k\left(\sum_{j=0}^{n}\frac{1}{j!}\left(\frac{\lambda}{\mu}\right)^j\right)^{-1}, \qquad k = 0,\ldots,n.$$

10.83.

$$\begin{cases}\frac{\lambda-\mu}{\mu}, & \lambda > \mu,\\ 0, & \lambda \le \mu.\end{cases}$$

10.84. Introduce the states: A = "the request A is processed"; B= "the request B is processed"; 0 = "the system is free". The transition intensities $0 \to A$, $0 \to B$, $A \to 0$, $B \to 0$ are equal to λ_1, λ_2, μ_1, μ_2, respectively. The probability for a request to be rejected equals $(\lambda_1\mu_2 + \mu_1\lambda_2)/(\lambda_1\mu_2 + \mu_1\lambda_2 + \mu_1\mu_2)$.

10.85.

$$\pi_A = \frac{\lambda_1}{\lambda_1+\mu_1}, \quad \pi_B = \frac{\mu_1\lambda_2}{(\lambda_1+\lambda_2+\mu_2)(\lambda_1+\mu_1)}.$$

The probability that the request B is accepted and processed is equal to $(1-\pi_B)\mu_2/(\lambda_1+\mu_2)$.

10.87. Let $X(t) \in \{0,1,\ldots,n+1\}$ be the number of requests in a system (either processing or waiting in the queue) at moment t. To find the stationary distribution $\{\pi_k\}$ observe that $X(t)$ is the birth-and-death process with the birth intensity $\lambda_k = \lambda, k=0,\ldots,n$ and the death intensity $\mu_j = \mu + j\nu$. The probability that the request is accepted and processed is equal to $\sum_{k=0}^{n} \pi_k \rho_k$, where ρ_k is the probability that the request has k requests in a queue before it will be processed.

It order to find ρ_k, consider the additional construction. Let $\{\tau_j, j \geq 1\}$ be the time moments when the requests, in the queue before the initial request (inclusively), leave the system (i.e., they either have been processed or lost), $\tau_0 = 0$. Denote by Y_m the position of the initial request in the queue at the time moment τ_m. We assume that $Y_m = \infty$ if the initial request is lost and $Y_m = 0$ if it was processed at the moment τ_m. The sequence $Y_m, m \geq 0$ is the Markov chain with the transition probabilities

$$p_{i,i-1} = \frac{\mu+(i-1)\nu}{\mu+i\nu}, \quad p_{i,\infty} = \frac{\nu}{\mu+i\nu},$$

where $i = 1,\ldots,n+1$. Thus, ρ_k is the probability that the sequence $\{Y_m, m \geq 0\}$ hits 0 earlier than ∞ given $Y_0 = k$. This probability is equal to $p_{k,k-1}p_{k-1,k-2}\cdots p_{1,0} = \mu/(\mu+k\nu)$.

10.88. Let

$$A = \sum_{j=0}^{n-1} \frac{(\lambda/\mu)^j}{j!} + \frac{(\lambda/\mu)^n}{n!} \sum_{k=0}^{m} \left(\frac{\lambda}{n\mu}\right)^k.$$

The stationary distribution is

$$\pi_k = \begin{cases} A^{-1}(\lambda/\mu)^k/k!, & k=0,\ldots,n, \\ A^{-1}(\frac{\lambda}{\mu})^k/(n!n^{k-n}), & k=n+1,\ldots,n+m. \end{cases}$$

10.90. (1) States: 0 = "the system is free"; A = "there is only request A in the system"; B = "there is only request B in the system"; AB = "the request A is processing, the request B is waiting in the queue"; AA = "the request A is processing, another request A is waiting in the queue"; BB = "the request B is processing, another request B is waiting in the queue".

(2) This probability is equal to the probability of hitting 0 for the following discrete-time Markov chain. States set: 0, B, AB, C ("cemetery", meaning the request is lost). The transition probabilities are $p_{0,0} = p_{C,C} = 1, p_{B,0} = \mu_2/(\lambda_1+\mu_2), p_{B,AB} = \lambda_1/(\lambda_1+\mu_2), p_{A,B} = \mu_2/(\lambda_1+\mu_2), p_{A,C} = \lambda_1/(\lambda_1+\mu_2)$. The initial state $X_0 = B$.

(3) The required probability is equal to $\pi_2 = \lambda_1^2/(\lambda_1^2+\lambda_1\mu_1+\mu_1^2)$; see the system from Problem 10.80 if $n=1$.

11

Renewal theory. Queueing theory

Theoretical grounds

Queueing theory was founded by Danish scientist A. Erlang who worked for the Copenhagen Telephone Exchange for many years at the beginning of the twentieth century. The works of F. Pollaczek, A. Khinchin, L. Takacs, and also B. Gnedenko and his school had significant influence on the development of queueing theory.

The main model of queueing theory can be described as follows. Assume that one or several devices (machines, routers, cash desks) accept and process some requests (components, packets, clients) that arrive at random. The processing time of each request may also be random. There is a large variability of rules and orders of processing, which have an influence on the quality of service. Examples of such rules could be the following: existence or nonexistence of a queue, different priorities of the requests, and so on. We also have to know what happens to request processing at the current instant of time when a new request with higher priority arrives. In this case the first request may be discarded or may be added to the queue for later processing. Moreover, processing can start from the very beginning or may take into account the time of the initial processing.

The main objects of queueing theory investigations, among others, are the following characteristics: average time that a request spends from arrival until processing; average number of requests that have been processed during a certain period of time; distribution of time until the system becomes free from requests; the distribution of a busy period (i.e., the length of time from the instant the request arrives at an empty system until the instant when the system is empty again); probability that the arrived request is discarded (this happens if the queue is overfull); and the limit or stationary distributions for queueing systems.

Compound Poisson processes with jumps having the same sign are widely used to describe queueing systems in renewal theory:

$$\xi(t) = t - S(t), \quad S(t) = \sum_{k=1}^{N(t)} \xi_k,$$

D. Gusak et al., *Theory of Stochastic Processes*, Problem Books in Mathematics,
DOI 10.1007/978-0-387-87862-1_11, © Springer Science+Business Media, LLC 2010

$$\zeta(t) = S(t) - t, \quad \mathsf{P}(\xi_k > 0) = 1,$$

where $N(t)$ is a Poisson process with rate $\lambda > 0$, and $\{\xi_k,\ k \geq 1\}$ is a sequence of i.i.d. random variables with distribution function $F(x) = \mathsf{P}(\xi_k \leq x)$.

Processes $\xi(t)$ and $\zeta(t)$ have a temporal interpretation in queueing theory. Moments of the process $N(t)$ jumps are associated with moments of request arrivals, jumps ξ_k of a process $S(t)$ are associated with a time necessary for a service of the kth request, and $S(t) = \sum_{k=1}^{N(t)} \xi_k$ is a total time necessary for service of all requests arrived before t. A process $\zeta(t) = S(t) - t$ describes additional time necessary for service after t if the device is not idle in $[0,t]$. A process $\xi(t) = t - S(t)$ is called a controlling process.

A process $\alpha(t) = \xi^+(t) = \sup_{0 \leq t' \leq t} \xi(t')$ is called a nonbusy process or idle process. This is a continuous nondecreasing process. A process $\beta(t) = t - \alpha(t)$ is called a busy process.

Denote by τ_k the difference between the nth and $(n-1)$th jumps of the process $N(t)$, a sequence $\{\tau_k,\ k \geq 1\}$ is a sequence of independent exponential distributed random variables with rate λ.

Introduce virtual waiting time $w(t), t \geq 0$ or the virtual waiting process, as follows.

$$w(t) = \begin{cases} 0, 0 \leq t < \tau_1 = t_1, & w(t_1) = \xi_1, \\ (\xi_1 - t + \tau_1)^+, & t \in [t_1, t_2), \\ (w(t_n) - t + t_n)^+, & t \in [t_n, t_{n+1}), \quad n > 1, \end{cases} \tag{11.1}$$

where

$$t_n = \sum_{k \leq n} \tau_k, \quad w(t_1) = \xi_1, \quad \Delta_n w = w(t_n) - w(t_n - 0) = \xi_n, \quad n > 1.$$

Graphs of $\xi(t), \alpha(t), \beta(t)$, and $w(t)$ are shown on plot 1, page 360.

Processes $\alpha(t), \beta(t)$ are piecewise linear functions. Intervals of linear growth $\beta(t)$ are called busy periods $\{\widetilde{\theta}_k,\ k \geq 1\}$ (they have the same distribution). Intervals where a function $\beta(t)$ is constant form idle periods. Comparing graphs of the processes $\xi(t), \alpha(t)$, and $w(t)$, it is easy to see that at busy periods

$$w(t) = \alpha(t) - \xi(t) > 0, \quad \tau_k' < t < \tau_k' + \widetilde{\theta}_k, \quad (k \geq 1), \tag{11.2}$$

where τ_k' is the initial time of the kth busy period. Outside busy periods $w(t) \equiv 0$.

If $w(0) = 0$ and $\mathsf{E}\xi(1) = 1 - \lambda\mu < 0$ ($\mu = \mathsf{E}\xi_k > 0$) then the following formula holds for the generating function of $w(t)$,

$$\omega(z,t) = \mathsf{E}e^{-zw(t)} = e^{tk(z)} \left(1 - z \int_0^t e^{-uk(z)} P_0(u)\,du\right), \tag{11.3}$$

Here $k(z)$ is the cumulant function of a process $\zeta(t)$; $P_0(u) = \mathsf{P}(w(u) = 0)$ is a probability that a system is free of requests at the time moment u. The function $k(z)$ is defined by

$$k(z) = \ln \mathsf{E}e^{-z\zeta(1)} = z + \lambda\left(\mathsf{E}e^{-z\xi_1} - 1\right) = z\left(1 - \lambda \int_0^\infty e^{-zx}\overline{F}(x)\,dx\right), \tag{11.4}$$

where $\overline{F}(x) := \mathsf{P}(\xi_1 > x)$,

$$\widetilde{p}_0(s) = s\int_0^{\infty} e^{-sx} P_0(x)dx = s\left(s+\lambda-\lambda\widehat{\Pi}(s)\right)^{-1}, \quad \widehat{\Pi}(s) = \mathsf{E}e^{-s\widetilde{\theta}_1}, \tag{11.5}$$

and $\widehat{\Pi}$ is the moment generating function of the first busy period $\widetilde{\theta}_1$.

There are three possible modes for a queueing system with one device and a Poisson stream of requests depending on a sign of $\mathsf{E}\zeta(1) = \lambda\mu - 1$ $(\zeta(t) = \sum_{k=1}^{N(t)} \xi_k - t)$.

(1) $\mathsf{E}\zeta(1) = \lambda\mu - 1 > 0$ $(\lambda\mu > 1)$. Then average service time is greater than average time of request arrival: $\mu > 1/\lambda$. This mode is called above-critical.

(2) $\mathsf{E}\zeta(1) = 0$ $(\mu = 1/\lambda)$. This mode is called critical.

(3) $\mathsf{E}\zeta(1) < 0$ $(\lambda\mu < 1,\ \mu < 1/\lambda)$. In this case the equilibrium mode exists, and random variables $w(t)$ weakly converge as $t \to \infty$.

Note that the distribution function of busy period $\Pi(x) = \mathsf{P}(\widetilde{\theta}_1 < x) \to 1$ as $x \to \infty$ only in the last two cases. In the first case

$$\Pi(+\infty) = (\lambda\mu)^{-1} < 1.$$

Usually the third (equilibrium) case $\mu < \lambda^{-1}$ is the most interesting from the point of view of queueing theory.

Classical results of renewal theory cited below are widely used for queueing systems study.

Definition 11.1. *Assume that* $\{T_n, n \geq 1\}$ *are nonnegative i.i.d. random variables with a distribution function* F, $F(0) < 1$. *Let a sequence of nonnegative random variables* $\{S_n, n \geq 0\}$ *be constructed recurrently*

$$S_{n+1} = S_n + T_{n+1}, \quad n \geq 0,$$

and S_0 *does not depend on* $\{T_n, n \geq 1\}$.

The variables S_n *are called renewal epochs. A random sequence* $\{S_n, n \geq 0\}$ *is said to be a pure renewal process if* $S_0 = 0$, *and a delayed renewal process otherwise.*

For example, T_n may be successive service times of requests, intervals between request arrivals, and so on.

Let $z : \mathbb{R} \to \mathbb{R}$ be a measurable function that equals zero when $x < 0$. The *renewal equation* is an equation of the form

$$Z(x) = z(x) + \int_{[0;x]} Z(x-y)F(dy), \quad x > 0.$$

The renewal equation can be written briefly as

$$Z = z + F * Z, \tag{11.6}$$

where "*" is a convolution sign; that is, $F * Z(x) = \int_{[0;x]} Z(x-y)F(dy)$.

Let us introduce the renewal function

$$U = \sum_{n=0}^{\infty} F^{*n}, \tag{11.7}$$

where F^{*n} is a convolution of n distribution functions F, and F^{*0} a distribution function of a unit measure concentrated at 0.

Theorem 11.1. *If z is a bounded function then*

$$Z = U * z \tag{11.8}$$

is a solution of equation (11.6). *This solution is unique in a class of functions that are bounded on finite intervals and vanish on* $(-\infty, 0)$.

Theorem 11.2. *Assume that the conditions of the previous theorem are satisfied, z is a Riemann integrable function, and*

$$\sum_n \sup_{x\in[n,n+1]} |z(x)| < \infty.$$

If F is not arithmetic, that is,

$$\forall\, h > 0: \ \mathsf{P}(T_1 \in h\mathbb{N}) < 1,$$

then for each $h > 0$ the following equality holds,

$$\lim_{t\to+\infty} Z(t) = \mu^{-1} \int_0^\infty z(y)dy, \tag{11.9}$$

where $\mu \in (0,\infty]$ is the expectation of T_1.

If T_1 is arithmetic then

$$\forall\, x \geq 0: \ \lim_{n\to\infty} Z(x+nh) = \frac{h}{\mu} \sum_k z(x+kh),$$

for all h multiples of a span of a distribution, that is, multiples of the largest g such that $\mathsf{P}(T_1 \in g\mathbb{N}) = 1$.

Bibliography

[22] Volume 1, Chapter 13, Volume 2, Chapters 6, 11, 13, 14; [29]; [6]; [32]; [28]; [47].

Problems

For problems for Markov models in queueing theory see Chapter 10 (Problems 10.76–10.90).

11.1. (Renewal theorem). Assume that the conditions of Theorem 11.1 are satisfied. Prove that

(1) If F is a nonarithmetic distribution then for any $h > 0$

$$\lim_{t\to\infty} (U(t+h) - U(t)) = \frac{h}{\mu}. \tag{11.10}$$

(2) If F is an arithmetic distribution then (11.10) is satisfied when h is a multiple of the span of F.

11.2. Let $\{S_n, n \ge 0\}$ be a delayed renewal process, F_0 be a d.f. of S_0. Denote by $V(t)$ the expected number of renewal epochs until the moment of time t. Prove that
(a) $V(t) = \sum_{n=0}^{\infty} \mathsf{P}(S_n \le t)$.
(b) $V = F_0 + F * V$.
(c) $V = F_0 * U$, where U is given by (11.7).
(d) $\lim_{t\to\infty}(V(t)/t) = \mu^{-1}$.
(e) The renewal rate is constant; that is, $V(t) = \mu^{-1}t$, if and only if

$$F_0(t) = \mu^{-1} \int_0^t (1 - F(y))dy.$$

11.3. Let $\nu(t)$ be a number of renewal epochs within $[0,t]$. Prove that
(a) There exists $\theta > 0$ such that $\mathsf{E}e^{\theta\nu(t)} < \infty$.
(b) If $S_0 = 0$ then $\mathsf{E}\nu(t) = U(t)$, where the function U is given by (11.7).

11.4. A device accepts requests. The processing times for different requests are i.i.d. random variables with distribution function F, expectation μ, and variance σ^2. The processing of a new request starts immediately after the previous one has been processed. The first request starts processing at the moment of time $t = 0$. Let:

$H(t,x)$ be the distribution function of residual processing time for the request processed at the moment t.

$G(t,x)$ be the distribution function of the total time this request had been processing before t.

$L(t,x)$ be the distribution function of full time processing of the request.
(a) Express functions H, G, L in terms of F and U.
(b) Find limits for the distributions $H(t,x)$, $G(t,x)$, $L(t,x)$ as $t \to \infty$ in the case when the distribution of F is nonarithmetical.
(c) Express the first moments of the limit distributions in terms of the moments of F assuming $\mu < \infty$.

11.5. Let V be a renewal function for a delayed renewal process (see Problem 11.2). Assume that F is nonarithmetical, and a function z satisfies the conditions of Theorem 11.2. Consider $Z = V * z$. Prove that

$$\lim_{t\to\infty} Z(t) = \mu^{-1} \int_0^{\infty} z(y)dy.$$

11.6. Solve Problem 11.4 assuming that the first request was processing before moment 0, and its residual processing time has a distribution function F_0.

11.7. Let F be a distribution function of nonnegative random variable ξ. The function

$$\widehat{F}(s) := \int_0^{\infty} e^{-st} dF(t) = Me^{-s\xi}, \quad s \ge 0$$

is called the Laplace–Stieltjes transform of F. Prove that $M\xi^k < \infty$ if and only if the function $\widehat{F}(s)$ is k times differentiable at 0. Moreover $M\xi^k = (-1)^k (d^k\widehat{F}/ds^k)|_{s=0}$.

11.8. Assume that a machine processes details in consecutive order. The distribution function of a service time is F. Find the Laplace–Stieltjes transform of
(a) A moment of time when nth detail will be processed.
(b) A moment of time when a defective detail starts processing, if the probability that any detail is defective is p.
(c) Time duration until ξ details are processed, where ξ has a Poisson distribution with parameter λ.

Suppose that details are processed independently of each other, and independently on ξ.

11.9. Assume that $\zeta, \eta_1, \eta_2, \dots$ are independent random variables, ζ takes values in $\mathbb{Z}_+$, $\mathsf{P}(\zeta = k) = p_k$, and random variables $\{\eta_k, k \in \mathbb{N}\}$ are i.i.d. and nonnegative. Find the Laplace–Stieltjes transform of $\sum_{k=1}^{\zeta} \eta_k$.

11.10. A router has accepted n independent packets, and processes them in consecutive order. Assume that each packet has a type A with probability p and type B with probability $q = 1 - p$, and processing times have distribution functions F_A and F_B, respectively.

Find the Laplace–Stieltjes transform of a time duration necessary for all packets' service. Particularly, find answers in the cases when
(a) F_A and F_B have exponential distribution with parameters α and β, respectively;
(b) F_A and F_B are distribution functions of constant random variables equal to α and β, respectively.

11.11. Generalize a result of Problem 11.10 to the case when m types of packets arrive with probabilities $p_1, \dots, p_m$, and distribution functions of service times are $F_1, \dots, F_m$ respectively.

11.12. Customers in a store are conventionally divided into m categories. Probability that a customer is of the kth category is equal to p_k, and distribution function of his purchase equals F_k in this case.

Assume that a cash desk breaks down with probability α while serving each customer. Find the Laplace–Stieltjes transform of money that a cash desk obtains until first breakdown.

11.13. Service time of a detail is a random variable with distribution function F_1. After a detail is processed, the machine does not work during a random time with distribution function F_2. This time is needed for machine supervision and loading a new detail.

Let $Z(t)$ be the probability that a machine is processing some detail at the moment of time t.
(a) Express $Z(t)$ in terms of F_1, F_2. Find the Laplace transform $\widehat{Z}(s) = \int_0^\infty Z(t)$ $e^{-st}\,dt$.
(b) Find $\lim_{t \to \infty} Z(t)$.
(c) Let $L_1(t), L_2(t)$ be expectations of processed detail numbers and idle periods until the moment of time t, respectively. Express L_1, L_2 and their Laplace–Stieltjes

transforms $\widehat{L}_i(s) = \int_0^\infty e^{-st} dF_i(t)$ in terms of F_1, F_2 and their Laplace–Stieltjes transforms.

Suppose that the first detail started processing at the moment of time $t = 0$, intervals of service and idle times are independent, and F_1, F_2 are nonarithmetical.

11.14. Consider the queueing system described in Problem 11.13. Let:

$W_1(t,x)$ be the distribution function of the residual time from t until the first moment when some detail was completely processed,

$W_2(t,x)$ be the distribution function of time from t until the moment when the detail arrived first after t was completely served,

$W_3(t,x)$ be the distribution function of time from t until the first start of service after t.

(a) Express $W_i(t,x)$ in terms of F_1, F_2.
(b) Find $\lim_{t\to\infty} W_i(t,x)$.
(c) Find limit distributions in item (1) assuming that F_1 and F_2 have exponential distributions with parameters α_1, α_2, and in the case when $F_1(x) = \mathbb{I}_{x\ge\beta}$, $F_2(x) = (1 - e^{-\alpha x})\mathbb{I}_{x\ge 0}$.

11.15. Assume that $\alpha \in [0,1)$ and a function z satisfies conditions of Theorem 11.2. Prove that equation

$$Z(t) = z(t) + \alpha \int_0^t Z(t-u)F(du)$$

has a solution, this solution is unique in a class of locally bounded functions, and is of the form

$$Z(t) = \int_0^t z(t-u)U(du),$$

where $U = \sum_{n=0}^\infty \alpha^n F^{*n}$. Express the Laplace–Stieltjes transform of Z in terms of Laplace transform $\widehat{z}(s) = \int_0^\infty e^{-st} z(t)dt$ and the Laplace–Stieltjes transform $\widehat{F}(s) = \int_0^\infty e^{-st} dF(t)$.

11.16. A device processes details in consecutive order. The time of each detail processing is a random variable with distribution function F. A probability that the detail is defective equals $p \in (0,1)$. Denote by $H(t)$ a probability that a device has not processed any defective detail before a moment of time t. Let $L_1(t), L_2(t), L_3(t)$ be the expectations of the number of details processed by a device before t if defective detail: (a) stops a device, and new details are not accepted, (b) is eliminated immediately, (c) is processed (with the same distribution as a quality detail), respectively. Find functions H, L_1, L_2, L_3 and their Laplace transforms.

Suppose that quality and processing of different details are independent, and the processing of the first detail started at moment zero.

How the answer would change if the residual processing time of the first detail had distribution function F_1?

11.17. Requests of two types arrive at one device. The distribution functions of the service times for these types of requests are G_1 and G_2, respectively. At the instant when the device becomes free, either a type 1 request arrives with probability $p \in (0,1)$ or a type 2 request arrives with probability $q = 1 - p$.

Let:

$L(t)$ be an expectation of type 1 requests that have arrived before the moment of time t,

$H(t,x)$ be a distribution function of a time interval from t until the first arrival of a new request after t,

$H_1(t,x), H_2(t,x)$ be distribution functions of a time interval from t until the first arrival of a new request (type 1 or 2, respectively) after t.

Find L, H, H_1, H_2 and their Laplace–Stieltjes transforms. Suppose that the first request arrives at the moment of time 0. Type and service time are independent for different requests.

11.18. Service time of a detail has an exponential distribution with a parameter α. A processing of a new detail starts immediately after the previous detail was completely processed. However, if some detail is processing for more than a period of time β, then the machine overheats and stops.

Let ξ be a moment when the machine stops. Find $\mathsf{E}\xi, \mathsf{D}\xi$, if the service of the first detail starts at the moment $t = 0$.

11.19. Solve Problem 11.18 if a service time has uniform distribution on $[0, \alpha]$, $\alpha > \beta$.

11.20. Patients with diseases of types $1, \ldots, n$ come to the doctor and their arrivals form independent Poisson streams with rates $\lambda_1, \ldots, \lambda_n$. Service time of a patient with a kth disease is a random variable with expectation β_k.

Prove that

(a) if $\sum_{k=1}^{n} \lambda_k \beta_k < 1$, then with probability 1 there exists a moment of time when there are no patients in the queue (assume that the doctor is an "ideal" one, that is, works without breaks, days off, etc.)

(b) The answer remains the same if patients with some diseases have higher priority than others (if a new patient with an urgent diseases arrives then a patient with lower priority who already started the examination can either be examined completely or re-examined later taking into account initial examination time).

(c) Moments when no patients remain form a renewal process.

Generalize the problem to the case when there are several doctors with the same qualification and specialization at the hospital.

11.21. A Poisson stream of requests arrives at the device. The rate of the stream equals $\lambda > 0$. The distribution function of service time of each request equals F, and the expectation equals μ. Assume that $\lambda\mu < 1$. If the device is busy then the new request is added at the end of the queue.

Denote by Π a distribution function of the busy period, that is, the length of time from the instant the request arrives at an empty system until the instant when the system is empty again. Prove that

(a) Laplace–Stieltjes transform $\widehat{\Pi}(s) = \int_0^\infty e^{-st} d\Pi(t)$ satisfies the equation $\widehat{\Pi}(s) = \widehat{F}(s+\lambda - \lambda\widehat{\Pi}(s))$, $\mathrm{Re}\, s > 0$.
(b) $\int_0^\infty t\, d\Pi(t) = \mu/(1-\lambda\mu)$.
(c) $\int_0^\infty t^2 d\Pi(t) = \int_0^\infty t^2 dF(t)(1-\lambda\mu)^{-3}$.

11.22. Consider a queueing system from Problem 11.21, where service time is
(a) Nonrandom and equals μ.
(b) Exponentially distributed with parameter $1/\mu$.
Find the first three moments and variance of the busy period.

11.23. Patients' arrival to the doctor forms a Poisson stream with rate $\lambda > 0$. The expectation of the service time equals μ, $\lambda\mu < 1$. Find a probability that a new patient catches a doctor free, if the doctor started her work a long enough time ago.

11.24. Assume that a device accepts n independent Poisson streams of requests with rates $\lambda_1, \ldots, \lambda_n$ respectively. Assume that a distribution function of the service time of requests from the kth stream equals F_k, $\sum_{k=1}^n \lambda_k \mu_k < 1$. Suppose also that all requests have the same priority. Prove that Laplace–Stieltjes transformation of the busy period satisfies the equation

$$\lambda\widehat{\Pi}(s) = \sum_{k=1}^n \lambda_k \widehat{F}_k(s+\lambda-\lambda\widehat{\Pi}(s)),$$

where $\lambda = \lambda_1 + \cdots + \lambda_n$. Find the expectation of the busy period.

11.25. Consider a queueing system from Problem 11.24 under the assumption that requests from different streams have different priorities. Consider the following cases.
(a) A higher priority request interrupts a lower priority request, and the lower priority request is added to the queue. The service time of the lower priority request resumes at the point where it was interrupted.
(b) If the service of a lower priority request has been already started, then a higher priority request does not interrupt it. The service of a higher priority request starts before all lower priority requests from the queue.
Write an equation for the Laplace–Stieltjes transformation of the busy period distribution function. Find the expectation of the busy period.

11.26. Solve Problem 11.25 if the service time of a lower priority request does not resume at the point where it was interrupted, and service starts from the very beginning.

11.27. A telephone exchange has an infinite number of channels. Assume that request arrivals follow a Poisson process with rate λ. The service time for each request has distribution function F. Let $X(t)$ be the number of busy channels at the time moment t, $X(0) = 0$. Find $\mathsf{P}(X(t) = k)$, $\mathsf{E}z^{X(t)}, k \geq 0, |z| < 1$.

11.28. A device accepts a Poisson stream of requests with a rate $\lambda > 0$. The service time for each request has the distribution function F. If a device is busy then a new request is added at the end of the queue. Let $t_n, n \geq 1$ be the moment when the nth request is processed; X_n is a number of requests in the system just after t_n. Prove that $\{X_n, n \geq 1\}$ is a homogeneous Markov chain with transition matrix

$$P = \left\| \begin{array}{cccc} p_0 & p_1 & p_2 & \cdots \\ p_0 & p_1 & p_2 & \cdots \\ 0 & p_0 & p_1 & \cdots \\ 0 & 0 & p_0 & \cdots \\ \multicolumn{4}{c}{\dots\dots\dots} \end{array} \right\|,$$

where

$$p_k = \int_0^\infty \frac{(\lambda x)^k}{k!} e^{-\lambda x} dF(x).$$

Prove that if $\lambda\mu < 1$, where $\mu = \int_0^\infty x dF(x)$, then this Markov chain is ergodic and generating the function $\pi(z) = \sum_{n=0}^\infty \pi_n z^n$, $|z| < 1$ of the stationary distribution equals

$$\pi(z) = \frac{(1-\lambda\mu)(1-z)\widehat{F}(\lambda(1-z))}{\widehat{F}(\lambda(1-z)) - z}.$$

11.29. Denote by S_n the nth hitting moment of some state i by a Markov chain (possible $S_n = \infty$).
(a) Prove that $\{S_n\}$ is a delayed renewal process.
(b) Prove the ergodic theorem for the Markov chain (Theorem 10.3).

11.30. A device accepts a Poisson stream of requests with a rate λ. The expected service time is μ, $\lambda\mu < 1$. Prove that a generating function of virtual waiting time $w(\theta_s)$ stopped at the moment θ_s, where θ_s has exponential distribution with rate $s > 0$, is of the form:

$$\widetilde{\omega}(z,s) = \mathsf{E}e^{-zw(\theta_s)} = s\int_0^\infty e^{-st}\omega(z,t)dt = \frac{s - z\widetilde{p}_0(s)}{s - k(z)}.$$

Use (11.5) and prove identity

$$\widetilde{\omega}(z,s) = \frac{s}{s-k(z)} \frac{s - z + \lambda\left[1 - \widehat{\Pi}(s)\right]}{s + \lambda\left[1 - \widehat{\Pi}(s)\right]}, \qquad \mathsf{P}(w(\theta_s) = 0) = \widetilde{p}_0(s).$$

11.31. Assume that $\lambda\mu < 1$ in Problem 11.30. Denote by w_* the weak limit of virtual waiting time $\lim_{t\to\infty} w(t)$. Use the last identity of the previous problem and prove that the generating function of the limit virtual waiting time is

$$\widetilde{w}_*(z) := \mathsf{E}e^{-zw_*} = \frac{1-\lambda\mu}{1 - \lambda\int_0^\infty e^{-zx}\overline{F}(x)dx} = \frac{z(1-\lambda\mu)}{z - \lambda(1 - \int_0^\infty e^{-zx}dF(x))},$$

where $\overline{F}(x) = \mathsf{P}(\xi_k > x)$, $x > 0$.

Find probability $p_+ = \mathsf{P}(w_* = 0)$ and show that the last identity is equivalent to the classical Pollaczek–Khinchin formula

$$\mathsf{E}e^{-zw_*} = p_+\left(1 - q_+\mu^{-1}\int_0^\infty e^{-zx}\overline{F}(x)dx\right)^{-1}, \qquad q_+ = 1 - p_+ = \lambda\mu.$$

11.32. A server accepts a Poisson stream of packets with rate $\lambda > 0$. The service time of each packet is nonrandom and equals τ, $\lambda\tau < 1$. If the server is busy then the new packet is added at the end of the queue. Find the Laplace–Stieltjes transformation and expectation of the waiting time of an arbitrary packet in stationary conditions.

11.33. A router accepts a Poisson stream of requests with rate $\lambda > 0$. Each request contains 1 symbol with probability p and $(m+1)$ symbols with probability $q = 1-p$. The service time of one symbol is nonrandom and equals τ. Find the Laplace–Stieltjes transformation and expectation of the waiting time experienced by an arbitrary request in the stationary mode. It is assumed that the capacity of the queue is infinite.

11.34. Find a cumulant function $\psi(\alpha) = \ln \mathsf{E}e^{i\alpha\zeta(1)}$ for a service process $\zeta(t) = \sum_{k=1}^{N(t)} \xi_k - t$, where the characteristic function of service time equals $\varphi(\alpha) = 1/(1-i\alpha)^2$. Find $\mathsf{E}\zeta(1)$. For which λ does the stationary mode exist?

11.35. Consider the queueing system with a process $\zeta(t)$ from the previous problem. Use the Pollaczek–Khinchin formula (see Problem 11.31) to find the generating function of stationary virtual waiting time ω_* when $\lambda < \frac{1}{2}$, $(\overline{F}(x) = \frac{1}{2}(1+x)e^{-x}$, $x > 0)$. If $\lambda = \frac{1}{4}$ then expand $\tilde{w}_*(z)$ into linear-fractional functions, find Laplace transformations, and find the distribution of w_*.

Hints

11.1. Take $z = \mathbb{I}_{[0,h]}$ in Theorem 11.2.
11.2. Use item (c) and previous problem in a solution of item (d).
(e) If $V(t) = \mu^{-1}t$, then (b) implies

$$F_0(t) = \frac{t}{\mu} - \frac{1}{\mu}\int_0^t (t-y)F(dy).$$

Now integrate by parts.
11.5. Use the Lebesgue dominated convergence theorem.
11.6. (a) All formulas have the same form as in Problem 11.4 where a function U is replaced by V.
(b) Use Problem 11.5.
11.15. Use the Banach fixed point theorem. $\widehat{Z} = \widehat{z}(1-\alpha\widehat{F})^{-1}$.
11.19. See solution of Problem 11.18.
11.20. Use the strong law of large numbers.
11.21. Prove that $\Pi(+\infty) = 1$ (see Problem 11.20), and prove that Π^{*n} is a distribution function of a busy period if service of one request just started and there are $n-1$ requests in the queue.

Assume that the service of the first request has finished at the time moment u. A probability that n requests arrived during this period of time is equal to

$e^{-\lambda u}((\lambda u)^n/n!)$. Thus $\Pi(x) = \sum_{n=0}^{\infty} \int_0^x e^{-\lambda u}((\lambda u)^n/n!)\Pi^{*n}(x-u)F(du)$. All summands are nonnegative, so we may change the order of sum and integration:

$$\begin{aligned} \widehat{\Pi}(s) &= \int_0^{\infty} \sum_{n=0}^{\infty} e^{-su} e^{-\lambda u} \frac{(\lambda u)^n}{n!} \widehat{\Pi}^n(s) F(du) \\ &= \int_0^{\infty} e^{-u(s+\lambda-\lambda\widehat{\Pi}(s))} F(du) = \widehat{F}\left(\lambda + s - \lambda \widehat{\Pi}(s)\right). \end{aligned} \tag{11.11}$$

Note that we have not proved uniqueness of the solution (in fact, such uniqueness also holds).

Functions $\widehat{\Pi}$ and $\widehat{F}$ are analytical when $\operatorname{Re} s > 0$ and equalities (11.11) hold true for all real positive s. Therefore (11.11) holds true for all s, $\operatorname{Re} s > 0$. Differentiate (11.11), pass to a limit as $s \to 0+$ and find the moments of Π.

11.24. Use the reasoning of Problem 11.21 and the following fact. The probability that the request from the kth stream arrived earlier than other requests equals $\lambda_k/(\lambda_1 + \cdots + \lambda_n)$.

11.29. Let $z(n) = f_{ij}^{(n)}$, F be a distribution function of the return time from a point j to itself. Use Theorem 11.2 with $x = 0$, $h = 1$.

11.30. $\mathsf{P}(w(\theta_s) = 0) = \lim_{z \to \infty} \widetilde{\omega}(z,s)$.

11.31. To find the generating function for w_*, pass to a limit as $s \to 0$ in a relation for $\widetilde{\omega}(z,s)$.

11.32, 11.33. Apply the Pollaczek–Khinchin formula (Problem 11.31).

Answers and Solutions

11.3. (a) Let $\varepsilon > 0$ and $\delta > 0$ be such that $\mathsf{P}(T_1 \geq \delta) > \varepsilon$. Use a sequence of i.i.d. random variables $\widetilde{T}_n = \delta \mathbb{I}_{\{T_n \geq \delta\}}$ and construct a renewal process $\widetilde{S}_n$ ($\widetilde{S}_0 = 0$) and process $\widetilde{\nu}(t)$. Then

$$\mathsf{P}(\widetilde{\nu}(m\delta) = k) = C_k^m \varepsilon^{m+1}(1-\varepsilon)^{k-m} \leq \text{const} \cdot k^m \cdot (1-\varepsilon)^k.$$

So the expectation $\mathsf{E}e^{\theta\widetilde{\nu}(m\delta)}$ is finite when $\theta < -\ln(1-\varepsilon)$. To complete the proof, observe that $\widetilde{\nu}(t)$ is a monotone process and $\widetilde{\nu}(t) \geq \nu(t)$.

11.4. (a) Assume that the last request before time moment t was served at the moment u. Then the conditional probability that the next request is served on the time interval $(t, t+x]$ equals $F(t+x-u) - F(t-u)$. So

$H(t,x) = \sum_n \int_0^t (F(t+x-u) - F(t-u))\mathsf{P}(S_n \in du) = \int_0^t (F(t+x-u) - F(t-u))U(du)$,

$G(t,x) = \int_{t-x}^t (1 - F(t-u))U(du)$, $L(t,x) = \int_{t-x}^t (F(x) - F(t-u))U(du)$.

(b) Due to Theorem 11.2 with $z(u) = F(u+x) - F(u)$, $z(u) = (1 - F(u))\mathbb{I}_{u \in [0,x)}$, $z(u) = (F(x) - F(u))\mathbb{I}_{u \in [0,x)}$, we obtain the following limit distributions for H, G, L respectively.

$$\mu^{-1} \int_0^x (1 - F(u))du; \quad \mu^{-1} \int_0^x (1 - F(u))du; \quad \mu^{-1} \int_0^x u dF(u).$$

(c) $(\mu^2+\sigma^2)/(2\mu)$; $(\mu^2+\sigma^2)/(2\mu)$; $(\mu^2+\sigma^2)/\mu$.
11.8. (a) $(\widehat{F}(s))^n$.
(b) $p(1-q\widehat{F}(s))^{-1}$ with $q=1-p$.
(c) $\exp\{\lambda(\widehat{F}(s)-1)\}$.
11.9. $\sum_{k=0}^{\infty} p_k(\widehat{F}(s))^k = \widehat{G}(\ln \widehat{F}(s))$, where F is the distribution function of η_k, G is the distribution function of ζ.
11.10. $(p\widehat{F_A}(s)+q\widehat{F_B}(s))^n$.
(a) $\left(\frac{p\alpha}{\alpha+s}+\frac{q\beta}{\beta+s}\right)^n$.
(b) $(pe^{-s\alpha}+qe^{-s\beta})^n$.
11.11. $\left(\sum_{k=1}^m p_k\widehat{F}_k(s)\right)^n$.
11.12. Profit equals $\sum_{j=1}^{\xi}\eta_j$, $(\sum_1^0=0)$, where η_j has the distribution function $\sum_k p_kF_k$, $\mathsf{P}(\xi=n)=\alpha\beta^n, n\in\mathbb{Z}_+$ $\beta=1-\alpha$. Therefore

$$Me^{-s\sum_{j=1}^{\xi}\eta_s} = \sum_{n=0}^{\infty} Me^{-s\sum_{j=1}^{n}\eta_j}\alpha\beta^n = \frac{\alpha}{1-\beta\sum_k p_k\widehat{F}_k(s)}.$$

11.13. (a) Let $F=F_1*F_2$ be the distribution function of the time between starts (or endings) of processing of two successive details, $U=\sum_{n=0}^{\infty}F^{*n}$. Then

$$Z(t)=\int_0^t(1-F_1(t-u))U(du),\quad \widehat{Z}(s)=\frac{1-\widehat{F}_1(s)}{s(1-\widehat{F}_1(s)\widehat{F}_2(s))}.$$

(b) Take $z(x)=1-F_1(x)$ in Theorem 11.2 and obtain

$$\lim_{t\to\infty}Z(t)=(\mu_1+\mu_2)^{-1}\int_0^{\infty}(1-F_1(u))du=\frac{\mu_1}{\mu_1+\mu_2}.$$

(c) A function L_1 is a renewal function of a pure renewal process with $F=F_1*F_2$; that is, $L_1=\sum_{n=0}^{\infty}F^{*n}$. A function L_2 is a renewal function for a delayed renewal process, where the distribution function of delay equals F_1.

$$\widehat{L}_1=\frac{1}{1-\widehat{F}_1\widehat{F}_2};\quad \widehat{L}_2=\frac{\widehat{F}_1}{1-\widehat{F}_1\widehat{F}_2}.$$

11.14. Epochs when details end their processing form a delayed renewal process. The first renewal epoch is the moment when the first detail finished processing, its distribution function equals F_1. Thus

$$W_1(t,x)=\int_0^t(F(t+x-u)-F(t-u))V(du),$$

where $V=F_1*U$, U and F are defined in a solution of Problem 11.13 (see also Problems 11.4, 11.6).

Analogously, $W_3(t,x)=\int_0^t(F(t+x-u)-F(t-u))U(du)$.

$$\lim_{t\to\infty}W_1(t,x)=\lim_{t\to\infty}W_3(t,x)=(\mu_1+\mu_2)^{-1}\int_0^x(1-F(u))du=:W(x).$$

It is easy to see that $W_2=W_3*F_1$. So, $\lim_{t\to\infty}W_2(t,x)=W*F_1(x)$.
11.16. Denote $U(t):=\sum_{n=0}^{\infty}q^nF^{*n}$ (see Problem 11.15). Then

$$H(t)=q\int_0^t(1-F(t-u))U(du),\quad L_1(t)=1-H(t).$$

Let $\{S_n, n \geq 0\}$ be epochs when the processing of details starts.

Assume that u is the last instant prior to t when the processing of some detail started. Then the conditional expectation of a number of defective details arrived at instant u equals $\sum_{n=0}^{\infty} np^n q = p/q$ (for $L_2(t)$), or p (for $L_3(t)$). So $L_2(t) = (p/q)\int_0^t (1-F(t-u))\widetilde{U}(du)$, $L_3(t) = p\int_0^t (1-F(t-u))\widetilde{U}(du)$, where $\widetilde{U} = \sum_{n=0}^{\infty} F^{*n}$ is a renewal function for a process with nondefective details only.

11.17. Let G be the distribution function of the first arrival of the type 1 request. Then $G(t) = p + q\int_0^t G(t-s)G_2(ds)$; that is, $G = p\sum_{n=0}^{\infty} q^n G_2^{*n}, \widehat{G} = p/(1-q\widehat{G_2})$. Denote by $F = G_1 * G$ a distribution function of the time between two successive arrivals of type 1 requests, $U = \sum_{n=0}^{\infty} F^{*n}$. A problem to calculate $L(t)$ is equivalent to the following one. Service time of a request has a distribution function F. After the service is complete a device is idle during a time with distribution function G. Then (see solution of Problem 11.13)

$$L = G * U, \quad \widehat{L} = \frac{p}{1-p\widehat{G_1}-q\widehat{G_2}}.$$

Analogously, to find H_1 or H_2 use the solution of Problem 11.14. Function H is defined by the formula from Problem 11.4 (b) with distribution function between successive requests equal to $F = pG_1 + qG_2$.

11.18. Probability that a machine overheats during the first detail processing is equal to $e^{-\alpha\beta}$. Therefore the distribution function of the machine's stopping time satisfies the equation

$$\begin{aligned} Z(x) &= z(x) + \int_0^x Z(x-y)\alpha e^{-\alpha y} \mathbb{I}_{y\in[0,\beta)} dy \\ &= z + (1-e^{-\alpha\beta}) Z * \widetilde{F}, \end{aligned} \tag{11.12}$$

where $z(x) = e^{-\alpha\beta}\mathbb{I}_{x\geq\beta}$, $d\widetilde{F}(x) = \alpha e^{-\alpha x}(1-e^{-\alpha\beta})^{-1}\mathbb{I}_{x\in[0,\beta)}dx$.

Let us integrate the left-hand and right-hand sides of (11.12) with respect to x. Then

$$\mathsf{E}X = \frac{e^{\alpha\beta}-1}{\alpha}, \quad DX = \frac{e^{2\alpha\beta}-1-2\alpha\beta e^{\alpha\beta}}{\alpha^2}.$$

11.22. (a) $\mu/(1-\lambda\mu)$; $\mu^2/(1-\lambda\mu)^3$; $(2\lambda\mu+1)\mu^3/(1-\lambda\mu)^5$;$\sigma = \lambda\mu^3/(1-\lambda\mu)^3$.
(b) $\mu(1-\lambda\mu)^{-1}$; $2\mu^2(1-\lambda\mu)^{-3}$; $6\mu^3(1+\lambda\mu)(1-\lambda\mu)^{-5}$;
$\sigma^2 = (1+\lambda\mu)\mu^2(1-\lambda\mu)^{-3}$.

11.23. One can use the result of Problem 11.13 (b), where F_1 has exponential distribution with a rate λ, and F_2 is a busy interval. Then the corresponding probability equals (see also Problem 11.21 (b))

$$\frac{\int_0^\infty x dF_1(x)}{\int_0^\infty x dF_1(x) + \int_0^\infty x dF_2(x)} = \frac{\frac{1}{\lambda}}{\frac{1}{\lambda} + \frac{\mu}{1-\lambda\mu}} = 1 - \lambda\mu.$$

11.24. $(\sum_{k=1}^n \lambda_k\mu_k)(\sum_{k=1}^n \lambda_k)^{-1}(1-\sum_{k=1}^n \lambda_k\mu_k)^{-1}$.

11.25. The answer is the same as in Problem 11.24.

11.27. Let $N(t)$ be the number of requests that arrived before t.

It can be proved (see Problem 5.17) that conditional distribution of call moments $\tau_1, \ldots, \tau_n$ given $N(t) = n$ is equal to the distribution of division points of $[0,t]$ by a system of n independent uniformly distributed in $[0,t]$ random variables.

The probability that a call arrived at the moment $u \in [0,t]$ will be serving at moment t equals $1 - F(t-u)$. So

$$\mathsf{P}(X(t) = k/N(t) = n) = C_n^k t^{-n} \left(\int_0^t (1 - F(u))du \right)^k \left(\int_0^t F(u)du \right)^{n-k},$$

$n \geq k$. This implies that

$$\mathsf{P}(X(t) = k) = \frac{\left(\int_0^t (1 - F(u))du \right)^k}{k!} e^{-\lambda \int_0^t (1-F(u))du};$$

$$\mathsf{E} z^{X(t)} = e^{-\lambda (1-z) \int_0^t (1-F(u))du}.$$

11.34. $\psi(\alpha) = \lambda(\varphi(\alpha) - 1) - i\alpha = i\alpha\left(\lambda(2 - i\alpha)(1 - i\alpha)^{-2} - 1\right)$; $E\zeta(1) = 2\lambda - 1$.

11.35. $Ee^{-z\omega_*} = (1 - 2\lambda)\left[1 - \lambda(2+z)(1+z)^{-2}\right]^{-1}$, $q_+ = 1 - p_+ = 2\lambda$ as $\lambda < \frac{1}{2}$.

If $\lambda = \frac{1}{4}$, then $p_+ = \frac{1}{2}$,

$$\mathsf{P}(\omega_* > x) = \frac{1}{4}\left[\left(1 + \frac{5}{\sqrt{17}}\right) e^{-x(7-\sqrt{17})/8} + \left(1 - \frac{5}{\sqrt{17}}\right) e^{-x(7+\sqrt{17})/8}\right] =$$

$$= e^{-7x/8}\left(\frac{1}{2}\operatorname{ch}\frac{\sqrt{17}x}{8} + \frac{5\sqrt{17}}{34}\operatorname{sh}\frac{\sqrt{17}x}{8}\right), \quad x > 0.$$

12

Markov and diffusion processes

Theoretical grounds

Let $\mathbb{T} \subset \mathbb{R}$, $(\Omega, \mathcal{F}, \{\mathcal{F}_t\}_{t \in \mathbb{T}}, \mathsf{P})$ be a probability space with complete filtration. Let $X = \{X(t), t \in \mathbb{T}\}$ be an adapted stochastic process taking values in some metric space $(\mathbb{X}, \mathcal{X})$, which sometimes is called the phase space of the process X.

Definition 12.1. *Process X is called a* Markov process *(MP) with respect to filtration $\{\mathcal{F}_t\}_{t \in \mathbb{T}}$ if for any $s \in \mathbb{T}$ and for any event $A \in \mathcal{F}_{\geq s} := \sigma\{X(u), u \geq s, u \in \mathbb{T}\}$ the following equality holds*

$$\mathsf{P}(A/\mathcal{F}_s) = \mathsf{P}(A/X(s)) \ \mathsf{P}\text{-}a.s. \tag{12.1}$$

If equality (12.1) holds for some filtration then it holds for natural filtration $\{\mathcal{F}_t^X\}_{t \in \mathbb{T}}$ as well. The sets from σ-algebras $\mathcal{F}_s^X$ and $\mathcal{F}_{\geq s}^X$ can be interpreted as the events from the "past" and from the "future" of the process X (with respect to the "present" moment s of time). There exist some equivalent characterizations of the Markov property (12.1).

Theorem 12.1. *A stochastic process X is a Markov process if and only if one of the following properties holds.*

(1) For any $s \in \mathbb{T}$ and for any bounded $\mathcal{F}_{\geq s}$-measurable real random variable $F : \Omega \to \mathbb{R}$ P-a.s. the following equality holds,

$$\mathsf{E}(F/\mathcal{F}_s) = \mathsf{E}(F/X(s)).$$

(2) For any $s \leq t$, $s, t \in \mathbb{T}$ and for any bounded $\mathcal{B}(\mathbb{X})$-measurable function $f : \mathbb{X} \to \mathbb{R}$ P-a.s. the following equality holds,

$$\mathsf{E}(f(X(t))/\mathcal{F}_s) = \mathsf{E}(f(X(t))/X(s)).$$

(3) For any $s \leq t$, $s, t \in \mathbb{T}$ and for any $B \in \mathcal{B}(\mathbb{X})$, P-a.s. the following equality holds,

$$\mathsf{P}(X(t) \in B/\mathcal{F}_s) = \mathsf{P}(X(t) \in B/X(s)).$$

The stochastic process X is Markov with respect to natural filtration if and only if one of the following equivalent properties holds,

D. Gusak et al., *Theory of Stochastic Processes*, Problem Books in Mathematics,
DOI 10.1007/978-0-387-87862-1_12, © Springer Science+Business Media, LLC 2010

(4) For any $m \geq 1$ and any $s_1 < \cdots < s_m < s \leq t$, $s_1, \ldots, s_m, s, t \in \mathbb{T}$, and for any bounded $\mathcal{X}$-measurable function f P-a.s. the following equality holds,

$$\mathsf{E}\big(f(X(t))/X(s_1), \ldots, X(s_m), X(s)\big) = \mathsf{E}\big(f(X(t))/X(s)\big).$$

(5) For any $s \leq t$, $s, t \in \mathbb{T}$ and for any set $B \in \mathcal{X}$ P-a.s. the following equality holds,

$$\mathsf{P}\big(X(t) \in B/X(s_1), \ldots, X(s_m), X(s)\big) = \mathsf{P}\big(X(t) \in B/X(s)\big).$$

(6) For any $s \in \mathbb{T}$ and for any sets $A \in \mathcal{F}_s$, $B \in \mathcal{F}_{\geq s}$ P-a.s. the following equality holds,

$$\mathsf{P}(A \cap B/X(s)) = \mathsf{P}(A/X(s))\mathsf{P}(B/X(s)).$$

Theorem 12.2. *Let $\{X(t), t \in \mathbb{T}\}$ be an n-dimensional stochastic process with independent increments. Then it is a Markov process with respect to natural filtration.*

Definition 12.2. *A function $\mathsf{P}(s, x, t, B), s \leq t, x \in \mathbb{X}, B \in \mathcal{X}$ taking values in $[0, 1]$ is called a* transition function, *or a* Markov transition function, *if it satisfies the following conditions.*

(1) For s, x, t fixed the function $\mathsf{P}(s, x, t, \cdot)$ is a probability measure on $(\mathbb{X}, \mathcal{X})$,

(2) For s, t, B fixed the function $\mathsf{P}(s, \cdot, t, B)$ is $\mathcal{X} - \mathcal{B}(\mathbb{R})$-measurable,

(3) $\mathsf{P}(s, x, s, B) = \mathbb{I}_{x \in B}$ for any $s \in \mathbb{T}$, $x \in \mathbb{X}$, $B \in \mathcal{X}$,

(4) For any $s < u < t$ $(s, u, t \in \mathbb{T})$, $x \in \mathbb{X}$, $B \in \mathcal{X}$ this function satisfies the Kolmogorov–Chapman equations

$$\mathsf{P}(s, x, t, B) = \int_{\mathbb{X}} \mathsf{P}(s, x, u, dy)\mathsf{P}(u, y, t, B).$$

Remark 12.1. It follows from condition (3) that the transition function satisfies the Kolmogorov–Chapman equations for any $s \leq u \leq t$.

Definition 12.3. *We say that Markov process $X = \{X(t), t \in \mathbb{T}\}$ has transition function, or transition probability $\mathsf{P}(s, x, t, B)$, if for any $s \leq t$, $s, t \in \mathbb{T}$ and any $B \in \mathcal{X}$,*

$$\mathsf{P}(X(t) \in B/X(s)) = \mathsf{P}(s, X(s), t, B) \ \mathsf{P}\text{-a.s.},$$

or, what is the same,

$$\mathsf{P}(s, x, t, B) = \mathsf{P}(X(t) \in B/X(s) = x)$$

for almost all x with respect to the distribution $X(s)$.

In the case when the phase space is countable, the transition function exists for an arbitrary Markov process (see Problems 12.3 and 12.4).

Consider one more form of the Kolmogorov–Chapman equations: for $s \leq u \leq t$, $B \in \mathcal{X}$, and for almost all x with respect to the distribution of $X(s)$,

$$\mathsf{P}(s, x, t, B) = \mathsf{E}\big(\mathsf{P}(u, X(u), t, B)/X(s) = x\big).$$

Definition 12.4. *The Markov process $X = \{X(t), t \in \mathbb{T}\}$ with the values in the space $(\mathbb{X}, \mathcal{X})$ and with transition function $\mathsf{P}(s, x, t, B)$ is called a* homogeneous Markov process *(HMP) if $\mathsf{P}(s, x, t, B)$ depends only on the difference $t - s$ for any $B \in \mathcal{X}$, $x \in \mathbb{X}$.*

For homogeneous Markov processes with $\mathbb{T} = \mathbb{R}^+$ or $\mathbb{Z}^+$, it is sufficient to consider only the functions $\mathsf{P}(x,t,B) := \mathsf{P}(0,x,t,B)$, $t \in \mathbb{T}$.

If we have HMP, then the conditions (1)–(4) from Definition 12.2 of the transition function can be rewritten in the form

(1′) For any fixed x,t the function $\mathsf{P}(x,t,\cdot)$ is a probability measure on $\mathcal{X}$,

(2′) For any fixed t,B the function $\mathsf{P}(\cdot,t,B)$ is $\mathcal{X}-\mathcal{B}(\mathbb{R})$-measurable,

(3′) $\mathsf{P}(x,0,B) = \delta_x(B)$ for any x,B,

(4′) For any x,s,t,B this function satisfies the Kolmogorov–Chapman equations:

$$\mathsf{P}(x,s+t,B) = \int_{\mathbb{X}} \mathsf{P}(x,s,dy)\mathsf{P}(y,t,B).$$

Definition 12.5. *The transition probability* $\mathsf{P}(t,x,B)$ *of HMP X is called* measurable *if for any $B \in \mathcal{X}$ it is a measurable function of the pair of variables (t,x) on the product of the spaces $(\mathbb{R}^+, \mathcal{B}(\mathbb{R}^+)) \times (\mathbb{X}, \mathcal{X})$. HMP X is called* weakly measurable *if its transition function is measurable.*

Denote by $\mathbb{B} = \mathbb{B}(\mathbb{X})$ the Banach space of all real-valued Borel functions $f : \mathbb{X} \to \mathbb{R}$ with the norm $\|f\| = \sup_{x\in\mathbb{X}} |f(x)|$. Define on $\mathbb{B}$ the family of operators $\{T_t, t > 0\}$ via the formula

$$T_t f(x) = \int_{\mathbb{X}} f(y)\mathsf{P}(t,x,dy), \; T_0 f = f. \tag{12.2}$$

Then it follows from the measurability of $\mathsf{P}(t,x,\cdot)$ with respect to x that the inclusion holds $T_t f \in \mathbb{B}$, for $f \in \mathbb{B}$; that is, $T_t : \mathbb{B} \to \mathbb{B}$. The family of linear operators $\{T_t, t \geq 0\}$ creates the semigroup; that is, for any $t,s \geq 0$,

$$T_{t+s} = T_t T_s \tag{12.3}$$

(see Problem 12.31).

In what follows, we consider only weakly measurable HMP.

Define the *resolvent operator* $\{R_\lambda, \mathrm{Re}\,\lambda > 0\}$ of HMP X as the family of operators on $\mathbb{B}$ of the form

$$R_\lambda f(x) = \int_0^\infty e^{-\lambda t} \int_{\mathbb{X}} f(y)\mathsf{P}(t,x,dy)dt = \int_0^\infty e^{-\lambda t} T_t f(x)dt.$$

The measurability of the transition probability provides that the integral is well defined. If we denote $\mathbb{B}^{\mathbb{C}}$ the space of complex-valued bounded Borel functions $f : \mathbb{X} \to \mathbb{C}$ with the uniform norm $\|f\| = \sup_{x\in\mathbb{X}} |f(x)|$, then $R_\lambda : \mathbb{B} \to \mathbb{B}^{\mathbb{C}}$. The principal properties of resolvent operators are the following ones.

(1) For any $\lambda \neq \mu$, the *resolvent equation* holds:

$$R_\lambda R_\mu = \frac{R_\mu - R_\lambda}{\lambda - \mu},$$

(2) $R_\lambda \mathbb{I} = (1/\lambda)\mathbb{I}$ and for any $\lambda > 0$, here $\mathbb{I}$ stands for the function that equals 1 identically; for $f \in \mathbb{B}_+ := \{f \in \mathbb{B} \,|\, f \geq 0 \text{ on } \mathbb{X}\}$ we have $R_\lambda f \in \mathbb{B}_+$,

(3) $\|R_\lambda\| \leq 1/\mathrm{Re}\,\lambda$.

Denote $\mathbb{B}_0$ the set of those functions from the space $\mathbb{B}$, for which $\lim_{t\downarrow 0} \|T_t f - f\| = 0$.

Lemma 12.1. *(1) The set $\mathbb{B}_0$ is a closed linear subspace in the $\mathbb{B}$.*

(2) The operators T_t transfer $\mathbb{B}_0$ into $\mathbb{B}_0$.

(3) For any $f \in \mathbb{B}$ we have that $R_\lambda f \in \mathbb{B}_0$.

(4) For any $f \in \mathbb{B}_0$ we have that $\lim_{\lambda \to +\infty} \|\lambda R_\lambda f - f\| = 0$.

Definition 12.6. *The operator $A\varphi := \lim_{h \downarrow 0} h^{-1}(T_h \varphi - \varphi)$, defined for those functions φ, for which this limit exists in the sense of strong convergence in $\mathbb{B}$, is called the* infinitesimal operator, *or* generator *of the semigroup $\{T_t, t \geq 0\}$, related to HMP X.*

The set of the functions φ mentioned above is denoted as D_A. Evidently, $D_A \subset \mathbb{B}_0$.

Theorem 12.3. *For any $f \in \mathbb{B}_0$ the following equality holds, $AR_\lambda f = \lambda R_\lambda f - f$, and for any $\varphi \in D_A$ the following equality holds, $R_\lambda A\varphi = \lambda R_\lambda \varphi - \varphi$.*

Remark 12.2. Definition 12.6 is a general definition of the generator of a semigroup. The generator A of semigroup $\{T_t, t \geq 0\}$, defined on $\mathbb{B}$, satisfies the relation

$$\frac{dT_t f}{dt} = AT_t f, \; f \in \mathbb{B}, \tag{12.4}$$

$$\frac{dT_t f}{dt} = T_t A f, \; f \in D_A. \tag{12.5}$$

Theorem 12.4. *(Hille–Yosida) Let $\mathbb{B}$ be a real Banach space, $\mathbb{B}^{\mathbb{C}}$ be its complex extension, and on $\mathbb{B}^{\mathbb{C}}$ the family of operators R_λ ($Re\ \lambda > 0$) be defined that satisfy the following assumptions.*

(1) $\|R_\lambda\| \leq 1/\mathrm{Re}\,\lambda$,

(2) For $\lambda > 0$, $R_\lambda(\mathbb{B}) \subset \mathbb{B}$,

(3) For any λ, μ, the resolvent equation *holds: $R_\lambda - R_\mu = (\mu - \lambda) R_\lambda R_\mu$,*

(4) For some $\lambda > 0$ the set $R_\lambda(\mathbb{B})$ is dense in $\mathbb{B}$.

Then there exists the semigroup of linear operators $\{T_t, t \geq 0\}$ on $\mathbb{B}$, satisfying the conditions:

(1) $T_t T_s = T_{t+s}$, $t, s \geq 0$,

(2) For any $x \in \mathbb{B}$, $\|T_t x - x\| \to 0$, $t \downarrow 0$,

(3) $\|T_t\| \leq 1$,

(4) For any $x \in \mathbb{B}$, $R_\lambda x = \int_0^\infty e^{-\lambda t} T_t x dt$.

Remark 12.3. The Hille–Yosida theorem asserts that any family of operators which has resolvent properties corresponds to some semigroup. In the general case this semigroup does not correspond to the transition function of some MP. For the assumptions on phase space and on resolvent operators that are sufficient for some semigroup to create the transition function of MP, see, for example [24], Vol. 2.

Now, let $\mathbb{T} = \mathbb{R}^+$ or $\mathbb{Z}^+$, $(\mathbb{X}, \mathfrak{X})$ be measurable space, and $\mathsf{P}(t, x, B)$ be the function satisfying the assumptions (1′)–(3′). Let us have the stochastic process $X(t, \omega)$, defined on Ω and with values in $\mathbb{X}$. Denote σ-algebras $\mathcal{F}_{\geq 0}$ $\mathcal{F}_{\leq t}$, generated by the sets of the form $\{X(s) \in B\}$, $s \in \mathbb{T}$, and $s \leq t$, correspondingly. Let $\{\mathsf{P}_x, x \in \mathbb{X}\}$ be the family of probability measures defined on the σ-algebra $\mathcal{F} = \mathcal{F}_{\geq 0}$.

Definition 12.7. *The pair* $(X, \mathsf{P}.)$ *is called the* homogeneous Markov family *(HMF) with transition function* $\mathsf{P}(t,x,B)$, *if for any* $t,h \in \mathbb{T}$, $x \in \mathbb{X}$, $B \in \mathcal{X}$ P_x*-a.s. we have that*

$$\mathsf{P}_x(X(t+h) \in B/\mathcal{F}_{\le t}) = \mathsf{P}(h, X(t), B).$$

Kolmogorov–Chapman equations (4′) are a consequence of the property mentioned above.

For a given HMF $(X, \mathsf{P}.)$ and every $x_* \in \mathbb{X}$, the process X, considered on the probability space $(\Omega, \mathcal{F}, \mathsf{P}_{x_*})$, is a HMP with the transition function $\mathsf{P}(t,x,B)$ such that $X(0) = x_*$ a.s. Moreover, for a given probability measure μ on $\mathcal{X}$, define the measure P_μ by

$$\mathsf{P}_\mu(A) = \int_{\mathbb{X}} \mathsf{P}_x(A)\mu(dx),\ A \in \mathcal{F}.$$

Then the process X, considered on the probability space $(\Omega, \mathcal{F}, \mathsf{P}_\mu)$, is a HMP with the transition function $\mathsf{P}(t,x,B)$ such that the distribution of $X(0)$ is equal to μ. Therefore, the notion of HMF allows one to consider sets of Markov processes with the same transition function and various probability laws of the initial value $X(0)$. When HMF has the same transition function with some HMP X, we say that this HMF corresponds to the process X. The expectation w.r.t. P_x or P_μ is denoted by E_x or E_μ, respectively.

In general, it is not a trivial problem to construct, for a given HMP, a HMF with the same transition function. Fortunately, for typical and most important processes this problem can be solved explicitly (see, e.g., Problem 12.23). In particular, the construction described in Problem 12.23 allows one to extend the definition of a Wiener process and consider a Wiener process with arbitrary distribution of its initial value $W(0)$.

Let $(X, \mathsf{P}.)$ be HMF, and let filtration $\{\mathcal{F}_t, t \in \mathbb{T}\}$, $\mathcal{F}_t \subseteq \mathcal{F}_{\ge 0}$ be defined on the space Ω.

Definition 12.8. *The stochastic process* X *is called* progressively measurable, *if for any* $s \in \mathbb{T}$ *the restriction of* $X(\cdot, \omega)$ *on* $[0,s] \times \Omega$ *is* $\mathcal{B}([0,s]) \times \mathcal{F}_s$*-measurable. (See Problem 3.7.)*

Definition 12.9. *HMF* $(X, \mathsf{P}.)$ *is called a* strong Markov family *with respect to the filtration* $\{\mathcal{F}_t, t \in \mathbb{T}\}$, *if the following conditions hold.*

(1) Stochastic process X *is progressively measurable.*

(2) For any stopping time τ *and for any* $t \in \mathbb{T}$, *any* $x \in \mathbb{X}$ *and* $B \in \mathcal{X}$ *we have that*

$$\mathsf{P}_x(X(\tau+t) \in B/\mathcal{F}_\tau) = \mathsf{P}(t, X(\tau), B).$$

If, for a Markov process X, the corresponding HMF is a strong Markov family, then the process X is said to have a strong Markov property.

Theorem 12.5. *Let the process* $X(t)$ *with HMF* $(X, \mathsf{P}.)$ *have the trajectories that are continuous from the right, and the function* $F(t,x) := \mathsf{E}_x f(X(t))$ *satisfy the relation:* $F(s,y) \to F(t,x)$ *for* $s \uparrow t$, $y \to x$ *and for any continuous bounded function* f. *Then HMF* $(X, \mathsf{P}.)$ *is a strong Markov family.*

For example, m-dimensional Wiener process $W(t), t \in \mathbb{R}^+$ has a strong Markov property (see Problem 12.24).

Definition 12.10. *HMF* $(X, \mathsf{P}.)$ *in a phase space* $(\mathbb{R}^m, \mathcal{B}(\mathbb{R}^m))$ *is called a* diffusion process *if the following conditions hold.*

(1) For the semigroup $\{T_t, t \geq 0\}$*, generated by the transition probability* $\mathsf{P}(t,x,B)$ *of the process, the generator* A *is well defined on the space* $C^2_{fin}(\mathbb{R}^m)$ *of doubly continuously differentiable functions with compact supports in* $\mathbb{R}^m$,

*(2) There exist such subjects as continuous vector function (*drift vector, *or* drift coefficient*)* $b(x) = (b^i(x), 1 \leq i \leq m)$ *and continuous, symmetric, and positively definite matrix function (*diffusion matrix, *or* diffusion coefficient*)* $a(x) = (a^{ij}(x), 1 \leq i, j \leq m)$ *for which*

$$Af(x) = \mathcal{L}f(x) := \frac{1}{2}\sum_{i,j=1}^{m} a^{ij}(x)\frac{\partial^2 f(x)}{\partial x_i \partial x_j} + \sum_{i=1}^{m} b^i(x)\frac{\partial f(x)}{\partial x_i},\ f \in C^2_{fin}(\mathbb{R}^m),$$

(3) The trajectories of X *are continuous a.s.*

Theorem 12.6. *Let* $(X, \mathsf{P}.)$ *be such a HMF on* $(\mathbb{R}^m, \mathcal{B}(\mathbb{R}^m))$ *that for any* $\varepsilon > 0$ *uniformly in* $x \in \mathbb{R}^m$ *the following conditions hold.*

(1) $\mathsf{P}(t,x,\mathbb{R}^m \setminus B(x,\varepsilon)) = o(t),\ t \to 0,$

(2) There exists a continuous vector function $b(x) = (b^i(x), 1 \leq i \leq m)$ *for which* $\int_{B(x,\varepsilon)} (y^i - x^i)\mathsf{P}(t,x,dy) = b^i(x)t + o(t),\ t \to 0,$

(3) There exists a continuous matrix function $a(x) = (a^{ij}(x), 1 \leq i, j \leq m)$ *for which* $\int_{B(x,\varepsilon)} (y^i - x^i)(y^j - x^j)\mathsf{P}(t,x,dy) = a^{ij}(x)t + o(t),\ t \to 0.$

(Here $B(x,\varepsilon)$ *denotes a ball of radius* ε *with the center at the point* x*.) Then the process* X *is a diffusion process, its generator* A *is defined on the space* $C^2_{uni}(\mathbb{R}^m)$ *of the bounded functions, uniformly continuous together with their partial derivatives of the first and the second order, and on this space the generator equals* $\mathcal{L}f$.

Theorem 12.7. *Let HMF* $(X, \mathsf{P}.)$ *satisfies conditions (1)–(3) of Theorem 12.6 uniformly in* x *from any bounded set, and for any bounded set* $B \subset \mathbb{R}^m$ *there exists a bounded set* $B' \supset B$ *such that* $\mathsf{P}(t,x,B) = o(t)$ *as* $t \to 0$ *uniformly in* $x \in \mathbb{R}^m \setminus B'$*. Then* X *is a diffusion process.*

Theorem 12.8. *Assume the generator* A *of some diffusion process is well defined on the subset of functions* $f \in C^2(\mathbb{R}^m)$ *for which*

$$|f(x)| + \left|\frac{\partial f(x)}{\partial x_i}\right| + \left|\frac{\partial^2 f(x)}{\partial x_i \partial x_j}\right| \leq \varphi(x),$$

where $\varphi(x) \to 0$ *as* $\|x\| \to \infty$, $1 \leq i,\ j \leq m$*, and coincide on this set with* $\mathcal{L}$*. Also, assume the transition probability to have a form* $\mathsf{P}(t,x,B) = \int_B p(t,x,y)dy$*, where the density* $p : (0,\infty) \times \mathbb{R}^m \times \mathbb{R}^m \to \mathbb{R}^+$ *is continuous in all its variables together with* $\partial p/\partial t$, $\partial p/\partial x_i$ *and* $\partial^2 p/\partial x_i \partial x_j$*. Let, moreover,*

$$|p(t,x,y)| + \left|\frac{\partial p}{\partial t}(t,x,y)\right| + \left|\frac{\partial p}{\partial x_i}(t,x,y)\right| + \left|\frac{\partial^2 p}{\partial x_i \partial x_j}(t,x,y)\right| \leq c(t,y)\varphi(x),$$

where $c : (0,\infty)\times\mathbb{R}^m \to \mathbb{R}^+$ *is a continuous positive function. Then the transition probability density p satisfies the equation*

$$\frac{\partial p(t,x,y)}{\partial t} = \frac{1}{2}\sum_{i,j=1}^{m} a^{ij}(x)\frac{\partial^2 p(t,x,y)}{\partial x_i \partial x_j} + \sum_{i=1}^{m} b^i(x)\frac{\partial p(t,x,y)}{\partial x_i}, \tag{12.6}$$

or, in a brief form, $\partial p/\partial t = \mathcal{L}_x p$. *Here index x means that the operator* $\mathcal{L}$ *is applied to the density p considered as a function of x with t, y fixed.*

Now, let the functions $a^{ij}\in C^2(\mathbb{R}^m)$, $b^i\in C^1(\mathbb{R}^m)$. Define the operator $\mathcal{L}^*$, formally conjugate to $\mathcal{L}$, by the formula

$$\mathcal{L}^* g(x) := \frac{1}{2}\sum_{i,j=1}^{m} \frac{\partial^2}{\partial x_i \partial x_j}\left(a^{ij}(x)g(x)\right) - \sum_i \frac{\partial}{\partial x_i}\left(b_i(x)g(x)\right),$$

and this equality relates to such smooth functions g for which $\mathcal{L}^*$ is correctly defined.

Theorem 12.9. *Let* $(X,\mathsf{P}_\cdot)$ *be a diffusion process with the generator* $\mathcal{L}$. *Also, let the density* $p(t,x,y)$ *of some transition probability have continuous partial derivatives* $\partial p/\partial t$, $\partial p/\partial y_i$, *and* $\partial^2 p/\partial y_i\partial y_j$. *Then this density satisfies the equation*

$$\frac{\partial p(t,x,y)}{\partial t} = \frac{1}{2}\sum_{i,j=1}^{m} \frac{\partial^2\left(a^{ij}(y)p(t,x,y)\right)}{\partial y_i \partial y_j} - \sum_{i=1}^{m} \frac{\partial}{\partial y_i}\left(b^i(y)p(t,x,y)\right), \tag{12.7}$$

or, in other terms, $\partial p/\partial t = \mathcal{L}_y^* p$.

Remark 12.4. Equation (12.6) is called the *backward Kolmogorov equation*; the equation (12.7) is called the *forward Kolmogorov equation*, or *Fokker–Planck equation*. Equation (12.6) is in fact equation (12.4), and (12.7) is in fact equation (12.5), rewritten in terms of parabolic partial differential equations.

Now, consider the m-dimensional stochastic differential equation

$$dX(t) = b(X(t))dt + \sigma(X(t))dW(t),\ X(0) = x\in\mathbb{R}^m,\ t\in\mathbb{R}^+, \tag{12.8}$$

and suppose that its (homogeneous in t) coefficients satisfy conditions of Theorem 14.1, that are sufficient for existence and uniqueness of its strong solution (the elements of the theory of stochastic differential equations are given in Chapter 14).

Let $X = X(t,\omega,x)$ be the solution of equation (12.8). Define the function

$$\mathsf{P}(t,x,B) := \mathsf{P}(X(t,\omega,x)\in B),\quad B\in\mathcal{B}(\mathbb{R}^m).$$

Theorem 12.10. *(1) For any* $t,h\in\mathbb{R}^+$, $x\in\mathbb{R}^m$, *and* $B\in\mathcal{B}(\mathbb{R}^m)$ P-*a.s. the following equality holds,* $\mathsf{P}\big(X(t+h,\omega,x)\in B/\mathcal{F}_t\big) = \mathsf{P}\big(h, X(t,\omega,x), B\big)$.

(2) If we put $\Omega' := \Omega\times\mathbb{R}^m$, $X(t) = X(t,\omega') = X(t,\omega,x)$, $\mathcal{F}_{\geq 0} := \sigma\{X(t), t\geq 0\}$, $\mathcal{F}_{\leq t} := \sigma\{X(s), s\leq t\}$, *and define for measurable sets* $C\subset\mathbb{R}^m\times\Omega$, $C\in\mathcal{F}_{\geq 0}$ *the function* $\mathsf{P}_x(C) := \mathsf{P}\big(\omega\big|\ (x,\omega)\in C\big)$, *then the pair* $(X,\mathsf{P}_\cdot)$ *is a HMF with transition probability* $\mathsf{P}(t,x,B)$.

(3) Let A be the infinitesimal operator of a semigroup generated by HMF $(X,\mathsf{P}_\cdot)$. *Then* $C^2_{fin}(\mathbb{R}^m)\subset D_A$, *and for* $f\in C^2_{fin}(\mathbb{R}^m)$ *we have the equality*

$$Af(x) = \mathscr{L}f(x) = \frac{1}{2}\sum_{i,j=1}^{m} a^{ij}(x)\frac{\partial^2 f(x)}{\partial x_i \partial x_j} + \sum_{i=1}^{m} b^i(x)\frac{\partial f(x)}{\partial x_i},$$

where $a^{ij}(x) = \sigma(x)\sigma^T(x)$.

Therefore, our HMF is a diffusion process.

Bibliography

[9], Chapter VI; [38], Chapter IV, §5–6; [90], Chapters 8–11; [24], Volume 2; [25], Chapter I, §4, Chapter VIII; [27], Chapter 5; [18]; [19]; [40]; [51], Chapters 14 and 15; [70]; [79], Chapters 19, 25, 26, 29; [22], Chapter X; [46], Chapter 12, §12.1; [61], Chapters VII–VIII; [68], Chapters 8 and 13; [87].

Problems

12.1. Prove that a process, being a Markov one with respect to some filtration to which it is adapted, is also a Markov process with respect to its natural filtration (see Definition 3.4).

12.2. Prove that real-valued stochastic process $X = \{X(t), \mathfrak{F}_t^X, t \in \mathbb{T}\}$ which consists of independent random variables (process with independent values) is a Markov process with respect to natural filtration.

12.3. Let phase space $\mathbb{X}$ of Markov process $\{X_t, t \geq 0\}$ be at most a countable set; that is, let X_t be a Markov chain with continuous time. For such a process, put $\mathbb{X}_t := \{i \in \mathbb{X} | \mathsf{P}(X_t = i) \neq 0\}, t \geq 0$; $p_{ij}(s,t) = \mathsf{P}(X_t = j / X_s = i), s \leq t, i \in \mathbb{X}_s, j \in \mathbb{X}_t$. Prove that its transition probabilities $p_{ij}(s,t)$ satisfy the conditions:

(a) $p_{ij}(s,t) \geq 0, i \in \mathbb{X}_s, j \in \mathbb{X}_t, s \leq t$.

(b) For any $i \in \mathbb{X}_s, 0 \leq s \leq t$ $\sum_{j \in \mathbb{X}_t} p_{ij}(s,t) = 1$.

(c) $p_{ij}(s,s) = \delta_{ij}$, $i, j \in \mathbb{X}_s$.

(d) For any $i \in \mathbb{X}_s, j \in \mathbb{X}_t, 0 \leq s \leq u \leq t$ the equations hold,

$$p_{ij}(s,t) = \sum_{k \in \mathbb{X}_u} p_{ik}(s,u) p_{kj}(u,t).$$

12.4. In the framework of Problem 12.3 put $p_{ij}(s,t) = 0$ for $i \in \mathbb{X}_s, j \notin \mathbb{X}_t$ and $p_{ij}(s,t) = p_{i_0 j}(s,t)$ for $i \notin \mathbb{X}_s, j \in \mathbb{X}_t$, where $i_0 \in \mathbb{X}_s$ is chosen arbitrarily. Prove that the properties (a)–(d) of transition probabilities mentioned in Problem 12.3 will still hold for such an "extension" of $p_{ij}(s,t)$.

12.5. Consider the "random broken line" $Y(t) = (t-k)X(k) + (k+1-t)X(k+1)$, where $\{X(k), k \geq 0\}$ is a real-valued Markov chain. Is it a Markov process?

12.6. Prove that the Wiener process with values in $\mathbb{R}^m$ is MP having a transition function and determine this function.

12.7. Prove that the Poisson process is MP having a transition function and determine this function.

12.8. Let $\{X(t), t \geq 0\}$ be a real-valued MP. Are the processes (a)$\{X([t]), t \geq 0\}$; (b) $\{[X(t)], t \geq 0\}$ Markov ones?

12.9. Let $\{X(t), t \geq 0\}$ be a real-valued MP with a transition function. Prove that $Y(t) := (X(t), t)$ is a homogeneous MP. Determine the transition function of $Y(t)$.

12.10. Prove that Gaussian process $X = \{X(t), t \geq 0\}$ is a Markov process if and only if for arbitrary $0 \leq s_1 < s_2 < t$ conditional distributions of $X(t)$ with respect to σ-algebras $\sigma(X(s_1))$ and $\sigma(X(s_1), X(s_2))$ coincide.

12.11. Prove that Gaussian process $X = \{X(t), t \geq 0\}$ is a Markov process if and only if its covariance function satisfies the following conditions.

(i) $R_X(s_1, s_2)R_X(s_2, s_3) = R_X(s_1, s_3)R_X(s_2, s_2)$ for arbitrary $0 \leq s_1 < s_2 < s_3$,

(ii) If $R_X(s, s) = 0$ for some $s \geq 0$, then $R_X(s_1, s_2) = 0$ for arbitrary $s_1 \leq s \leq s_2$.

12.12. Let the homogeneous MP take its values in $\mathbb{R}^m$ and its transition function be such that $\mathsf{P}(x, t, B) = \mathsf{P}(x + y, t, B + y)$ for any $x, y \in \mathbb{R}^m$, $t \geq 0$, $B \in \mathcal{B}(\mathbb{R}^m)$. Prove that this process has independent increments.

12.13. Let the transition function of HMP be equal to

$$\mathsf{P}(s, x, A) = \frac{1}{\sqrt{2\pi}\sigma(s)} \int_A \exp\left\{\frac{-(y - m(s)x)^2}{2\sigma^2(s)}\right\} dy.$$

Prove that Kolmogorov–Chapman equations hold if and only if $m(t+s) = m(t)m(s)$ and $\sigma^2(t+s) = m^2(t)\sigma^2(s) + \sigma^2(t)$. (Note that if m, σ are continuous then either $m(t) = e^{at}$, $\sigma(t) = b(e^{2at} - 1)$, $a \neq 0$, $b > 0$, or $m(t) = 1$, $\sigma(t) = bt$, $b > 0$).

12.14. Let $\{\xi_i, i \geq 0\}$ be i.i.d.r.v., $\xi_i = \pm 1$ with probabilities $\frac{1}{2}$, $S_0 = 0$, $S_n = \sum_{i=1}^n \xi_i$, $X_n = \max_{0 \leq k \leq n} S_k$. Prove that X is *not* a Markov process with respect to the filtration generated by the process S.

12.15. Let $\{W(t), t \geq 0\}$ be a Wiener process, $\alpha, \beta > 0$. The *Ornstein–Uhlenbeck process* is defined as

$$V_t := e^{-\beta t} W(\alpha e^{2\beta t}),\ t \geq 0.$$

Is it a Markov process? A homogeneous Markov process? As for the Ornstein–Uhlenbeck process see also Example 6.9, Problems 6.12, 14.7, and 14.8.

12.16. Prove that the Brownian bridge is a nonhomogeneous Markov process and determine its transition function.

12.17. (*Telegraph signal*) Let $\{N(t), t \geq 0\}$ be the Poisson process with intensity $\lambda > 0$. Let the random variable ζ be independent of N and $\mathsf{P}\{\zeta = -1\} = \mathsf{P}\{\zeta = 1\} = \frac{1}{2}$. Put $X(t) = \zeta(-1)^{N(t)}$, $t \geq 0$. Is this process Markov?

12.18. Let N be the Poisson process. Prove that the process $X(t) = N(-t), t \in (-\infty, 0]$ is Markov and determine its transition function.

12.19. Let W be the Wiener process. Prove that the process $X(t) = W(-t), t \in (-\infty, 0]$ is Markov and determine its transition function.

12.20. Let $\{X(t), t \geq 0\}$ be a Markov process. For $T > 0$ define the process $\{Y(s) = X(T-s), s \in [0,T]\}$. Is the process Y Markov? Is Y homogeneous if X is homogeneous? Compare with Problem 5.37.

12.21. Let $\{X(t), t \geq 0\}$ and $\{Y(t), t \geq 0\}$ be real-valued Markov processes. Are the following processes Markov: $\{X(t)+Y(t), t \geq 0\}$ $\{X(t)Y(t), t \geq 0\}$?

12.22. Give an example of a martingale that is not a Markov process and an example of a Markov process that is not a martingale.

12.23. Let W be an m-dimensional Wiener process. Put $\Omega = C(\mathbb{R}^+, \mathbb{R}^m)$, $\mathcal{F} = \mathcal{B}(C(\mathbb{R}^+, \mathbb{R}^m))$, and for $x \in \mathbb{R}^m$ define the measure P_x as the distribution in $(\Omega, \mathcal{F})$ of the random element $x + W(\cdot)$ (see Definition 16.1 and Problem 16.1). Verify that $(W, P.)$ is a HMF with the transition function

$$\mathsf{P}(t,x,B) = (2\pi t)^{-(m/2)} \int_B e^{-((\|y-x\|^2)/2t)}\, dy.$$

12.24. (1) Prove the strong Markov property of the Wiener process.

(2) Prove the following stronger version of the previous statement: if τ is a stopping time w.r.t. natural filtration $\{\mathcal{F}_t^W\}$ generated by the Wiener process W, then the stochastic process $\{V(t) := W(t+\tau) - W(\tau),\ t \geq 0\}$ is a Wiener process as well, and it is independent of σ-algebra $\mathcal{F}_\tau^W$.

12.25. Let $W(t) = (W_1(t), \dots, W_m(t))$ be an m-dimensional Wiener process, $m \geq 1$, and $X_m(t) := \left(\sum_{k=1}^m W_k^2(t)\right)^{1/2}$ be a Bessel process, that is, the radial component of the process W. Is this process Markov?

12.26. Let W be a Wiener process.

(1) Assume that $g : \mathbb{R} \to \mathbb{R}$ is a bijection, and both g and g^{-1} are measurable functions. Prove that the process $X(t) = g(W(t)),\ t \geq 0$ is a homogeneous Markov process and determine its transition function.

(2) Prove that $X(t) = |W(t)|, t \geq 0$ is a homogeneous Markov process and determine its transition function.

(3) Are the processes from the previous two items strong Markov processes?

12.27. Let

$$\mathsf{P}(t,x,A) = \begin{cases} \mathsf{P}_W(t,x,A), & x \neq 0 \\ \delta_0(A), & x = 0 \end{cases}, \quad t \geq 0,\ A \in \mathcal{B}(\mathbb{R}),$$

where P_W is the transition function of Wiener process. Check that P satisfies the Kolmogorov–Chapman equations. Prove that the Markov process with transition function P is *not* a strong Markov process.

12.28. Let X be a homogeneous Markov process, and $\phi : \mathbb{X} \to \mathbb{R}$ be a bounded measurable function. Prove that double-component process $Y(t) = (X(t), \int_0^t \phi(X(s))\,ds)$, $t \geq 0$ is also a homogeneous Markov process.

12.29. Let X be a homogeneous Markov process, and $\phi : \mathbb{X} \to \mathbb{R}$ be a nonnegative measurable bounded function,

$$Q(t,x,A) := \mathsf{E}_x \exp\left(-\int_0^t \phi(X(s))\,ds\right) \mathbb{I}_A(X(t)),\ x \in \mathbb{X},\ A \in \mathcal{B}(\mathbb{X}).$$

Prove that Q is a *substochastic transition function*; that is, for Q all the conditions (1′)–(4′) from the definition of a transition function are satisfied except the condition $Q(t,x,\mathbb{X}) = 1, t \geq 0, x \in \mathbb{X}$, instead of which the following condition holds, $Q(t,x,\mathbb{X}) \leq 1, t \geq 0, x \in \mathbb{X}$.

12.30. Let $\mathbb{X} = \mathbb{R}^+$. Prove the following statements.

(1) The function

$$Q_+(t,x,A) = \frac{1}{\sqrt{2\pi t}} \int_A \left[e^{-(((y-x)^2)/2)} + e^{-(((y+x)^2)/2)}\right] dy,\ t \in \mathbb{R}^+,\ x \in \mathbb{X},\ A \in \mathcal{B}(\mathbb{X})$$

is a transition function,

(2) The function

$$Q_-(t,x,A) = \frac{1}{\sqrt{2\pi t}} \int_A \left[e^{-(((y-x)^2)/2)} - e^{-(((y+x)^2)/2)}\right] dy,\ t \in \mathbb{R}^+,\ x \in \mathbb{X},\ A \in \mathcal{B}(\mathbb{X})$$

is a substochastic transition function.

The functions Q_+, Q_- are transition functions for the processes, called the *Brownian motion on* $\mathbb{R}^+$ *with reflection* and the *Brownian motion on* $\mathbb{R}^+$ *with absorption* at zero point, correspondingly.

12.31. Prove that the family of operators $\{T_t, t \geq 0\}$ defined by the equality (12.2) creates a semigroup, and moreover, $\|T_t\| = 1$.

12.32. (1) Prove that the resolvent R_λ is an operator-valued analytical function of λ for $\mathrm{Re}\,\lambda > 0$, and

$$\frac{d}{d\lambda} R_\lambda f = -R_\lambda^2 f.$$

(2) Prove that for any λ with $\mathrm{Re}\,\lambda > 0$ and any $n \geq 1$,

$$\frac{d^n}{d\lambda^n} R_\lambda f = (-1)^n n! R_\lambda^{n+1} f.$$

12.33. Prove Lemma 12.1.

12.34. Determine the semigroup generated by the m-dimensional Wiener process. Prove that the space $C^2_{\mathrm{uni}}(\mathbb{R}^m)$ is contained in D_A, and for those functions $Af = \frac{1}{2}\Delta f$, where Δ is the Laplace operator (Laplacian).

12.35. Let N be the Poisson process, $X(t) = i^{N(t)}$, $t \geq 0$, where $i = \sqrt{-1}$. Prove that X is a homogeneous Markov process. Determine its semigroup, infinitesimal operator, and resolvent.

12.36. Determine the resolvent of the Poisson process.

12.37. Determine the resolvent of the Wiener process.

12.38. Let the diffusion process $\{X(t),\mathcal{F}_t,t\in\mathbb{R}^+\}$ with continuous drift coefficient $\mu(x)$ and continuous diffusion coefficient $\sigma(x)\neq 0$ start from the point $x\in(a,b)$, and let $\tau_a\,(\tau_b)$ be the hitting time of the point a (point b) by this process. Denote $u(x)=\mathsf{P}(\tau_b<\tau_a\big|X_0=x),\ \ a<x<b$. We know that $u\in C^2(\mathbb{R})$. Prove that the function u satisfies the following differential equation.

$$\frac{\sigma^2(x)}{2}u''(x)+u'(x)\mu(x)=0,\ a<x<b,$$

with boundary conditions $u(a)=0$, $u(b)=1$.

12.39. Consider a HMF corresponding to the m-dimensional Wiener process W (see discussion after Definition 12.7 and Problem 12.23).

(1) Prove that for any random variable ξ for which the expectations below are correctly defined, the equality holds $\mathsf{E}_\mu\xi=\int_{\mathbb{R}^m}(\mathsf{E}_x\xi)\,\mu(dx)$.

(2) Let $B(x,r)$ be the ball with the center $x\in\mathbb{R}^m$ and of radius r. Prove that the Wiener process starting from the point x exits the ball $B(x,r)$ with probability 1.

(3) Let τ_r denote the exit time for the process W from the ball $B(x,r)$. Find the distribution of random variable $W(\tau_r)$ assuming that $W(0)=x$, and prove that $\mathsf{E}_x\tau_r<\infty$.

(4) Prove that $\mathsf{E}_x\tau_r=r^2/m$.

(5) Let φ be a bounded measurable function, $f(x)=\mathsf{E}_x\phi(W(\tau_r))$. Using the strong Markov property of the Wiener process (Problem 12.24), prove that $f(x)=\int_{S(x,r)}f(y)\mu(dy)$, where $S(x,r)$ is the spherical surface of the ball $B(x,r)$, and μ is the unit mass uniformly distributed on $S(x,r)$.

12.40. (Example of a local martingale that is not a martingale). Let $\{W(t),t\in\mathbb{R}^+\}$ be a Wiener process with values in $\mathbb{R}^3$, and $W(0)=x_0\neq 0$. Also, let the function $h:\mathbb{R}^3\setminus\{0\}\to\mathbb{R}$, $h(x)=\|x\|^{-1}, x\in\mathbb{R}^3\setminus\{0\}$.

(1) Prove that $h(W(t))$ is a local martingale but *is not* a martingale with respect to natural filtration.

(2) Prove that $\mathsf{E}_{x_0}(h(W(t)))^2<\infty$ for all $t\in\mathbb{R}^+$ and $x_0\in\mathbb{R}^3$, and moreover, $\sup_{t\geq 0}\mathsf{E}_{x_0}(h(W(t)))^2<\infty$, where the expectation is under the condition that the process starts from the point x_0.

Hints

12.1. Use Definition 12.1.
12.2. Use item (4) or (5) of Theorem 12.1.
12.6, 12.7. Use independence of increments of corresponding processes.
12.9. Consider the case when $\Omega=\mathbb{X}^{\mathbb{T}}$ is the set of all the functions defined on $\mathbb{T}$ and taking their values in $\mathbb{X}$, $\mathcal{F}$ is the σ-algebra generated by cylinder sets. Define a

new phase space $\widetilde{\mathbb{X}} := \mathbb{R}^+ \times \mathbb{X}$, and a new measurable space $\widetilde{\Omega} = \{\omega = \widetilde{x}(t) \,|\, \widetilde{x}(t) = (t_0 + t, x(t))\}$, where $x \in \mathbb{X}^{\mathbb{T}}$, $t_0 \in \mathbb{R}^+$, with the corresponding σ-algebra $\widetilde{\mathfrak{F}}$. Define on $\widetilde{\mathfrak{F}}$ the probability function $\widetilde{\mathsf{P}}_{\widetilde{x}}(\widetilde{B})$, $\widetilde{x} = (t_0, x)$ as the product of measures λ_{t_0} and $\mathsf{P}_{t_0,x}$, where λ_{t_0} is a probability measure on the space of functions of the form $g_s(t) = s + t$, $s, t \in \mathbb{R}^+$, concentrated on the function $g_{t_0}(t)$, and $\mathsf{P}_{t_0,x}$ are the measures on $\mathfrak{F}$ corresponding to the original nonhomogeneous process that starts from the point x at t_0. Now it is necessary to check that the constructed process is homogeneous and determine its transition probabilities.

12.11. Use Problems 12.10 and 6.20.

12.15, 12.16. Use Problems 12.11 and 6.18.

12.18, 12.19. To prove the Markov property, use the symmetry of the property from item (6) of Theorem 12.1 with respect to time reversion.

12.24. In order to shorten the notation, consider the one-dimensional case; the multidimensional case can be treated quite analogously.

(1) Use Theorem 12.5: the function $F(t,x)$ from the formulation of this theorem can be given explicitly, $F(t,x) = (2\pi t)^{-(1/2)} \int_{\mathbb{R}} f(y) e^{-(((y-x)^2)/2t)}\, dy$. Prove that this function is continuous in $(t,x) \in (0,+\infty) \times \mathbb{R}$ for every continuous bounded function f (in fact, even for every measurable bounded f).

(2) At first, prove that $\mathsf{E}\mathbb{I}_A g(\overline{V}) = \mathsf{E}\mathbb{I}_A \mathsf{E} g(\overline{V})$ for any $k \geq 1$, $0 \leq t_1 \leq t_2 \leq \cdots \leq t_k$, continuous function $g : \mathbb{R}^k \to \mathbb{R}$, event $A \in \mathscr{F}_\tau$, and vector $\overline{V} := (V(t_1), V(t_2), \ldots, V(t_k))$. In this order consider the discrete approximations of the stopping time τ by the stopping times $\tau_n(\omega) = \sum_{j=1}^\infty (j/2^n) \mathbb{I}_{A_{j,n}}$, where $A_{j,n} = \{(j-1)2^{-n} \leq \tau < j2^{-n})\}$. Deduce from here that for any event $B \in \mathscr{B}(\mathbb{R}^m)$ the equality $\mathsf{P}(A \cap \{\overline{V} \in B\}) = \mathsf{P}(A)\mathsf{P}(\overline{V} \in B)$ holds. Check that the proof follows from the last equality.

12.27. Consider Markov moment $\tau = \inf\{t \,|\, X(t) = 0\}$.

12.28. For arbitrary $s \leq t$ we can decompose the variable $I(t) := \int_0^t \phi(X(r))\, dr$ into the sum $I(s) + I(s,t)$, in which the second term $I(s,t) := \int_s^t \phi(X(r))\, dr$ is an $\mathfrak{F}_{\geq s}$-measurable variable.

12.29. Prove that the function Q admits the following probabilistic representation,

$$Q(t,x,A) = \mathsf{P}_x\left(X(t) \in A, \zeta > \int_0^t \phi(X(s))\, ds \right),$$

where ζ is a random variable independent of the process X and with a distribution $\mathrm{Exp}(1)$. Use this representation to prove the required statement.

12.30. The function Q_+ is a transition function of the process $|W|$; see Problem 12.26. The function Q_- admits the representation $Q_-(t,x,A) = \mathsf{P}_x(W(t) \in A, \tau_0 > t)$, where $\tau_0 = \inf\{r \,|\, W(r) = 0\}$ (prove this fact using the reflection principle for the Wiener process; see Problem 7.109). Use the above-mentioned probabilistic representations of the functions Q_+, Q_- in order to verify the required properties of these functions.

12.31. Use the Kolmogorov–Chapman equation (4′).

12.32. Use the resolvent equation.

12.33. Items (1), (2) follow directly from the definition of $\mathbb{B}_0$.

(3) Prove the following equality $T_h R_\lambda f - R_\lambda f = (e^{\lambda h} - 1) R_\lambda f - e^{\lambda h} \int_0^h T_s f e^{-\lambda s} ds$, whence $\|T_h R_\lambda f - R_\lambda f\| \leq (e^{\lambda h} - 1)\|R_\lambda f\| + h\|f\|$, $h \geq 0$.

(4) Use the Lebesgue dominated theorem and the fact that $\|T_t\| = 1$.
12.34. Use Theorem 12.6.
12.38. Let $h > 0$. Prove that

$$u(x) = \mathsf{E}\left(u(X_h)\middle|\, X_0 = x\right) + o(h),\ h \to 0+.$$

Then assume that $u \in C^2(\mathbb{R})$ and establish the following expansion with the help of the Taylor formula.

$$u(X_h) = u(x) + u'(x)\mu(x)h + \frac{\sigma^2(x)}{2}u''(x)h + \alpha(h),$$

where $\mathsf{E}(\alpha(h)|X_0 = x) \to 0,\ h \to 0+$.

12.39. (4) Check that it is enough to consider the case $x = 0$. At first, use the self-similar property of the Wiener process: a stochastic process $\widetilde{W}(t) := rW(t/r^2)$ is a Wiener process as well. Consider the moment τ_1 of the first exit of the trajectory of the process W from the ball $B(0,1)$ and the moment $\widetilde{\tau}_r$ of the first exit of the trajectory of the process $\widetilde{W}$ from the ball $B(0,r)$. Verify that $\widetilde{\tau}_r = r^2\tau_1$, whence obtain the relation $\mathsf{E}_0\widetilde{\tau}_r = C \cdot r^2$, where $C = \mathsf{E}_0\tau_1$ is some constant. This constant is finite; it follows from item (3). Calculate it for $m = 2$; for higher m the proof is similar but more technical. Let σ_1 be the moment of the first hit of the surface $x_1^2 + x_2^2 = r^2$ by the process W, and σ_2 be the moment of the first hit of one of the lines $x_1 = \pm 1$ by the process W. Prove that $\mathsf{E}_0\sigma_1 = \mathsf{E}_0\sigma_2 - \mathsf{E}_\mu\sigma_2$, where μ is the uniform distribution on the surface $x_1^2 + x_2^2 = r^2$. Use the fact that σ_2 is the moment of the first hit of the boundary of the interval $[-r,r]$ by the one-dimensional Wiener process (more exactly, by the second coordinate of W). Prove that $\mathsf{E}_x\sigma_2 = r^2 - x^2$. Use item (1) and prove that $C = \frac{1}{2}$ for $m = 2$.

Answers and Solutions

12.3. We prove only item (d). We have the following equalities

$$\begin{aligned} p_{ij}(s,t) &= \mathsf{P}(X_t = j/X_s = i) = \tfrac{\mathsf{P}(X_t=j,X_s=i)}{\mathsf{P}(X_s=i)} \\ &= \textstyle\sum_{k\in\mathbb{X}_u} \tfrac{\mathsf{P}(X_t=j,X_u=k,X_s=i)}{\mathsf{P}(X_s=i)} \\ &= \textstyle\sum_{k:\mathsf{P}(X_u=k,X_s=i)>0} \mathsf{P}(X_t = j/X_u = k, X_s = i)\tfrac{\mathsf{P}(X_u=k,X_s=i)}{\mathsf{P}(X_s=i)} \\ &= \textstyle\sum_{k\in\mathbb{X}_u} \mathsf{P}(X_t = j/X_u = k)\mathsf{P}(X_u = k/X_s = i) = \sum_{k\in\mathbb{X}_u} p_{ik}(s,u)p_{kj}(u,t). \end{aligned}$$

12.5. No.
12.6. $\mathsf{P}(t,x,A) = \mathsf{P}(x + W(t) \in A) = (1/\sqrt{2\pi t})\int_A e^{-(((y-x)^2)/2t)}\,dy$.
12.7. $\mathsf{P}(t,x,A) = \mathsf{P}(x + N(t) \in A) = \sum_{k:k+x\in A} e^{-\lambda t}(((\lambda t)^k)/k!)$.
12.8. (a) Yes; (b) no.
12.10. Necessity follows from the definition of a Markov property. Let's prove sufficiency. Denote for the fixed $s < t$ the random variable $\Delta(s,t) = X(t) - \mathsf{E}[X(t)|X(s)]$. The theorem on normal correlation provides that $\mathsf{E}[X(t)|X(s)]$ is a linear function of $X(s)$; that is, for arbitrary $s_1, \ldots, s_m \in [0,s)$ the random variables $\Delta(t,s), X(s_1), \ldots$

$X(s_m)$ create a Gaussian system. Moreover, by our assumption, for arbitrary $k = 1,\ldots,m$ the random variables $\Delta(t,s)$ and $X(s_k)$ are noncorrelated. At last, $\Delta(s,t)$ and $X(s)$ are also noncorrelated. Therefore (Proposition 6.3) the random variable $\Delta(s,t)$ is independent of the vector $(X(s_1),\ldots,X(s_m),X(s))$. So, the random variable $X(t)$ can be presented as the sum of random variable $\Delta(s,t)$, independent of $X(s_1),\ldots, X(s_m),X(s)$, and the random variable $\mathsf{E}[X(t)|X(s)]$, which is measurable with respect to the σ-algebra $\sigma(X(s_1),\ldots,X(s_m),X(s))$. It means that the conditional distribution of $X(t)$ with respect to $\sigma(X(s_1),\ldots,X(s_m),X(s))$ is a Gaussian distribution with the mean value $\mathsf{E}[X(t)|X(s)] + \mathsf{E}\Delta(s,t)$ and variance $\mathsf{D}\Delta(s,t)$. This distribution equals the conditional distribution of $X(t)$ with respect to σ-algebra $\sigma(X(s))$.

12.12. Let's write the sequence of equalities that hold true for all the necessary values of s,t,A: $\mathsf{P}(X(t+s)-X(t) \in A/\mathfrak{F}_t) = \mathsf{P}(X(t+s)-X(t) \in A/X(t)) = \mathsf{P}(X(t+s)-x \in A/X(t)=x)\big|_{\{X(t)=x\}} = \mathsf{P}(X(t+s) \in A+x/X(t)=x)\big|_{\{X(t)=x\}} = \mathsf{P}(x,s,x+A)\big|_{\{X(t)=x\}} = \mathsf{P}(0,s,A)$, and the last probability is constant, whence the proof follows.

12.13. Denote $p(s,x,y)$ the density of transition probability and obtain the following relations.

$$
\begin{aligned}
I &:= \int_{\mathbb{R}} p(s,x,z)p(t,z,y)dz \\
&= \frac{1}{2\pi\sigma(s)\sigma(t)} \int_{\mathbb{R}} \exp\left\{-\frac{(z-m(s)x)^2}{2\sigma^2(s)}\right\} \exp\left\{-\frac{(y-m(t)z)^2}{2\sigma^2(t)}\right\} dz \\
&= \frac{1}{m(t)} \frac{1}{2\pi\sigma(s)\frac{\sigma(t)}{m(t)}} \int_{\mathbb{R}} \exp\left\{-\frac{(z-m(s)x)^2}{2\sigma^2(s)}\right\} \exp\left\{-\frac{\left(\frac{y}{m(t)}-z\right)^2}{2\frac{\sigma^2(t)}{m^2(t)}}\right\} dz.
\end{aligned}
$$

The right-hand side, up to the term $1/m(t)$, is the density at the point 0 of the sum of two independent Gaussian random variables

$$
N\big(m(s)x,\sigma^2(s)\big) \ \text{and} \ N\left(-\frac{y}{m(t)}, \frac{\sigma^2(t)}{m^2(t)}\right).
$$

Therefore

$$
\begin{aligned}
I &= \frac{1}{m(t)} \frac{1}{\sqrt{2\pi}\sqrt{\sigma^2(s)+\frac{\sigma^2(t)}{m^2(t)}}} \exp\left\{-\frac{\left(\frac{y}{m(t)}-m(s)x\right)^2}{\sigma^2(s)+\frac{\sigma^2(t)}{m^2(t)}}\right\} \\
&= \frac{1}{\sqrt{2\pi}\sqrt{m^2(t)\sigma^2(s)+\sigma^2(t)}} \exp\left\{-\frac{(y-m(s)m(t)x)^2}{m^2(t)\sigma^2(s)+\sigma^2(t)}\right\},
\end{aligned}
$$

whence the required statement follows immediately.

12.14. It is sufficient to check the following relations: $\mathsf{P}(X_3 = 1/X_2 = 0, S_2 = 0) = \frac{1}{2} \neq \mathsf{P}(X_3 = 1/X_2 = 0, S_2 = -2) = 0$, but for a Markov process these probabilities must be equal.

12.15. The Ornstein–Uhlenbeck process is a homogeneous Markov process.

12.16. The transition probability density equals

$$
p(s,x,t,y) = \exp\left\{-\frac{(x(1-s)-y(1-t))^2}{2(1-t)(1-s)(t-s)}\right\}\left(\frac{1-s}{2\pi(1-t)(t-s)}\right)^{1/2}.
$$

12.17. The telegraph signal is a Markov process.
12.18. The transition probability density equals

$$p(s,x,t,y) = \sqrt{\frac{s}{2\pi t(t-s)}} \exp\left\{ -\frac{s(y-\frac{t}{s}x)^2}{2t(t-s)} \right\}, \ s<t<0, \ x,y \in \mathbb{R}.$$

12.19.

$$\mathsf{P}(s,x,t,\{y\}) = \left(\frac{t}{s}\right)^y \left(1-\frac{t}{s}\right)^{x-y} \frac{x!}{y!(x-y)!}, \ s<t<0, \ x,y \in \mathbb{Z}^+, \ x \geq y.$$

12.20. The answer for the first question is positive, and for the second one it is, generally speaking, negative.
12.21. The answers for both questions are, generally speaking, negative.
12.22. The example of a martingale that is not a Markov process can be described as $X_0 = X_1 = 1$, $X_{n+1} = X_n + X_{n-1} \cdot (\varepsilon_n / 2^n)$, where ε_n are i.i.d. Bernoulli random variables; the example of a Markov process that is not a martingale can be described as $W(t)+t$, where $W(t)$ is a Wiener process.
12.25. Yes.
12.26. Denote by $\{\mathcal{F}_t\}$ the filtration generated by the process W. For the arbitrary measurable function g we have the following equalities.

$$\mathsf{E}[\mathbb{I}_A(X(t))/\mathcal{F}_s] = \mathsf{E}[\mathbb{I}_{g^{-1}(A)}(W(t)/\mathcal{F}_s] = \mathsf{P}_W(t-s, W(s), g^{-1}(A)),$$

where P_W is the transition function of a Wiener process (see Problem 12.6). In item (1), we have that $W(s) = g^{-1}(X(s))$, therefore the transition function of the process X equals

$$\mathsf{P}(x,t,A) = \mathsf{P}_W(g^{-1}(x), t, g^{-1}(A)), \ x \in \mathbb{R}, \ t \geq 0, \ A \in \mathcal{B}(\mathbb{R}).$$

In item (2), the function $g(x) = |x|$ is not a bijection, nevertheless, for any Borel set $A \subset \mathbb{R}^+$ and arbitrary $t \in \mathbb{R}^+$ the values of transition probability satisfy the relations

$$\begin{aligned} \mathsf{P}_W(t, W(s), g^{-1}(A)) &= \mathsf{P}_W(t, W(s), A) + \mathsf{P}_W(t, W(s), -A) \\ &= \frac{1}{\sqrt{2\pi t}} \int_A e^{-(((y-W(s))^2)/2t)}\, dy + \frac{1}{\sqrt{2\pi t}} \int_{-A} e^{-(((y-W(s))^2)/2t)}\, dy \\ &= \frac{1}{\sqrt{2\pi t}} \int_A \left[e^{-(((y-W(s))^2)/2t)} + e^{-(((y+W(s))^2)/2t)} \right] dy \end{aligned}$$

and thus depend only on $|W(s)|$. It means that the transition function of the process $|W|$ equals

$$\mathsf{P}(t,x,A) = \frac{1}{\sqrt{2\pi t}} \int_A \left[e^{-(((y-x)^2)/2t)} + e^{-(((y+x)^2)/2t)} \right] dy, \ x \in \mathbb{R}^+, \ t \geq 0, \ A \in \mathcal{B}(\mathbb{R}^+).$$

The processes $g(W)$ and $|W|$ are strong Markov processes.
12.35. Let λ be the intensity of Poisson process, then

$$T_t f(i^k) = \sum_{n=0}^{\infty} e^{-\lambda t} \frac{(\lambda t)^n}{n!} f(i^{k+n}), \ t \geq 0, \ Af(i^k) = \lambda [f(i^{k+1}) - f(i^k)],$$

$$\begin{aligned} R_\mu f(i^k) = {} & [(\lambda+\mu)^4 - \lambda^4]^{-1} \Big((\lambda+\mu)^3 f(i^k) + (\lambda+\mu)^2 \lambda f(i^{k+1}) \\ & + (\lambda+\mu)\lambda^2 f(i^{k+2}) + \lambda^3 f(i^{k+3}) \Big), \ \mu > 0, \ k = 0,1,2,3. \end{aligned}$$

12.36. Let λ be the intensity of Poisson process, then

$$R_{\mu}f(k) = \sum_{n=0}^{\infty} \frac{\lambda^n}{(\lambda+\mu)^{n+1}} f(k+n),\ \mu > 0,\ k \in \mathbb{Z}^+.$$

12.37. $R_\lambda f(x) = \int_{\mathbb{R}} Q_\lambda(x,y) f(y)\, dy$,

$$Q_\lambda(x,y) = \int_0^\infty e^{-\lambda t} \frac{e^{-(((y-x)^2)/2t)}}{\sqrt{2\pi t}}\, dt = \frac{1}{\sqrt{2\lambda}} e^{-\sqrt{2\lambda}|y-x|}.$$

12.39. (2) If the trajectory of a Wiener process stays inside the ball $B(x,r)$ till the moment $t=n$, then all the increments $W(1)-W(0), W(2)-W(1), \dots, W(n)-W(n-1)$ do not exceed $2r$ in absolute value. These increments are jointly independent and Gaussian. Therefore, if τ_r is the moment of the first exit from the ball $B(x,r)$, then $\mathsf{P}_x(\tau_r \geq n) \leq (\mathsf{P}(|W(1)-W(0)| < 2r))^n = p^n$, where $p < 1$. It means that $\mathsf{P}(\tau_r = \infty) = 0$.

(3) It follows with evidence from item (1) that the random point $W(\tau_r)$ is situated on the spherical surface $S(x,r)$ that restricts the ball $B(x,r)$. Because the distribution density of any increment $W(t)-W(s)$ depends only on $t-s$ and on the modulus of the vector $W(t)-W(s)$, we have that the distribution of the Wiener trajectory does not change under any rotation of the whole space around the center of the ball for any angle. So, the distribution of the random point $W(\tau_r)$ is invariant with respect to all the rotations of the spherical surface $S(x,r)$. The unique distribution with this property is the uniform distribution for which the probability for $W(\tau_r)$ to hit within some domain of the spherical surface is proportional to the area of this domain. It means that the distribution of $W(\tau_r)$ is uniform on $S(x,r)$. Let $F(t)$ be the distribution function of τ_r. Then, according to item (2),

$$\begin{aligned}\mathsf{E}_x\tau_r &= \textstyle\int_0^\infty t dF(t) \leq \sum_{n=1}^\infty \int_{n-1}^n t dF(t)\\ &\leq \textstyle\sum_{n=1}^\infty n \int_{n-1}^n dF(t) \leq \sum_{n=1}^\infty n\mathsf{P}_x(\tau_r \geq n-1) \leq \sum_{n=1}^\infty np^{n-1} < \infty.\end{aligned}$$

(5) Let a Wiener process start from the point x. At the moment τ_r, according to item (2), the distribution of $W(\tau_r)$ is uniform on $S(x,r)$. According to the strong Markov property, the process W can be considered as a Wiener process with uniform initial distribution. Therefore, due to item (1), we obtain that $\mathsf{E}_x\varphi(W(\tau)) = \mathsf{E}_\mu\varphi(W(\tau)) = \int_{S(x,r)} \mathsf{E}_y\varphi(W(\tau))\mu(dy)$.

12.40. (1) Denote the increasing sequence of stopping times $\tau_k := \inf\{t > 0 |\ \|W(t)\| \leq 1/k\}$, $k \geq 1$. Then $\tau_k \to \infty$ a.s., because in $\mathbb{R}^3$ $\mathsf{P}(W(t) = 0 \text{ for some } t > 0) = 0$. The function h is harmonic in $\mathbb{R}^3 \setminus \{0\}$ and for any $k \geq 1$ $D_k := \{x \in \mathbb{R}^3 |\ \|x\| > 1/k\} \subset D(h)$. Consider the closure $\overline{D}_k$ and define on it the function $g_k(x) := \mathsf{E}_x h(W(\tau_k))$, where E_x is an expectation under initial condition $W(0) = x$. Using the strong Markov property (12.6) of the Wiener process and Problem 12.39, item (6), we obtain that $g_k(x) = \int_{S(x,r)} g_k(y) dm(y)$ for all sufficiently small balls $B(x,r)$ with the centers in the point $x \in D_k$ and spherical surfaces $S(x,r)$, where the measure m has unit mass and is concentrated on $S(x,r)$. Therefore, g_k is harmonic in D_k. Moreover, it is continuous in $\overline{D}_k$ and $g_k|\partial D_k = h$, where $\partial D_k = \overline{D}_k \setminus D_k$. From the principle of the maximum for harmonic functions we obtain that $g_k = h$ in $\overline{D}_k$ for any k. Furthermore, for any $k \geq 1$, $x \in D_k$ and $t > 0$

$$\mathsf{E}_x(h(W(\tau_k))/\mathfrak{F}_t) = \mathbb{I}_{\tau_k \le t} h(W(t \wedge \tau_k)) + \mathbb{I}_{\tau_k > t} \mathsf{E}_x(W(\tau_k)/\mathfrak{F}_t).$$

(Here $\{\mathfrak{F}_t\}$ is the filtration generated by the process W.) It follows from the strong Markov property that on the set $\{\tau_k > t\}$ we have the equality $\mathsf{E}_x(h(W(\tau_k))/\mathfrak{F}_t) = \mathsf{E}_{W(t)}(h(W(\tau_k))) = g_k(W(t))$. Because $g_k = h$ in $\overline{D}_k$, then, combining previous relations, we obtain that $\mathsf{E}_x(W(\tau_k)/\mathfrak{F}_t) = h(W(t \wedge \tau_k))$. Furthermore, $W(0) = x_0$, and $x_0 \in D_k$ for all sufficiently large k, and we obtain that $\{h(W(t \wedge \tau_k)), t \in \mathbb{R}^+\}$ is a bounded martingale, whence $\{h(W(t)), t \in \mathbb{R}^+\}$ is a local martingale. But it is not a martingale. Indeed, for $t > 0$ and $R > 2\|x_0\|$ we have that

$$\begin{aligned}
\mathsf{E}_{x_0} h(W(t)) &= (2\pi t)^{-3/2} \int_{\mathbb{R}^3} \|y\|^{-1} e^{-((\|y-x_0\|^2)/2t)} dy \\
&\le (2\pi t)^{-3/2} \Big(\int_{\|y\| \le R} \|y\|^{-1} dy + \int_{\|y\| > R} \|y\|^{-1} e^{-((\|y\|^2)/8t)} dy \Big) \\
&\le \frac{C_1 R^2}{(2\pi t)^{3/2}} + \frac{C_2}{R} \to 0,
\end{aligned}$$

if, at first, $t \to \infty$, and then $R \to \infty$. It means that $\mathsf{E}_{x_0} h(W(t)) \ne \mathsf{E}_{x_0}(h(W(0))) = \|x_0\|^{-1}$ for large t.

(2) Please, prove this statement on your own, similarly to previous estimates.

13

Itô stochastic integral. Itô formula. Tanaka formula

Theoretical grounds

Let $\{W(t), t \in \mathbb{R}^+\}$ be a Wiener process, and $\{g(t), \mathscr{F}_t^W, t \in \mathbb{R}^+\}$ a stochastic process (recall that the previous notation means that g is adapted to a natural filtration $\{\mathscr{F}_t^W\}$ of the Wiener process). Let $\mathscr{F}^W = \sigma\{W_t, t \geq 0\}$. A process g is said to belong to the class $\hat{\mathscr{L}}_2([a,b])$ if it is measurable and $\mathsf{E}\int_a^b g^2(s)ds < \infty$. A process g belongs to the class $\hat{\mathscr{L}}_2$ if it belongs to $\hat{\mathscr{L}}_2([0,t])$ for all $t \in \mathbb{R}^+$.

Assume process $g \in \hat{\mathscr{L}}_2([a,b])$ is simple, that is, has the form $g(s) = g(t_k)$, $s \in (t_k, t_{k+1}]$, $a = t_0 < t_1 < \cdots < t_n = b$, $g(a) = g_0$. Then *the stochastic integral* of g on $[a,b]$ with respect to the Wiener process is defined as $\int_a^b g(s)dW(s) := \sum_{k=0}^{n-1} g(t_k)[W(t_{k+1}) - W(t_k)]$. If $g \in \hat{\mathscr{L}}_2([a,b])$ then there exists a sequence of simple processes $g_n \in \hat{\mathscr{L}}_2([a,b])$ such that $\mathsf{E}\int_a^b [g(s) - g_n(s)]^2 ds \to 0, n \to \infty$. The *Itô stochastic integral* of a process g with respect to the Wiener process is defined by the formula $I_{[a,b]}(g) := \int_a^b g(s)dW(s) := \text{l.i.m.} \int_a^b g_n(s)dW(s)$.

Let $g \in \hat{\mathscr{L}}_2([0,T])$. Then $g(\cdot)1\!\!1_{\cdot\in[0,t]} \in \hat{\mathscr{L}}_2([0,T])$ for every $t \in [0,T]$. Define a collection of stochastic integrals $I_t(g) = \int_0^t g(s)dW(s)$ when $t \in [0,T]$ by a formula $I_t(g) = I_{[0,T]}(g(\cdot)1\!\!1_{\cdot\in[0,t]})$. Denote $I_{[s,t]}(g) = I_{[0,T]}(g(\cdot)1\!\!1_{\cdot\in[s,t]})$, $[s,t] \subset [0,T]$. (See Problem 13.7.)

Theorem 13.1. *Stochastic integral $\{I_t(g), t \in [0,T]\}$ of a process $g \in \hat{\mathscr{L}}_2([0,T])$ with respect to a Wiener process has the following properties.*

(1) It is linear with respect to the integrand: $I_t(af + bg) = aI_t(f) + bI_t(g), f, g \in \hat{\mathscr{L}}_2([0,T])$.

(2) It is additive with respect to the interval of integration: $I_t(f) = I_u(f) + I_{[u,t]}(f)$ for all $0 \leq u \leq t$.

(3) It has a modification continuous in $t \in [0,T]$.

(4) It is a square integrable martingale with respect to a flow $\{\mathscr{F}_t, t \in \mathbb{R}^+\}$, and is isometric $\|I_t(g)\|_{L_2(\mathsf{P})} = \|g\|_{L_2([0,t])}$, $g \in \hat{\mathscr{L}}_2$.

D. Gusak et al., *Theory of Stochastic Processes*, Problem Books in Mathematics,
DOI 10.1007/978-0-387-87862-1_13,

It is known ([60] Chapter IV, §3) that each measurable adapted process has a progressively measurable modification. In the following we always consider a progressive modification for measurable adapted processes.

The Itô integral can be extended to the class of measurable adapted processes with square integrable trajectories by the locality principle (Problem 13.10). We also call this extension the Itô integral. It keeps properties (1)–(3) of Theorem 13.1.

Let process X be of the form

$$X(t) = X(0) + \int_0^t \mu(s)ds + \int_0^t \sigma(s)dW(s),\ t \in \mathbb{R}^+,$$

where σ is a measurable adapted process such that $\mathsf{P}(\int_0^t \sigma^2(s)ds < \infty) = 1,\ t \in \mathbb{R}^+$ and μ is measurable and adapted, and $\mathsf{P}(\int_0^t |\mu(s)|ds < \infty) = 1,\ t \in \mathbb{R}^+$. Then the process $X(t)$ is said to have a stochastic differential $dX(t) = \mu(t)dt + \sigma(t)dW(t)$.

Theorem 13.2. *(The Itô formula) Let function $F : [0,\infty) \times \mathbb{R} \to \mathbb{R}$ belong to the class $C^1([0,\infty)) \times C^2(\mathbb{R})$ of the functions, continuously differentiable in $t \in [0,\infty)$ and twice continuously differentiable in $x \in \mathbb{R}^+$. Suppose that process $\{X(t), \mathscr{F}_t, t \in \mathbb{R}^+\}$ has a stochastic differential $dX(t) = f(t)dW(t) + g(t)dt$. Then the following formula holds true:* P*-a.s. for all $t \in \mathbb{R}^+$,*

$$F(t,X(t)) = F(0,X_0) + \int_0^t \frac{\partial F(s,X(s))}{\partial x} f(s)dW(s)$$

$$+ \int_0^t \left(\frac{\partial F(s,X(s))}{\partial s} + \frac{\partial F(s,X(s))}{\partial x} g(s) + \frac{1}{2}\frac{\partial^2 F(s,X(s))}{\partial x^2} f^2(s) \right) ds.$$

The statement of Theorem 13.2 can be written in the following form. The process $Y(t) = \{F(t,X(t)), t \in \mathbb{R}^+\}$ has a stochastic differential

$$dY(t) = \frac{\partial F}{\partial t}(t,X(t))dt + \frac{\partial F}{\partial x}(t,X(t))dX(t) + \frac{1}{2}\frac{\partial^2 F}{\partial x^2}(t,X(t))(dX(t))^2,$$

where $(dX(t))^2$ is defined via the following formal "operating with differentials" rules: $dt \cdot dt = dt \cdot dW(t) = dW(t) \cdot dt = 0$, $dW(t) \cdot dW(t) = dt$.

Theorem 13.3. *(The multidimensional Itô formula) Assume that $X(t) = (X_1(t), \ldots, X_m(t))$ is a stochastic process, and each coordinate $X_i(t)$ has a stochastic differential $dX_i(t) = \mu_i(t)dt + \sigma_i(t)dW_i(t)$, $W_i(t)$ are correlated Wiener processes: $\langle W_i, W_j\rangle(t) = \int_0^t \rho_{ij}(s)ds$, and $F : [0,\infty) \times \mathbb{R}^m \to \mathbb{R}$ belongs to the class $C^1([0,\infty)) \times C^2(\mathbb{R}^m)$. Then*

$$F(t,X(t)) = F(0,X_0) + \int_0^t \left[\frac{\partial F(s,X(s))}{\partial t} + \sum_{i=1}^m \frac{\partial F(s,X(s))}{\partial x_i}\mu_i(s) \right.$$

$$\left. + \frac{1}{2}\sum_{i=1}^m \frac{\partial^2 F(s,X(s))}{\partial x_i \partial x_j}\sigma_i(s)\sigma_j(s)\rho_{ij}(s) \right] ds + \sum_{i=1}^m \int_0^t \frac{\partial F(s,X(s))}{\partial x_i}\sigma_i(s)dW_i(s).$$

Definition 13.1. *Let* $\{W(t),\ t\in\mathbb{R}^+\}$ *be a one-dimensional Wiener process. The limit in probability*

$$\begin{aligned} L(t,x) &:= \lim_{\varepsilon\downarrow 0}(2\varepsilon)^{-1}\int_0^t \mathrm{I\!I}_{W(s)\in(x-\varepsilon,x+\varepsilon)}ds \\ &= \lim_{\varepsilon\downarrow 0}(2\varepsilon)^{-1}\lambda^1\{s\in[0,t] \mid W(s)\in(x-\varepsilon,x+\varepsilon)\} \end{aligned} \tag{13.1}$$

is called the local time of the process W at the point x on the time interval $[0,t]$. *The limit in* (13.1) *in fact exists both with probability 1 and in the mean square (see Problems* 13.47 and 13.52).

The notion of a local time was introduced by Lévy [56]; see also [13], [87]. Often it is said that the local time describes the *relative time* spent by the trajectory of the process W near the point x. Recall (see Problem 3.21) that the total time spent by this trajectory at the point x is a.s. equal to zero.

Theorem 13.4. *(The Tanaka formula) For every* $(t,x)\in\mathbb{R}^+\times\mathbb{R}$,

$$|W(t)-x|-|W(0)-x| = \int_0^t \mathrm{sign}(W(s)-x)dW(s)+L(t,x) \ \ a.s.$$

The Tanaka formula can be considered as a generalization of the Itô formula for the function $f(x)=|x|$, which is not twice differentiable (see also [66] for further generalizations).

Theorem 13.5. *(The Fubini theorem for stochastic integrals) Let* $\{W(t),\ t\in\mathbb{R}^+\}$ *be a Wiener process, a function* $f=f(x,s,\omega):[a,b]\times[c,d]\times\Omega\to\mathbb{R}$ *be jointly measurable with respect to all arguments,* $f(x,s)\in\hat{\mathscr{L}}_2([a,b])$ *for all* $s\in[c,d]$, $f(x,s)\in\hat{\mathscr{L}}_2([c,d])$ *for all* $x\in[a,b]$, *and* $\int_a^b f^2(x,s)dx\in\hat{\mathscr{L}}_1([c,d])$, *that is,* $\mathsf{E}\int_c^d\int_a^b f^2(x,s)dxds<\infty$. *Then the stochastic integral* $\int_c^d f(x,s)dW(s)$ *exists for each* $x\in[a,b]$, *and being considered as a function of x it is Lebesgue integrable on* $[a,b]$ *with probability 1. Also stochastic integral* $\int_c^d\left(\int_a^b f(x,s)dx\right)dW(s)$ *exists, and a.s. the following equality holds true.*

$$\int_a^b\left(\int_c^d f(x,s)dW(s)\right)dx = \int_c^d\left(\int_a^b f(x,s)dx\right)dW(s).$$

Assume that $\{M(t),\mathfrak{F}_t,\ t\in[0,T]\}$ is a square integrable martingale with trajectories in $\mathbb{D}([0,T])$ (see Theoretical grounds in Chapter 3). Let $\langle M\rangle(t)$ be its quadratic characteristic. Denote by $\Pi_b([0,T])$ a space of simple functions g such that $g(s)=g(t_k)$, $s\in(t_k,t_{k+1}]$ with some partition $\{0=t_0<t_1<\cdots<t_n=T\}$ and $g(t_k)$ being a bounded $\mathfrak{F}_{t_k}$-measurable random variable for every $k=0,\dots,n-1$.

For $g\in\Pi_b([0,T])$, the Itô stochastic integral of g w.r.t. martingale M is defined as $\int_0^T g(s)dM(s):=\sum_{k=0}^{n-1}g(t_k)(M(t_{k+1})-M(t_k))$.

Introduce a norm $\|g\|_{\langle M\rangle}:=(\mathsf{E}\int_0^T g^2(s)d\langle M\rangle(s))^{1/2}$ in $\Pi_b([0,T])$. The just-defined correspondence $I:g\to\int_0^T g(s)dM(s)$ is a linear and isometric mapping from $\Pi_b([0,T])$ to $L_2(\Omega,\mathfrak{F},\mathsf{P})$. Define the space $\hat{\mathscr{L}}_2([0,T],\langle M\rangle)$ as the completion of $\Pi_b([0,T])$ in the norm $\|\cdot\|_{\langle M\rangle}$, and extend the correspondence I to the whole space $\hat{\mathscr{L}}_2([0,T],\langle M\rangle)$. This extension is called the Itô stochastic integral w.r.t. M and is denoted by I. For a process $g\in\hat{\mathscr{L}}_2([0,T],\langle M\rangle)$ the Itô stochastic integral of g w.r.t. M is defined as the image of g under I, and is denoted by $I(g)$.

A process $I_t(g)$, $t \in [0,T]$ can be introduced similarly to the Wiener case. This process satisfies properties (1) and (2) of Theorem 13.1. In addition,

$$\mathsf{E}I_t(g_1)I_t(g_2) = \mathsf{E}\int_0^T g_1(s)g_2(s)d\langle M\rangle(s),\ g_1,g_2 \in \hat{\mathscr{L}}_2([0,T],\langle M\rangle),$$
$$\mathsf{E}(I_t^2(g) - I_s^2(g)/\mathfrak{F}_s) = \mathsf{E}(\int_s^t g^2(u)d\langle M\rangle(u)/\mathfrak{F}_s),$$

and if $M \in \mathscr{M}^{2,c}$ then $I_\cdot(g) \in \mathscr{M}^{2,c}$.

For continuous square integrable martingales, the following version of the Itô formula holds true.

Theorem 13.6. *Let $M \in \mathscr{M}^{2,c}$, and $f \in C^2(\mathbb{R})$. Then*

$$f(M(t)) = f(M(0)) + \int_0^t f'(M(s))dM(s) + \frac{1}{2}\int_0^t f''(M(s))d\langle M\rangle(s).$$

Bibliography

[9], Chapter VIII; [38], Chapter II; [64]; [13]; [90], Chapter 12, §12.1–12.3; [24], Volume 3, Chapter I, §2–3; [25], Chapter VIII, §1; [51], Chapter 19; [57], Chapters 4 and 5; [79], Chapters 30 and 32; [20], Chapters 11 and 12; [8], Chapter 7, §7.1–7.4; [46], Chapter 12, §12.4 and §12.6; [54], Chapter 3, §3.4; [61], Chapters III–IV; [68], Chapter 13, §13.1.1; [85], Chapters 6–8; [66].

Problems

13.1. Let $f \in L_2([0,T])$ be a real-valued nonrandom function, and $\{W(t),\ t \in [0,T]\}$ be the Wiener process.

(1) Prove that $I_{[0,T]}(f) = \int_0^T f(s)dW(s)$ has a Gaussian distribution. Find its mean and variance. The integral of a nonrandom function is also called the Wiener integral, or the Itô–Wiener integral.

(2) Prove that $\mathsf{E}\exp\{iuI_{[0,T]}(f)\} = \exp\{-(u^2/2)\int_0^T f^2(t)dt\}$.

(3) Verify that for a nonrandom $f \in L_2([0,T])$ the Wiener integral $\int_0^T f(s)dW(s)$ coincides with the stochastic integral w.r.t. the process W considered as a process with orthogonal increments (Chapter 8, Problems 8.13 and 8.14) if $f \in L_2([0,T])$.

13.2. (1) Let $\{W(s), s \in [0,t]\}$ be a Wiener process. Prove the following integration by parts formula. If a function $f(\cdot,\omega) \in C^1[0,T]$ for a.a. $\omega \in \Omega$, and $f \in \mathscr{L}_2$, then

$$\int_0^t f(s,\omega)dW(s) = f(t,\omega)W(t) - \int_0^t W(s)f'(s,\omega)ds.$$

(2) Prove that $\int_0^t s dW(s) = tW(t) - \int_0^t W(s)ds$.

13.3. Prove that the random variables $\int_0^1 t dW(t)$ and $\int_0^1 (2t^2-1)dW(t)$ are independent.

13.4. (1) Find $\int_0^t W(s)dW(s)$, where $\{W(s), s \in [0,t]\}$ is a Wiener process.

(2) Find variances of the integrals $\int_0^t |W(s)|^{1/2}dW(s)$ and $\int_0^t (W(s)+s)^2 dW(s)$.

(3) Prove that $\int_0^t W^2(s)dW(s) = \frac{1}{3}W^3(t) - \int_0^t W(s)ds$.

13.5. Assume that process $g \in \mathscr{L}_2([0,T])$ is continuous in the mean square sense. Prove that

$$\int_0^T g(t)dW(t) = \underset{n\to\infty}{\text{l.i.m.}} \sum_{i=0}^{n-1} g\left(t_i^{(n)}\right)\left(W\left(t_{i+1}^{(n)}\right) - W\left(t_i^{(n)}\right)\right),$$

where a sequence of partitions $\pi_n := \{0 = t_0^{(n)} < \cdots < t_n^{(n)} = T\}$ is such that $|\pi_n| \to 0$ as $n \to \infty$.

13.6. Assume that process $g \in \hat{\mathscr{L}}_2([0,T])$. Prove that

$$\int_0^T g(t)dW(t) = \underset{n\to\infty}{\text{l.i.m.}} \sum_{k=1}^{n-1} n \int_{(k-1)T/n}^{kT/n} g(s)ds(W((k+1)T/n) - W(kT/n)).$$

13.7. Let $g \in \hat{\mathscr{L}}_2([0,T])$.

(1) Prove that for each $t \in [0,T]$ a restriction of the process g on $[0,t]$ lies in a class $\hat{\mathscr{L}}_2([0,t])$.

(2) Prove that $\int_0^t g(s)dW(s) = \int_0^T \mathrm{I\!I}_{[0,t]}(s)g(s)\,dW(s)$, $t \in [0,T]$ a.s.

13.8. (Localization of the stochastic integral) Let $f, g \in \hat{\mathscr{L}}_2([0,T])$. Denote $A = \{\omega \in \Omega \mid \int_0^T (f(t,\omega) - g(t,\omega))^2 dt = 0\}$. Prove that $\int_0^t f(s)dW(s) = \int_0^t g(s)dW(s)$, $t \in [0,T]$ almost surely on the set A.

13.9. Let $f \in \hat{\mathscr{L}}_2([0,T])$, τ be a stopping time.

(1) Prove that the process $f_\tau := f(\cdot)\,\mathrm{I\!I}_{[0,\tau]}$ belongs to $\hat{\mathscr{L}}_2([0,T])$.

(2) Denote $J(t) = \int_0^t f(s)dW(s)$, $t \in [0,T]$. Prove that $J(\tau \wedge T) = \int_0^T \mathrm{I\!I}_{t\le\tau} f(t)dt := I_{[0,T]}(f_\tau)$ a.s. (in the left-hand side, the stopping time $\tau \wedge T$ is substituted into the continuous stochastic process J).

13.10. Let the measurable adapted stochastic process $f(t)$, $t \in [0,T]$ have square integrable trajectories. Prove that there exists a unique process $I(t)$, $t \in [0,T]$ such that for any $g \in \hat{\mathscr{L}}_2([0,T])$ the equality $I(t) = \int_0^t g(s)dW(s)$, $t \in [0,T]$ holds almost surely on the set $\{\omega \in \Omega \mid \int_0^T (f(t,\omega) - g(t,\omega))^2 dt = 0\}$.

Stochastic process $I(t)$, $t \in [0,T]$ is called the stochastic Itô integral of f w.r.t. W, and is denoted $I(t) = \int_0^t f(s)dW(s)$.

13.11. Assume that $g(t)$, $t \in [0,T]$ is a measurable adapted process with square integrable trajectories. Prove that $\sum_{k=1}^{n-1} n \int_{(k-1)T/n}^{kT/n} g(s)ds(W((k+1)T/n) - W(kT/n))$ converges in probability as $n \to \infty$ to $\int_0^T g(t)dW(t)$. Compare with Problem 13.6.

13.12. Assume that $\{f(t),\ t \in [0,T]\}$ is a continuous adapted process. Prove that the following convergence in probability holds

$$\int_0^T f(t)dW(t) = \underset{n\to\infty}{\text{P-lim}} \sum_{i=0}^{n-1} f\left(t_i^{(n)}\right)\left(W\left(t_{i+1}^{(n)}\right) - W\left(t_i^{(n)}\right)\right),$$

where a sequence of partitions $\pi_n := \{0 = t_0^{(n)} < \cdots < t_n^{(n)} = T\}$ is such that $|\pi_n| \to 0$ as $n \to \infty$. Compare with Problem 13.5.

13.13. Let $\pi_n = \{0 = t_{n,0} < t_{n,1} < \cdots < t_{n,n} = 1\}$ be a sequence of partitions such that $|\pi_n| \to 0$ $n \to \infty$.

Find limits in probability.

(a) $\lim_{n\to\infty} \sum_{k=0}^{n-1} W(t_{n,k})(W(t_{n,k+1}) - W(t_{n,k}))$.

(b) $\lim_{n\to\infty} \sum_{k=0}^{n-1} W(t_{n,k+1})(W(t_{n,k+1}) - W(t_{n,k}))$.

(c) $\lim_{n\to\infty} \sum_{k=0}^{n-1} W\left((t_{n,k+1} + t_{n,k})/2\right)(W(t_{n,k+1}) - W(t_{n,k}))$.

13.14. Assume that a sequence of progressively measurable stochastic processes $b_n(t)$, $t \in [0,T]$ is such that

(1) $\mathsf{P}\left(\int_0^T b_n^2(t)dt < \infty \right) = 1, n \ge 0$.

(2) $\mathsf{P}\left(\lim_{n\to\infty} \int_0^T (b_n(t) - b_0(t))^2 dt = 0 \right) = 1$.

Prove that

$$\int_0^T b_n(t)dW(t) \xrightarrow{\mathsf{P}} \int_0^T b_0(t)dW(t), \quad n \to \infty.$$

13.15. Let $\{W_n(t),\ t \in [0,T], n \ge 0\}$ be a sequence of Wiener processes such that $W_n(t) \xrightarrow{\mathsf{P}} W_0(t), n \to \infty$ for each $t \in [0,T]$. Assume that stochastic processes $\xi_n(t)$, $t \in [0,1]$ are $\sigma\{W_n(s),\ s \le t\}$-adapted, respectively, $\mathsf{P}\left(\int_0^1 \xi_n^2(t)dt < \infty \right) = 1, n \ge 0$ and $\lim_{n\to\infty} \mathsf{E} \int_0^1 (\xi_n(t) - \xi_0(t))^2 dt = 0$. Prove the convergence of stochastic integrals

$$\lim_{n\to\infty} \int_0^1 \xi_n(t)dW_n(t) = \int_0^1 \xi_0(t)dW_0(t)$$

in probability.

13.16. Calculate $\mathsf{E}(\int_0^T W^n(t)dW(t))^2$, $n > 1$.

13.17. Apply the Itô formula and find stochastic differentials of the processes.

(a) $X(t) = W^2(t)$.

b) $X(t) = \sin t + e^{W(t)}$.

(c)$X(t) = t \cdot W(t)$.

(d) $X(t) = W_1^2(t) + W_2^2(t)$, where $W(t) = (W_1(t), W_2(t))$ is a two-dimensional Wiener process.

(e) $X(t) = W_1(t) \cdot W_2(t)$.

(f) $X(t) = (W_1(t) + W_2(t) + W_3(t))(W_2^2(t) - W_1(t)W_3(t))$, where $(W_1(t), W_2(t), W_3(t))$ is a three-dimensional Wiener process.

13.18. Let $f \in \mathscr{L}_2$, $X(t) = \int_0^t f(s)dW(s)$. Prove that process $M(t) := X^2(t) - \int_0^t f^2(s)ds$, $t \ge 0$ is a martingale.

13.19. Let $W(t) = (W_1(t), \ldots, W_m(t))$, $t \ge 0$ be an m-dimensional Wiener process, $f : \mathbb{R}^m \to \mathbb{R}, f \in C^2(\mathbb{R}^m)$. Prove that

$$f(W(t)) = f(0) + \int_0^t \nabla f(W(s))dW(s) + \frac{1}{2}\int_0^t \Delta f(W(s))ds,$$

where $\nabla = ((\partial/\partial x_1), \dots, (\partial/\partial x_m))$ is the gradient, and $\triangle = \sum_{i=1}^m (\partial^2/\partial x_i^2)$ is the Laplacian.

13.20. Find the stochastic differential $dZ(t)$ of a process $Z(t)$, if:
(a) $Z(t) = \int_0^t f(s)dW(s)$, where $f \in \hat{\mathscr{L}}_2([0,t])$.
(b) $Z_t = \exp\{\alpha W(t)\}$.
(c) $Z_t = \exp\{\alpha X(t)\}$, where process $\{X(t),\ t \in \mathbb{R}^+\}$ has a stochastic differential $dX(t) = \alpha dt + \beta dW(t)$, $\alpha, \beta \in \mathbb{R}$, $\beta \neq 0$.
(d) $Z_t = X^n(t)$, $n \in \mathbb{N}$, $dX(t) = \alpha X(t)dt + \beta X(t)dW(t)$, $\alpha, \beta \in \mathbb{R}$, $\beta \neq 0$.
(e) $Z_t = X^{-1}(t)$, $dX(t) = \alpha X(t)dt + \beta X(t)dW(t)$, $\alpha, \beta \in \mathbb{R}$, $\beta \neq 0$.

13.21. Let $h_n(x)$, $n \in \mathbb{Z}^+$ be the Hermite polynomial of nth order:

$$h_n(x) = (-1)^n e^{(x^2/2)} \frac{d^n}{dx^n}\left(e^{-(x^2/2)}\right).$$

(1) Verify that $h_0(x) = 1$, $h_1(x) = x$, $h_2(x) = x^2 - 1$, $h_3(x) = x^3 - 3x$.
(2) Prove that the multiply Itô stochastic integrals

$$I_n(t) := \left(\int_0^t \int_0^{t_1} \cdots \int_0^{t_{n-1}} dW(t_n) \cdots dW(t_2) dW(t_1) \right), \quad n \geq 1$$

are well-defined.
(3) Prove that

$$n! I_n(t) = t^{n/2} h_n\left(\frac{W(t)}{\sqrt{t}}\right), \quad t > 0.$$

13.22. Let $\Delta_n(T) = \{(t_1, \dots, t_n) |\ 0 \leq t_1 \leq \cdots \leq t_n \leq T\}$. For $f \in L_2(\Delta_n(T))$, denote

$$I_n^f(T) := \left(\int_0^T \int_0^{t_n} \cdots \int_0^{t_2} f(t_1, \dots, t_n) dW(t_1) \cdots dW(t_n) \right).$$

(1) Prove that $\mathsf{E} I_n^f(T) I_m^g(T) = 0$ for any $n \neq m, f \in L_2(\Delta_n(T))$, $g \in L_2(\Delta_m(T))$, and

$$\mathsf{E} I_n^f(T) I_n^g(T) = \int_0^T \int_0^{t_n} \cdots \int_0^{t_2} f(t_1, \dots, t_n) g(t_1, \dots, t_n) dt_1 \cdots dt_n$$

for any $n \geq 1, f, g \in L_2(\Delta_n(T))$.
(2) Let $0 = a_0 \leq a_1 \leq \cdots \leq a_m \leq T$, $k_i \in \mathbb{Z}^+$, $i = 0, \dots, m-1$, $n = k_0 + \cdots + k_{m-1}$. Put

$$f(t_1, \dots, t_n) = \prod_{i=0}^{m-1} \prod_{j=k_0+\cdots+k_i}^{k_0+\cdots+k_{i+1}-1} \mathbb{I}_{[a_i, a_{i+1})}(t_j).$$

Prove that

$$I_n^f(T) = \prod_{i=0}^{m-1} \frac{(a_{i+1} - a_i)^{k_i/2}}{k_i!} h_{k_i}\left(\frac{W(a_{i+1}) - W(a_i)}{\sqrt{a_{i+1} - a_i}}\right).$$

(3) Prove that for any $m \in \mathbb{N}$, $a_i \in [0, T]$, $k_i \in \mathbb{Z}^+$, $i = 1, \dots, m$ there exist functions $f_j \in L_2([0,T]^j)$, $0 \leq j \leq n = k_0 + \cdots + k_{m-1}$ and a constant c such that

$$W(a_1)^{k_1} \cdots W(a_m)^{k_m} = c + \sum_{j=0}^n I_j^{f_j}(T).$$

(4) Use the completeness of polynomial functions in any space $L_2(\mathbb{R}^n,\gamma)$ where γ is a Gaussian measure (see, e.g., [5]), and prove that for any random variable ξ measurable w.r.t. $\sigma(W(a_1),\ldots,W(a_m))$ there exists a unique sequence of functions $\{f_n \in L_2(\Delta_n(T)),\ n \in \mathbb{Z}^+\}$ such that

$$\xi = f_0 + \sum_{j=0}^{\infty} I_j^{f_j}(T). \tag{13.2}$$

Moreover

$$f_0 = \mathsf{E}\xi,\ \mathsf{E}\xi^2 = f_0^2 + \sum_{n\geq 1} \|f_n\|^2_{L_2(\Delta_n(T))}. \tag{13.3}$$

(5) Prove that for any random variable $\xi \in L_2(\Omega,\mathscr{F}_T^W,\mathsf{P})$ there exists a unique sequence of functions $\{f_n \in L_2(\Delta_n(T)),\ n \in \mathbb{Z}^+\}$ such that (13.2), (13.3) are satisfied.

13.23. Let $X(t)$ be a.s. a positive stochastic process, $dX(t) = X(t)(\alpha dt + \beta dW(t))$, $\alpha,\beta \in \mathbb{R}$. Prove that $d\ln X(t) = (\alpha - \frac{1}{2}\beta^2)dt + \beta dW(t),\ t \in \mathbb{R}^+$.

13.24. Assume that processes $X(t)$ and $Y(t)$, $t \in \mathbb{R}^+$ have stochastic differentials $dX(t) = a(t)dt + b(t)dW(t)$, $dY(t) = \alpha(t)dt + \beta(t)dW(t)$. Prove that

$$\begin{aligned} d(XY)_t &= X(t)dY(t) + Y(t)dX(t) + dX(t)dY(t) \\ &= (\alpha(t)X(t) + a(t)Y(t) + b(t)\beta(t))dt + (\beta(t)X(t) + b(t)Y(t))dW(t). \end{aligned}$$

13.25. Assume that processes $X(t)$ and $Y(t)$, $t \in \mathbb{R}^+$ have stochastic differentials $dX(t) = adt + bdW(t)$, $dY(t) = Y(t)(\alpha dt + \beta dW(t))$, $\alpha,\beta,a,b \in \mathbb{R}$, and $Y(0) > 0$ a.s. Prove that $dY^{-1}(t) = Y^{-1}(t)\left((-\alpha+\beta^2)dt - \beta dW(t)\right)$, $d\left(X(t)Y^{-1}(t)\right) = Y^{-1}(t)\left[(a - b\beta + X(t)(\beta^2 - \alpha))dt + (b - \beta X(t))dW(t)\right]$. (Note that $Y(t) > 0, t \geq 0$ a.s. if $Y(0) > 0$, see Problem 14.2.)

13.26. Let $\{W(t),\mathfrak{F}_t,\ t \in [0,T]\}$ be a Wiener process, $\{f(t),\mathfrak{F}_t^W,\ t \in [0,T]\}$ be a bounded process, $|f(t)| \leq C$, $t \in [0,T]$. Prove that $\mathsf{E}|\int_0^t f(s)dW(s)|^{2m} \leq C^{2m}t^m (2m-1)!!$

13.27. Assume that a process X has a stochastic differential $dX(t) = U(t)dt + dW(t)$, where U is bounded process. Define a process $Y(t) = X(t)M(t)$, where $M(t) = \exp\{-\int_0^t U(s)dW(s) - \frac{1}{2}\int_0^t U^2(s)ds\}$. Apply the Itô formula and prove that $Y(t)$ is an $\mathfrak{F}_t$-martingale where $\mathfrak{F}_t = \sigma\{U(s),W(s),s \leq t\}$.

13.28. Let $\{W(t),\ t \in [0,T]\}$ be a Wiener process. Assume process $\{f(t),\mathfrak{F}_t^W,\ t \in [0,T]\}$ is such that $\int_0^T \mathsf{E}|f(t)|^{2m}dt < \infty$. Prove that $\mathsf{E}|\int_0^t f(s)dW(s)|^{2m} \leq [m(2m-1)]^m t^{m-1} \int_0^t \mathsf{E}|f(s)|^{2m}ds$.

13.29. Prove that for any $p \geq 1$ there exist positive constants c_p, C_p such that

$$c_p\mathsf{E}|\max_{t\in[0,T]}\int_0^t f(s)dW(s)|^{2p} \leq \mathsf{E}|\int_0^T f^2(s)ds)|^p \leq C_p\mathsf{E}|\max_{t\in[0,T]}\int_0^t f(s)dW(s)|^{2p}.$$

13.30. Let $\{f(t),\mathfrak{F}_t^W,\ t \in \mathbb{R}^+\}$ be a bounded stochastic process, $f \in \mathscr{L}_2$, and a process $X(t)$ be the stochastic exponent, that is, have the form $X(t) = \exp\{\int_0^t f(s)dW(s) - \frac{1}{2}\int_0^t f^2(s)ds\}$, where $\{W(t),\mathfrak{F}_t,\ t \in \mathbb{R}^+\}$ is a Wiener process. Prove that $dX(t) = X(t)f(t)dW(t)$.

13.31. Let $(\Omega, \mathcal{F}, \{\mathcal{F}_t\}_{t\in[0,T]}, \mathsf{P})$ be a filtered probability space, $\{W(t), \mathcal{F}_t,\ t \in [0,T]\}$ be a Wiener process, $\{\gamma(t), \mathcal{F}_t,\ t \in [0,T]\}$ be a progressively measurable stochastic process such that $\mathsf{P}\{\int_0^T \gamma^2(s)ds < \infty\} = 1$, and $\{\xi(t), \mathcal{F}_t,\ t \in [0,T]\}$ be a stochastic process of the form $\xi(t) = 1 + \int_0^t \gamma(s)dW(s)$, $t \in [0,T]$ (its stochastic integrability was grounded in Problem 13.10). Assume that $\xi(t) \ge 0$, $t \in [0,T]$ P-a.s. Prove that ξ is a nonnegative supermartingale and $\mathsf{E}\xi(t) \le 1$, $t \in [0,T]$.

13.32. Assume that a process $\{X(t),\ t \in \mathbb{R}^+\}$ has a stochastic differential $dX(t) = \alpha X(t)dt + \sigma(t)dW(t)$, $X(0) = X_0 \in \mathbb{R}$, where $\alpha \in \mathbb{R}, \sigma \in \mathcal{L}_2([0,t])$ for all $t \in \mathbb{R}^+$. Find $m(t) := \mathsf{E}X(t)$.

13.33. Assume that a process $\{X(t),\ t \in \mathbb{R}^+\}$ has a stochastic differential $dX(t) = \mu(t)dt + \sigma(t)dW(t)$, where $\mu \in L_1([0,t]), \sigma \in \mathcal{L}_2([0,t])$ for all $t \in \mathbb{R}^+$, and $\mu(t) \ge 0$ for all $t \in \mathbb{R}^+$. Prove that $\{X(t), \mathcal{F}_t^W,\ t \in \mathbb{R}^+\}$ is a submartingale, where $\mathcal{F}_t^W$ is a flow of σ-algebras generated by a Wiener process $\{W(t),\ t \in \mathbb{R}^+\}$.

13.34. Let $X(t) = \exp\{\int_0^t \mu(s)ds + \int_0^t \sigma(s)dW(s)\}$, and $\mu \in L_1([0,t]), \sigma \in \mathcal{L}_2([0,t])$ for all $t \in \mathbb{R}^+$. Find conditions on functions μ and σ under which the process $X(t)$ is (a) a martingale; (b) a submartingale; (c) a supermartingale.

13.35. Apply the Itô formula and prove that the following processes are martingales with respect to natural filtration:

(a) $X_t = e^{t/2}\cos W(t)$.

(b) $X_t = e^{t/2}\sin W(t)$.

(c) $X_t = (W(t)+t)\exp\{-W(t) - \frac{1}{2}t\}$.

13.36. Let $\{W(t), \mathcal{F}_t,\ t \in \mathbb{R}^+\}$ be the Wiener process. Prove that $W^4(1) = 3 + \int_0^1 (12(1-t)W(t) + 4W^3(t))dW(t)$.

13.37. Let $\{W(t), \mathcal{F}_t,\ t \in \mathbb{R}^+\}$ be the Wiener process, and τ be a stopping time such that $\mathsf{E}\tau < \infty$. Prove that $\mathsf{E}W(\tau) = 0$, $\mathsf{E}W^2(\tau) = \mathsf{E}\tau$.

13.38. Let $a < x < b$, $\{X(t) = x + W(t),\ t \in \mathbb{R}^+\}$ where W is the Wiener process. Denote $\varphi(x) := \mathsf{P}(X(\tau) = a)$ where τ is the exit moment of the process X from the interval (a,b). Prove that $\varphi(x) \in C^\infty[a,b]$ and that φ satisfies differential equation $\varphi''(x) = 0$, $x \in [a,b]$, and $\varphi(a) = 1$, $\varphi(b) = 0$. Prove that $\mathsf{P}(X(\tau) = a) = (b-x)/(b-a)$. This is a particular case of Problem 14.31.

13.39. Let $\{H(t), \mathcal{F}_t^W,\ t \in [0,T]\}$ be a stochastic process with $\int_0^T H^2(t)dt < \infty$ a.s. Put $M(t) := \int_0^t H(s)dW(s)$.

(1) Prove that $M(t)$ is a local square integrable martingale.

(2) Let $\mathsf{E}\ \sup_{t\le T} M^2(t) < \infty$. Prove that $\mathsf{E}\int_0^T H^2(t)dt < \infty$ and the process $M(t)$ is a square integrable martingale.

13.40. Let

$$p(t,x) = \frac{1}{\sqrt{1-t}}\exp\left\{-\frac{x^2}{2(1-t)}\right\},\quad 0 \le t < 1,\ \ x \in \mathbb{R},\ \text{ and } p(1,x) = 0.$$

Put $M(t) := p(t, W(t))$, where $W(t)$ is a Wiener process.

(a) Prove that $M(t) = M(0) + \int_0^t (\partial p/\partial x)(s, W(s))dW(s)$.

(b) Let $H(t) = (\partial p/\partial x)(t, W(t))$. Prove that $\int_0^1 H^2(t)dt < \infty$ a.s., and $\mathsf{E}\int_0^1 H^2(t)dt = +\infty$.

13.41. (1) Let $\{W(t) = (W_1(t), \dots, W_m(t)), \mathfrak{F}_t, t \geq 0\}$ be an m-dimensional Wiener process. Put $M(t) := \sum_{i=1}^m (W_i(t))^2$. Prove that $\{M(t) - mt, \mathfrak{F}_t, t \geq 0\}$ is a martingale and its quadratic characteristic equals

$$\langle M \rangle (t) = \int_0^t 4M(u)\,du.$$

(2) Let $N(t) = (M(t))^{1/2}$,

$$\varphi(x) = \begin{cases} \ln|x|, & m = 2, \\ |x|^{2-m}, & m \geq 3. \end{cases}$$

Prove that $\{\varphi(N(t)), \mathfrak{F}_t, t \geq 0\}$ is a local martingale.

13.42. Let $\{M_t, \mathfrak{F}_t, t \in [0,T]\}$ be a martingale of the form $\int_0^t H(s)dW(s) + \int_0^t K(s)ds$, where $\int_0^t H^2(s)ds < \infty$ a.s, $\int_0^t |K(s)|ds < \infty$ a.s. Prove that $K(t) = 0 \ \lambda^1|_{[0,T]} \times \mathsf{P}$ a.s.

13.43. Let $M \in \mathscr{M}^{2,c}$. Prove that

$$M^2(t) - M^2(s) = 2\int_0^t M(s)dM(s) + \langle M \rangle (t) - \langle M \rangle (s), \ s < t.$$

13.44. Let $\{M(t), \mathfrak{F}_t, \ t \in \mathbb{R}^+\}$ be a continuous local martingale, $M(0) = 0$, and $\lim_{t \to \infty} \langle M \rangle (t) = \infty$ a.s. Denote

$$\tau_t := \inf\{s > 0 | \ \langle M \rangle (s) > t\}, \ t > 0.$$

Prove that a stopped stochastic process $\{M(\tau_t), \ t \geq 0\}$ is a Wiener process with respect to the filtration $\{\mathfrak{F}_{\tau_t}, \ t \in \mathbb{R}^+\}$.

13.45. Let $M \in \mathscr{M}^{2,c}_{\text{loc}}$, $M(0) = 0$ and $\mathsf{E}\langle M \rangle (t) < \infty$ for all $t > 0$. Prove that $M \in \mathscr{M}^{2,c}$.

13.46. (The Dynkin formula)

(1) Assume that a stochastic process $\{X(t), \mathfrak{F}_t, \ t \in \mathbb{R}^+\}$ has a stochastic differential $dX(t) = \mu(t, X(t))dt + \sigma(t, X(t))dW(t)$, $X(0) = x \in \mathbb{R}$, where μ and σ are continuous bounded functions. Let a function f be bounded and $f = f(t,x) \in C^1(\mathbb{R}^+) \times C^2(\mathbb{R})$, τ be a bounded stopping time, and a second-order differential operator $\mathscr{L}$ be of the form

$$\mathscr{L} f(t,x) = \mu(t,x)\frac{\partial f}{\partial x} + \frac{1}{2}\sigma^2(t,x)\frac{\partial^2 f}{\partial x^2}.$$

(This operator is similar to the operator $\mathscr{L}$ introduced in Definition 12.16, but now its coefficients are nonhomogeneous in time.) Prove that

$$\mathsf{E}f(\tau, X_\tau) = f(0, X_0) + \mathsf{E}\int_0^\tau \left(\frac{\partial f}{\partial t} + \mathscr{L} f\right)(u, X_u)du.$$

(2) Prove the following version of the Dynkin formula for unbounded stopping times. Let $C_0^2(\mathbb{R}^n)$ be the class of twice continuously differentiable functions on $\mathbb{R}^n$ with compact support, and τ be a stopping time such that $\mathsf{E}_x\tau < \infty$. Then

$$\mathsf{E}_x f(X_\tau) = f(x) + \mathsf{E}\int_0^\tau \mathscr{L} f(X_u)du.$$

(3) Let n-dimensional stochastic process $\{X(t), \mathfrak{F}_t,\ t \in \mathbb{R}^+\}$ have a stochastic differential of the form $dX_i(t) = \mu_i(t, X(t))dt + \Sigma_{j=1}^m b_{i,j}(t, X(t))dW_j(t)$, $X(0) = x \in \mathbb{R}^n$, where $W(t) = (W_1(t), W_2(t), \ldots, W_m(t))$ is an m-dimensional Wiener process. Assume that all components of μ and b are bounded continuous functions, a function f is bounded, $f = f(t,x) \in C^1(\mathbb{R}^+) \times C^2(\mathbb{R}^n)$, and τ is a bounded stopping time. Write down and prove the multidimensional Dynkin formula for X.

13.47. Let $\{W(t),\ t \in \mathbb{R}^+\}$ be a Wiener process. Prove that the limit (13.1) exists in $L_2(\mathsf{P})$ and for all $t \in \mathbb{R}^+$ and $x \in \mathbb{R}$,

$$(W(t) - x)^+ - (W(0) - x)^+ = \int_0^t \mathbb{I}_{W(s)\in[x,\infty)} dW(s) + \frac{1}{2}L(t,x) \text{ a.s.}$$

13.48. Prove Theorem 13.4 (the Tanaka formula).

13.49. Prove that there exists continuous in (t,x) a modification of $L(t,x)$.

13.50. Let $\{W(t),\ t \in \mathbb{R}^+\}$ be a Wiener process. Prove that for all $t \in \mathbb{R}^+$ and $a \le b$,

$$\int_a^b L(t,x)dx = \int_0^t \mathbb{I}_{W(s)\in(a,b)} ds \text{ a.s.}$$

13.51. Let $f \in L_1(\mathbb{R})$. Prove that for each $t \in \mathbb{R}^+$,

$$\int_{-\infty}^{\infty} L(t,x) f(x)dx = \int_0^t f(W(s))ds \text{ a.s.}$$

13.52. Prove that a limit in (13.1) exists with probability 1.

13.53. (1) Let $\{L(t,y), t \ge 0, y \in \mathbb{R}\}$ be a local time of a Wiener process. Prove that $\mathsf{E}_x(L(\tau_a \wedge \tau_b, y)) = 2u(x,y)$, $a \le x \le y \le b$ where $\tau_z := \inf\{s \ge 0 | W(s) = z\}$, $z = a, b$, and $u(x,y) = ((x-a)(b-y))/(b-a)$.

(2) Prove that

$$\mathsf{E}_x \int_0^{\tau_a\wedge\tau_b} f(W(s))ds = 2\int_a^b u(x,y)f(y)dy$$

for any nonnegative bounded Borel function $f : \mathbb{R} \to \mathbb{R}^+$.

13.54. Let $W(t) = (W_1(t), \ldots, W_n(t))$ be an n-dimensional Wiener process, $X(t) = a + W(t)$ where $a = (a_1, \ldots, a_n) \in \mathbb{R}^n$ $(n \ge 2)$, and let $R > \|a\|$. Denote $\tau_R := \inf\{t \ge 0 \mid X(t) \notin B(0,R)\}$ where $B(0,R) = \{x \in \mathbb{R}^n \mid \|x\| < R\}$. Find $\mathsf{E}\tau_R$.

13.55. Let X be the process from Problem 13.54, but $\|a\| > R$. Find the probability for this process to hit the ball $B(0,R)$.

13.56. Find some function $f=(f_1,f_2)$ such that for any process $X(t)=(X_1(t),X_2(t))$ satisfying the equation

$$\begin{cases} dX_1(t) = -X_2(t)dW(t) + f_1(X_1(t),X_2(t))dt, \\ dX_2(t) = X_1(t)dW(t) + f_2(X_1(t),X_2(t))dt, \end{cases}$$

the equality $X_1^2(t)+X_2^2(t)=X_1^2(0)+X_2^2(0), t\geq 0$ a.s. holds true.

13.57. (1) Prove the *Lévy theorem*. A stochastic process $\{W(t),\mathcal{F}_t,t\geq 0\}$ is a Wiener process if and only if it is a square integrable martingale, $W(0)=0$, and $\mathsf{E}((W(t)-W(s))^2/\mathcal{F}_s)=t-s$ for any $s<t$.

(2) Prove the following generalization of the Lévy theorem. For any square integrable martingale $\{M(t),\mathcal{F}_t,t\geq 0\}$ with the quadratic characteristic (see Definition 7.10) of the form $\langle M\rangle(t)=\int_0^t \alpha(s)\,ds$ where $\alpha\in L_1([0,t])$ for any $t>0$, and $\alpha(s)>0$ for all $s>0$, there exists a Wiener process $\{W(t),\mathcal{F}_t,t\geq 0\}$ such that $M(t)=\int_0^t(\alpha(s))^{1/2}\,dW(s)$.

(3) Prove the multidimensional version of the Lévy theorem. An n-dimensional stochastic process $\{W(t)=(W_1(t),W_2(t),\dots,W_n(t)),\mathcal{F}_t,t\geq 0\}$ is an n-dimensional Wiener process if and only if it is a square integrable martingale, $W(0)=0$, and the processes $W_i(t)W_j(t)-\delta_{ij}t$ are martingales for any $1\leq i,j\leq n$.

Another formulation for the latter claim is that, for any $1\leq i,j\leq n$, the joint quadratic characteristic $\langle W_i,W_j\rangle(t)$ equals $\delta_{ij}t$.

13.58. Prove that stochastic process $Y(t):=\int_0^t \operatorname{sign}(W(s))dW(s)$, $t\geq 0$ is a Wiener process.

13.59. Let $W(t)=(W_1(t),\dots,W_m(t))$ be a Wiener process. Assume that a progressively measurable process $U(t),t\geq 0$ takes values in a space of $n\times m$ matrices and $U(t)U^*(t)=id_n$ a.s., where id_n is the identity $n\times n$ matrix. Prove that $\widetilde{W}(t)=\int_0^t U(s)dW(s)$ is an n-dimensional Wiener process.

13.60. Assume that a progressively measurable stochastic process $\beta(t),t\geq 0$ satisfies the condition:

$$\exists\, c,C>0\ \forall\, t\geq 0:\ c\leq \beta(t)\leq C \text{ a.s.}$$

Let

$$A(t)=\int_0^t \beta(s)ds,\ A^{-1}(t):=\inf\left\{s\geq 0 \,\middle|\, \int_0^s \beta(z)dz=t\right\}.$$

Prove that $A^{-1}(t)$ is a stopping time. Consider filtration $\{\mathcal{F}_t^A:=\mathcal{F}_{A^{-1}(t)},t\geq 0\}$. Prove that the process

$$\widetilde{W}(t)=\int_0^{A^{-1}(t)} \beta^{1/2}(s)dW(s)$$

is an $\{\mathcal{F}_t^A,t\geq 0\}$-Wiener process. Prove the following change of variables formula,

$$\int_0^{A^{-1}(t)} b(s)dW(s)=\int_0^t b(A^{-1}(z))\beta^{-1/2}(A^{-1}(z))d\widetilde{W}(z),$$

where $b(t), t \geq 0$ is a $\sigma\{W(s), s \leq t\}$-adapted process such that

$$\mathsf{P}\left(\int_0^T b^2(s)ds < \infty\right) = 1$$

for each $T > 0$.

Hints

13.1. Represent the integral as a mean-square limit of the corresponding integral sums.
13.2. See the hint to Problem 13.1. The Itô formula can also be applied.
13.3. Prove that these random variables are jointly Gaussian (see Problem 13.1), and calculate their covariance.
13.4. (3) Apply the Itô formula to $W^3(t)$.
13.5. Apply the definition of a stochastic integral and properly approximate a continuous process $f \in \hat{\mathscr{L}}_2([0,T])$.
13.6. Prove that process $g_n = \sum_{k=1}^{n-1} n \int_{(k-1)T/n}^{kT/n} g(s)ds\, \mathbb{I}_{[kT/n,(k+1)T/n)}$ belongs to $\hat{\mathscr{L}}_2([0,T])$, and $g_n \to g, n \to \infty$ in $\hat{\mathscr{L}}_2([0,T])$. In order to prove this verify that linear operator $P_n:\ f \to \sum_{k=1}^{n-1} n \int_{(k-1)T/n}^{kT/n} f(s)ds\, \mathbb{I}_{[kT/n,(k+1)T/n)}$ in $\hat{\mathscr{L}}_2([0,T])$ has a norm 1, and a sequence $\{P_n,\ n \geq 1\}$ strongly converges to the identity operator in $\hat{\mathscr{L}}_2([0,T])$.
13.8. Prove that a sequence of simple processes

$$f_n(t) = \sum_k n \int_{(k-1)/n}^{k/n} f(s)ds\, \mathbb{I}_{t \in [k/n,(k+1)/n)}$$

converges to a process f in $\hat{\mathscr{L}}_2([0,T])$. Construct a similar sequence g_n for g and observe that the Itô integrals of these processes $\int_0^t f_n(s)dW(s)$, $\int_0^t g_n(s)dW(s)$ coincide on the set A for any fixed $t \in [0,T]$. In order to prove that

$$\mathsf{P}\left(\omega \in A \mid \exists t \in [0;T]\ \int_0^t f(s)dW(s) \neq \int_0^t g(s)dW(s)\right) = 0$$

notice that processes $\int_0^t f(s)dW(s)$, $\int_0^t g(s)dW(s)$ are continuous in t.
13.9. Prove the required statements for a stopping time that take values in a finite set. Then approximate arbitrary stopping time τ by a sequence of stopping times $\tau_n = \sum_{k \geq 1} k/n\, \mathbb{I}_{\tau \in [(k-1)/n,\, k/n)}$. See also the solution to Problem 13.37.
13.10. Put $\tau_n = \inf\{t \geq 0 \mid \int_0^t f^2(s)ds = n\} \wedge T$. Then τ_n is a stopping time, a sequence $\{\tau_n\}$ is nondecreasing, and for a.a. ω there exists $n_0 = n_0(\omega)$ such that $\tau_n(\omega) = T,\ n \geq n_0$. Set $f_n(t) = f(t)\, \mathbb{I}_{\tau_n > t}$. Then $\int_0^T (f_n(t,\omega) - f(t,\omega))^2 dt = 0$ if $\tau_n(\omega) = T$. Denote $I(t) = \int_0^t f_n(s)dW(s)$ for $\{\omega \mid \tau_n(\omega) = T\}$ and apply the results of the previous two problems.
13.11. Let

$$\tau_m = \inf\{t \geq 0 \mid \int_0^t g^2(s)ds = m\} \wedge T,\ g_n(t) = \sum_{k=1}^{n-1} n \int_{(k-1)T/n}^{kT/n} g(s)ds\, \mathbb{I}_{[kT/n,(k+1)T/n)}(t).$$

Then (see Problems 13.6, 13.10) $\int_0^T g_n^2(t)dt \le \int_0^T g^2(t)dt$ for all $\omega \in \Omega$, $n \in \mathbb{N}$, and

$$\int_0^T g(s)dW(s) = \int_0^T g(s)\mathbb{1}_{\tau_m = T}dW(s),$$

$$\sum_{k=1}^{n-1} n\int_{(k-1)T/n}^{kT/n} g(s)ds(W((k+1)T/n) - W(kT/n))$$

$$= \int_0^T g_n(t)\mathbb{1}_{\tau_m=T}dW(t)$$

for a.a. ω such that $\tau_m(\omega) = T$. Tend $n \to \infty$ and apply Problems 13.6, 13.10.

13.12. Apply results of Problems 13.5–13.10.

13.13. The first sum converges to $\int_0^1 W(t)dW(t) = (W^2(1) - 1)/2$ in $L_2(\mathsf{P})$. Compare the sums in (b), (c) with the sum in (a) and apply the reasoning of Problem 3.19 to estimate differences between the sums.

13.14. Introduce stopping times

$$\tau_{N,n} = \inf\{t \ge 0 | \int_0^t b_n^2(s)ds \ge N\} \cup \{T\}.$$

Then

$$\int_0^T \left(b_n(t)\mathbb{1}_{t \le \tau_{N,n}} - b_0(t)\mathbb{1}_{t\le\tau_{N,0}}\right)^2 dt \to 0, \quad n \to \infty.$$

Therefore $\int_0^T b_n(t)\mathbb{1}_{t\le\tau_{N,n}}dW(t)$ converges in $L_2(\mathsf{P})$ to $\int_0^T b_0(t)\mathbb{1}_{t\le\tau_{N,n}}dW(t)$ as $n \to \infty$, and so converges in probability. The localization property of the stochastic integral (Problem 13.8) implies that

$$\int_0^T b_n(t)\mathbb{1}_{t\le\tau_{N,n}}dW(t) = \int_0^T b_n(t)dW(t)$$

for a.a. ω from the set $\{\tau_{N,n} = T\}$. So, to complete the proof, it suffices to observe that

$$\forall\, \varepsilon > 0\, \exists\, N: \ \overline{\lim_{n\to\infty}}\, \mathsf{P}(\tau_{N,n} < T) < \varepsilon.$$

13.15. Introduce auxiliary processes

$$\xi_{n,m}(t) = \sum_{k=0}^{m-2} m\int_{k/m}^{k+1/m} \xi_n(s)ds\mathbb{1}_{[((k+1)/m),((k+2)/m))}(t).$$

Then

$$\left(\int_0^1 \xi_n(t)dW_n(t) - \int_0^1 \xi_0(t)dW_0(t)\right)^2 \le 3\Bigg(\left(\int_0^1 (\xi_n(t) - \xi_{n,m}(t))dW_n(t)\right)^2 + \left(\int_0^1 \xi_{n,m}(t)dW_n(t) - \int_0^1 \xi_{0,m}(t)dW_0(t)\right)^2 + \left(\int_0^1 (\xi_0(t) - \xi_{0,m}(t))dW_0(t)\right)^2\Bigg). \tag{13.4}$$

It is easy to see that for each m we have the following convergence in probability,

$$\lim_{n\to\infty}\int_0^1 \xi_{n,m}(t)dW_n(t) = \int_0^1 \xi_{0,m}(t)dW_0(t).$$

It can be proved that $P_m:\ f \to \sum_{k=0}^{m-2} m\int_{k/m}^{k+1/m} f(s)ds\,\mathrm{1\!I}_{[((k+1)/m),((k+2/m))}(t)$ in $\mathscr{L}_2([0,T])$ has a norm 1, and a sequence $\{P_m,\ m\geq 1\}$ strongly converges to the identity operator in $\mathscr{L}_2([0,T])$. So,

$$\lim_{m\to\infty}\mathsf{E}\left(\int_0^1(\xi_0(t)-\xi_{0,m}(t))dW_0(t)\right)^2$$
$$=\lim_{m\to\infty}\mathsf{E}\int_0^1(\xi_0(t)-\xi_{0,m}(t))^2dt = 0.$$

Observe that

$$\mathsf{E}\left(\int_0^1(\xi_n(t)-\xi_{n,m}(t))dW_n(t)\right)^2$$
$$=\mathsf{E}\int_0^1(\xi_n(t)-\xi_{n,m}(t))^2dt \leq 3\Big(\mathsf{E}\int_0^1(\xi_n(t)-\xi_0(t))^2dt$$
$$+\mathsf{E}\int_0^1(\xi_0(t)-\xi_{0,m}(t))^2dt+\mathsf{E}\int_0^1(\xi_{0,m}(t)-\xi_{n,m}(t))^2dt\Big)$$
$$\leq 6\mathsf{E}\int_0^1(\xi_n(t)-\xi_0(t))^2dt+3\mathsf{E}\int_0^1(\xi_0(t)-\xi_{0,m}(t))^2dt.$$

Therefore the first term in the right-hand side of (13.4) converges in the mean square to 0 as $n,m\to\infty$. So, the left-hand side of (13.4) converges to 0 in probability as $n\to\infty$.

13.19. Apply the multidimensional Itô formula.

13.20. Apply the Itô formula.

13.21. (2), (3) Use the method of mathematical induction. In particular, in item (3), let $a(s) := (W(s))/\sqrt{s}$. Then $da(s) = s^{-1/2}dW(s) - s^{-3/2}W(s)ds/2$. Due to the Itô formula

$$\begin{aligned} d\big(s^{n/2}h_n(a)\big) &= \tfrac{n}{2}s^{n/2-1}h_n(a)+s^{n/2}h_n'(a(s))\big(s^{-1/2}dW(s)-s^{-3/2}W(s)ds/2\big)\\ &+\tfrac{1}{2}s^{n/2}h_n''(a(s))s^{-1}ds = \tfrac{s^{n/2-1}}{2}\big(nh_n(a(s))-a(s)h_n'(a(s))\\ &+h_n''(a(s))\big)ds+s^{(n-1)/2}h_n'(a(s))dW(s).\end{aligned}$$

The following property of Hermite polynomials is well known: $nh_n(x)-xh_n'(x)+h_n''(x)=0$, and also $h_n'(x)=nh_{n-1}(x)$. That is

$$d\big(s^{n/2}h_n(a(s))\big) = ns^{(n-1)/2}h_{n-1}(a(s))dW(s),$$

that which was to be demonstrated (this is the inductive step).

13.22. (1) Use properties of stochastic integrals.

(2) Apply Problem 13.21.

(4) Uniqueness follows from item (1). To prove existence, use items (1) and (3).

(5) Random variables $I_n^{f_n}(T)$, $I_m^{f_m}(T)$ are orthogonal in $L_2(\Omega,\mathscr{F}_T^W,\mathsf{P})$ if $m\neq n$ (see item (1)), so $\{I_n^{f_n}(T)|\ f_n\in L_2(\Delta_n(T))\}$ are orthogonal subspaces of $L_2(\Omega,\mathscr{F}_T^W,\mathsf{P})$. Therefore, it suffices to verify that linear combinations of multiple Itô integrals are dense in $L_2(\Omega,\mathscr{F}_T^W,\mathsf{P})$. As mentioned, the set of polynomials is dense in any space $L_2(\mathbb{R}^n,\gamma)$, where γ is a Gaussian measure. So any square integrable random variable

of the form $g(W(s_1),\dots,W(s_n))$ can be approximated in $L_2(\mathsf{P})$ by linear combinations of multiple stochastic integrals. Prove now that a set of square integrable random variables having the form $g(W(s_1),\dots,W(s_n))$, $s_k \in [0,T]$, $n \in \mathbb{N}$ is dense in $L_2(\Omega,\mathscr{F}_T^W,\mathsf{P})$.
13.23.–13.25. Apply the Itô formula. In Problem 13.24 this can be made straightforwardly; in Problems 13.23 and 13.25 an additional limit procedure should be used because the functions $x \mapsto \ln x$ and $(x,y) \mapsto x/y$ do not belong to the classes $C^2(\mathbb{R})$ and $C^2(\mathbb{R}^2)$, respectively. For instance, for any given $c > 0$ there exists a function $F_c \in C^2(\mathbb{R})$ such that $F_c(x) = \ln x, x \geq c$. Write the Itô formula for $F_c(X(t))$ and then tend $c \to 0+$.
13.26. Let $X(t) := \int_0^t f(s)dW(s)$. Put $\tau_N := \inf\{t \in \mathbb{R}^+ \,|\, \sup_{0\leq s\leq t}|X(s)| \geq N\} \wedge T$, apply the Itô formula to $|X(t\wedge\tau_N)|^{2m}$, obtain the estimate $\mathsf{E}|X(t)|^{2m} \leq C^2 m(2m-1)\int_0^t \mathsf{E}|X(s)|^{2m-2}ds$, and apply mathematical induction.
13.28. Apply the Itô formula and obtain the equality $\mathsf{E}|X(t\wedge\tau_N)|^{2m} = m(2m-1)\mathsf{E}\int_0^{t\wedge\tau_N}|X(s)|^{2m-2}f^2(s)ds$, where $X(t)$ and τ_N are the same as in Problem 13.26. Apply the Hölder inequality with $p = m/(m-1), q = m$ to the right-hand side. Verify and then use the following facts: $\mathsf{E}|X(t\wedge\tau_N)|^{2m} < \infty$ and $\mathsf{E}|X(t\wedge\tau_N)|^{2m}$ is nondecreasing in t.
13.29. Prove that the quadratic characteristic of the square integrable martingale $\int_0^t f(s)dW(s)$, $t\in[0,T]$ equals $\int_0^t f^2(s)ds$, $t\in[0,T]$. Apply the Burkholder–Davis inequality for a continuous-time martingale (Theorem 7.17).
13.30. Apply the Itô formula.
13.31. Consider the following stopping times: $\tau_n = \inf\{t\in[0,T] \mid \int_0^t \gamma^2(s)ds \geq n\}$, where we set $\tau_n = T$ if $\int_0^T \gamma^2(s)ds < n$. Use properties of the Itô integral and verify that process $\xi(t\wedge\tau_n)$ is a continuous nonnegative $\mathfrak{F}_t$-martingale. Check that $\tau_n \to T$ a.s. as $n\to\infty$. Apply the Fatou lemma and deduce both statements.
13.32. Write the equation for $m(t)$.
13.33. Prove that $\int_0^t \sigma(t)dW(t)$ is a martingale with respect to the indicated flow and apply the definition of a submartingale.
13.34. Apply the Itô formula.
13.39. (1) Consider a sequence of stopping times $\tau_n = \inf\{t\in\mathbb{R}^+ \mid \int_0^t H^2(s)ds = n\}\wedge T$ and prove that $\mathsf{E}M^2(\tau_n) = \mathsf{E}\int_0^{\tau_n} H^2(s)ds$.
(2) Apply the Lebesgue monotone convergence theorem and relation $\mathsf{E}\int_0^{\tau_n} H^2(s)ds = \mathsf{E}M^2(\tau_n) \leq \mathsf{E}\sup_{t\leq T} M^2(t)$.
13.41. (1) To calculate a quadratic characteristic apply the Itô formula to the difference $M^2(t) - M^2(s), 0 \leq s < t$.
(2) Apply the multidimensional version of the Itô formula.
13.42. Consider a sequence of stopping times $\tau_n = \inf\{t\in\mathbb{R}^+ \,|\, \int_0^t H^2(s)ds = n\}\wedge T$.
13.43. Apply the generalized Itô formula.
13.46. Apply the Itô formula and Doob's optional sampling theorem.
13.49. Apply Theorem 3.7.
13.51. Use approximation and Problem 13.50.
13.53. (1) Due to Problem 13.38, $\mathsf{P}\{\tau_a < \tau_b\} = (b-x)/(b-a)$. Prove that

$$\mathsf{E}_x \int_0^{\tau_a\wedge\tau_b} \mathrm{sign}(W(s)-y)dW(s) = 0$$

and apply Theorem 13.4 to deduce the identity $\mathsf{E}_x(L(\tau_a \wedge \tau_b, y)) = |b-y| - |x-y| + (b-x)/(b-a)(|a-y| - |b-y|)$ and then the required statement.
(2) Follows from item (1) and Problem 13.51.
13.57. (1) Apply the generalization of the Itô formula to a function $F(W(t)) = \exp\{iuW(t)\}$. Denote $I(t) := \mathsf{E}(F(W(t))/\mathfrak{F}_s)$. Then $I(t)$ satisfies the equation $I(t) = F(W(s)) - (u^2/2)\int_s^t I(\theta)\,d\theta$. Therefore, $I(t) = F(W(s))\exp\{-(u^2/2)(t-s)\}$, and so $\mathsf{E}(\exp\{iu(W(t)-W(s))\}/\mathfrak{F}_s) = \exp\{-(u^2/2)(t-s)\}$.
(2) Put $W(t) := \int_0^t ((dM(s))/(\alpha(s)))$. Verify that $\{W(t),\, t \geq 0\}$ is a Wiener process.
13.58. A process Y_t is a square integrable martingale with quadratic characteristic $\langle Y\rangle_t = \int_0^t \operatorname{sign}^2(W(s))ds$. Use Problem 3.21 and prove that $\langle Y\rangle_t = t$. Finally apply Problem 13.57 (Lévy theorem).
13.59. See hint to Problem 13.58.
13.60. Due to Problem 7.44, a random variable $A^{-1}(t)$ is a stopping time. Check that $\widetilde{W}(t)$ is a continuous $\mathscr{F}_{A^{-1}(t)}$-martingale, $\mathsf{E}\widetilde{W}(t) = 0, \mathsf{E}\widetilde{W}^2(t) = t$, and use the Lévy theorem (Problem 13.57). To prove the change of variable formula, approximate a process $b(t)$ by simple processes, prove the formula for them and apply the result of Problem 13.14.

Answers and Solutions

13.13. (a)
$$\frac{W^2(1)-1}{2},$$
(b)
$$\frac{W^2(1)+1}{2},$$
(c)
$$\frac{W^2(1)}{2}.$$
13.16.
$$\frac{(2n-1)!!}{n+1}T^{n+1}.$$
13.32. $m(t) = X_0 e^{\alpha t}$.
13.34. For almost all $s \in \mathbb{R}^+$ and $\omega \in \Omega$, the value $\mu(s,\omega) - ((\sigma^2(s,\omega))/2)$ should (a) equal 0; (b) ≥ 0: (c) ≤ 0.
13.36. Put $X(t) := \mathsf{E}(W^4(1)/\mathfrak{F}_t^W), 0 \leq t \leq 1$. Due to the Markov property of a Wiener process, $X(t) = E(W^4(1)/W(t))$. The conditional distribution of $W(1)$ given $W(t)$ is Gaussian, $\mathcal{N}(W(t), 1-t)$, thus
$$\begin{aligned} X(t) &= \mathsf{E}\left((W(1)-W(t)+W(t))^4/W(t)\right) = \mathsf{E}\left((W(1)-W(t))^4/W(t)\right) \\ &\quad + 4\mathsf{E}\left((W(1)-W(t))^3 W(t)/W(t)\right) + 6\mathsf{E}\left((W(1)-W(t))^2 W^2(t)/W(t)\right) \\ &\quad + 4\mathsf{E}\left((W(1)-W(t))W^3(t)/W(t)\right) + W^4(t) = 3(1-t)^2 + 6(1-t)W^2(t) \\ &\quad + W^4(t). \end{aligned}$$

Hence, due to the Itô formula $X(s) = X(0) + \int_0^s (12(1-t)W(t) + 4W(t)^3)dW(t)$ (check this). Finally, $X(0) = \mathsf{E}W^4(1) = 3$.

13.37. Consider the adapted function $f(s,\omega) := \mathbb{I}_{\tau(\omega)\geq s}$. Then $\mathsf{P}\left(\int_0^\infty f^2(s)ds < \infty\right) = \mathsf{P}(\tau < \infty) = 1$. Let us show that $\int_0^t f(s)dW(s) = W(t\wedge\tau)$ a.s. Put $\tau_n = k/2^n$ if $(k-1)/2^n \leq \tau \leq k/2^n$, $\tau_n = \infty$ if $\tau = \infty$. Consider integrals $\int_0^t \mathbb{I}_{\tau_n\geq s}dW(s) = \int_0^\infty \mathbb{I}_{\tau_n\wedge t\geq s}dW(s)$. If $t = i/2^n$ for some i, then $\int_0^t \mathbb{I}_{\tau_n\geq s}dW(s) = \int_0^\infty \mathbb{I}_{\tau_n\wedge t\geq s}dW(s) = W_{\tau_n\wedge t}$. Due to continuity of the stochastic integral and Wiener process, the last equality is satisfied for all t. Next, $\int_0^\infty \mathsf{E}(\mathbb{I}_{\tau_n\geq s} - \mathbb{I}_{\tau\geq s})^2 ds = \int_0^\infty (\mathsf{P}(s\leq\tau_n) - \mathsf{P}(s\leq\tau))ds = \mathsf{E}\tau_n - \mathsf{E}\tau \leq 1/2^n \to 0$, $n\to\infty$, so $\int_0^t \mathbb{I}_{s\leq\tau}dW(s) = \text{l.i.m.}_{n\to\infty}\int_0^t \mathbb{I}_{s\leq\tau_n}dW(s) = \text{l.i.m.}_{n\to\infty} W(\tau_n\wedge t) = W(\tau\wedge t)$. Then P-a.s. $W(\tau) = \int_0^\tau \mathbb{I}_{s\leq\tau}dW(s) = \int_0^\infty \mathbb{I}_{s\leq\tau}dW(s)$, and $\mathsf{E}\int_0^\infty \mathbb{I}^2_{s\leq\tau}ds = \mathsf{E}\tau < \infty$. So, $\mathsf{E}W(\tau) = \mathsf{E}\int_0^\infty \mathbb{I}_{s\leq\tau}dW(s) = 0$, and $\mathsf{E}W^2(\tau) = \mathsf{E}(\int_0^\infty \mathbb{I}_{s\leq\tau}dW(s))^2 = \int_0^\infty \mathsf{E}\mathbb{I}^2_{s\leq\tau}ds = \mathsf{E}\tau$.

13.44. One can verify that random variable τ_t is a stopping time for each $t > 0$. One can also check that $\langle M\rangle_{\tau_t} = t$ a.s. Let us show that $\{M_{\tau_t}, \mathcal{F}_{\tau_t}, t\in\mathbb{R}^+\}$ is a square integrable martingale. Define a localizing sequence

$$\sigma_k := \inf\{t > 0 | \, |M_t| > k\}.$$

Then $\{M_{t\wedge\sigma_k}, \mathcal{F}_t, t\in\mathbb{R}^+\}$ is a bounded martingale, and hence, by Theorem 7.5 a process $\{M_{\tau_t\wedge\sigma_k}, \mathcal{F}_{\tau_t}, t\in\mathbb{R}^+\}$ is a bounded martingale as well. According to Problem 13.43

$$M^2(\tau_t\wedge\sigma_k) = 2\int_0^{\tau_t\wedge\sigma_k} M(s)dM(s) + \langle M\rangle(\tau_t\wedge\sigma_k).$$

It was mentioned in Theoretical grounds in Chapter 7 that for $M\in\widetilde{\mathcal{M}}^2$ it holds that $M^2(t) - \langle M\rangle(t)$ is a martingale. Therefore, $M^2(t\wedge\sigma_k) - \langle M\rangle(t\wedge\sigma_k)$, and so $M^2(\tau_t\wedge\sigma_k) - \langle M\rangle(\tau_t\wedge\sigma_k)$ are martingales. That is, a process $\int_0^{\tau_t\wedge\sigma_k} M(s)dM(s)$ is a martingale, moreover a bounded martingale. Thus its expectation is zero and $\mathsf{E}M^2(\tau_t\wedge\sigma_k) = \mathsf{E}\langle M\rangle(\tau_t\wedge\sigma_k) \leq \mathsf{E}\langle M\rangle(\tau_t) \leq t$. Due to Fatou's lemma $\{M(\tau_t), \mathcal{F}_{\tau_t}, t\in\mathbb{R}^+\}$ is a square integrable martingale. Due to Problem 13.57, it suffices now to prove that $M(\tau_t)$ is a continuous process and $\mathsf{E}\left((M(\tau_t) - M(\tau_s))^2/\mathcal{F}_{\tau_s}\right) = t - s$. Let us first prove the last relation. It is easy to see that $\mathsf{E}\left((M(\tau_t) - M(\tau_s))^2/\mathcal{F}_{\tau_s}\right) = \langle M^\tau\rangle(t) - \langle M^\tau\rangle(s)$, where $\langle M^\tau\rangle$ is a quadratic characteristic of a martingale $M(\tau_t)$. Due to the generalization of the Itô formula, $M^2(\tau_t) = 2\int_0^{\tau_t} M(s)dM(s) + t$. We can consider the last relation as a Doob–Meyer decomposition for supermartingale $M^2(\tau_t)$, where $\int_0^{\tau_t} M(s)dM(s)$ is a local martingale, and $A(t) = t$ is nonrandom and thus a predictable nondecreasing process. Because of uniqueness of such decomposition $\langle M^\tau\rangle(t) = t$.

Finally, let us prove the continuity $M(\tau_t)$. A function τ_t is right continuous. Therefore $M(\tau_t)$ has right continuous trajectories a.s.

Note that $\langle M\rangle(\tau_t-) = \langle M\rangle(\tau_t) = t$ and M is a continuous process. Denote by $A\subset\Omega$ the set on which the trajectories are not continuous. Then $A = \{\omega\in\Omega|$ there exists $t > 0$ such that $\tau_t- \neq \tau_t$, and $M(\tau_t-) \neq M(\tau_t)\} \subset \bigcup_{r,s\in\mathbb{Q}}\{\omega\in\Omega|$ there exists $t > 0$ such that $\tau_t- < r < s < \tau_t$, $\langle M\rangle(r) = \langle M\rangle(s)$, $M(r)\neq M(s)\} \subset \bigcup_{r,s\in\mathbb{Q}, 0<r<s}\{\langle M\rangle(r) = \langle M\rangle(s), M(r)\neq M(s)\}$. Fix $0 < r < s$, $r,s\in\mathbb{Q}$. Denote $\sigma = \inf\{u\geq r | \, \langle M\rangle(u) > \langle M\rangle(r)\}$. Then $\{\langle M\rangle(r) = \langle M\rangle(s), M(r)\neq M(s)\} = \{\sigma\geq s, M(r)\neq M(\sigma\wedge s)\}$ and $\langle M\rangle(\sigma\wedge s\wedge\sigma_k) - \langle M\rangle(r\wedge\sigma_k) = 0$. Then $(M(\sigma\wedge s\wedge\sigma_k))^2 -$

$(M(r\wedge\sigma_k))^2 = \int_{r\wedge\sigma_k}^{\sigma\wedge s\wedge\sigma_k} M(u)dM(u)$, and the right-hand side has zero expectation, and the expectation of the left-hand side equals $\mathsf{E}(M(\sigma\wedge s\wedge\sigma_k)-M(r\wedge\sigma_k))^2$. Tend $k\to\infty$ and apply the Fatou lemma: $\mathsf{E}(M(\sigma\wedge s)-M(r))^2=0$. It is easy to deduce now that $\mathsf{P}(A)=0$.

13.45. Let us write down a generalization of the Itô formula for a localizing sequence $\{\tau_n, n\ge 1\}$ (see also Problem 13.43),

$$M^2(t\wedge\tau_n) = 2\int_0^{t\wedge\tau_n} M(s)dM(s) + \langle M\rangle(t\wedge\tau_n),$$

and $\int_0^{t\wedge\tau_n} M(s)dM(s)$ is a martingale (see Theoretical grounds to Chapter 7). So $\mathsf{E}\int_0^{t\wedge\tau_n} M(s)dM(s)=0$, hence $\mathsf{E}M^2(t\wedge\tau_n)\le\mathsf{E}\langle M\rangle(t)<\infty$. The application of Fatou's lemma completes the proof.

13.46. (3) $\mathsf{E}f(\tau,X_\tau)=f(0,X_0)+\mathsf{E}\int_0^\tau(\mathscr{L}f+\partial f/\partial t)(u,X_u)du$, where $\mathscr{L}=\Sigma_{i=1}^n\mu_i(t,x)(\partial f(t,x))/\partial x_i+\frac{1}{2}\Sigma_{i,j=1}^n\sigma_{ij}(t,x)(\partial^2 f(t,x))/\partial x_i\partial x_j$ with $\sigma_{ij}:=\Sigma_{k=1}^m b_{ik}b_{jk}$.

13.47. Put $f_x(y)=(y-x)^+$. Define approximations $f_{x\varepsilon}(y)(\varepsilon>0)$ of the function $f_x(y)$:

$$f_{x\varepsilon}(y)=\begin{cases}0, & \text{if } y\le x-\varepsilon,\\ (y-x+\varepsilon)^2/4\varepsilon, & \text{if } x-\varepsilon\le y\le x+\varepsilon,\\ y-x, & \text{if } y\ge x+\varepsilon.\end{cases}$$

There exists a sequence $\varphi_n\in C^\infty(\mathbb{R})$ of functions, with compact supports that contract to $\{0\}$, and such that $g_n:=\varphi_n * f_{x\varepsilon}$ (i.e., $g_n(y)=\int_{\mathbb{R}} f_{x\varepsilon}(y-z)\varphi_n(z)dz$) satisfy relations: $g_n\to f_{x\varepsilon}$ and $g_n'\to f_{x\varepsilon}'$ uniformly on $\mathbb{R}$, and $g_n''\to f_{x\varepsilon}''$ pointwise except at points $x\pm\varepsilon$. Notice that $g_n\in C^\infty(\mathbb{R})$. For example, we can put $\varphi_n(y)=n\varphi(ny)$, where $\varphi(y)=c\exp\{-(1-y^2)^{-1}\}$ for $|y|<1$ and $\varphi(y)=0$ for $|y|\ge 1$; and a constant c is such that $c\int_{-1}^1\varphi(y)dy=1$. Apply the Itô formula to g_n:

$$g_n(W(t))-g_n(W(0))=\int_0^t g_n'(W(s))dW(s)+\frac{1}{2}\int_0^t g_n''(W(s))ds.$$

For all t a sequence $\mathbb{I}_{s\in[0,t]}g_n'(W(s))$ converges as $n\to\infty$ to $\mathbb{I}_{s\in[0,t]}f_{x\varepsilon}'(W(s))$ uniformly on $\mathbb{R}^+\times\Omega$. So, $\int_0^t g_n'(W(s))dW(s)$ converges in $L_2(\mathsf{P})$ to $\int_0^t f_{x\varepsilon}'(W(s))dW(s)$. Observe that $\lim_{n\to\infty}g_n''(W(s))=f_{x\varepsilon}''(W(s))$ a.s. for each $s\in\mathbb{R}^+$ because $\mathsf{P}(W(s)=x\pm\varepsilon)=0$ for all $\varepsilon>0$. Due to Fubini's theorem (applyied to a product of measures $\mathsf{P}\times\lambda^1$) this limit relation holds a.s. for almost all $s\in\mathbb{R}^+$ with respect to Lebesgue measure λ^1. Because $|g_n''|\le(2\varepsilon)^{-1}$, the Lebesgue dominated theorem implies the convergence $\int_0^t g_n''(W(s))ds\to\int_0^t f_{x\varepsilon}''(W(s))ds$ in $L_2(\mathsf{P})$ and a.s. So, for each x and t

$$f_{x\varepsilon}(W(t))-f_{x\varepsilon}(W(0))=\int_0^t f_{x\varepsilon}'(W(s))dW(s)+\frac{1}{2}\int_0^t\frac{1}{2\varepsilon}\mathbb{I}_{W(s)\in(x-\varepsilon,x+\varepsilon)}ds.\ \text{a.s.} \tag{13.5}$$

A sequence $f_{x\varepsilon}(W(t))-f_{x\varepsilon}(W(0))$ converges $L_2(\mathsf{P})$ to $(W(t)-x)^+-(W(0)-x)^+$ as $\varepsilon\downarrow 0$ because $|f_{x\varepsilon}(W(t))-f_{x\varepsilon}(W(0))|\le|W(t)-W(0)|$. Moreover $\mathsf{E}\Big(\int_0^t(f_{x\varepsilon}'(W(s))-\mathbb{I}_{W(s)\in[x,\infty)})^2ds\Big)\le\mathsf{E}\int_0^t\mathbb{I}_{W(s)\in(x-\varepsilon,x+\varepsilon)}ds\le\int_0^t(2\varepsilon/\sqrt{2\pi s})ds\to 0$ as $\varepsilon\to 0$. Therefore $\int_0^t f_{x\varepsilon}'(W(s))dW(s)$ converges in $L_2(\mathsf{P})$ to $\int_0^1\mathbb{I}_{W(s)\in[x,\infty)}dW(s)$. This implies the required statement.

13.48. A process $(-W)$ is also a Wiener process. So it possesses a local time at the point $(-x)$. Let us denote it by $L^-(t,-x)$. Applying Definition 13.1 to $L^-(t,-x)$ we obtain that $L^-(t,-x) = L(t,x)$ a.s. Combine the last equality and the application of the result of Problem 13.47 to $(-W)$ and $(-x)$ instead of W and x. Then we get

$$(W(t)-x)^- - (W(0)-x)^- = -\int_0^t \mathbb{I}_{W(s)\in(-\infty,x]}dW(s) + \frac{1}{2}L(t,x)$$

(check the last equality). Add this equality to the equality from Problem 13.47, and obtain the required statement because $\int_0^1 \mathbb{I}_{W(s)=x}dW(s) = 0$ a.s.

13.50. Without loss of generality we may assume that $L(t,x)$ is continuous in (t,x) (see Problem 13.49). Denote $I(t,x) = \int_0^t \mathbb{I}_{W(s)\in[x,\infty)}dW(s)$, $I(t,x)$ is continuous in (t,x). Then (Problem 13.47)

$$\frac{1}{2}L(t,x) = (W(t)-x)^+ - (W(0)-x)^+ - I(t,x). \tag{13.6}$$

Let $f'_{x\varepsilon}(z) = (1/2\varepsilon)\int_{x-\varepsilon}^{x+\varepsilon} \mathbb{I}_{z\in[y,\infty)}dy$ (see solution of Problem 13.47). Due to the stochastic Fubini theorem (Theorem 13.5), we obtain

$$\begin{aligned}\int_0^t f'_{x\varepsilon}(W(s))dW(s) &= \tfrac{1}{2\varepsilon}\int_0^t\int_{x-\varepsilon}^{x+\varepsilon} \mathbb{I}_{W(s)\in[y,\infty)}dy\, dW(s)\\ &= \tfrac{1}{2\varepsilon}\int_{x-\varepsilon}^{x+\varepsilon}\int_0^t \mathbb{I}_{W(s)\in[y,\infty)}dW(s)\, dy = \tfrac{1}{2\varepsilon}\int_{x-\varepsilon}^{x+\varepsilon} I(t,y)dy \text{ a.s.}\end{aligned}$$

(Verify that conditions of the stochastic Fubini theorem are really satisfied.)

Substituting the received identity into formula (13.5) we obtain that a.s.

$$f_{x\varepsilon}(W(t)) - f_{x\varepsilon}(W(0)) - \frac{1}{2\varepsilon}\int_{x-\varepsilon}^{x+\varepsilon} I(t,y)dy = \frac{1}{2}\int_0^t \frac{1}{2\varepsilon}\mathbb{I}_{W(s)\in(x-\varepsilon,x+\varepsilon)}ds. \tag{13.7}$$

Let us integrate the last equality with respect to x. Then

$$\begin{aligned}\int_a^b \Big\{f_{x\varepsilon}(W(t)) - f_{x\varepsilon}(W(0)) - \frac{1}{2\varepsilon}\int_{x-\varepsilon}^{x+\varepsilon} I(t,y)dy\Big\}dx&\\ = \frac{1}{4\varepsilon}\int_0^t\int_a^b \mathbb{I}_{x-\varepsilon<W(s)<x+\varepsilon}dxds \text{ a.s.}&\end{aligned} \tag{13.8}$$

For any $z\in\mathbb{R}$

$$\lim_{\varepsilon\downarrow 0}\frac{1}{2\varepsilon}\int_a^b \mathbb{I}_{(x-\varepsilon,x+\varepsilon)}(z)dx = \mathbb{I}_{z\in(a,b)} + \mathbb{I}_{z=a} + \mathbb{I}_{z=b}. \tag{13.9}$$

Make $\varepsilon\downarrow 0$ in (13.8). A function I is continuous, so identity (13.9) implies that

$$\int_a^b \Big((W(t)-x)^+ - (W(0)-x)^+ - I(t,x)\Big)dx = \frac{1}{2}\int_0^1 \mathbb{I}_{W(s)\in(a,b)}ds \text{ a.s.}$$

The required statement follows from the last identity and (13.6).

13.52. Let us apply equality (13.8). Its left-hand side is continuous in $\varepsilon>0$ and the right-hand side is left-continuous in $\varepsilon>0$. So, identity (13.8) holds for all $\varepsilon>0$ simultaneously, for a.a. ω. The left-hand side converges to $\frac{1}{2}L(t,x)$ as $\varepsilon\downarrow 0$ (see (13.6)) because $I(t,\cdot)$ is continuous. That is what was to be demonstrated.

13.54. Let $m\in\mathbb{N}$. Apply the multidimensional Dynkin formula to the process X, $\tau=\tau_m=\tau_R\wedge m$ and a bounded function $f\in C^2(\mathbb{R})$ such that $f(x)=\|x\|^2$ as $\|x\|\le R$. Observe that $\mathscr{L}f(x) = \frac{1}{2}\Delta f(x) = n$ when $\|x\|\le R$. Therefore

$$\begin{aligned}\mathsf{E}f(X(\tau_m)) &= f(a)+\tfrac{1}{2}\mathsf{E}\int_0^{\tau_m}\triangle f(X(s))ds\\ &= \|a\|^2+\mathsf{E}\int_0^{\tau_m} nds = \|a\|^2+n\mathsf{E}\tau_m.\end{aligned}$$

So $\mathsf{E}\tau_m \le (1/n)(R^2-\|a\|^2)$ for all $m\in\mathbb{N}$. Thus $\tau_R = \lim_{m\to\infty}\tau_m < \infty$ a.s. and $\mathsf{E}\tau_R = (1/n)(R^2-\|a\|^2)$.

13.55. Let σ_k be the first exit time from the ring $A_k = \{x\in\mathbb{R}^n \mid R\le\|x\|\le 2^kR\}$, $k\in\mathbb{N}$. Let also $f_{n,k}\in C^2(\mathbb{R})$ have a compact support and

$$f_{n,k}(x) = \begin{cases} -\ln|x|, & n=2,\\ |x|^{2-n}, & n>2,\end{cases}$$

as $R\le\|x\|\le 2^kR$. Because $\triangle f_{n,k}=0$ on a set A_k, Dynkin's formula implies

$$\mathsf{E}f_{n,k}(X(\sigma_k)) = f_{n,k}(a) \quad \text{for all } k\in\mathbb{N}. \tag{13.10}$$

We put $p_k := \mathsf{P}(|X(\sigma_k)|=R)$, $q_k=\mathsf{P}(|X(\sigma_k)|=2^kR)$, and consider cases $n=2$ and $n>2$ separately.

If $n=2$ then due to (13.10)

$$-\ln R\cdot p_k - (\ln R + k\ln 2)q_k = -\ln\|a\|,\ k\in\mathbb{N}.$$

So $q_k\to 0$ as $k\to\infty$, and $\mathsf{P}(\sigma_k<\infty)=1$.

If $n>2$, then (13.10) implies $p_k\cdot R^{2-n} + q_k(2^k\cdot R)^{2-n} = \|a\|^{2-n}$. Because $0\le q_k\le 1$,

$$\lim_{k\to\infty} p_k = \mathsf{P}(\sigma_k<\infty) = \left(\frac{\|a\|}{R}\right)^{2-n}.$$

13.56. $f_1(x_1,x_2) = -0,5x_1$; $f_2(x_1,x_2) = -0,5x_2$.

14

Stochastic differential equations

Theoretical grounds

Consider a complete filtration $\{\mathcal{F}_t, t \in [0,T]\}$ and an m-dimensional Wiener process $\{W(t), t \in [0,T]\}$ with respect to it. By definition, a stochastic differential equation (SDE) is an equation of the form

$$dX(t) = b(t,X(t))dt + \sigma(t,X(t))dW(t),\ 0 \le t \le T, \tag{14.1}$$

with $X_0 = \xi$, where ξ is an $\mathscr{F}_0$-measurable random vector, $b = b(t,x) : [0,T] \times \mathbb{R}^n \to \mathbb{R}^n$, and $\sigma = \sigma(t,x) : [0,T] \times \mathbb{R}^n \to \mathbb{R}^{n\times m}$ are measurable functions. Equality (14.1) is simply a formal writing of the stochastic integral equation

$$X(t) = X_0 + \int_0^t b(s,X(s))ds + \int_0^t \sigma(s,X(s))dW(s),\ 0 \le t \le T. \tag{14.2}$$

Definition 14.1. *A strong solution to stochastic differential equation* (14.1) *on the interval* $0 \le t \le T$ *is an* $\mathcal{F}_t$*-adapted* $\mathbb{R}^m$*-value process* $\{X(t), 0 \le t \le T\}$ *with a.s. continuous paths and such that after its substitution into the left- and right-hand sides of relation* (14.2)*, for each* $0 \le t \le T$ *the equality holds with probability 1.*

Definition 14.2. *Equation* (14.1) *has a unique strong solution in the interval* $[0,T]$ *if the fact that processes* $X = X(t)$ *and* $Y = Y(t)$*,* $t \in [0,T]$*, are strong solutions to the given equation with the same initial condition, implies that* X *is a modification of* Y *(then these continuous processes do not differ, i.e.,* $\mathsf{P}(X(t) = Y(t), t \in [0,T]) = 1$*).*

Theorem 14.1. *Assume both Lipschitz and linear growth conditions*

$$|b(t,x) - b(t,y)| + |\sigma(t,x) - \sigma(t,y)| \le L|x-y|,\ x,y \in \mathbb{R}^n, t \in [0,T];$$
$$|b(t,x)|^2 + |\sigma(t,x)|^2 \le L(1+|x|^2),\ t \in [0,T], x \in \mathbb{R}^n,$$

where $L > 0$ *is a constant. Then the stochastic differential equation has a unique strong solution. Here we denote Euclidean norm of both vector and matrix by the symbol* $|\cdot|$*; that is,*

DOI 10.1007/978-0-387-87862-1_14,

$$|b(t,x)| = \Big(\sum_{k=1}^{n} |b_k(t,x)|^2\Big)^{1/2}, \ |\sigma(t,x)| = \Big(\sum_{k=1,j=1}^{n,m} |\sigma_{kj}(t,x)|^2\Big)^{1/2}.$$

Theorem 14.1 gives the simplest conditions for existence and uniqueness of a strong solution; those are the classical Lipschitz and linear growth conditions. There exist generalizations of the theorem to the case where the Lipschitz condition is replaced by a weaker one.

Theorem 14.2. *Consider a scalar equation with homogeneous coefficients*

$$X(t) = X_0 + \int_0^t b(X(s))ds + \int_0^t \sigma(X(s))dW(s),\ 0 \le t \le T, \tag{14.3}$$

and assume that its coefficients satisfy the next conditions.

(1) The functions $b(x)$ and $\sigma(x)$ are bounded.

(2) There exists a strictly increasing function $\rho(u)$ on $[0,\infty)$ such that $\rho(0) = 0$, $\int_{0+} \rho^{-2}(u) = \infty$, and $|\sigma(x) - \sigma(y)| \le \rho(|x-y|)$ for all $x, y \in \mathbb{R}$ (it is Yamada's condition [91], see also [38]).

(3) There exists an increasing convex function $\varsigma(u)$ on $[0,\infty)$, such that $\varsigma(0) = 0$, $\int_{0+} \varsigma^{-1}(u) = \infty$, and $|b(x) - b(y)| \le \varsigma(|x-y|)$ for all $x, y \in \mathbb{R}$.

Then equation (14.3) has a unique strong solution.

In particular, one may take $\rho(u) = u^{\alpha}$, $\alpha \ge \frac{1}{2}$, and $\varsigma(u) = Cu$.

Now, we pass to the definition of a weak solution. Assume that only nonrandom coefficients $b(t,x)$ and $\sigma(t,x)$ are given, and at the moment there is no stochastic object at hand.

Theorem 14.3. *Consider a scalar equation* (14.3) *with homogeneous coefficients and assume that its coefficients satisfy the Lipschitz and linear growth conditions. Then a process X has a strong Markov property.*

Definition 14.3. *If on a certain probability space $(\Omega, \mathfrak{F}, \mathsf{P})$ one can construct a filtration $\{\mathfrak{F}_t, t \in [0,T]\}$ and two processes $\{(\widetilde{X}(t), \widetilde{W}(t)), t \in [0,T]\}$, which are adapted to the filtration and such that $\{\widetilde{W}(t), t \in [0,T]\}$ is a Wiener process, and $\widetilde{X}(t)$ is a solution to equation* (14.1) *in which W is changed for $\widetilde{W}$, then it is said that equation* (14.1) *has a weak solution.*

Theorem 14.4. *Let coefficients $b(t,x)$ and $\sigma(t,x)$ be measurable locally bounded functions, continuous in x for each $t \in [0,T]$, and moreover*

$$|b(t,x)|^2 + |\sigma(t,x)|^2 \le L(1+|x|^2),\ t \in [0,T],\ x \in \mathbb{R}^n.$$

Then equation (14.1) *has a weak solution.*

Remark 14.1. Throughout this and the next chapters by a Wiener process we mean a process W that satisfies the usual definition of a Wiener process, except, maybe, for the condition $W(0) = 0$. However, if no initial condition is specified, it is assumed that $W(0) = 0$.

Bibliography

[9], Chapter VIII; [38], Chapter IV; [90], Chapter 12, §§12.4–12.5; [24], Volume 3, Chapters II and III; [25], Chapter VIII, §§2–4; [26]; [51], Chapter 19; [57], Chapter 4; [79], Chapter 31; [20], Chapter 14; [8], Chapter 7, §7.5; [46], [49], Chapter 21; Chapter 12, §12.5; [54], Chapter 3, §3.5; [61], Chapter V; [68], Chapter 13, §13.1; [85], Chapters 9, 10, and 15.

Problems

14.1. Let $\{W(t), t \in \mathbb{R}^+\}$ be a one-dimensional Wiener process. Prove that the next processes are the solutions to corresponding stochastic differential equations.

(a) $X(t) = e^{W(t)}$ is a solution to an SDE $dX(t) = \frac{1}{2}X(t)dt + X(t)dW(t)$.

(b) $X(t) = (W(t))/(1+t)$ with $W(0) = 0$ is a solution to an SDE $dX(t) = -(1/(1+t))X(t)dt + (1/(1+t))dW(t), X(0) = 0$.

(c) $X(t) = \sin W(t)$ with $W(0) = a \in (-(\pi/2).(\pi/2))$ is a solution to an SDE $dX(t) = -\frac{1}{2}X(t)dt + (1-X^2(t))^{1/2}dW(t)$ for $t < \tau(\omega) = \inf\{s > 0 | W(s) \notin [-(\pi/2), (\pi/2)]\}$.

(d) $(X_1(t), X_2(t)) = (t, e^t W(t))$ is a solution to an SDE

$$\begin{pmatrix} dX_1(t) \\ dX_2(t) \end{pmatrix} = \begin{pmatrix} 1 \\ X_2(t) \end{pmatrix} dt + \begin{pmatrix} 0 \\ e^{X_1(t)} \end{pmatrix} dW(t),$$

(e) $(X_1(t), X_2(t)) = (\operatorname{ch} W(t), \operatorname{sh} W(t))$ is a solution to an SDE

$$\begin{pmatrix} dX_1(t) \\ dX_2(t) \end{pmatrix} = \frac{1}{2}\begin{pmatrix} X_1(t) \\ X_2(t) \end{pmatrix} dt + \begin{pmatrix} X_2(t) \\ X_1(t) \end{pmatrix} dW(t).$$

14.2. Prove that the process $X(t) = X_0 \exp\{(r - (\sigma^2/2))t + \sigma W(t)\}$ is a strong solution to an SDE $dX(t) = rX(t)dt + \sigma X(t)dW(t), X(0) = X_0$, $t \in \mathbb{R}^+$, and find an equation that is satisfied by the process $X(t) = X_0 \exp\{rt + \sigma W(t)\}$.

14.3. Let the process $\{X(t), t \in \mathbb{R}^+\}$ be a solution to an SDE

$$dX(t) = (\mu_1 X(t) + \mu_2)dt + (\sigma_1 X(t) + \sigma_2)dW(t),\ X_0 = 0, t \in \mathbb{R}^+.$$

(1) Find an explicit form for $X(t)$.

(2) Let $S(t) = \exp\{(\mu_1 - (\sigma_1^2/2))t + \sigma_1 W(t)\}$, where W is the same Wiener process that is written in the equation for X.

(a) Prove that the process $\{S(t)\}$ is a strong solution to an SDE $dS(t) = \mu_1 S(t)dt + \sigma_1 S(t)dW(t)$, $t \in \mathbb{R}^+$.

(b) Find a stochastic differential equation that is satisfied by the process $\{S^{-1}(t)\}$.

(3) Prove that $d(X(t)S^{-1}(t)) = S^{-1}(t)((\mu_2 - \sigma_1\sigma_2)dt + \sigma_2 dW(t))$.

14.4. Let $\{W(t) = (W_1(t), \dots, W_n(t)), t \in \mathbb{R}^+\}$ be an n-dimensional Wiener process. Find a solution to SDE

$$dX(t) = rX(t)dt + X(t)\Big(\sum_{k=1}^{n} \alpha_k dW_k(t)\Big),\ X(0) > 0.$$

14.5. (1) Prove that the process

$$X(t) = \alpha(1 - t/T) + \beta t/T + (T - t) \int_0^t \frac{dW(s)}{T - s}, \quad 0 \le t \le T,$$

is a solution to SDE

$$dX(t) = \frac{\beta - X(t)}{T - t} dt + dW(t), \; t \in [0, T], \; X_0 = \alpha.$$

(2) Prove that $X(t) \to \beta$ as $t \to T-$, a.s.

(3) Prove that the process X is a Brownian bridge over the interval $[0, T]$ with fixed endpoints $X(0) = \alpha$ and $X(T) = \beta$. (A standard Brownian bridge is obtained with $T = 1$, $\alpha = \beta = 0$, see Example 6.1 and Problems 6.13, 6.21.)

14.6. Solve the next stochastic differential equations.

(a) $\begin{pmatrix} dX_1(t) \\ dX_2(t) \end{pmatrix} = \begin{pmatrix} 1 \\ 0 \end{pmatrix} dt + \begin{pmatrix} 1 & 0 \\ 0 & X_1(t) \end{pmatrix} \begin{pmatrix} dW_1(t) \\ dW_2(t) \end{pmatrix},$

where $W(t) = (W_1(t), W_2(t))$ is a two-dimensional Wiener process.

(b) $dX(t) = X(t)dt + dW(t)$.

(c) $dX(t) = -X(t)dt + e^{-t}dW(t)$.

14.7. (1) Solve the Ornstein–Uhlenbeck equation (or the Langevin equation)

$$dX(t) = \mu X(t)dt + \sigma dW(t), \quad \mu, \sigma \in \mathbb{R}.$$

The solution is called the Ornstein–Uhlenbeck process (cf. Problem 6.12).

(2) Find $\mathsf{E}X(t)$ and $\mathsf{D}X(t)$ (cf. Example 6.2).

14.8. (1) Solve the mean-reverting Ornstein–Uhlenbeck equation

$$dX(t) = (m - X(t))dt + \sigma dW(t), \quad \mu, \sigma \in \mathbb{R}.$$

Here the coefficient m is the "mean value". Respectively, a solution to this equation is called the mean-reverting Ornstein–Uhlenbeck process.

(2) Find $\mathsf{E}X(t)$ and its asymptotic behavior as $t \to \infty$, and also find $\mathsf{D}X(t)$.

14.9. Solve an SDE

$$dX(t) = r dt + \alpha X(t)dW(t), \; r, \alpha \in \mathbb{R}.$$

14.10. Solve the next stochastic differential equations:

$$(1)\; dX(t) = \left(\sqrt{1 + X^2(t)} + 1/2X(t)\right)dt + \sqrt{1 + X^2(t)}dW(t),$$

$$(2)\; dX(t) = \left(\frac{2}{1+t}X(t) - a(1+t)^2\right)dt + a(1+t)^2 dW(t).$$

14.11. Consider a linear SDE

$$dX(t) = A(t)x(t)dt + \sigma dW(t), \; X(0) = x,$$

where $A(t)$ is a nonrandom continuous on $\mathbb{R}^+$ function, moreover $A(t) \le -\alpha < 0$ for all $t \ge 0$. Prove that

$$\mathsf{E}X^2(t) \le e^{-2\alpha t}\left(x^2 + \frac{\sigma^2}{2\alpha}(1 - e^{-2\alpha t})\right) \le \frac{\sigma^2}{2\alpha} + \left(x^2 - \frac{\sigma^2}{2\alpha}\right)e^{-2\alpha t}.$$

14.12. Solve a two-dimensional SDE,

$$\begin{aligned} dX_1(t) &= X_2(t)dt + \alpha dW_1(t), \\ dX_2(t) &= X_1(t)dt + \beta dW_2(t), \end{aligned}$$

where $(W_1(t), W_2(t))$ is a two-dimensional Wiener process and $\alpha, \beta \in \mathbb{R}$.

14.13. (1) Solve a system of stochastic differential equations

$$dX(t) = Y(t)dt,\ dY(t) = -\beta X(t)dt - \alpha Y(t)dt + \sigma dW(t),$$

where α, β, and σ are positive constants.

(2) Show that in the case where the vector $(X(0), Y(0))$ has a joint Gaussian distribution, the vector process (X, Y) is Gaussian. Find its covariance function.

14.14. (Feynman–Kac formula) Let $\{X(t), \mathcal{F}_t, t \in \mathbb{R}^+\}$ be a diffusion process that admits a stochastic differential

$$dX(t) = \mu(t, X(t))dt + \sigma(t, X(t))dW(t),\ 0 \le t \le T,$$

and let there exist a solution $f(t,x)$, $(t,x) \in [0,T] \times \mathbb{R}$, to a partial differential equation

$$\frac{\partial f}{\partial t} + \mathscr{L} f(t,x) = r(t,x) f(t,x),\ 0 \le t \le T,$$

with boundary condition $f(T,x) = g(x)$. (Here $r(t,x) \in C([0,T] \times \mathbb{R})$, $r \ge 0$, and the operator $\mathscr{L}$ is introduced in Problem 13.46.) Prove that

$$f(t,x) = \mathsf{E}\left(g(X(T))e^{-\int_t^T r(u,X(u))du} / X(t) = x\right).$$

14.15. Prove that the next one-dimensional SDE has a unique strong solution:

$$dX(t) = \log(1 + X^2(t))dt + \mathbb{1}_{X(t)>0} X(t)dW(t),\ X_0 = a \in \mathbb{R}.$$

14.16. Let a, c, d be real constants and $a > 0$. Consider a one-dimensional SDE,

$$dX(t) = (cX(t) + d)dt + (2aX(t) \vee 0)^{1/2} dW(t). \tag{14.4}$$

(1) Prove that for any initial value $X(0)$, the equation has a unique strong solution.

(2) Prove that in the case $d \ge 0$ and $X(0) \ge 0$, the solution is nonnegative; that is, $X(t) \ge 0$ for all $t \ge 0$, a.s.

14.17. (Gronwall–Bellman lemma) Let $\{x(t), 0 \le t \le T\}$ be a nonnegative continuous on $[0,T]$ function that satisfies the inequality $x(t) \le a + b\int_0^t x(s)ds$, $t \in [0,T]$, $a, b \ge 0$. Prove that $x(t) \le ae^{bt}$, $t \in [0,T]$.

14.18. Let the assumptions of Theorem 14.1 hold. Prove that under an additional assumption $\mathsf{E}|\xi|^2 < \infty$, the unique solution X to equation (14.1) satisfies the inequality

$$\mathsf{E}|X(t)|^2 \le k_1 \exp(k_2 t),\ 0 \le t \le T$$

with some constants k_1 and k_2.

14.19. Let $\{X_{s,x}(t), t \ge s\}, x \in \mathbb{R}^d$ be a solution to SDE (14.1) for $t \ge s$, with initial condition $X_{s,x}(s) = x$. Assume that the coefficients of the equation satisfy both Lipschitz and linear growth conditions. Prove that for any $T > 0$ and $p \ge 2$, there exists a constant c such that for $X_{s,x}(t)$ the following moment bounds are valid.

(1) $\forall\, s,t \in [0,T], s \le t\ \forall\, x \in \mathbb{R}^d$: $\mathsf{E}|X_{s,x}(t)|^p \le c(1+|x|^p)$.
(2) $\forall\, s,t,\ 0 \le s \le t \le T\ \forall\, x_1,x_2$: $\mathsf{E}|X_{s,x_1}(t) - X_{s,x_2}(t)|^p \le c|x_1 - x_2|^p$.
(3) $\forall\, x\ \forall\, s,t_1,t_2 \in [0,T],\ s \le t_1 \wedge t_2$:

$$\mathsf{E}|X_{s,x}(t_1) - X_{s,x}(t_2)|^p \le c(1+|x|^p)|t_1 - t_2|^{p/2}.$$

(4) $\forall\, x\ \forall\, s_1,s_2,t \in [0,T],\ t \ge s_1 \vee s_2$:

$$\mathsf{E}|X_{s_1,x}(t) - X_{s_2,x}(t)|^p \le c(1+|x|^p)|s_1 - s_2|^{p/2}.$$

14.20. Assume the conditions of Problem 14.19. Prove that the process $\{X_{s,x}(t), t \ge s\}$, $x \in \mathbb{R}^n$ has a modification that is continuous in (s,t,x).

14.21. Let $(\Omega, \mathcal{F}, \{\mathcal{F}_t\}_{t\in[0,T]}, \mathsf{P})$ be a filtered probability space, $\{W(t), \mathcal{F}_t, t \in [0,T]\}$ be a Wiener process, and $\{\gamma(t), \mathcal{F}_t, t \in [0,T]\}$ be a progressively measurable stochastic process with $\mathsf{P}\{\int_0^T \gamma^2(s)ds < \infty\} = 1$. Consider an SDE $dX(t) = \gamma(t)X(t)dW(t)$, $X(0) = 1$.

(1) Prove that there exists a nonnegative continuous solution to the equation, which is unique and given by the formula $X(t) = \exp\{\int_0^t \gamma(s)dW(s) - \frac{1}{2}\int_0^t \gamma^2(s)ds\}$. The process $X(t)$ is called a stochastic exponent (see Problem 13.30).

(2) Prove that the process X is a supermartingale and $\mathsf{E}X(t) \le 1$.

14.22. (Novikov's condition for martingale property of stochastic exponent) Let the conditions of Problem 14.21 hold. Then under $\mathsf{E}\exp\{\frac{1}{2}\int_0^T \gamma^2(s)ds\} < \infty$, a supermartingale $X(t,\gamma) := \exp\{\int_0^t \gamma(s)dW(s) - \frac{1}{2}\int_0^t \gamma^2(s)ds\}$ is a martingale and $\mathsf{E}X(t,\gamma) = 1$.

14.23. Assume the conditions of Problem 14.21 and for a certain $\delta > 0$ let it hold that

$$\sup_{t\le T} \mathsf{E}\exp\{\delta\gamma^2(t)\} < \infty.$$

Prove that $\mathsf{E}X(T,\gamma) = 1, t \in [0,T]$.

14.24. Let under the conditions of Problem 14.21, γ be a Gaussian process with $\sup_{t\le T} \mathsf{E}|\gamma(t)| < \infty$ and $\sup_{t\le T} \mathsf{D}\gamma(t) < \infty$. Prove that $\mathsf{E}X(T,\gamma) = 1$.

14.25. (Girsanov theorem for continuous time) Let under the conditions of item (1) of Problem 14.21, it hold $\mathsf{E}X(t,\gamma) = 1$. Define a probability measure Q by an equality $d\mathsf{Q} = X(t,\gamma)d\mathsf{P}$. Prove that on the probability space $(\Omega, \mathcal{F}, \mathsf{Q})$ a stochastic process $\widetilde{W}(t) := W_t - \int_0^t \gamma(s)ds$ is Wiener with respect to the filtration $\{\mathcal{F}_t\}_{t\in[0,T]}$.

14.26. Let $a : \mathbb{R}^n \to \mathbb{R}^n$ be a bounded measurable function. Construct a weak solution $\{X(t) = X_x(t)\}$ to an SDE $dX(t) = a(X(t))dt + dW(t)$, $X(0) = x \in \mathbb{R}^n$.

14.27. Let W be a Wiener process. Prove that $X(t) := W^2(t)$ is a weak solution to an SDE $dX(t) = dt + 2\sqrt{|X(t)|}d\widetilde{W}(t)$ where $\widetilde{W}$ is another Wiener process.

14.28. Let
$$X(t) = x_0 + \int_0^t a(s)dL(s) + \int_0^t b(s)dW(s),$$
where $L(t), t \geq 0$ is a continuous nondecreasing adapted process. Consider the processes β, A^{-1}, and $\widetilde{W}$ from Problem 13.60 and introduce processes $\widetilde{X}(t) = X(A^{-1}(t))$ and $\widetilde{L}(t) = L(A^{-1}(t))$. Prove that
$$d\widetilde{X}(t) = a(A^{-1}(t))d\widetilde{L}(t) + b(A^{-1}(t))\beta^{-1/2}(A^{-1}(t))d\widetilde{W}(t).$$
In particular, if $\beta(t) = c(X(t))$, $a(t) = \alpha(X(t))$, and $b(t) = \sigma(X(t))$, where α, σ, c are nonrandom functions, $c(x) > 0$, and $L(t) = t$, then the process $\widetilde{X}(t)$ is a solution to SDE,
$$d\widetilde{X}(t) = \alpha(\widetilde{X}(t))c^{-1}(\widetilde{X}(t))dt + \sigma(\widetilde{X}(t))c^{-1/2}(\widetilde{X}(t))d\widetilde{W}(t).$$

14.29. Let measurable processes $\alpha(t)$ and $\beta(t)$, $t \geq 0$ be adapted to the σ-algebra generated by a Wiener process and
$$\exists\, c, C > 0\ \forall t \geq 0: \ |\alpha(t)| \leq C,\ c \leq \beta(t) \leq C.$$
Prove that for a process X with
$$dX(t) = \alpha(t)dt + \beta(t)dW(t),\ \geq 0,$$
it holds with probability 1 that
$$\limsup_{t \to \infty} |X(t)| = +\infty \text{ a.s.}$$

14.30. Let $X(t), t \geq 0$ satisfy SDE,
$$dX(t) = a(X(t))dt + b(X(t))dW(t),\ t \geq 0,$$
where $a, b : \mathbb{R} \to \mathbb{R}$ satisfy the Lipschitz condition. Let $b(x) > 0$ for all $x \in [x_1, x_2]$ and $X(0) = x_0 \in [x_1, x_2]$. Prove that with probability 1 the exit time $\tau = \inf\{t \geq 0 |\, X(t) \notin (x_1, x_2)\}$ of the process X from the interval $[x_1, x_2]$ is finite and $\mathsf{E}\tau^m < \infty$ for all $m > 0$.

14.31. Let $\{X(t), t \geq 0\}$ be a solution to SDE (14.3) with initial condition $X(0) = x$, where b, $\sigma : \mathbb{R} \to \mathbb{R}$ satisfy the Lipschitz condition and $\sigma(x) \neq 0, x \in \mathbb{R}$. For the process X, prove that the probability $p_{ab}(x)$, $x \in (a, b)$, to hit the point a before the point b equals $(s(b) - s(x))/(s(b) - s(a))$, where
$$s(x) = \int_{c_1}^x \exp\left\{ - \int_{c_2}^y \frac{2b(z)}{\sigma^2(z)} dz \right\} dy$$
with arbitrary constants c_1 and c_2.

14.32. Find the probability $p_{ab}(x), x \in (a, b)$ to hit the point a before the point b for a process X that satisfies the next stochastic differential equations with initial condition $X(0) = x$.

(a) $dX(t) = dW(t)$.
(b) $dX(t) = dW(t) + Kdt$.
(c) $dX(t) = (2 + \sin X(t))dW(t)$.
(d) $dX(t) = AX(t)dt + BX(t)dW(t)$ with $B \neq 0$ and $A > 0$.
(e) $dX(t) = (A/(X(t)))dt + dW(t)$, where $A > 0$.

14.33. Find the probability that a path of a Wiener process intersects a straight line $y = kt + l$ with $t \geq 0$.

14.34. Prove that with probability 1 the process $X(t)$ from Problem 14.32 (e) does not hit the origin in finite time for $A \geq 1/2$.

14.35. Let W be a Wiener process on $\mathbb{R}^n$ with $n > 1$, and $x \in \mathbb{R}^n \setminus \{0\}$. Prove that the stochastic process $X(t) = \|x + W(t)\|$ satisfies SDE,

$$dX(t) = \frac{n-1}{2X(t)}dt + dB(t),$$

where $B(t)$ is a one-dimensional Wiener process. Use Problem 14.34 and check that $\mathsf{P}(\exists t > 0 |\ X(t) = 0) = 0$.

14.36. Let coefficients of an SDE and a function s satisfy the conditions of Problem 14.31. Prove that:

(a) If $\lim_{x\to+\infty} s(x) = +\infty$ and $\lim_{x\to-\infty} s(x) = -\infty$, then

$$\mathsf{P}(\sup_{t\geq 0} X(t) = +\infty) = \mathsf{P}(\inf_{t\geq 0} X(t) = -\infty) = 1.$$

(b) If $\lim_{x\to+\infty} s(x) = +\infty$ and $\lim_{x\to-\infty} s(x) = c \in \mathbb{R}$, then

$$\mathsf{P}(\sup_{t\geq 0} X(t) < \infty) = \mathsf{P}(\inf_{t\geq 0} X(t) = -\infty) = \mathsf{P}(\lim_{t\to+\infty} X(t) = -\infty) = 1.$$

(c) If there exist finite limits $\lim_{x\to-\infty} s(x) = s(-\infty)$ and $\lim_{x\to+\infty} s(x) = s(+\infty)$, then

$$\mathsf{P}(\sup_{t\geq 0} X(t) < \infty / X(0) = x) = \mathsf{P}(\inf_{t\geq 0} X(t) = -\infty / X(0) = x) = \tfrac{s(+\infty) - s(x)}{s(+\infty) - s(-\infty)}.$$

14.37. Assume that coefficients of an SDE satisfy the conditions of Problem 14.31. Prove that:

(a) $\mathsf{P}(\limsup_{t\to\infty} |X(t)| = \infty) = 1$.

(b) $\mathsf{P}(\liminf_{t\to\infty} X(t) = -\infty,\ \limsup_{t\to\infty} X(t) \in \mathbb{R}) = 0$.

(c) With probability 1 one of the three disjoint events occurs: either $\lim_{t\to\infty} X(t) = +\infty$, or $\lim_{t\to\infty} X(t) = -\infty$, or $\{\liminf_{t\to\infty} X(t) = -\infty, \limsup_{t\to\infty} X(t) = +\infty\}$; that is,

$$\mathsf{P}(\liminf_{t\to\infty} X(t) = -\infty,\ \limsup_{t\to\infty} X(t) = +\infty) + \mathsf{P}(\lim_{t\to\infty} X(t) = -\infty) + \mathsf{P}(\lim_{t\to\infty} X(t) = +\infty) = 1.$$

14.38. Let coefficients of an SDE satisfy the conditions of Problem 14.31. Denote by $\tau_{[a,b]} = \inf\{t \geq 0 |\ X(t) \notin [a,b]\}$ the first exit time of the process X from an interval $[a,b]$. Prove that the function $v(x) = \mathsf{E}(\tau_{[a,b]} / X(0) = x)$ is finite and equal to

$$v(x) = -\int_a^x 2\varphi(y) \int_a^y \frac{dz}{\sigma^2(z)\varphi(z)} dy + \int_a^b 2\varphi(y) \int_a^y \frac{dz}{\sigma^2(z)\varphi(z)} dy \frac{\int_a^x \varphi(z)dz}{\int_a^b \varphi(z)dz},$$

where

$$\varphi(x) = \exp\left\{ -\int_a^x \frac{2b(z)}{\sigma^2(z)} dz \right\}.$$

14.39. Let τ be the exit time of a Wiener process from an interval $[-a,b]$ with $a>0$ and $b>0$. Find $\mathsf{E}\tau$.

Hints

14.1. Apply the Itô formula.
14.2. Apply the Itô formula to $X(t)=(r-(\sigma^2/2))t+\sigma W(t)$ and $F(x) = e^x$.
14.3. (1) Note that it is a linear heterogeneous equation. Apply a method similar to the variation of constants method.
(2) Apply the Itô formula.
14.4. Look for a solution in the form $X(t)=X(0)\exp\{at+\sum_{k=1}^n b_k W_k(t)\}$, apply the Itô formula, and take into account the independence of components of W_k.
14.5. In order to prove that $\lim_{t\to T-}(T-t)\int_0^t((dW(s))/(T-s))=0$ a.s., set $M(t)=\int_0^t((dW(s))/(T-s))$, apply modified Theorem 7.16 (some analogue of Doob's martingale inequality for continuous time), and prove that $\mathsf{P}(\sup_{T(1-2^{-n})\le t\le T(1-2^{-n-1})}(T-t)|M(t)|>\varepsilon)\le 2\varepsilon^{-2}2^{-n}$. Apply the Borel–Cantelli lemma and obtain that for a.a. ω there exists $n(\omega)<\infty$ such that for all $n\ge n(\omega)$ it holds $\omega\notin A_n$, where

$$A_n=\{\omega|\sup_{T(1-2^{-n})\le t\le T(1-2^{-n-1})}(T-t)|M(t)|>2^{-(n/4)}\}.$$

14.6. (b) Multiply both sides of the equation by e^{-t} and compare with $d(e^{-t}W(t))$.
14.7, 14.8. Use Problem 14.1.
14.9. Multiply both sides by $\exp\{-\alpha W(t)+\frac{1}{2}\alpha^2 t\}$.
14.10. (1) First solve an SDE,

$$dX(t)=\sqrt{1+X^2(t)}dW(t)+\frac{1}{2}X(t)dt.$$

For this purpose write the Itô formula for a function f which is unknown at the moment:

$$\begin{aligned}f(X(t))=f(X(0))&+\int_0^t f'(X(s))\sqrt{1+X^2(s)}dW(s)\\&+\int_0^t f'(X(s))X(s)\,ds+1/2\int_0^t f''(X(s))(1+X^2(s))ds,\end{aligned}$$

and find the function f such that the integrand expression in the Lebesgue integral in the latter equality is identical zero. Then use the fact that in the initial equation the first summands in the drift and diffusion coefficients coincide.
(2) For the most part, the reasoning is similar.
14.11. Use the Itô formula for $X^2(t)$, compose an ordinary differential equation for $\mathsf{E}X^2(t)$, and solve it.
14.15. Use Theorem 14.1.
14.16. (1) Check that the conditions of Theorem 14.2 hold true.
(2) Separately consider the cases $d=0$ and $d>0$. In the first case, based on uniqueness of the solution to equation (14.4), prove that $X(t)\equiv 0$ if $X(0)=0$; and if $X(0)>0$ then set $\sigma=\inf\{t|\,X(t)=0\}$ and show that $X(t)=X(t\wedge\sigma)$. Let $d>0$. We set $\sigma_{-\varepsilon}=\inf\{t|\,X(t)=-\varepsilon\}$ where $\varepsilon>0$ satisfies $-c\varepsilon+d>0$. Suppose that $\mathsf{P}(\sigma_{-\varepsilon}<\infty)>0$. Then with probability 1, if to choose any $r<\sigma_{-\varepsilon}$ such that $X(t)<0$

for $t \in (r, \sigma_{-\varepsilon})$, we have $dX(t) = (cX(t) + d)dt$ in the interval $(r, \sigma_{-\varepsilon})$; that is, $X(t)$ is growing in this interval, which is impossible.

14.18. Use the Gronwall–Bellman lemma.

14.20. Use the results of the previous problem and check that $\forall\, R > 0, T > 0, p \geq 2\ \exists c > 0\ \forall\, s_1, s_2, t_1, t_2 \in [-T, T], s_1 \leq t_1,\ s_2 \leq t_2\ \forall\, x_1, x_2 \in \mathbb{R}^n,\ \|x_2\| \leq R, \|x_2\| \leq R:$

$$\mathsf{E}\|X_{s_1,x_1}(t_1) - X_{s_2,x_2}(t_2)\|^p \leq c(|s_1 - s_2|^{p/2} + |t_1 - t_2|^{p/2} + \|x_1 - x_2\|^p).$$

Use Problem 3.12 and Theorem 3.7.

14.21. (1) Existence of the solution of the given form can be derived from the Itô formula. Let Y be another continuous solution. Prove by the Itô formula that $d((Y(t))/(X(t))) = 0$.

(2) It follows from Problem 13.31.

14.24. Use Problem 14.23.

14.25. Similarly to Problem 7.96, check that the process $\{M_t\}$ is a Q-local martingale if and only if $M_t X(t, \gamma)$ is a local P-martingale. Next, use the Itô formula and obtain that $\widetilde{W}(t) \cdot X(t, \gamma) = \int_0^t X(s, \gamma) dW(s) + \int_0^t \widetilde{W}(s) X(s, \gamma) f(s) dW(s)$; that is, this process is a local P-martingale. Thus, $\widetilde{W}(t)$ is a local Q-martingale. Now, because the quadratic variation $[\widetilde{W}]_t = t$, obtain, based on Theorem 7.17, that $\widetilde{W}$ is a square-integrable Q-martingale.

14.27. Write the Itô formula for X and compare it with the required equation. Use Problem 13.58.

14.28. Make an ordinary change of variables in the first integral:

$$\int_0^{A^{-1}(t)} a(s) dL(s) = \int_0^t a(A^{-1}(z)) d\widetilde{L}(z).$$

For another integral, use Problem 13.60.

14.33. The desired probability equals the probability that the process satisfying $dX(t) = dW(t) - k\,dt$ and $X(0) = -l$ hits the point 0. Denote $\tau = \inf\{t \geq 0 | X(t) = 0\}$. If $k = 0$ then $\mathsf{P}(\tau < \infty) = 1$. Let, to be specific, $k < 0$. Then with probability 1 (see Problem 3.18), $\lim_{t \to +\infty} X(t) = +\infty$. Therefore, $\mathsf{P}(\tau < \infty) = 1$ if $l \geq 0$. For $l < 0$,

$$\begin{aligned}\mathsf{P}(\tau < \infty) &= \lim_{n \to \infty} \mathsf{P}(X(t) \text{ hits 0 before the point } n) \\ &= \lim_{n \to \infty} \tfrac{e^{-kn} - e^{-kl}}{e^{-kn} - 1} = e^{-kl}.\end{aligned}$$

14.34. Use the reasoning from Problem 14.36.

14.35. Let $\zeta = \inf\{t \geq 0 |\ X(t) = 0\} \cup \{+\infty\}$. Then by the Itô formula

$$dX(t) = \frac{n-1}{2X(t)} dt + \sum_{k=1}^{n} \frac{W_k(t)}{\|W(t)\|} dW_k(t),\ \ t \leq \zeta.$$

From Problem 13.59 it follows that the process

$$B(t) = \sum_{k=1}^{n} \int_0^t \frac{W_k(s)}{\|W(s)\|} dW_k(s)$$

is Wiener.

14.36. (a) Denote by P_x the distribution of X under the condition $X(0) = x$. Then for any $x_1, x_2, x_1 \leq x \leq x_2 : \mathsf{P}_x(\sup_{t \geq 0} X(t) \geq x_2) \geq \mathsf{P}_x(\text{the process } X \text{ hits } x_2 \text{ before } x_1) = (s(x) - s(x_1))/(s(x_2) - s(x_1))$ (see Problem 14.31). Let $x_1 \to -\infty$.

(b) For any $x_2 > x$ $\mathsf{P}_x(\sup_{t\geq 0} X(t) > x_2) \leq \mathsf{P}_x$(there exists $x_1 \leq x$ such that the process X hits x_2 before x_1) $= \lim_{x_1\to -\infty} \mathsf{P}_x(X(t)$ hits x_2 before $x_1) = (s(x) - s(-\infty))/(s(x_2) - s(-\infty))$. Let $x_2 \to +\infty$ and use the result of Problem 14.37.

(c) Use reasoning of item (b) and the result of Problem 14.37.

14.37. (a) With probability 1 the process $X(t)$ exits from any interval (see Problem 14.30).

(b) Use the strong Markov property of a solution to SDE (see Theorem 14.3), the Borel–Cantelli lemma, and Problem 14.31, and check the following. If for some $c \in \mathbb{R}$ there exists a sequence of Markov times $\{\tau_n\}$ such that $\lim_{n\to\infty} \tau_n = +\infty$ and $X(\tau_n) = c$, then $\mathsf{P}(\liminf_{t\to\infty} X(t) = -\infty) = \mathsf{P}(\limsup_{t\to\infty} X(t) = +\infty) = 1$. If $\mathsf{P}(\liminf_{t\to\infty} X(t) = -\infty$ and $\limsup_{t\to\infty} X(t) > c) > 0$, then take $\tau_n = \inf\{t \geq \sigma_n : X(t) = c\}$ with $\sigma_{n+1} = \inf\{t \geq \tau_n : X(t) = c-1\}$ and $\sigma_0 = 0$.

(c) See (a) and (b).

14.38. Notice that the function v is twice continuously differentiable and

$$\begin{aligned} Lv(x) &= -1, x \in [a,b], \\ v(a) &= v(b) = 0, \end{aligned}$$

where $Lv(x) = b(x)v'(x) + \frac{1}{2}\sigma^2(x)v''(x)$. Then by the Itô formula and the properties of a stochastic Itô integral one has that for any $n \in \mathbb{N}$,

$$\begin{aligned} &\mathsf{E}(v(X(n\wedge\tau_{[a,b]}))/X(0) = x) \\ &= v(x) + \mathsf{E}\left(\int_0^{n\wedge\tau_{[a,b]}} Lv(X(s))ds/X(0) = x\right) \\ &= v(x) - \mathsf{E}(n\wedge\tau_{[a,b]}/X(0) = x). \end{aligned}$$

Because $\tau_{[a,b]} < \infty$ a.s. (see Problem 14.30), by the dominated convergence theorem the left-hand side of the equality tends to 0, whereas by the monotone convergence theorem the expectation on the right-hand side converges to $\mathsf{E}(\tau_{[a,b]}/X(0) = x)$.

14.39. Use the result of Problem 14.38 with the process $dX(t) = dW(t)$.

Answers and Solutions

14.14. By the Itô formula

$$df(t,X(t)) = \left(\frac{\partial f}{\partial t} + \mathscr{L}f(t,X(t))\right)dt + dP(t),$$

where $dP(t) = (\partial f/\partial x)\sigma(t,x)dW(t)$. Taking into account that f is a solution to the given partial differential equation, we obtain

$$df(t,X(t)) = r(t,X(t))f(t,X(t))dt + dP(t),\ 0 \leq t \leq T.$$

Because this SDE is linear in f, its solution has a form

$$f(T,X(T)) = e^{\int_t^T r(u,X(u))du}\left(f(t,X(t)) + \int_t^T e^{-\int_t^s r(u,X(u))du} dP(s)\right).$$

Taking the expectation under the condition $X(t) = x$, accounting for the boundary condition, and using the martingale property of the stochastic process P, then we obtain the desired statement.

14.17. Let $a > 0$. Then for $t > 0$

$$\left(\log\left(a+b\int_0^t x(s)ds\right)\right)' = \frac{bx(t)}{a+b\int_0^t x(s)ds} \le b.$$

Integrating from 0 up to t we obtain

$$\log\left(a+b\int_0^t x(s)ds\right) - \log a \le bt, \ t \in [0,T],$$

and $a+b\int_0^t x(s)ds \le ae^{bt}$. Consider the case $a = 0$ yourself.

14.19. (1) Let $\tau_N = \inf\{t \ge s | \ \|X_{s,x}(t)\| \ge N\} \wedge T$. For simplicity we consider only the case $m = n = 1$. The Itô formula implies that

$$|X_{s,x}(t\wedge\tau_N)|^p = |x|^p + p\int_s^{t\wedge\tau_N} |X_{s,x}(z)|^{p-1}\text{sign}\,(X_{s,x}(z))\,(b(z,X_{s,x}(z))dz$$
$$+\sigma(z,X_{s,x}(z))dW(z)) + \frac{1}{2}p(p-1)\int_s^{t\wedge\tau_N} |X_{s,x}(z)|^{p-2}\sigma^2(z,X_{s,x}(z))dz.$$

Therefore,

$$\begin{aligned}\mathsf{E}|X_{s,x}(t\wedge\tau_n)|^p &\le K_1\left(|x|^p + \mathsf{E}\int_s^{t\wedge\tau_N}|X_{s,x}(z)|^{p-2}(1+|X_{s,x}(z)|+\xi_{sz}^2)dz\right)\\ &\le K_2\left(|x|^p + \mathsf{E}\int_s^{t\wedge\tau_N}(1+|X_{s,x}(z)|^p)dz\right)\\ &\le K_3\left(|x|^p + \int_s^t(1+\mathsf{E}|X_{s,x}(z\wedge\tau_N)|^p)dz\right)\\ &\le K_3\left(|x|^p + T + \int_s^t \mathsf{E}|X_{s,x}(z\wedge\tau_N)|^p dz\right).\end{aligned}$$

Now, the Gronwall–Bellman lemma implies the inequality

$$\mathsf{E}|X_{s,x}(t\wedge\tau_N)|^p \le c(1+|x|^p), \ \ 0\le s\le t\le T, x\in\mathbb{R},$$

where the constant c does not depend on N. Use Fatou's lemma to prove the desired inequality.

Items (2) to (4) are proven similarly to (1) based on the Itô formula and Gronwall–Bellman lemma.

14.22. Let $a > 0$, and $\sigma_a = \inf\{t\in[0,T] \mid \int_0^t\gamma(s)dW(s) - \int_0^t\gamma^2(s)ds = -a\}$, $\sigma_a = T$, if $\inf_{t\in[0,T]}\left(\int_0^t\gamma(s)dW(s) - \int_0^t\gamma^2(s)ds\right) > -a$. Let also $\lambda\le 0$. Show that $\mathsf{E}X(\sigma_a,\lambda\gamma) = 1$. According to Problem 14.21,

$$X(\sigma_a,\lambda\gamma) = 1+\lambda\int_0^{\sigma_a} X(s,\lambda\gamma)\gamma(s)dW(s).$$

Thus, it is enough to show that $\mathsf{E}\int_0^{\sigma_a} X^2(s,\lambda\gamma)\gamma^2(s)ds < \infty$. But this relation is implied by the next two bounds: $\mathsf{E}\int_0^{\sigma_a}\gamma^2(s)ds \le \mathsf{E}\int_0^T\gamma^2(s)ds \le \mathsf{E}\exp\{\frac{1}{2}\int_0^T\gamma^2(s)ds\} < \infty$, and

$$X(s,\lambda\gamma) = \exp\left\{\lambda(\int_0^s\gamma(u)dW(u) - \int_0^s\gamma^2(u)du\right\}$$
$$\times\exp\left\{(\lambda-\frac{\lambda^2}{2})\int_0^s\gamma^2(u)du\right\} \le \exp\{|\lambda|a\}, \text{ for all } \lambda\le 0 \text{ and } s\le\sigma_a.$$

Now, we show that $\mathsf{E}X(\sigma_a,\lambda\gamma)=1$ for $0<\lambda\le 1$. Define

$$\rho(\sigma_a,\lambda\gamma)=e^{\lambda a}X(\sigma_a,\lambda\gamma),\ A(\omega)=\int_0^{\sigma_a}\gamma^2(s)ds,$$

$$B(\omega)=\int_0^{\sigma_a}\gamma(s)dW(s)-\int_0^{\sigma_a}\gamma^2(s)ds+a\ge 0,$$

and let $u(z)=\rho(\sigma_a,\lambda\gamma)$, where $\lambda=1-\sqrt{1-z}$. It is clear that $0\le\lambda\le 1$ if and only if $0\le z\le 1$. Besides, $u(z)=\exp\{(z/2)A(\omega)+(1-\sqrt{1-z})B(\omega)\}$. For $0\le z<1$, one can P-a.s. expand the function $u(z)$ into a series: $u(z)=\sum_{k=0}^{\infty}(z^k/k!)p_k(\omega)$, where $p_k(\omega)\ge 0$ P-a.s., for all $k\ge 0$. Problem 13.31 implies that $\mathsf{E}u(z)\le e^{a(1-\sqrt{1-z})}<\infty$. If $0\le z_0<1$ and $|z|\le z_0$, then $\mathsf{E}\sum_{k=0}^{\infty}(|z|^k)/(k!)p_k(\omega)\le \mathsf{E}u(z_0)<\infty$. Therefore, due to the Fubini Theorem for any $|z|<1$ we have $\mathsf{E}u(z)=\sum_{k=0}^{\infty}(z^k/k!)\mathsf{E}p_k(\omega)$.

On the other hand, for $-\infty\le z<1$ we have $e^{a(1-\sqrt{1-z})}=\sum_{k=0}^{\infty}(z^k/k!)c_k$, where $c_k\ge 0$, $k\ge 0$. From this, and also from the equality $\mathsf{E}\rho(\sigma_a,\lambda\gamma)=e^{\lambda a}$, $\lambda\le 0$, we obtain for $-1<z\le 0$ that $\sum_{k=0}^{\infty}(z^k/k!)\mathsf{E}p_k(\omega)=\sum_{k=0}^{\infty}(z^k/k!)c_k$, thus, $\mathsf{E}p_k(\omega)=c_k$, $k\ge 0$, and then for $0\le z<1$ we have $\mathsf{E}u(z)=\sum_{k=0}^{\infty}(z^k/k!)c_k=e^{a(1-\sqrt{1-z})}$. Because $A(\omega)$ and $B(\omega)$ are nonnegative P-a.s., then $\rho(\sigma_a,\lambda\gamma)\uparrow\rho(\sigma_a,\gamma)$ for $\lambda\uparrow 1$. Due to the monotone convergence theorem,

$$e^a=\lim_{\lambda\uparrow 1}\mathsf{E}\rho(\sigma_a,\lambda\gamma)=\mathsf{E}\rho(\sigma_a,\gamma).$$

Evidently, $1=\mathsf{E}X(\sigma_a,\gamma)=\mathsf{E}X(\sigma_a,\gamma)\mathbb{I}_{\sigma_a<T}+\mathsf{E}X(T,\gamma)\mathbb{I}_{\sigma_a=T}$. Therefore

$$\mathsf{E}X(T,\gamma)=1-\mathsf{E}X(\sigma_a,\gamma)\mathbb{I}_{\sigma_a<T}+\mathsf{E}X(T,\gamma)\mathbb{I}_{\sigma_a<T}.$$

We have that $\lim_{a\to\infty}\mathbb{I}_{\sigma_a<T}=0$ P-a.s. In addition, we know that $\mathsf{E}X(T,\gamma)\le 1$. Therefore, $\lim_{a\to\infty}\mathsf{E}X(T,\gamma)\mathbb{I}_{\sigma_a<T}=0$. Moreover, on the set $\{\sigma_a<T\}$ we have $X(\sigma_a,\gamma)=\exp\{-a+\frac{1}{2}\int_0^{\sigma_a}\gamma^2(s)ds\}\le\exp\{-a+\frac{1}{2}\int_0^T\gamma^2(s)ds\}$, whence

$$\mathsf{E}X(\sigma_a,\gamma)\mathbb{I}_{\sigma_a<T}\le e^{-a}\mathsf{E}\exp\left\{\frac{1}{2}\int_0^T\gamma^2(s)ds\right\}\to 0,\ a\to\infty.$$

Thus, $\mathsf{E}X(T,\gamma)=1$.

14.23. By the Jensen inequality, $\exp\{\frac{1}{2}\int_0^T\gamma^2(t)dt\}\le(1/T)\int_0^T\exp\{(T\gamma^2(t))/2\}dt$, hence for $T\le 2\delta$ it holds

$$\mathsf{E}\exp\left\{\frac{1}{2}\int_0^T\gamma^2(t)dt\right\}\le\sup_{t\le T}\mathsf{E}\exp\{\delta\gamma^2(t)\}<\infty,$$

and the proof follows from Problem 14.22. Now, let $T>2\delta$. Represent $X(T,\gamma)$ as a product: $X(T,\gamma)=\prod_{j=0}^{n-1}X(t_j,t_{j+1},\gamma)$, where $0=t_0<t_1<\cdots<t_n=T$ and $X(t_j,t_{j+1},\gamma)=\exp\{\int_{t_j}^{t_{j+1}}\gamma(t)dW(t)-\frac{1}{2}\int_{t_j}^{t_{j+1}}\gamma^2(t)dt\}$, $t_{j+1}-t_j\le 2\delta$, $0\le j\le n-1$. Then $\mathsf{E}X(t_j,t_{j+1},\gamma)=1$ $\mathsf{E}(X(t_j,t_{j+1},\gamma)/\mathfrak{F}_{t_j})=1$ P-a.s. Thus,$\mathsf{E}X(T,\gamma)=\mathsf{E}(\mathsf{E}(X(T,\gamma)/\mathfrak{F}_{t_{n-1}}))=\mathsf{E}X(t_{n-1},\gamma)=\cdots=\mathsf{E}X(t_1,\gamma)=1$.

14.26. Define a martingale $M(t)=\exp\{\int_0^t a(W(s))dW(s)-\int_0^t a^2(W(s))ds/2\}$, where W is a Wiener process. Fix $T>0$ and define a probability measure Q by relation $dQ=M_Td\mathsf{P}$ on $\mathfrak{F}_T^W$. Then, due to the Girsanov theorem, the stochastic process $\hat{W}(t):=-\int_0^t a(W(s))ds+W(t)$ is a Wiener process with respect to the measure Q for $t \le T$, at that $dW(t) = a(W(t))dt + d\hat{W}(t)$. If one sets $W(0) = x$, then the couple $(W, \hat{W})$ forms the desired weak solution.

14.29. Perform the time change $\xi(t) = X(A^{-1}(t))$, where $A(t) = \int_0^t \beta^2(s)ds$ (see Problem 14.28). Then $d\xi(t) = \widetilde{\alpha}(t)dt + d\widetilde{W}(t)$, where $\widetilde{\alpha}(t) = \alpha(A^{-1}(t)) \times \beta^{-2}(A^{-1}(t))$, $|\widetilde{\alpha}(t)| \le K := C/c^2$. Use Problem 5.39 with $a = 1$ and $b = 2K + 2m + 1$.

14.30. Let bounded functions $\widetilde{a}$ and $\widetilde{b}$ satisfy the Lipschitz condition, $\widetilde{b}(x) \ge c > 0$, $x \in \mathbb{R}$, where c is a constant, and $\widetilde{a}(x) = a(x)$, $\widetilde{b}(x) = b(x)$ for all $x \in [x_1, x_2]$.

Denote by $\widetilde{X}(t)$ a solution to SDE with coefficients $\widetilde{a}$ and $\widetilde{b}$. Then $\mathsf{P}(X(t) = \widetilde{X}(t),\ t \le \tau) = 1$ and the process $\widetilde{X}$ satisfies the condition of Problem 14.29, whence $\mathsf{P}(\tau < \infty) = 1$. One can see from the solution of Problems 14.29 and 5.39 that the probabilities $\mathsf{P}(\tau \ge n)$ do not exceed $\alpha\beta^n$, where $\alpha > 0$ and $\beta \in (0;1)$ are constants. This implies that all moments of τ are finite.

14.31. Notice that the function s satisfies the equation $Ls(x) = 0, x \in \mathbb{R}$, where $L = b(x)(d/dx) + \frac{1}{2}\sigma^2(x)(d^2/dx^2)$. Therefore, the Itô formula implies that $ds(X(t)) = f(X(t))dW(t)$, where $f(x) = \exp\{-\int_{c_2}^x ((2b(z))/(\sigma^2(z)))dz\}\sigma(x)$. Because f is a bounded in (a,b) function and the exit time ζ_{ab} of the process $X(t)$ from the interval (a,b) is finite with probability 1, we have $E\zeta_{ab}^2 < \infty$ (see Problem 14.30), then $\mathsf{E}s(X(\zeta_{ab})) = s(x)$. Thus, $s(a)\mathsf{P}(X(\zeta_{ab}) = a) + s(b)\mathsf{P}(X(\zeta_{ab}) = b) = s(x)$, and therefore, $\mathsf{P}(X(\zeta_{ab}) = a) = (s(b) - s(x))/(s(b) - s(a))$.

14.32. $p_{ab}(x) = (s(b) - s(x))/(s(b) - s(a))$, where (see Problem 14.31):

(a) $s(x) = x$.

(b) $s(x) = -e^{-kx}$.

(c) $s(x) = x$.

(d) $s(x) = \begin{cases} x^{-\frac{2A}{B^2}+1}, & 2A \ne B^2 \\ \log x, & 2A = B^2. \end{cases}$

(e) $s(x) = \begin{cases} x^{-2A+1}, & A \ne 1/2 \\ \log x, & A = 1/2. \end{cases}$

14.33. $\begin{cases} e^{-kl}, & kl > 0 \\ 1, & kl \le 0. \end{cases}$

14.39. $\mathsf{E}\tau = ab$.

15

Optimal stopping of random sequences and processes

Theoretical grounds

The optimal stopping problem can be considered for nearly any stochastic process, and its formulation will be similar in each case. But its solution will be relatively simple only for a few processes. One class of such processes consists of discrete-time Markov chains. Let$\{X_n, n \in \mathbb{Z}^+\}$ be a Markov chain with a finite or countable phase space $(\mathbb{X}, \mathfrak{X})$ and one-step transition probabilities $\{p_{xy}, x, y \in \mathbb{X}\}$. Consider also a bounded function $f : \mathbb{X} \to \mathbb{R}^+$. Let $X_\infty := x_0$ and $f(x_0) = 0$. We need the following: (1) to compute $v(x) := \sup_\tau \mathsf{E}_x f(X_\tau)$, where sup is taken over all Markov moments τ, and the chain starts from the point $x \in \mathbb{X}$; and (2) to find a Markov moment τ_0, for which $v(x) = \mathsf{E}_x f(X_{\tau_0})$. As in game theory, $f(x)$ is called a payoff function, or a premium function, $v(x)$ is a price of the game, and τ_0 is an optimal strategy. Often only stopping times τ are considered and in this case τ_0 is said to be an optimal stopping time.

Definition 15.1. *A function $g : \mathbb{X} \to \mathbb{R}^+$ is said to be excessive (with respect to a chain X) if for any $x \in \mathbb{X}$ it holds*

$$Pg(x) \le g(x),$$

where P is the operator defined by $Pg(x) = \sum_{y \in \mathbb{X}} p_{xy} g(y)$.

Lemma 15.1. *If a function f is excessive, then for any Markov moment τ the inequality $f(x) \ge \mathsf{E}_x f(X_\tau)$ holds.*

Remark 15.1. In fact, an optimal strategy for an excessive function is to stop immediately.

Lemma 15.2. *The price of the game is an excessive function. Furthermore, the price of the game $v(x)$ is the least among all excessive functions $h(x)$ with $h(x) \ge f(x)$ for any $x \in \mathbb{X}$.*

The price of the game is said to be the excessive majorant for a function f or the least excessive majorant for f.

D. Gusak et al., *Theory of Stochastic Processes*, Problem Books in Mathematics, 229
DOI 10.1007/978-0-387-87862-1_15, © Springer Science+Business Media, LLC 2010

Corollary 15.1. *If $\#\mathbb{X} = n$, then the price of the game $v(x)$ is a minimal function satisfying the system of $3n$ inequalities*

$$v(x) \geq \sum_{y \in \mathbb{X}} p_{xy} v(y), \; v(x) \geq f(x), \; v(x) \geq 0, \; x \in \mathbb{X}.$$

Definition 15.2. *A set Γ of points x where the payoff $f(x)$ is equal to its excessive majorant $v(x)$, is called a supporting set or a stopping set. A set Γ^c of points x where the payoff function $f(x)$ is less than its excessive majorant $v(x)$, is said to be a continuation set.*

Theorem 15.1. *If the phase space $\mathbb{X}$ is finite, the time τ_0 of the first hit of a stopping set by a chain X visits a stopping set at the first time is an optimal strategy (and it is also an optimal stopping).*

Let now $\{X(t), t \geq 0\}$ be a diffusion process defined on $\mathbb{R}^n$. That is, $X(t)$ satisfies an SDE $dX(t) = b(t,X(t))dt + \sigma(t,X(t))dW(t)$, where $X(t) \in \mathbb{R}^n$, $b(t,x) : \mathbb{R}^+ \times \mathbb{R}^n \to \mathbb{R}^n$, $\sigma(t,x) : \mathbb{R}^+ \times \mathbb{R}^n \to \mathbb{R}^{n \times m}$ and $\{W(t), t \in \mathbb{R}^+\}$ is an m-dimensional Wiener process. Let also $f : \mathbb{R}^n \to \mathbb{R}^+$ be a continuous payoff function. The optimal stopping problem in this case consists in finding a stopping time τ_0 with respect to the filtration $\{\mathcal{F}_t^W, t \in \mathbb{R}^+\}$ generated by W, such that $\mathsf{E}_x f(X(\tau_0)) = \sup_\tau \mathsf{E}_x f(X(\tau))$, $x \in \mathbb{R}^n$ where sup is taken over all Markov stopping times is considered and the process starts at $t = 0$ from x. We also have to find $v_f(x) := \mathsf{E}_x f(X(\tau_0))$.

Let us introduce some notions. All functions are supposed to be measurable.

Definition 15.3. *A function $f : \mathbb{R}^n \to \mathbb{R}^+$ is said to be lower semicontinuous if for any point $x \in \mathbb{R}^n$ and for any sequence $x_n \to x$ as $n \to \infty$, it holds*

$$f(x) \leq \liminf_{n \to \infty} f(x_n).$$

Definition 15.4. *A lower semicontinuous function $f : \mathbb{R}^n \to \mathbb{R}^+$ is said to be superharmonic with respect to a diffusion process X if*

$$f(x) \geq \mathsf{E}_x f(X_\tau)$$

for all $x \in \mathbb{R}^n$ and stopping times τ. If the lower semicontinuity is not required, then the function is called excessive (cf. Lemma 15.1).

Definition 15.5. *Let $h : \mathbb{R}^n \to \mathbb{R}^+$ be a measurable function and f be a superharmonic function with $f \geq h$. Then f is said to be a superharmonic majorant for h. If, in addition, for any superharmonic majorant f_1 for the function h it holds $f_1 \geq f$, then f is called the least superharmonic majorant for h and denoted by $\widehat{h}$.*

Concerning a construction of a superharmonic majorant, see Problem 15.14. If "superharmonic function" in Definition 15.5 is replaced by "excessive function", then one gets definitions of excessive majorant and the least excessive majorant.

Denote $\Gamma = \{x \mid f(x) = v_f(x)\}$, $\Gamma^c = \{x \mid f(x) < v_f(x)\}$, $\tau_\Gamma = \inf\{t \in \mathbb{R}^+ \mid X(t) \in \Gamma\}$, and $\Gamma_N = \{x \mid f(x) \wedge N = \widehat{f(x) \wedge N}\}$.

Theorem 15.2. *Let $f : \mathbb{R}^n \to \mathbb{R}^+$ be a continuous payoff function. Then*

(1) $v_f(x) = \widehat{f}(x)$.

(2) If $\tau_\Gamma < \infty$ P_x a.s. and the sequence $\{f(X_{\tau_{\Gamma_N}}), N \geq 1\}$ is uniformly integrable then $v_f(x) = \mathsf{E}_x f(X_{\tau_\Gamma})$ and $\tau_0 = \tau_\Gamma$ is an optimal stopping.

(3) If τ_1 is an optimal stopping, then $\tau_1 \geq \tau_\Gamma$, P_x-a.s.

Theorem 15.3. *(Construction of the least superharmonic majorant) Let g_0 be a non-negative lower semicontinuous function defined on $\mathbb{R}^n$. Define inductively*

$$g_n(x) = \sup_{t \in S_n} \mathsf{E}_x g_{n-1}(X(t)),$$

where $S_n := \{k2^{-n} | 0 \leq k \leq 4^n\}$, $n \in \mathbb{N}$. Then $g_n \uparrow \widehat{g_0}$, where $\widehat{g_0}$ is the least superharmonic majorant for the function g_0.

Now, let a premium function $g \in C^2(\mathbb{R}^n)$, $\mathscr{L}$ be the generator of a homogeneous diffusion process X satisfying an SDE $dX(t) = b(X(t))dt + \sigma(X(t))dW(t)$. That is,

$$\mathscr{L} f(x) = \sum_{i=1}^{n} \frac{\partial f}{\partial x_i} b_i(x) + \frac{1}{2} \sum_{i,j=1}^{n} \frac{\partial^2 f}{\partial x_i \partial x_j} \sigma_{ij}(x).$$

Introduce a set

$$U = \{x \in \mathbb{R}^n | \mathscr{L} g(x) > 0\}. \tag{15.1}$$

Then $U \subset \Gamma^c$.

Bibliography

[19], Chapter III; [67]; [23], Chapter 6; [81]; [54], Chapter 2; [61], Chapter X; [85].

Problems

15.1. Prove Lemma 15.1.

15.2. Prove: if a function f is excessive and $\tau \geq \sigma$ are two Markov moments, then

$$\mathsf{E}_x f(X_\sigma) \geq \mathsf{E}_x f(X_\tau).$$

15.3. Let a function f be excessive, and τ_Γ be a time of the first entry into a set $\Gamma \subset \mathbb{X}$ by a Markov chain $\{X_n, n \in \mathbb{Z}^+\}$. Prove that the function $h(x) := \mathsf{E}_x f(X_{\tau_\Gamma})$ is excessive as well.

15.4. Prove: if a payoff function f is excessive, then the price of the game v equals f.

15.5. Prove: if an excessive function g dominates a payoff function f then it also dominates the price of the game v.

15.6. Prove Lemma 15.2.

15.7. Prove the following properties of excessive functions defined on $\mathbb{X}$.

(1) If a function f is excessive and $\alpha > 0$, then the function αf is excessive as well.

(2) If functions f_1, f_2 are excessive, then the sum $f_1 + f_2$ is excessive as well.

(3) If $\{f_\alpha,\ \alpha \in \mathfrak{A}\}$ is a family of excessive functions, then the function $f(x) := \inf_{\alpha\in\mathfrak{A}} f_\alpha(x)$ is excessive as well.

(4) If $\{f_n, n \geq 1\}$ are excessive functions and $f_n \uparrow f$ pointwise, then f is an excessive function as well.

15.8. Let a Markov chain $\{X_n,\ n \in \mathbb{Z}^+\}$ be a symmetric random walk with absorption points 0 and N. That is, $\mathbb{X} = \{0, 1, \ldots, N\}$, $p_{xy} = \frac{1}{2}$ if $x = 1, 2, \ldots, N-1$, $y = x \pm 1$, $p_{NN} = 1$, and $p_{00} = 1$. Prove: the class of functions $f : \mathbb{X} \to \mathbb{R}^+$, which are excessive w.r.t. this random walk, coincides with the class of concave functions.

15.9. Let a Markov chain $\{X_n,\ n \in \mathbb{Z}^+\}$ be the same as in Problem 15.8.

(1) Prove that the price of the game $v(x)$ is the least concave function for which $v(x) \geq f(x)$, $x \in \mathbb{X} = \{0, 1, \ldots, N\}$.

(2) Prove that to stop at time τ_0, when the chain first enters any of the points x with $f(x) = v(x)$, is an optimal strategy.

15.10. Let a Markov chain $\{X_n,\ n \in \mathbb{Z}^+\}$ be a random walk over the set $\mathbb{X} = \{0, 1, \ldots, N\}$ and $p_{x(x+1)} = p$, $p_{x(x-1)} = q = 1 - p \neq p$ if $x = 1, 2, \ldots, N-1$ and $p_{NN} = 1$, $p_{00} = 1$ (nonsymmetric random walk with absorption at points 0 and N).

(1) Describe the class of functions $f : \mathbb{X} \to \mathbb{R}^+$ which are excessive w.r.t. the random walk.

(2) Let a premium function $f(x) = x$. Find an optimal strategy and calculate the price of the game if: (a) $q > p$; (b) $q < p$.

15.11. Let a Markov chain $\{X_n,\ n \in \mathbb{Z}^+\}$ be a random walk over a set $\mathbb{X} = \{0, 1, \ldots, N\}$ and $p_{x(x+1)} = p$, $p_{x(x-1)} = q = 1 - p$ if $x = 1, 2, \ldots, N-1$ and $p_{N(N-1)} = 1$, $p_{01} = 1$ (nonsymmetric random walk with reflection at points 0 and N). Describe the class of functions $f : \mathbb{X} \to \mathbb{R}^+$ that are excessive w.r.t. the random walk.

15.12. (Optimal stopping for a Wiener process with absorption) Consider a Wiener process $\{W(t), t \in \mathbb{R}^+\}$ with $W(0) = x \in [0, a]$. Furthermore, if the process visits point 0 or a, then it stays there forever; that is, $\mathbb{X} = [0, a]$ (such a process is called a Wiener process with absorption at the points 0 and a). Let a function $f \in C([0, a])$ and is nonnegative. Find the price of the game $v(x) = \sup_\tau \mathsf{E}_x f(W(\tau))$ and construct a Markov moment τ_0, for which $v(x) = \mathsf{E}_x f(W(\tau_0))$.

15.13. (Optimal stopping for a two-dimensional Wiener process) Let $W(t) = \{(W_1(t), W_2(t)), t \in \mathbb{R}^+\}$ be a Wiener process in $\mathbb{R}^2$.

(1) Prove that only constant functions are superharmonic nonnegative functions with regard to W.

(2) Prove that there is no optimal stopping for an unbounded function f.

(3) Prove that the continuation set is $\Gamma^c = \{x |\ f(x) < ||f||_\infty\}$, where $||f||_\infty = \sup_{x\in\mathbb{R}^2} |f(x)|$.

(4) Prove that if the logarithmic capacity $cap\,(\partial\Gamma^c)=0$, then $\tau_\Gamma=\infty$ a.s. (The logarithmic capacity of a planar compact set is the value $\gamma(E)=\exp\{-V(E)\}$, where $V(E)=\inf_{\mathsf{P}}\int_{E\times E}\ln|u-v|^{-1}d\mathsf{P}(u)d\mathsf{P}(v)$ and the infimum is taken over all probabilistic measures on E. The value $V(E)$ is called the Robbins constant for the set E and the set E is called polar if $V(E)=+\infty$ or, the equivalent, if $\gamma(E)=0$.)

(5) Prove that if $cap\,(\partial\Gamma^c)>0$ then $\tau_\Gamma<\infty$ a.s. and $v_f(x)=||f||_\infty=\mathsf{E}_x f(W_{\tau_\Gamma})$. It means that τ_Γ is the optimal stopping.

15.14. Let us suppose that there exists a Borel set H such that $g_H(x):=\mathsf{E}_x g(X(\tau_H))$ dominates a function g and, at the same time, $g_H(x)\ge\mathsf{E}_x g(X(\tau))$ for all stopping times τ and all $x\in\mathbb{R}^n$. Prove that in this case $v_g(x)=g_H(x)$; that is, τ_H is an optimal stopping.

15.15. (Optimal stopping for an n-dimensional Wiener process if $n\ge 3$) Let $W(t)=\{(W_1(t),\dots,W_n(t)),t\in\mathbb{R}^+\}$ be a Wiener process in $\mathbb{R}^n$ with $n\ge 3$.

(1) Let a premium function be

$$g(x)=\begin{cases}||x||^{-1}, & \text{if } ||x||\ge 1,\\ 1, & \text{if} ||x||<1,\end{cases}$$

$x\in\mathbb{R}^3$. Prove that this function is superharmonic in $\mathbb{R}^3$. Furthermore, it is such that $v_g=g$, and it is optimal to stop immediately regardless on an initial point.

(2) Let a premium function be

$$h(x)=\begin{cases}||x||^{-\alpha}, & \text{if} ||x||\ge 1,\\ 1, & \text{if} ||x||<1,\end{cases}$$

$\alpha>1$, a set $H=\{x\in\mathbb{R}^n|\ ||x||\le 1\}$, and a function $\widetilde{h}(x)=\mathsf{E}_x h(W(\tau_H))$ (remember that τ_H is the moment of the first hit on a set H).

(a) Show that $\widetilde{h}(x)=\mathsf{P}_x(\tau_H<\infty)$.

(b) Show that

$$\widetilde{h}(x)=\begin{cases}1, & \text{if } ||x||<1,\\ ||x||^{-1}, & \text{if } ||x||\ge 1,\end{cases}$$

It means that the function $\widetilde{h}$ coincides with the function g from item 1), which is the superharmonic majorant for h.

(c) Prove that $v_h=\widetilde{h}=g$ and the moment τ_H is an optimal stopping.

15.16. Let $\{W(t),t\in\mathbb{R}^+\}$ be a one-dimensional Wiener process. Find v_g and an optimal stopping τ_0 where it exists, if:

(a) $v_g(x)=\sup_\tau\mathsf{E}_x|W(t)|^p$ with $p>0$.

(b) $v_g(x)=\sup_\tau\mathsf{E}_x e^{-W^2(\tau)}$.

(c) $v_g(s,x)=\sup_\tau\mathsf{E}_{s,x}(e^{-\rho\tau}\operatorname{ch}W(\tau))$, where $\rho>0$ and $\operatorname{ch}z:=(e^z+e^{-z})/2$. (Here the expectation is taken under the condition $W(s)=x$.)

15.17. Prove that $U\subset\Gamma$ (see formula (15.1)).

15.18. (Optimal stopping for a time-dependent premium function) Let a premium function g be of the following form: $g = g(t,x) : \mathbb{R} \times \mathbb{R}^n \to \mathbb{R}^+$, $g \in C(\mathbb{R} \times \mathbb{R}^n)$. Find $g_0(x)$ and τ_0 such that

$$g_0(x) = \sup_{\tau} \mathsf{E}_x g(\tau, X(\tau)) = \mathsf{E}_x g(\tau_0, X(\tau_0)),$$

where X is a diffusion process with $dX(t) = b(X(t))dt + \sigma(X(t))dW(t)$, $t \in \mathbb{R}_+$ and $X(0) = x$, $b : \mathbb{R}^n \to \mathbb{R}^n$, $\sigma : \mathbb{R}^n \to \mathbb{R}^{n \times m}$ are measurable functions, W is an m-dimensional Wiener process.

15.19. Let $X(t) = W(t)$, $t \geq 0$ be a one-dimensional Wiener process and a premium function $g(t,x) = e^{-\alpha t + \beta x}$, $x \in \mathbb{R}$, where $\alpha, \beta \geq 0$ are some constants.

(1) Prove that the generator $\widetilde{\mathscr{L}}$ of the process

$$Y_{(s,x)}(t) = \begin{pmatrix} s+t \\ W(t)+x \end{pmatrix}$$

is given by

$$\widetilde{\mathscr{L}} f(s,x) = \frac{\partial f}{\partial s} + \frac{1}{2}\frac{\partial^2 f}{\partial x^2}, \; f \in C^2(\mathbb{R}).$$

(2) Deduce that $\widetilde{\mathscr{L}} g = (-\alpha + \frac{1}{2}\beta^2)g$ and the identity $v_g = g$ holds true when $\beta^2 \leq 2\alpha$ and the optimal strategy is the instant stopping; if $\beta^2 > 2\alpha$ then optimal moment τ_0 does not exist and $v_g = +\infty$.

15.20. Let $\{Y_n, \mathfrak{F}_n, 0 \leq n \leq N\}$ be the Snell envelope for a nonnegative process $\{X_n, \mathfrak{F}_n, 0 \leq n \leq N\}$ (see Problem 7.22). Let $\mathscr{T}_{n,N}$ be the set of stopping times taking values in the set $\{n, n+1, \dots, N\}$.

(1) Prove that the r.v.

$$\tau_0 := \inf\{0 \leq n \leq N \,|\, Y_n = X_n\}$$

is a stopping time and the stopped stochastic process $\{Y_n^{\tau_0} = Y_{n \wedge \tau_0}, \mathfrak{F}_n, 0 \leq n \leq N\}$ is a martingale.

(2) Prove that $Y_0 = \mathsf{E}(X_{\tau_0}/\mathfrak{F}_0) = \sup_{\tau \in \mathscr{T}_{0,N}} \mathsf{E}(X_\tau/\mathfrak{F}_0)$ (i.e., in this sense τ_0 is an optimal stopping).

(3) Generalizing the statements (1) and (2), prove that the r.v.

$$\tau_n := \inf\{n \leq j \leq N \,|\, Y_j = X_j\}$$

is a stopping time and

$$Y_n = \sup_{\tau \in \mathscr{T}_{n,N}} \mathsf{E}(X_\tau/\mathfrak{F}_n) = \mathsf{E}(X_{\tau_n}/\mathfrak{F}_n).$$

15.21. Prove that a stopping τ is optimal if and only if $Y_\tau = X_\tau$ and the stopped process $\{Y_n^\tau, \mathfrak{F}_n, 0 \leq n \leq N\}$ is a martingale. This statement means that τ_0 from Problem 15.20 is the least optimal stopping time.

15.22. Let $Y_n = M_n - A_n$ be the Doob–Meyer decomposition for a Snell envelope (see Problem 7.62). Put $\tau_1 = \inf\{0 \le n \le N | A_{n+1} \ne 0\} \wedge N$. Prove that τ_1 is an optimal stopping and $\tau_1 \ge \tau$ for any optimal stopping τ. It means that τ_1 is the largest optimal stopping.

15.23. Let a sequence $\{X_n, 0 \le n \le N\}$ be a homogeneous Markov chain taking values in a finite set $\mathbb{X}$ with transition matrix P. Also let a function $\varphi = \varphi(n,x) : \{0,1,\dots,N\} \times \mathbb{X} \to \mathbb{R}$ be measurable. Prove that the Snell envelope for the sequence $Z_n := \psi(n,X_n)$ is determined by the formula $Y_n = y(n,Z_n)$ where the function y is given by relations: $y(N,x) = \psi(N,x)$ for any $x \in \mathbb{X}$ and for $0 \le n \le N-1$ $u(n,\cdot) = \max(\psi(n,\cdot), Pu(n+1,\cdot))$.

15.24. Let $\{Y_n, 0 \le n \le N\}$ be a Snell envelope for a sequence $\{X_n, 0 \le n \le N\}$. Prove that for every n $\mathsf{E}Y_n = \sup_{\tau \in \mathcal{T}_{n,N}} \mathsf{E}X_\tau$. In particular, $\mathsf{E}Y_0 = \sup_{\tau \in \mathcal{T}_{0,N}} \mathsf{E}X_\tau$.

15.25. Prove that τ_0 is optimal if and only if $\mathsf{E}X_{\tau_0} = \sup_{\tau \in \mathcal{T}_{0,N}} \mathsf{E}X_\tau$.

Hints

15.2. Write formula (15.3) (see solution to Problem 15.1) for σ and τ. Deduce that $\mathsf{E}_x \alpha^\sigma f(X_\sigma) \ge \mathsf{E}_x \alpha^\tau f(X_\tau)$. Next, let $\alpha \to 1$.

15.4. Derive from the definition of $v(x)$ that $v(x) \ge f(x)$ and from Problem 15.20 (or Lemma 15.1) that $f(x) \ge v(x)$.

15.7. Use the definition of excessive function or Lemma 15.1.

15.8. Use the definition of excessive function.

15.9. Use Theorem 15.1 and Problem 15.8.

15.12. *I method.* At first, one can prove that for the Wiener process with absorption, the class of functions $g : [0,a] \to \mathbb{R}^+$ satisfying the condition $g(x) \ge \mathsf{E}_x g(W(\tau))$ for all Markov moments τ coincides with the class of all nonnegative concave functions. The proof of the statement that any concave function satisfies this inequality is rather technically complicated (see, e.g., [19]). To prove the concavity of the function g satisfying the inequality $g(x) \ge \mathsf{E}_x g(W(\tau))$, establish the inequality

$$\mathsf{E}_x g(W(\tau)) = g(x_1)\frac{x_2 - x}{x_2 - x_1} + g(x_2)\frac{x - x_1}{x_2 - x_1},$$

where τ is the moment of the first exit from the interval $[x_1, x_2] \in [0,a]$. In this order check that the probability $\mathsf{P}(x, x_1, x_2)$ to start from a point x and to get to x_1 earlier than to x_2 for $0 \le x_1 \le x_2 \le a$ is equal to $(x_2 - x)/(x_2 - x_1)$. (See Problem 14.31.) Next, prove that the price of the game is the least concave dominant of the function f. Consider a strategy τ that consists in waiting till the moment when process W visits point x_1 or point x_2 and then the strategies τ_1 or τ_2, respectively, are used. Here

τ_i, $i = 1,2$ are strategies leading under initial states x_1 and x_2 to the average payoff which is more than $v(x_1) - \varepsilon$ or $v(x_2) - \varepsilon$. (The existence of such strategies follows from the definition of supremum.) Derive that

$$\mathsf{E}_x f(W(\tau)) \geq \frac{x_2 - x}{x_2 - x_1} v(x_1) + \frac{x - x_1}{x_2 - x_1} v(x_2) - \varepsilon,$$

and that

$$v(x) \geq \frac{x_2 - x}{x_2 - x_1} v(x_1) + \frac{x - x_1}{x_2 - x_1} v(x_2).$$

Thus, the function v is concave. It means that the price of the game is the least nonnegative, concave dominant for the function f. Indeed, v obviously dominates f and is concave. Moreover, it follows from the above considerations that for any other concave dominant z of f we have that $z(x) \geq \mathsf{E}_x z(W(\tau)) \geq \mathsf{E}_x f(W(\tau)) = v(x)$. And finally, prove that for $f \in C([0,a])$ the optimal Markov moment equals $\tau_0 = \inf\{t | \, W(t) \in \Gamma\}$, where $\Gamma = \{x \in [0,a] | \, f(x) = v(x)\}$.

II method. Use the fact that a Wiener process is a diffusion process. Take into account the absorption at 0 and a.

15.13. (1) Suppose that f is a nonnegative superharmonic function with regard to W and there exist two points $x, y \in \mathbb{R}^2$ with $f(x) < f(y)$. Consider $\mathsf{E}_x f(W(\tau))$, where τ is the time of the first visit by the Markov process W_t small disk with center at y. Use a multidimensional version of the Dynkin formula (Problem 13.46) and Problem 13.55.

(2) Follows directly from item (1).

(3) Follows from item (1) and the definition of continuation set.

(4) and (5) follow from item (3). (See also [63] and [43].) Indeed, according to [43] and [63], if a set B is compact and $\tau_B = \inf\{t > 0 | \, W(t) \in B\}$ then $\mathsf{P}(\tau_B < \infty)$ equals 0 or 1 depending on whether the set B has zero or positive logarithmic capacity; in our case $B = \mathbb{R}^2 \backslash \mathfrak{D}$ is compact.

15.15. (1) Follows obviously from the definitions.

(2)(b) Use Problems 13.54 and 13.55.

(c) Use item (b) and Problem 15.14.

15.19. The first statement is evident. For the second one consider only a case where $\beta^2 > 2\alpha$. First, the set $U := \{(s,x) | \widetilde{\mathscr{L}} g(s,x) > 0\} = \mathbb{R}^2$ in this case, thus, $\Gamma^c = \mathbb{R}^2$. It means that τ_0 does not exist. Second, use Theorem 15.3 in order to construct the least superharmonic majorant:

$$\begin{aligned} E s_{s,x} g(Y(t)) &= \sup\nolimits_{t \in S_n} \mathsf{E} e^{-\alpha(s+t) + \beta W^x(t)} = \sup\nolimits_{t \in S_n} e^{-\alpha(s+t)} \cdot e^{\beta x + (1/2)\beta^2 t} \\ &= \sup\nolimits_{t \in S_n} g(s,x) e^{(-\alpha + (1/2)\beta^2)t} = g(s,x) \exp((-\alpha + \tfrac{1}{2}\beta^2) 2^n), \end{aligned}$$

therefore, $g_n(s,x) \to \infty$ as $n \to \infty$.

15.23. Use the definition of a Snell envelope and the fact that for a bounded and measurable function $f : \mathbb{X} \to \mathbb{R}$ and a homogeneous Markov chain $\{Z_n, 0 \leq n \leq N\}$ it holds

$$\mathsf{E}(f(Z_{n+1})/\mathfrak{F}_n) = Pf(Z_n),$$

where $Pf(x) = \sum_{y\in\mathbb{X}} p_{xy} f(y)$, $\{p_{xy}\}_{x,y\in\mathbb{X}}$ are entries of the transition matrix P.

15.24. Use Problem 15.20.

15.25. Let $\mathsf{E}(X_{\tau_0}/\mathfrak{F}_0) = \sup_{\tau\in\mathscr{T}_{0,N}} \mathsf{E}(X_\tau/\mathfrak{F}_0)$. Calculate the mathematical expectation for both parts and prove that $\mathsf{E}X_{\tau_0} = \sup_{\tau\in\mathscr{T}_{0,N}} \mathsf{E}X_\tau$. On the contrary, let $\mathsf{E}X_{\tau_0} = \sup_{\tau\in\mathscr{T}_{0,N}} \mathsf{E}X_\tau$. Prove that in this case $Y_\tau = X_\tau$ and Y^τ are martingales.

Answers and Solutions

15.1. Put $\varphi(x) = f(x) - \alpha Pf(x)$, $0<\alpha<1$. Then $f(x) = (\varphi + \alpha P\varphi + \cdots + \alpha^n P^n\varphi + \alpha^{n+1}P^{n+1}f)(x)$ and $\varphi(x)\ge 0$, $x\in E$. Furthermore, $0 \le P^n f = P^{n-1}(Pf) \le P^{n-1}f$ which means that $\alpha^n P^n f \to 0$ as $n\to\infty$. This implies that $f(x) = \sum_{n=0}^{\infty} \alpha^n P^n \varphi(x)$ where $P^0 = I$ is the identity operator. Check that $P^n\varphi(x) = \mathsf{E}_x\varphi(X_n)$. Then

$$f(x) = \mathsf{E}_x\left(\sum_{n=0}^{\infty} \alpha^n \varphi(X_n)\right). \tag{15.2}$$

And again, similarly to (15.2), prove that

$$\mathsf{E}_x \alpha^\tau f(X_\tau) = \mathsf{E}_x(\alpha^\tau\varphi(X_\tau) + \alpha^{\tau+1}\varphi(X_{\tau+1}) + \cdots). \tag{15.3}$$

Comparing (15.2) with (15.3) we can conclude that

$$f(x) \ge \mathsf{E}_x \alpha^\tau f(X_\tau).$$

Now, let α to 1.

15.3. Let $\tau'_\Gamma = \inf\{n\ge 1 \mid X_n\in\Gamma\}$. Then, $\tau'_\Gamma \ge \tau_\Gamma$. It follows from Problem 15.2 that $\mathsf{E}_x f(X_{\tau'_\Gamma}) \le \mathsf{E}_x f(X_{\tau_\Gamma}) = h(x)$. But, if the first step leads the Markov chain from x to y then $\mathsf{E}_x f(X_{\tau'_\Gamma}) = \mathsf{E}_y f(X_{\tau_\Gamma}) = h(y)$. So, $\mathsf{E}_x f(X_{\tau'_\Gamma}) = \sum_{y\in\mathbb{X}} p_{xy} h(y) = Ph(x)$. Thus, $Ph(x) \le h(x)$, $x\in\mathbb{X}$.

15.5. If $g\ge f$ and g is an excessive function, then for any strategy τ

$$\mathsf{E}_x f(X_\tau) \le \mathsf{E}_x g(X_\tau) \le g(x),$$

which implies that $v(x) = \sup_\tau \mathsf{E}_x f(X_\tau) \le g(x)$.

15.6. Because for $\tau=\infty$ the relation $\mathsf{E}_x f(X_\infty) = 0$ holds true, then $v(x)\ge 0$. Fix $\varepsilon>0$. It follows from the definition of supremum that for every $y\in\mathbb{X}$ there exists a strategy $\tau_{\varepsilon,y}$ such that

$$\mathsf{E}_y f(X_{\tau_{\varepsilon,y}}) \ge v(y) - \varepsilon.$$

Now let the strategy τ consist in making one step and then, if this step leads to the point y, we continue with strategy $\tau_{\varepsilon,y}$. It is evident that τ is a Markov moment (check this), and for this τ

$$\mathsf{E}_x f(X_\tau) = \sum_{y\in\mathbb{X}} p_{xy}\mathsf{E}_y f(X_{\tau_{\varepsilon,y}}) \geq \sum_{y\in\mathbb{X}} p_{xy}(v(y)-\varepsilon) \geq Pv(x)-\varepsilon.$$

Now pass to the limit as $\varepsilon \to 0$.

15.10. (1) Put $y_k = (q/p)^k$, $k = 0,\dots,N$, $Y = \{y_0,\dots,y_N\}$. Put a function $f : \{0,\dots,N\} \to \mathbb{R}^+$. It is excessive if and only if the function $g : Y \to \mathbb{R}^+$ determined by the identities $g(y_k) = f(k), k \in \{0,\dots,N\}$, is concave.

(2) For $f(x) = x$ the function g defined above is equal to $g(y) = \log_{q/p} y$. It is concave if $q > p$ and convex if $q < p$. For the first case a concave majorant for g is this function itself, and for the second case such majorant is a linear function w with $w(y_0) = g(y_0) = 0, w(y_N) = g(y_N) = N$; that is, $w(y) = ((y-y_0)/(y_N-y_0))N$. So, (a) an optimal strategy consists in the immediate stopping, and $v(x) = x$.

(b) An optimal strategy consists in the stopping at the first of one of the points 0 or N, and

$$v(x) = w\left(\left(\frac{q}{p}\right)^x\right) = \frac{\left(\frac{q}{p}\right)^x - 1}{\left(\frac{q}{p}\right)^N - 1}N.$$

15.11. Prove that every excessive function f is constant. For any x consider the moment τ_x of the first entry into the state x. Because all states are connected with each other and the total number of states is finite, then for any y it holds $\mathsf{P}_y(\tau_x < +\infty) = 1$, and thus, $\mathsf{E}_y f(\tau_x) = f(x)$. Considering f as a payoff function, we obtain that the corresponding price of the game $v(y) \geq f(x)$. Because f is excessive, then $v = f$, and it means that $f(y) \geq f(x)$. Changing the roles for the points x and y we obtain the opposite inequality, $f(x) \geq f(y)$. So, $f(x) = f(y)$.

15.14. It is evident that if $\overline{g}(x)$ is the least excessive majorant for the function g, then $\overline{g}(x) \leq g_H(x)$. On the other hand, $g_H(x) \leq \sup_\tau \mathsf{E}_x g(X(\tau)) = v_g(x)$, and it follows from Theorems 15.2 (1) and (15.3) that $v_g = g_H$.

15.16. (a) $v_g(x) = +\infty$, and τ_0 does not exist.

(b) $v_g(x) = 1$, and $\tau_0 = \inf\{t > 0 | W(t) = 0\}$.

(c) if $\rho < \frac{1}{2}$, then $v_g(s,x) = +\infty$, and τ_0 does not exist; if $\rho \geq \frac{1}{2}$, then $v_g(s,x) = g(s,x)$.

15.17. Let a point $x \in V \subset U$ and τ_0 be the time of the first exit from a bounded open set V. According to the Dynkin formula (see Problem 13.46, the multidimensional version) for any $v > 0$

$$\mathsf{E}_x g(X(\tau_0 \wedge v)) = g(x) + \mathsf{E}_x \int_0^{\tau_0\wedge v} \mathcal{L}g(X_s)ds > g(x).$$

It means that $g(x) < v_g(x)$; that is, $x \in \Gamma^c$.

15.18. Reduce the nonhomogeneous in time problem to a homogeneous one, the solution of which is determined by Theorem 15.2, as follows. Define a new diffusion process

$$Y(t) = Y_{s,x}(t) \text{ as } Y(t) = \begin{pmatrix} s+t \\ X_x(t) \end{pmatrix}, \ t \in \mathbb{R}^+,$$

where $X_x(t)$ is a diffusion starting from the point x, and $s \in \mathbb{R}^+$. Then

$$dY(t) = \begin{pmatrix} 1 \\ b(X(t)) \end{pmatrix} dt + \begin{pmatrix} 0 \\ \sigma(X(t)) \end{pmatrix} dW(t) = \tilde{b}(Y(t))dt + \tilde{\sigma}(Y(t))dW(t),$$

with

$$\tilde{b}(y) = \tilde{b}(t,x) = \begin{pmatrix} 1 \\ b(x) \end{pmatrix} \in \mathbb{R}^{n+1}, \tilde{\sigma}(y) = \tilde{\sigma}(t,x) = \begin{pmatrix} 0 \dots 0 \\ \dots\dots \\ \sigma(x) \end{pmatrix} \in \mathbb{R}^{(n+1)\times m},$$

where $y = (t,x) \in \mathbb{R} \times \mathbb{R}^n$. We see that Y is the diffusion process starting from the point $y = (s,x)$. Let $P_y = P_{s,x}$ be the distribution of Y and $\mathsf{E}_y = \mathsf{E}_{(s,x)}$ mean the expectation with respect to the measure P_y. The problem can be written in terms of $Y(t)$ as follows: to find g_0 and τ_0 such that

$$g_0(x) = v_g(0,x) = \sup_\tau \mathsf{E}_{0,x} g(Y(\tau)) = \mathsf{E}_{0,x} g(Y(\tau_0))$$

which is a particular case for the problem of finding $v_g(s,x)$ and τ with

$$v_g(s,x) = \sup_\tau \mathsf{E}_{s,x} g(Y(\tau)) = \mathsf{E}_{s,x} g(Y(\tau_0)).$$

This problem is standard if we replace $X(t)$ with $Y(t)$.

15.20. (1) Because $Y_N = X_N$, τ_0 is correctly defined. Besides, the events $\{\tau_0 = 0\} = \{Y_0 = X_0\} \in \mathcal{F}_0$, $\{\tau_0 = n\} = \{Y_0 > X_0\} \cap \cdots \cap \{Y_{n-1} > X_{n-1}\} \cap \{Y_n = X_n\} \in \mathcal{F}_n$, $1 \leq n \leq N$; that is, τ_0 is a stopping time. Furthermore, $Y_{(n+1)\wedge\tau_0} - Y_{n\wedge\tau_0} = (Y_{n+1} - Y_n)\mathbb{1}_{\tau_0 > n}$. Because $Y_n > X_n$ on the set $\{\tau_0 > n\}$ then $Y_n = \mathsf{E}(Y_{n+1}/\mathcal{F}_n)$. So, $\mathsf{E}(Y_{n+1}^{\tau_0} - Y_n^{\tau_0}/\mathcal{F}_n) = \mathsf{E}(Y_{(n+1)\wedge\tau_0} - Y_{n\wedge\tau_0}/\mathcal{F}_n) = \mathbb{1}_{\tau_0 > n}\mathsf{E}(Y_{n+1} - \mathsf{E}(Y_{n+1}/\mathcal{F}_n)/\mathcal{F}_n) = 0$. This implies that $\{Y_n^{\tau_0}, \mathcal{F}_n, 0 \leq n \leq N\}$ is a martingale.

(2) Because Y^{τ_0} is a martingale then $Y_0 = Y_0^{\tau_0} = \mathsf{E}(Y_N^{\tau_0}/\mathcal{F}_0) = \mathsf{E}(Y_{\tau_0}/\mathcal{F}_0) = \mathsf{E}(X_{\tau_0}/\mathcal{F}_0)$. From the other side, if $\tau \in \mathcal{T}_{0,N}$, then the stopped process Y^τ is a supermartingale. So, $Y_0^\tau = Y_0 \geq \mathsf{E}(Y_N^\tau/\mathcal{F}_0) = \mathsf{E}(Y_\tau/\mathcal{F}_0) \geq \mathsf{E}(X_\tau/\mathcal{F}_0)$. Thus, $\mathsf{E}(X_{\tau_0}/\mathcal{F}_0) = \sup_{\tau \in \mathcal{T}_{0,N}} \mathsf{E}(X_\tau/\mathcal{F}_0)$.

The proof of item (3) is similar to the proofs of (1) and (2), if we replace 0 with n.

15.21. Let the stopped process Y^τ be a martingale. Prove, taking into account the first condition, that $Y_0 = \mathsf{E}(X_\tau/\mathcal{F}_0)$, and deduce the optimality of τ from item 2) of Problem 15.20. And vice versa, let a stopping τ be optimal. Prove the following sequence of inequalities using Problem 15.20 (2).

$$Y_0 = \mathsf{E}(X_\tau/\mathcal{F}_0) \leq \mathsf{E}(Y_\tau/\mathcal{F}_0) \leq Y_0.$$

Based on this and the inequality $X_\tau \leq Y_\tau$ derive that $X_\tau = Y_\tau$. Then prove the inequalities $\mathsf{E}(Y_\tau/\mathcal{F}_0) = Y_0 \geq \mathsf{E}(Y_{\tau\wedge n}/\mathcal{F}_0) \geq \mathsf{E}(Y_\tau/\mathcal{F}_0)$ and deduce that $\mathsf{E}(Y_{\tau\wedge n}/\mathcal{F}_0) =$

$\mathsf{E}(Y_\tau/\mathfrak{F}_0) = \mathsf{E}(\mathsf{E}(Y_\tau/\mathfrak{F}_n)/\mathfrak{F}_0)$. Because $Y_{\tau\wedge n} \geq \mathsf{E}(Y_\tau/\mathfrak{F}_n)$ then $Y_{\tau\wedge n} = \mathsf{E}(Y_\tau/\mathfrak{F}_n)$ and this means that $\{Y_n^\tau, \mathfrak{F}_n, 0 \leq n \leq N\}$ is a martingale.

15.22. Because the process $\{A_n, 1 \leq n \leq N\}$ is predictable, then τ_1 is a stopping time. It is evident that $Y^{\tau_1} = M^{\tau_1}$ as $A_n^{\tau_1} = A_{n\wedge\tau_1} \equiv 0$. It means that Y^{τ_1} is a martingale. According to Problem 15.21, we need to prove that $Y_{\tau_1} = X_{\tau_1}$. For ω such that $\tau_1 = N$ the identities $Y_{\tau_1} = Y_N = X_N = X_{\tau_1}$ hold true. If $\{\tau_1 = n\}$, $0 \leq n < N$, then the following identities take place: $Y_{\tau_1}\mathbb{I}_{\tau_1=n} = Y_n\mathbb{I}_{\tau_1=n} = M_n\mathbb{I}_{\tau_1=n} = \mathsf{E}(M_{n+1}/\mathfrak{F}_n)\mathbb{I}_{\tau_1=n} = \mathsf{E}(Y_{n+1} + A_{n+1}/\mathfrak{F}_n)\mathbb{I}_{\tau=n} > \mathsf{E}(Y_{n+1}/\mathfrak{F}_n)\mathbb{I}_{\tau=n}$; that is, $Y_{\tau_1}\mathbb{I}_{\tau_1=n} = Y_n\mathbb{I}_{\tau_1=n} = X_n\mathbb{I}_{\tau_1=n} = X_{\tau_1}\mathbb{I}_{\tau_1=n}$. Thus, τ_1 is an optimal stopping, $\tau \geq \tau_1$, and $\mathsf{P}(\tau > \tau_1) > 0$. So, $\mathsf{E}A_\tau > 0$ and $\mathsf{E}Y_\tau = \mathsf{E}M_\tau - \mathsf{E}A_\tau = \mathsf{E}M_0 - \mathsf{E}A_\tau = \mathsf{E}Y_0 - \mathsf{E}A_\tau < \mathsf{E}Y_0$, and the stopped process M^τ is a martingale (check why it is true), and thus, τ is not an optimal stopping.

16

Measures in a functional spaces. Weak convergence, probability metrics. Functional limit theorems

Theoretical grounds

In this chapter, we consider random elements taking values in metric spaces and their distributions. The definition of a random element taking values in $\mathbb{X}$ involves the predefined σ-algebra $\mathcal{X}$ of subsets of $\mathbb{X}$. The following statement shows that in a *separable* metric space, in fact, the unique natural choice for the σ-algebra $\mathcal{X}$ is the Borel σ-algebra $\mathcal{B}(\mathbb{X})$.

Lemma 16.1. *Let $\mathbb{X}$ be a separable metric space, and $\mathcal{X}$ be a σ-algebra of subsets of $\mathbb{X}$ that contains all open balls. Then $\mathcal{X}$ contains every Borel subset of $\mathbb{X}$.*

Further on, while dealing with random elements taking values in a separable metric space $\mathbb{X}$, we assume $\mathcal{X} = \mathcal{B}(\mathbb{X})$. For a nonseparable space, a σ-algebra $\mathcal{X}$ is given explicitly.

Theorem 16.1. *(The Ulam theorem) Let $\mathbb{X}$ be a Polish space and μ be a finite measure on the Borel σ-algebra $\mathcal{B}(\mathbb{X})$. Then for every $\varepsilon > 0$ there exists a compact set $K_\varepsilon \subset \mathbb{X}$ such that $\mu(\mathbb{X}\backslash K_\varepsilon) < \varepsilon$.*

Generally, we deal with *functional spaces* $\mathbb{X}$; that is, spaces of the functions or the sequences with a given parametric set. Let us give an (incomplete) list of such spaces with the corresponding metrics. In all the cases mentioned below we consider $x, y \in \mathbb{X}$.

(1) $\mathbb{X} = C([0,T])$, $\rho(x,y) = \max_{t\in[0,T]} |x(t) - y(t)|$.
(2) $\mathbb{X} = C([0,+\infty))$, $\rho(x,y) = \sum_{k=1}^{\infty} 2^{-k} \left[\max_{t\in[0,k]} |x(t) - y(t)| \wedge 1\right]$.
(3) $\mathbb{X} = L_p([0,T]), p \in [1,+\infty)$, $\rho(x,y) = \left[\int_0^T |x(t) - y(t)|^p \, dt\right]^{1/p}$.
(4) $\mathbb{X} = L_\infty([0,T])$, $\rho(x,y) = \text{ess sup}_{t\in[0,T]} |x(t) - y(t)|$.
(5) $\mathbb{X} = \ell_p, p \in [1,+\infty)$, $\rho(x,y) = [\sum_{k=1}^{\infty} |x_k - y_k|^p]^{1/p}$.
(6) $\mathbb{X} = \ell_\infty$, $\rho(x,y) = \sup_{k\in\mathbb{N}} |x_k - y_k|$.
(7) $\mathbb{X} = c_0 = \{(x_k)_{k\in\mathbb{N}} : \exists \lim_{k\to\infty} x_k = 0\}$, $\rho(x,y) = \sup_{k\in\mathbb{N}} |x_k - y_k|$.

D. Gusak et al., *Theory of Stochastic Processes*, Problem Books in Mathematics,
DOI 10.1007/978-0-387-87862-1_16,

(8) $\mathbb{X} = \mathbb{D}([a,b])$ (Skorokhod space, see Chapter 3, Remark 3.2), the metrics in this space are given below.

Definition 16.1. *Let $X = \{X(t), t \in \mathbb{T}\}$ be a real-valued random process with $\mathbb{T} = [0,T]$ or $\mathbb{T} = [0,+\infty)$. If there exists a modification $\widetilde{X}$ of this process with all its trajectories belonging to some functional space $\mathbb{X}$ (e.g., one of the spaces from items (1)–(4), (8) of the above–given list), and the mapping $\hat{X} : \Omega \ni \omega \mapsto \widetilde{X}(\cdot,\omega)$ is $\mathcal{F}-\mathcal{X}$ measurable for a certain σ-algebra $\mathcal{X}$ in $\mathbb{X}$, then the process X is said to generate the random element $\hat{X}$ in $(\mathbb{X},\mathcal{X})$.*

If process X generates a random element $\hat{X}$ in $(\mathbb{X},\mathcal{X})$, then its distribution *in $(\mathbb{X},\mathcal{X})$ is the probability measure $\mu_X \equiv \mathsf{P} \circ \hat{X}^{-1}$, $\mu_X(A) = \mathsf{P}(\hat{X} \in A), A \in \mathcal{X}$. The notions of the random element generated by a random sequence and the corresponding distribution are introduced analogously.*

Example 16.1. The Wiener process generates a random element in $C([0,T])$ (see Problem 16.1). The distribution of the Wiener process in $C([0,T])$ is called the *Wiener measure*.

Definition 16.2. *A sequence of probability measures $\{\mu_n\}$ defined on the Borel σ-algebra of the metric space $\mathbb{X}$* weakly converges *to measure μ if*

$$\int_{\mathbb{X}} f d\mu_n \to \int_{\mathbb{X}} f d\mu, \quad n \to \infty \tag{16.1}$$

for arbitrary continuous bounded function $f : \mathbb{X} \to \mathbb{R}$ (notation: $\mu_n \Rightarrow \mu$). If the sequence of distributions of random elements $\hat{X}_n$ converges weakly to the distribution of the element $\hat{X}$, then the sequence of the random elements $\hat{X}_n$ converges weakly *or* by distribution *to $\hat{X}$ (notation: $\hat{X}_n \Rightarrow \hat{X}$ or $\hat{X}_n \xrightarrow{d} \hat{X}$). If processes X_n generate random elements in a functional space $\mathbb{X}$ and these elements converge weakly to the element generated by a process X, then the sequence of the processes X_n is said to converge to X* by distribution in $\mathbb{X}$.

A set $A \in \mathcal{B}(\mathbb{X})$ is called a *continuity set* for the measure μ if $\mu(\partial A) = 0$ (∂A denotes the boundary of the set A).

Theorem 16.2. *All the following statements are equivalent.*

(1) $\mu_n \Rightarrow \mu$.
(2) Relation (16.1) *holds for every bounded function f satisfying the Lipschitz condition: there exists L such that $|f(x) - f(y)| \le L\rho(x,y), x,y \in \mathbb{X}$ (here ρ is the metric in $\mathbb{X}$).*
(3) $\limsup_{n\to\infty} \mu_n(F) \le \mu(F)$ *for every closed set $F \subset \mathbb{X}$.*
(4) $\liminf_{n\to\infty} \mu_n(G) \ge \mu(G)$ *for every open set $G \subset \mathbb{X}$.*
(5) $\lim_{n\to\infty} \mu_n(A) = \mu(A)$ *for every continuity set A for the measure μ.*

Theorem 16.3. *(1) Let $\mathbb{X},\mathbb{Y}$ be metric spaces and $F : \mathbb{X} \to \mathbb{Y}$ be an arbitrary function. Then the set $\mathcal{D}_F$ of the discontinuity points for this function is a Borel set (moreover, a countable union of closed sets).*

(2) Let random elements $\hat{X}_n, n \geq 1$ in $(\mathbb{X}, \mathcal{B}(\mathbb{X}))$ converge weakly to a random element $\hat{X}$ and $\mathsf{P}(\hat{X} \in \mathcal{D}_F) = 0$. Then random elements $\hat{Y}_n = F(\hat{X}_n), n \geq 1$ in $(\mathbb{X}, \mathcal{B}(\mathbb{X}))$ converge weakly to the random element $\hat{Y} = F(\hat{X})$.

Consider the important partial case $\mathbb{X} = C([0,T])$. If X is a random process that generates a random element in $C([0,T])$, then for every $m \geq 1, t_1, \dots, t_m \in [0,T]$ the finite-dimensional distribution $\mathsf{P}^X_{t_1,\dots,t_m}$ can be represented as the image of the distribution of X in $C([0,T])$ under the mapping

$$\pi_{t_1,\dots,t_m} : C([0,1]) \ni x(\cdot) \mapsto (x(t_1), \dots, x(t_m)) \in \mathbb{R}^m.$$

("the finite-dimensional projection"). Because every function $\pi_{t_1,\dots,t_m}$ is continuous, Theorem 16.3 yields that, for every sequence of random processes X_n that converge by distribution in $C([0,T])$ to a process X, for every $m \geq 1, t_1, \dots, t_m \in [0,T]$ finite-dimensional distributions $\mathsf{P}^{X_n}_{t_1,\dots,t_m}$ converge weakly to the finite-dimensional distribution $\mathsf{P}^X_{t_1,\dots,t_m}$.

It should be mentioned that the inverse implication does not hold true and convergence of the finite-dimensional distributions of random processes does not provide their convergence by distribution in $C([0,T])$ (see Problem 16.13).

Definition 16.3. *(1) A family of measures $\{\mu_\alpha, \alpha \in \mathcal{A}\}$ is called* weakly *(or* relatively*)* compact *if each of its subsequences contains a weakly convergent subsequence.*

(2) A family of measures $\{\mu_\alpha, \alpha \in \mathcal{A}\}$ is called tight *if for every $\varepsilon > 0$ there exists a compact set $K_\varepsilon \subset \mathbb{X}$ such that $\mu_\alpha(\mathbb{X} \backslash K_\varepsilon) < \varepsilon, \alpha \in \mathcal{A}$.*

Theorem 16.4. *(The Prokhorov theorem) (1) If a family of measures $\{\mu_\alpha, \alpha \in \mathcal{A}\}$ is tight, then it is weakly compact.*

(2) If a family of measures $\{\mu_\alpha, \alpha \in \mathcal{A}\}$ is weakly compact and $\mathbb{X}$ is a Polish space, then this family is tight.

It follows from the definition that a sequence of measures $\{\mu_n\}$ converges weakly if and only if (a) this family if weakly compact, and (b) this family has at most one weak partial limit (i.e., if two of its subsequences converge weakly to measures μ' and μ'', then $\mu' = \mu''$). The statements given above provide the following criteria which are very useful for investigation of the convergence of the random processes by the distribution in $C([0,T])$.

Proposition 16.1. *In order for random processes X_n to converge by distribution in $C([0,T])$ to a random process X it is necessary and sufficient that*
(a) The sequence of their distributions in $C([0,T])$ is tight.
(b) All the finite-dimensional distributions of the processes X_n converge weakly to the corresponding finite-dimensional distributions of the process X.

For a tightness of a sequence of distributions in $C([0,T])$ of stochastic processes, a wide choice of sufficient conditions is available (see [25], Chapter 9; [4], Chapter 2; [9], Chapter 5). Here we formulate one such condition that is analogous to the Kolmogorov theorem (Theorem 3.12).

Theorem 16.5. *Let a sequence of random processes* $X_n = \{X_n(t), t \in [0,T]\}, n \ge 1$ *be such that, for some constants* $\alpha, \beta, C > 0$,

$$\mathsf{E}|X_n(t) - X_n(s)|^\alpha \le C|t-s|^{1+\beta}, \quad t,s \in [0,T], \quad n \in \mathbb{N}.$$

Then the sequence of the distributions of these processes in $C([0,T])$ *is tight.*

A *random walk* is a sequence of sums $S_n = \sum_{k=1}^n \xi_k, n \ge 1$, where $\{\xi_k k \in \mathbb{N}\}$ are independent random variables (in general, these variables can have various distributions, but we assume further they are identically distributed. For random walks, see also Chapters 10, 11, and 15).

Theorem 16.6. *(The Donsker theorem) Let* $\{S_n, n \ge 1\}$ *be a random walk and* $\mathsf{E}\xi_k^2 < +\infty$. *Then the random processes*

$$X_n(t) = \frac{S_{[nt]} - [nt]\mathsf{E}\xi_1}{\sqrt{n\mathsf{D}\xi_1}} + (nt - [nt])\frac{\xi_{[nt]+1} - \mathsf{E}\xi_1}{\sqrt{n\mathsf{D}\xi_1}}, \quad t \in [0,1], \quad n \ge 1$$

converge by distribution in $C([0,1])$ *to the Wiener process.*

Corollary 16.1. *Let* $F : C([0,1]) \to \mathbb{R}$ *be a functional with its discontinuity set* $\mathcal{D}_F$ *having zero Wiener measure. Then* $F(X_n) \Rightarrow F(W)$.

Note that Corollary 16.1 does not involve any assumptions on the structure of the laws of the summands ξ_k, and thus the Donsker theorem is frequently called *the invariance principle*: the limit distribution is invariant w.r.t. choice of the law of ξ_k. Another name for this statement is the *functional limit theorem*.

Processes with continuous trajectories have a natural interpretation as random elements valued in $C([0,T])$. This allows one to study efficiently the limit behavior of the distributions of the functionals of such processes. In order to extend this construction for the processes with càdlàg trajectories, one has to endow the set $\mathbb{D}([0,T],\mathbb{Y})$ with a structure of a metric space, and this space should be separable and complete (see Problem 16.17, where an example of an inappropriate metric structure is given). Below, we describe the metric structure on $\mathbb{D}([0,T],\mathbb{Y})$ introduced by A. V. Skorokhod.

Let $\mathbb{Y}$ be a Polish space with the metric ρ. Denote by Λ the class of strictly monotone mappings $\lambda : [0,T] \to [0,T]$ such that $\lambda(0) = 0, \lambda(T) = T$. Denote

$$\|\lambda\| = \sup_{s \ne t} \left| \ln\left(\frac{\lambda(t) - \lambda(s)}{t-s}\right) \right|.$$

Definition 16.4. *For* $x, y \in \mathbb{D}([0,T],\mathbb{Y})$, *denote*

$$d(x,y) = \inf\{\varepsilon \mid \exists \lambda \in \Lambda, \sup_{t\in[0,T]} \rho(x(\lambda(t)), y(t)) \le \varepsilon, \sup_{t\in[0,T]} |\lambda(t) - t| \le \varepsilon\},$$

$$d_0(x,y) = \inf\{\varepsilon \mid \exists \lambda \in \Lambda, \sup_{t\in[0,T]} \rho(x(\lambda(t)), y(t)) \le \varepsilon, \|\lambda\| \le \varepsilon\}.$$

Theorem 16.7. *(1) The functions d, d_0 are metrics on $\mathbb{D}([0,T],\mathbb{Y})$.*

(2) The space $(\mathbb{D}([0,T],\mathbb{Y}),d)$ is separable but is not complete.

(3) The space $(\mathbb{D}([0,T],\mathbb{Y}),d_0)$ is both separable and complete.

(4) A sequence $\{x_n\} \subset \mathbb{D}([0,T],\mathbb{Y})$ converges to some $x \in \mathbb{D}([0,T],\mathbb{Y})$ in the metric d if and only if this sequence converges to x in the metric d_0.

The last statement in Theorem 16.15 shows that the classes of the closed sets in the spaces $(\mathbb{D}([0,T],\mathbb{Y}),d)$ and $(\mathbb{D}([0,T],\mathbb{Y}),d_0)$ coincide, and therefore the definitions of the weak convergence in these spaces are equivalent.

For a tightness of a sequence of distributions in $\mathbb{D}([0,T])$ of stochastic processes, a wide choice of sufficient conditions is available (see [25], Chapter 9, [4], Chapter 3). We formulate one of them.

Theorem 16.8. *Let a sequence of random processes $X_n = \{X_n(t), t \in [0,T]\}, n \geq 1$ be such that, for some constants $\alpha, \beta, C > 0$,*

$$\mathsf{E}|X_n(t) - X_n(s)|^{\alpha}|X_n(r) - X_n(t)|^{\alpha} \leq C|r-s|^{1+\beta}, \quad s < t < r, \quad n \in \mathbb{N}.$$

Then the sequence of the distributions of these processes in $\mathbb{D}([0,T])$ is tight.

Consider a *triangular array* of random variables $\{\xi_{nk}, 1 \leq k \leq n\}$, where for every n random variables $\xi_{n1}, \dots, \xi_{nn}$ are independent and identically distributed. Consider the random walk $S_{nk} = \sum_{j=1}^{k} \xi_{nj}, 1 \leq k \leq n$ corresponding to this triangular array.

Theorem 16.9. *Let the central limit theorem hold for the array $\{\xi_{nk}, 1 \leq k \leq n\}$; that is there exists random variable η such that $S_{nn} \Rightarrow \eta$. Then the random processes*

$$X_n(t) := S_{[nt]}, \quad t \in [0,1], \quad n \geq 1$$

converge by distribution in $\mathbb{D}([0,1])$ to the stochastically continuous homogeneous process with independent increments Z such that

$$Z(1) \stackrel{d}{=} \eta.$$

Note that, under an appropriate choice of the array $\{\xi_{nk}\}$, any infinitely divisible distribution can occur as the distribution of the variable η. Correspondingly, any Lévy process can occur as the limiting process Z.

Together with a qualitative statement about convergence of a sequence of distributions, frequently (especially in applications) explicit estimates for the rate of convergence are required. The rate of convergence for a sequence of probability distributions can be naturally controlled by a distance between the prelimit and limit distributions w.r.t. some *probability metric*; that is, a metric on the family of probability measures. Below, we give a list of the most important and frequently used probability metrics.

The class of probability measures on a measurable space $(\mathbb{X},\mathcal{X})$ will be denoted by $\mathcal{P}(\mathbb{X})$. Consider first the case $\mathbb{X} = \mathbb{R}, \mathcal{X} = \mathcal{B}(\mathbb{R})$. In this case, every measure $\mu \in \mathcal{P}(\mathbb{R})$ is uniquely defined by its distribution function F_{μ}.

Definition 16.5. The uniform metric *(or* the Kolmogorov metric*) is the function*

$$d_U(\mu,\nu) = \sup_{x\in\mathbb{R}} |F_\mu(x) - F_\nu(x)|, \quad \mu,\nu \in \mathcal{P}(\mathbb{R}).$$

Definition 16.6. The Lévy metric *is the function*

$$d_L(\mu,\nu) = \inf\{\varepsilon \,|\, F_\nu(x-\varepsilon) - \varepsilon \le F_\mu(x) \le F_\nu(x+\varepsilon) + \varepsilon, x \in \mathbb{R}\}, \quad \mu,\nu \in \mathcal{P}(\mathbb{R}).$$

Definition 16.7. The Kantorovich metric *is the function*

$$d_K(\mu,\nu) = \int_{\mathbb{R}} |F_\mu(x) - F_\nu(x)|\,dx, \quad \mu,\nu \in \mathcal{P}(\mathbb{R}). \tag{16.2}$$

Note that the integral in the right-hand side of (16.2) can diverge, and thus the Kantorovich metric can take value $+\infty$.

Next, let $(\mathbb{X},\rho)$ be a metric space, $\mathcal{X} = \mathcal{B}(\mathbb{X})$.

Definition 16.8. The Lévy-Prokhorov metric *is the function*

$$d_{LP}(\mu,\nu) = \inf\{\varepsilon \,|\, \mu(A) \le \nu(A^\varepsilon) + \varepsilon, A \in \mathcal{B}(\mathbb{X})\}, \quad \mu,\nu \in \mathcal{P}(\mathbb{X}),$$

where $A^\varepsilon = \{y \,|\, \rho(y,A) < \varepsilon\}$ is the open ε-neighborhood of the set A.

It requires some effort to prove that d_{LP} is a metric indeed; see Problem 16.52.

For a Lipschitz function $f : \mathbb{X} \to \mathbb{R}$, denote by $\mathrm{Lip}(f)$ its Lipschitz constant; that is, the infimum of L such that $|f(x) - f(y)| \le L\rho(x,y), x,y \in \mathbb{X}$.

Definition 16.9. The Lipschitz metric *is the function*

$$d_{\mathrm{Lip}}(\mu,\nu) = \sup_{f:\,\mathrm{Lip}(f)\le 1} \left| \int_{\mathbb{X}} f\,d\mu - \int_{\mathbb{X}} f\,d\nu \right|, \quad \mu,\nu \in \mathcal{P}(\mathbb{X}).$$

The Lipschitz metric is closely related to the Kantorovich metric; see Theorem 16.12. Some authors use term “Kantorovich metric” for the metric d_{Lip}.

For $\mu,\nu \in \mathcal{P}(\mathbb{X})$, denote by $C(\mu,\nu)$ the class of all random elements $Z = (X,Y)$ in $(\mathbb{X}\times\mathbb{X}, \mathcal{X}\otimes\mathcal{X})$ such that the first component X has distribution μ and the second component Y has distribution ν. Such a random element is called *a coupling* for the measures μ,ν.

Definition 16.10. The Wasserstein metric *of the power $p \in [1,+\infty)$ is the function*

$$d_{W,p}(\mu,\nu) = \inf_{(X,Y)\in C(\mu,\nu)} [\mathsf{E}\rho^p(X,Y)]^{1/p}, \quad \mu,\nu \in \mathcal{P}(\mathbb{X}).$$

In general, the Wasserstein metric can take value $+\infty$.

We remark that some authors insist that, from the historical point of view, the correct name for $d_{W,p}$ is the Kantorovich metric (or the Kantorovich-Rubinstein metric). We keep the term "Wasserstein metric" which is now used more frequently.

The Wasserstein metric is a typical example of a *coupling* (or *minimal*) probability metric. The general definition for the coupling metric has the form

$$\inf_{(X,Y)\in C(\mu,\nu)} H(X,Y), \tag{16.3}$$

where H is some metric on the set of random elements (see [92], Chapter 1). In the definition of the Wasserstein metric, H is equal to the L_p-distance $H(X,Y) = \|\rho(X,Y)\|_{L_p}$.

Under quite general assumptions, the infimum in the definition of the Wasserstein metric is attained; that is, for a given $\mu,\nu\in\mathcal{P}(\mathbb{X}), p\in[1,+\infty)$ there exists an element $Z^* = (X^*,Y^*)\in C(\mu,\nu)$ such that

$$\mathsf{E}\rho^p(X^*,Y^*) = d^p_{W,p}(\mu,\nu) \tag{16.4}$$

(see Problem 16.55). Any element Z^* satisfying (16.4) is called an *optimal coupling* for the measures μ,ν w.r.t. metric $d_{W,p}$.

In the important particular cases, explicit formulae are available both for the Wasserstein metric and for corresponding optimal couplings.

Proposition 16.2. *Let $\mathbb{X}=\mathbb{R}$, $\rho(x,y)=|x-y|$. For arbitrary $\mu,\nu\in\mathcal{P}(\mathbb{X})$ define the vector*

$$Z_{\mu,\nu} = (F_\mu^{[-1]}(U), F_\nu^{[-1]}(U)),$$

where U is the random variable uniformly distributed on $[0,1]$ and $F^{[-1]}(x) = \inf\{y|\, F(y)>x\}$ is the quantile transformation for the function F.

Then for every $p\in[1,+\infty)$ the random vector $Z_{\mu,\nu}$ is an optimal coupling for the measures μ,ν w.r.t. to the metric $d_{W,p}$. In particular (see Problem 16.57),

$$d^p_{W,p}(\mu,\nu) = \int_0^1 \left|F_\mu^{[-1]}(x) - F_\nu^{[-1]}(x)\right|^p dx, \quad p\in[1,+\infty). \tag{16.5}$$

Now, let $(\mathbb{X},\mathcal{X})$ be an arbitrary measurable space. Recall that, by *the Hahn theorem,* for any σ-finite signed measure $\varkappa$ there exists a set $C\in\mathcal{X}$ such that $\varkappa(A)\ge 0$ for any $A\in\mathcal{X}, A\subset C$ and $\varkappa(B)\le 0$ for any $B\in\mathcal{X}, B\subset\mathbb{X}\backslash C$. The measure $|\varkappa|(\cdot) := \varkappa(\cdot\cap C) - \varkappa(\cdot\cap(\mathbb{X}\backslash C))$ is called *the variation* of the signed measure $\varkappa$, and $|\varkappa|(\mathbb{X})$ is called *the total variation* of $\varkappa$.

Definition 16.11. The total variation metric *(or* the total variation distance*) is the function*

$$d_V(\mu,\nu) = \|\mu-\nu\|_{\mathrm{var}}, \quad \mu,\nu\in\mathcal{P}(\mathbb{X}),$$

where $\|\mu-\nu\|_{\mathrm{var}}$ is the total variation of the signed measure $\mu-\nu$.

Definition 16.12. The Hellinger metric *is the function*

$$d_H(\mu,\nu)=\left[\int_{\mathbb{X}}\left(\sqrt{\frac{d\mu}{d\lambda}}-\sqrt{\frac{d\nu}{d\lambda}}\right)^2 d\lambda\right]^{1/2}, \quad \mu,\nu\in\mathcal{P}(\mathbb{X}),$$

where λ is an arbitrary σ-finite measure such that $\mu\ll\lambda,\nu\ll\lambda$. The value $d_H(\mu,\nu)$ does not depend on the choice of λ (see Problem 16.65).

The Hellinger metric is closely related to the Hellinger integrals.

Definition 16.13. The Hellinger integral *of the power $\theta\in[0,1]$ is the function*

$$H_\theta(\mu,\nu)=\int_{\mathbb{X}}\left(\frac{d\mu}{d\lambda}\right)^{\theta}\left(\frac{d\nu}{d\lambda}\right)^{1-\theta} d\lambda, \quad \mu,\nu\in\mathcal{P}(\mathbb{X}).$$

Here, as in the previous definition, λ is a measure such that $\mu\ll\lambda,\nu\ll\lambda$. For $\theta=0$ or 1, the notational convention $0^0=1$ is used.

The Hellinger integral $H_{1/2}(\mu,\nu)$ is also called the Hellinger affinity.

The values $H_\theta(\mu,\nu),\theta\in[0,1]$ do not depend on the choice of λ (see Problem 16.65). Hellinger integrals appear to be a useful tool for investigating the properties of absolute continuity and singularity of the measures μ and ν (see Definitions 17.1, 17.2, and Problems 16.68,16.69).

In order to estimate how close two probability distributions are each to other, some "distance" functions are also used, not being the metrics in the true sense. These functions can be nonsymmetric w.r.t. μ,ν, fail to satisfy the triangle inequality, and so on. Here we give one such function that is used most frequently.

Definition 16.14. *For $\mu,\nu\in\mathcal{P}(\mathbb{X})$, let λ be a σ-finite measure such that $\mu\ll\lambda,\nu\ll\lambda$. Denote $f=d\mu/d\lambda, g=d\nu/d\lambda$.* The relative entropy *(or* the Kullback–Leibler distance*) for the measures μ,ν is defined by*

$$\mathcal{E}(\mu\|\nu)=\int_{\mathbb{X}} f\ln\left(\frac{f}{g}\right)d\lambda.$$

Here, the notational conventions $0\ln(0/p)=0, p\geq 0$, and $p\ln(p/0)=+\infty, p>0$ are used.

The relative entropy can take value $+\infty$. Its value for a given μ,ν does not depend on the choice of λ (see Problem 16.65).

Let us formulate the most important properties of the probability metrics introduced above. Let $\mathbb{X}$ be a Polish space.

Theorem 16.10. *(1) A sequence $\{\mu_n\}\subset\mathcal{P}(\mathbb{X})$ converges weakly to $\mu\in\mathcal{P}(\mathbb{X})$ if and only if $d_{LP}(\mu_n,\mu)\to 0, n\to+\infty$.*

(2) The set $\mathcal{P}(\mathbb{X})$ with the Lévy–Prokhorov metric d_{LP} forms a Polish metric space.

Theorem 16.11. *Assume the metric ρ on the set $\mathbb{X}$ is bounded. Then for every $p \in [1,+\infty)$,*

(1) The set $\mathcal{P}(\mathbb{X})$ with the Wasserstein metric $d_{W,p}$ forms a Polish metric space.

(2) The sequence $\{\mu_n\} \subset \mathcal{P}(\mathbb{X})$ converges weakly to $\mu \in \mathcal{P}(\mathbb{X})$ if and only if $d_{W,p}(\mu_n,\mu) \to 0, n \to +\infty$.

Theorem 16.12. *(The Kantorovich–Rubinstein theorem) The Wasserstein metric $d_{W,1}$ coincides with the Lipschitz metric d_{Lip}. Furthermore, in the case $\mathbb{X} = \mathbb{R}, \rho(x,y) = |x-y|$ both these metrics are equal to the Kantorovich metric d_K.*

Convergence of a sequence of measures w.r.t. total variation metric d_{TV} is called *convergence in variation* (notation: $\mu_n \stackrel{\mathrm{var}}{\to} \mu$). This convergence is stronger than the weak convergence; that is, $\mu_n \stackrel{\mathrm{var}}{\to} \mu$ implies $\mu_n \Rightarrow \mu$, but inverse implication, in general, does not hold. The following statement, in particular, shows that convergence in the Hellinger metric is equivalent to convergence in variation.

Proposition 16.3. *For the Hellinger metric d_H and the total variation metric d_{TV}, the following relations hold.*

$$d_H^2 \le d_{TV} \le 2d_H.$$

Let us give one more property of the total variation metric, which has a wide range of applications in ergodic theory for Markov processes with a general phase space. The following statement, by different authors, is named *the coupling lemma* or *the Dobrushin lemma.*

Proposition 16.4. *For any $\mu,\nu \in \mathcal{P}(\mathbb{X})$,*

$$d_{TV}(\mu,\nu) = 2 \inf_{(X,Y)\in C(\mu,\nu)} \mathsf{P}(X \ne Y).$$

The coupling lemma states that the total variation metric, up to multiplier 2, coincides with the coupling metric that corresponds to the "indicator distance" $H(X,Y) = \mathsf{P}(X \ne Y)$.

The properties given above show that there exist close connections between various probability metrics. The variety of probability metrics used in the literature is caused by the fact that every such metric arises naturally from a certain class of models and problems. On the other hand, some of the metrics have several additional properties that appear to be useful because these properties provide more convenient and easy analysis involving these metrics. One such property is called *the tensorization property* and means that some characteristics of the metric are preserved under the operation of taking a tensor product. Let us give two examples of statements of such kind (see Problems 16.60, 16.67).

Proposition 16.5. *Let $\mathbb{X} = \mathbb{X}_1 \times \mathbb{X}_2$ and the metric ρ on $\mathbb{X}$ has the form*

$$\rho(x,y) = \left[\rho_1^p(x_1,y_1) + \rho_2^p(x_2,y_2)\right]^{1/p}, \quad x=(x_1,x_2), \quad y=(y_1,y_2) \in \mathbb{X},$$

where $p \in [1,+\infty)$ and ρ_1, ρ_2 are the metrics in $\mathbb{X}_1, \mathbb{X}_2$. Then the distance between arbitrary product-measures $\mu = \mu_1 \times \mu_2, \nu = \nu_1 \times \nu_2$ w.r.t. the Wasserstein metric of the power p is equal to

$$d_{W,p}(\mu,\nu) = \left[d^p_{W,p}(\mu_1,\nu_1) + d^p_{W,p}(\mu_2,\nu_2)\right]^{1/p}.$$

Proposition 16.6. *Let $\mathbb{X} = \mathbb{X}_1 \times \mathbb{X}_2$. Then the distance between arbitrary product-measures $\mu = \mu_1 \times \mu_2, \nu = \nu_1 \times \nu_2$ w.r.t. the Hellinger metric satisfies*

$$\frac{1}{2}d^2_H(\mu,\nu) = 1 - \left(1 - \frac{1}{2}d^2_H(\mu_1,\nu_1)\right)\left(1 - \frac{1}{2}d^2_H(\mu_2,\nu_2)\right).$$

In particular, $d^2_H(\mu;\nu) \le d^2_H(\mu_1,\nu_1) + d^2_H(\mu_2,\nu_2)$.

Bibliography

[4]; [9], Chapter V; [17]; [25], Chapter IX; [88]; [92], Chapter I.

Problems

16.1. Let $\{X(t), t \in [0,T]\}$ be a process that has a continuous modification. Prove that the process X generates a random element in $C([0,T])$.

16.2. Let $\{X(t), t \in [0,T]\}$ be a process that has a measurable modification and such that $\mathsf{E}\int_0^T X^2(t)\,dt < +\infty$. Prove that the process X generates a random element in $L_2([0,T])$.

16.3. Let $\{X(t), t \in [0,T]\}$ be a process that has a càdlàg modification. Prove that the process X generates a random element in $\mathbb{D}([0,T])$.

16.4. Let $\mathbb{X}$ be a metric space and μ be a finite measure on $\mathcal{B}(\mathbb{X})$. Prove that

(1) For any $A \in \mathcal{B}(\mathbb{X}), \varepsilon > 0$ there exist a closed set F_ε and open set G_ε such that $F_\varepsilon \subset A \subset G_\varepsilon$ and $\mu(G_\varepsilon \backslash F_\varepsilon) < \varepsilon$ (*the regularity property* for a measure on a metric space).

(2) If $\mathbb{X}$ is a Polish space then for any $A \subset \mathcal{B}(\mathbb{X}), \varepsilon > 0$ there exists a compact set $K_\varepsilon \subset A$ such that $\mu(A \backslash K_\varepsilon) < \varepsilon$ (a refinement of the Ulam theorem).

(3) For any $p \in [1,+\infty)$ the set $C_{b,u}(\mathbb{X})$ of all bounded uniformly continuous functions on $\mathbb{X}$ is dense in $L_p(\mathbb{X},\mu)$.

16.5. Let $(\mathbb{X},\rho)$ be a Polish space, and $\mu : \mathcal{B}(\mathbb{X}) \to [0,1]$ be an additive set function. Prove that μ is σ-additive (i.e., is a measure) if and only if $\mu(A) = \sup\{\mu(K) |\, K \subset A, K \text{ is a compact set}\}, A \in \mathcal{B}(\mathbb{X})$.

16.6. Let $\mathcal{X}$ be the σ-algebra of subsets of ℓ_∞ generated by the open balls, and $\{\xi_k, k \geq 1\}$ are i.i.d. random variables with $\mathsf{P}(\xi_k = \pm 1) = \frac{1}{2}$. Prove:

(1) The sequence $\{\xi_k\}$ generates a random element ξ in $(\ell_\infty, \mathcal{X})$.

(2) Every compact subset $K \subset \ell_\infty$ belongs to the σ-algebra $\mathcal{X}$, and for every such set $\mathsf{P}(\xi \in K) = 0$.

16.7. Let $\{\xi_k, k \geq 1\}$ be the sequence of i.i.d. random variables that have standard normal distribution. Prove:

(1) The sequence $\left\{\zeta_k = \xi_k / \sqrt{\ln(k+1)}\right\}$ does not generate a random element in c_0.

(2) The sequence $\{\zeta_k\}$ generates a random element ζ in the space ℓ_∞ with the σ-algebra $\mathcal{X}$ generated by the open balls.

(3) Every compact subset $K \subset \ell_\infty$ belongs to the σ-algebra $\mathcal{X}$, and for every such set $\mathsf{P}(\zeta \in K) = 0$.

Thereby, for the distributions of the element ζ and the element ξ introduced in the previous problem, the statement of the Ulam theorem fails.

16.8. Let X_n, X be the random variables, $X_n \Rightarrow X$, and the distribution function F_X be continuous in every point of some closed set K. Prove that $\sup_{x \in K} |F_{X_n}(x) - F_X(x)| \to 0, \quad n \to +\infty$.

16.9. Give an example of the random vectors $X_n = (X_n^1, X_n^2)$, $n \geq 1$, $X = (X^1, X^2)$ such that $X_n \Rightarrow X$ and

(a) For every n there exists a function $f_n \in C(\mathbb{R})$ such that $X_n^2 = f_n(X_n^1)$ a.s..

(b) There does not exist a measurable function f such that $X^2 = f(X^1)$ a.s.

Such an example demonstrates that *functional dependence is not preserved under weak convergence*.

16.10. Let $\{X_n\}$ be a sequence of random elements in $(\mathbb{X}, \mathcal{B}(\mathbb{X}))$ with a tight family of the distributions. Prove that for every $f \in C(\mathbb{X}, \mathbb{Y})$ the family of the distributions in $(\mathbb{Y}, \mathcal{B}(\mathbb{Y}))$ of the elements $Y_n = f(X_n)$ is also tight.

16.11. Let $\mathbb{X}, \mathbb{Y}$ be metric spaces, $\mathbb{X} \times \mathbb{Y}$ be their product, and $\{\mu_n\}$ be a sequence of measures on the Borel σ-algebra in $\mathbb{X} \times \mathbb{Y}$. Prove that the sequence $\{\mu_n\}$ is tight if and only if both the sequences $\{\mu_n^1\}, \{\mu_n^2\}$ of the *marginal distributions* for the measures $\{\mu_n\}$ are tight. The marginal distributions for a measure μ on $\mathcal{B}(\mathbb{X} \times \mathbb{Y})$ are defined by

$$\mu^1(A) = \mu(A \times \mathbb{Y}), \quad A \in \mathcal{B}(\mathbb{X}), \quad \mu^2(B) = \mu(\mathbb{X} \times B), \quad B \in \mathcal{B}(\mathbb{Y}).$$

16.12. Consider the following functional spaces:

(a) $\mathrm{Lip}([0,1]) = \left\{ f \mid \|f\|_{\mathrm{Lip}} := |f(0)| + \sup_{s,t \in [0,1], s \neq t} |f(t) - f(s)| / |t - s| < +\infty \right\}$ (Lipschitz functions).

(b) $\mathrm{H}_\gamma([0,1]) = \left\{ f \mid \|f\|_{H_\gamma} \equiv |f(0)| + \sup_{s,t \in [0,1], s \neq t} |f(t) - f(s)| / |t - s|^\gamma < +\infty \right\}$ (Hölder functions with the index $\gamma \in (0,1)$).

Are they Banach spaces w.r.t. norms $\|\cdot\|_{Lip}$ and $\|\cdot\|_{H_\gamma}$, respectively? Which of these spaces are separable?

16.13. Let

$$X_n(t) = nt\,\mathbb{I}_{[0,1/(2n))} + \left[1 - \frac{t}{n}\right]\mathbb{I}_{[1/(2n),1/n)}, \quad t \in [0,1], \quad n \geq 1, \quad X \equiv 0.$$

Prove that
(a) All the finite-dimensional distributions of the process X_n converge weakly to the corresponding finite-dimensional distributions of the process X.
(b) The sequence $\{X_n\}$ does not converge to X by distribution in $C([0,1])$.

16.14. Let $\{a,b,c,d\} \subset (0,1)$ and $a < b, c < d$. Calculate the distance between the functions $x = 1_{[a,b)}$ and $\mathbb{I}_{[c,d)}$, considered as elements of $\mathbb{D}([0,1])$, w.r.t. the metric (a) d; (b) d_0.

16.15. Let $x_n(t) = \mathbb{I}_{[1/2,1/2+1/n)}(t), t \in [0,1], n \geq 2$.

(1) Prove that the sequence $\{x_n\}$ is fundamental in $\mathbb{D}([0,1])$ w.r.t. the metric d, but this sequence does not converge.

(2) Check directly that this sequence is not fundamental in $\mathbb{D}([0,1])$ w.r.t. the metric d_0.

16.16. Is the set $\mathbb{D}([0,1])$ a closed subset of the space $\mathbb{B}([0,1])$ of all bounded functions on $[0,1]$ with the uniform metric?

16.17. Prove that the space $\mathbb{D}([0,1])$, endowed with the uniform metric, is complete but is not separable.

16.18. Prove that if $x_n \to x$ in the metrics d of the space $\mathbb{D}([0,1])$ and the function x is continuous, then $\|x_n - x\|_\infty \to 0$.

16.19. Prove that:

(1) $C([0,1])$ is a closed subset of $\mathbb{D}([0,1])$.

(2) $C([0,1])$ is a nowhere dense subset of $\mathbb{D}([0,1])$; that is, for every nonempty open ball $B \subset \mathbb{D}([0,1])$ there exists a nonempty open ball $B' \subset B$ such that $B' \cap C([0,1]) = \varnothing$.

16.20. Give an example of sequences $\{x_n\}, \{y_n\}$ in $\mathbb{D}([0,1])$ such that the sequences themselves converge in $\mathbb{D}([0,1])$ but the sequence of the $\mathbb{R}^2$-valued functions $\{z_n(t) = (x_n(t), y_n(t))\}$ does not converge in $\mathbb{D}([0,1],\mathbb{R}^2)$.

16.21. Is the mapping $S : (x,y) \mapsto x+y, x,y \in \mathbb{D}([0,1])$ continuous as a function $\mathbb{D}([0,1]) \times \mathbb{D}([0,1]) \to \mathbb{D}([0,1])$?

16.22. For a given $a < b$ consider the following functional on $C([0,1])$:

$$I_{ab}(x) = \int_0^1 \mathbb{I}_{x(s)\in[a,b]}\,ds, \quad x \in C([0,1]).$$

Prove that this functional is not continuous, but the set of its discontinuity points has zero Wiener measure.

16.23. Let $\{X(t) \in [0,1]\}$ be a process with continuous trajectories. For $a \in \mathbb{R}$, denote by D_a^X the set of $t \in [0,1]$ such that the corresponding one-dimensional distribution has an atom in the point a; that is, $D_a^X = \{t \in [0,1] | \mathsf{P}(X(t) = a) > 0\}$. Prove that:

(1) $D_a^X \in \mathcal{B}([0,1])$.

(2) The set of discontinuity points for the functional I_{ab} introduced in the previous problem has zero measure w.r.t. distribution of the process X if and only if the set $D_a^X \cup D_b^X$ has zero Lebesgue measure.

16.24. For a given $a < b$, consider the functional I_{ab} (Problem 16.22) on the space $\mathbb{D}([0,1])$. Prove that this functional is not continuous, but for every $c \neq 0$ the set of its discontinuity points has zero measure w.r.t. distribution of the process $X(t) = N(t) + ct, t \in [0,1]$, where N is the Poisson process. Does the last statement remain true for $c = 0$?

16.25. For a given $z \in \mathbb{R}$, consider the functional on $C(\mathbb{R}^+)$

$$\tau(z,x) = \begin{cases} \inf\{t \mid x(t) = z\}, & \{t \mid x(t) = z\} \neq \varnothing, \\ +\infty, & \{t \mid x(t) = z\} = \varnothing. \end{cases}$$

Prove that $\tau(z,\cdot)$ is not a continuous functional, but the set of its discontinuity points has zero Wiener measure.

16.26. On the space $C([0,1])$, consider the functionals

$$M(x) = \max_{t\in[0,1]} x(t), \quad \vartheta(x) = \min\{t \mid x(t) = M(x)\}, \quad x \in C([0,1]).$$

Prove that (a) the functional M is continuous; (b) the functional ϑ is not continuous, but the set of its discontinuity points has zero Wiener measure.

16.27. Prove that the following functionals on $C([0,1])$ are not continuous, but the sets of their discontinuity points have zero Wiener measure.

$$\varkappa(x) = \min\{t \in [0,1] | x(t) = x(1)\}, \quad \chi(x) = \max\{t \in [0,1] | x(t) = 0\},$$

$x \in C([0,1])$.

16.28. Let $a : [0,1] \to \mathbb{R}$ be a positive continuous function. Prove that the functional $M_a : \mathbb{D}([0,1]) \to \mathbb{R}, M_a(x) = \sup_{t\in[0,1]} x(t)/(a(t))$ is continuous.

16.29. Denote $T(x) = \inf\{t \in [0,1] \mid x(t-) \neq x(t)\}, x \in \mathbb{D}([0,1])$. Is $T(\cdot)$ a continuous functional on $\mathbb{D}([0,1])$?

16.30. For $c > 0$, denote $T_c(x) = \inf\{t \in [0,1] \mid |x(t-) - x(t)| > c\}$, $x \in \mathbb{D}([0,1])$, that is, the moment of the first jump of the function x with the jump size exceeding c. Prove that for any measure μ on $\mathbb{D}([0,1])$ there exists at most countable set A_μ such that for arbitrary $c \notin A_\mu$ the set of discontinuity points for T_c has zero measure μ.

16.31. Denote $\tau_a(x) = \inf\{t \in [0,1] | x(t) \geq a\}, x \in \mathbb{D}([0,1])$, that is, the moment of the first passage of x over the level a (if the set is empty, then put $\tau_a(x) = 1$). Describe the set of values $a \in \mathbb{R}$ such that the set of discontinuity points for τ_a has zero measure w.r.t. the distribution of the Poisson process in $\mathbb{D}([0,1])$.

16.32. Prove that for arbitrary $a \in \mathbb{R}$ the set of discontinuity points for the functional τ_a introduced in the previous problem has zero measure w.r.t. the distribution of the process $X(t) = N(t) - t, t \in [0,1]$, where N is the Poisson process.

16.33. Let $\{S_n = \sum_{k \leq n} \xi_k, n \in \mathbb{Z}^+\}$ be a random walk with $\mathsf{E}\xi_k = 0, \mathsf{E}\xi_k^2 = 1$. Prove that for any $a < b$

$$\frac{1}{N} \sum_{n \leq N} \mathsf{P}(S_n \in [a\sqrt{N}, b\sqrt{N}]) \to \int_0^1 \int_a^b \frac{1}{\sqrt{2\pi s}} e^{-y^2/(2s)}\, dyds, \quad N \to \infty.$$

16.34. Let $\{S_n, n \in \mathbb{Z}^+\}$ be as in the previous problem. Denote $H_S^N(z) = \#\{n \leq N | S_n \geq z \cdot \sqrt{N}\}, z \in \mathbb{R}$. Prove that

$$\mathsf{P}\left(\frac{1}{N} H_S^N(0) \leq \alpha\right) \to \frac{2}{\pi} \arcsin \sqrt{\alpha}, \quad N \to \infty, \quad \alpha \in (0,1).$$

16.35. (1) Let $\{S_n = \sum_{k \leq n} \xi_k, n \in \mathbb{Z}^+\}$ be the random walk with $\mathsf{P}(\xi_k = \pm 1) = \frac{1}{2}$. Prove that

$$\mathsf{P}(\max_{n \leq N} S_n \geq z) = 2\mathsf{P}(S_N > z) + \mathsf{P}(S_N = z), \quad z \in \mathbb{Z}^+.$$

(2) Let W be the Wiener process. Prove that

$$\mathsf{P}(\max_{s \leq t} W(s) \geq z) = 2\mathsf{P}(W(t) \geq z), \quad z \geq 0.$$

(3) Let $\{S_n = \sum_{k \leq n} \xi_k, n \in \mathbb{Z}^+\}$ be a random walk with $\mathsf{E}\xi_k = 0, \mathsf{E}\xi_k^2 = 1$. Prove that

$$\frac{\mathsf{P}(\max_{n \leq N} S_n \geq z \cdot \sqrt{n})}{\mathsf{P}(S_N \geq z \cdot \sqrt{n})} \to 2, \quad N \to \infty, \quad z \geq 0.$$

16.36. Let W be the Wiener process, $z > 0$.

(1) Find the distribution density of the random variable $\tau(z,W)$ (the functional $\tau(\cdot,\cdot)$ is defined in Problem 16.25).

(2) Prove that $\mathsf{E}\tau^\alpha(z,W) = +\infty$ for $\alpha \geq \frac{1}{2}$ and $\mathsf{E}\tau^\alpha(z,W) < +\infty$ for $\alpha \in (0, \frac{1}{2})$.

16.37. Prove that $\{Y(z) = \tau(z,W), z \in \mathbb{R}^+\}$ is a stochastically continuous homogeneous process with independent increments.

16.38. Find the cumulant and the Lévy measure of the process Y from the previous problem.

16.39. Let $\{S_n = \sum_{k\le n} \xi_k\}$ be the random walk with $\mathsf{P}(\xi_k = \pm 1) = \frac{1}{2}$, and H_S^N be the function defined in Problem 16.34. Prove that for $z < 0, \alpha \in (0,1)$,

$$\mathsf{P}\left(\frac{1}{N} H_S^N(z) \le \alpha\right) \to \int_0^{1-\alpha} \sqrt{\frac{2}{\pi^3}} \cdot \frac{|z|}{s^{\frac{3}{2}}} e^{-z^2/(2s)} \cdot \arcsin\sqrt{\frac{\alpha}{1-s}}\, ds, \quad N \to \infty.$$

Give the formula for $\lim_{N\to\infty} \mathsf{P}\left(N^{-1} H_S^N(z) \le \alpha\right)$ when $z > 0$.

16.40. Let $\{S_n\}$ be as in the previous problem.

(1) For a given n, N $(n < N)$ find $\mathsf{P}(S_m \le S_n, m \le N)$.

(2) Denote $\vartheta^N(S) \equiv \min\{m : S_m = \max_{n\le N} S_n\}$. Prove that

$$\mathsf{P}(\vartheta^N(S) \le \alpha \cdot N) \to \frac{2}{\pi} \arcsin\sqrt{\alpha}, \quad N \to \infty, \quad \alpha \in (0,1).$$

16.41. Let $\{S_n = \sum_{k\le n} \xi_k, n \in \mathbb{Z}^+\}$ be the random walk with $\mathsf{P}(\xi_k = \pm 1) = \frac{1}{2}$. Find $\mathsf{P}(\max_{n\le N} S_n = m, \vartheta^N(S) = k, S_N = r)$.

16.42. Find the joint distribution of the variables $\max_{t\in[0,1]} W(t), \vartheta(W), W(1)$, where W is a Wiener process, and the functional ϑ is defined in Problem 16.26.

16.43. Let $\{S_n = \sum_{k\le n} \xi_k, n \in \mathbb{Z}^+\}$ be the random walk with $\mathsf{P}(\xi_k = \pm 1) = \frac{1}{2}$. Find $\mathsf{P}(\max_{n\le N} S_n = m, \min_{n\le N} S_n = k, S_N = r)$.

16.44. Find the joint distribution of the variables $\max_{t\in[0,1]} W(t), \min_{t\in[0,1]} W(t), W(1)$ (W is a Wiener process). Compare with Problem 7.108.

16.45. Let $\{S_n = \sum_{k\le n} \xi_k, n \in \mathbb{Z}^+\}$ be a random walk with $\mathsf{E}\xi_k = 0, \mathsf{E}\xi_k^2 = 1$. Prove that

$$\mathsf{P}(\max_{n\le N} |S_n| \le z \cdot \sqrt{N}) \to \sum_{m\in\mathbb{Z}} (-1)^m \int_{(2m-1)z}^{(2m+1)z} \frac{1}{\sqrt{2\pi}} e^{-y^2/2}\, dy$$

$$= 1 - \frac{4}{\pi} \sum_{m=1}^{\infty} \frac{(-1)^m}{m+1} \exp\left\{-\frac{\pi^2(2m+1)^2}{8z^2}\right\}, \quad N \to \infty, \quad z > 0.$$

16.46. Without an explicit calculation of the distribution, show that $\varkappa(W) \stackrel{d}{=} \chi(W)$ (the functionals $\varkappa, \chi$ are defined in Problem 16.27).

16.47. Prove that

$$\mathsf{P}(\varkappa(W) \le \alpha) = \mathsf{P}(\chi(W) \le \alpha) = \frac{2}{\pi} \arcsin\sqrt{\alpha} = \lim_{N\to\infty} \mathsf{P}(S_n \ne 0, n \ge \alpha \cdot N),$$

$\alpha \in (0,1)$, where $\{S_n\}$ is the random walk with $\mathsf{P}(\xi_k = \pm 1) = \frac{1}{2}$.

16.48. Without passing to the limit, find the distribution of the variables $\varkappa(W), \chi(W)$. Compare with the previous problem.

16.49. In the array $\{\xi_{nk}, 1 \le k \le n\}$ let the random variables $\xi_{n1}, \dots, \xi_{nn}$ be i.i.d. with $\mathsf{P}(\xi_{n1} = 1) = \lambda/n, \mathsf{P}(\xi_{n1} = 0) = 1 - \lambda/n$. Prove that for any $a > 0$,

$$\mathsf{P}(S_{nk} \le ak, k = 1, \dots, n) \to \mathsf{P}(\max_{t \in [0,1]} (N(t) - at) \le 0), \quad n \to \infty,$$

where N is the Poisson process with intensity λ.

16.50. In the situation of the previous problem, prove that for arbitrary $a > 0$ the distributions of the variables $n^{-1}\#\{k : S_{nk} > ak\}$ weakly converge to the distribution of the variable $\int_0^1 \mathbb{I}_{N(t) > at}\, dt$.

16.51. Verify the following relations between the uniform metric d_U and and Lévy metric d_L.

(1) $d_U(\mu, \nu) \ge d_L(\mu, \nu), \mu, \nu \in \mathcal{P}(\mathbb{R})$.

(2) If the measure ν possesses a bounded density p_ν, then

$$d_U(\mu, \nu) \le \left(1 + \sup_{x \in \mathbb{R}} p_\nu(x)\right) d_L(\mu, \nu), \quad \mu \in \mathcal{P}(\mathbb{R}).$$

In particular, if $\nu \sim \mathcal{N}(0, 1)$, then

$$d_L(\mu, \nu) \le d_U(\mu, \nu) \le \left(1 + (2\pi)^{-1/2}\right) d_L(\mu, \nu), \quad \mu \in \mathcal{P}(\mathbb{R}).$$

16.52. Verify the metric axioms for the Lévy–Prokhorov metric d_{LP}:
(a) $d_{LP}(\mu, \nu) = 0 \Leftrightarrow \mu = \nu$.
(b) $d_{LP}(\mu, \nu) = d_{LP}(\nu, \mu)$.
(c) $d_{LP}(\mu, \nu) \le d_{LP}(\mu, \pi) + d_{LP}(\pi, \nu)$
for any $\mu, \nu, \pi \in \mathcal{P}$.

16.53. Prove the triangle inequality for the Wasserstein metric.

16.54. Let $\mathbb{X}$ be a Polish space, $\mu, \nu \in \mathcal{P}(\mathbb{X})$, and let $\{Z_n, n \ge 1\} \subset C(\mu, \nu)$ be an arbitrary sequence. Prove that the sequence of distributions of $Z_n, n \ge 1$ in $X \times \mathbb{X}$ is weakly compact.

16.55. Let $\mathbb{X}$ be a Polish space, $\mu, \nu \in \mathcal{P}(\mathbb{X}), p \in [1, +\infty)$. Prove the existence of an optimal coupling for the measures μ, ν, that is, of such an element $Z^* = (X^*, Y^*) \in C(\mu, \nu)$ that $\mathsf{E}\rho^p(X^*, Y^*) = d^p_{W,p}(\mu, \nu)$.

16.56. Let $\mathbb{X}$ be a Polish space, $p \in [1, +\infty)$. Prove that the class of optimal couplings for the measures μ, ν depends on μ, ν continuously in the following sense. For any sequences $\{\mu_n\} \subset \mathcal{P}(\mathbb{X})$ and $\{\nu_n\} \subset \mathcal{P}(\mathbb{X})$, convergent in the metric $d_{W,p}$ to measures μ and ν, respectively, and any sequence $Z^*_n, n \ge 1$ of optimal couplings for $\mu_n, \nu_n, n \ge 1$ weakly convergent to an element Z^*, the element Z^* is an optimal coupling for μ, ν.

16.57. Prove formula (16.5) (a) for discrete measures μ, ν; (b) in the general case.

16.58. Calculate the Wasserstein distance $d_{W,2}(\mu, \nu)$ for $\mu \sim \mathrm{U}(0,1), \nu \sim \mathrm{Exp}(\lambda)$. For what λ is this distance minimal; that is, which exponential distribution gives the best approximation for the uniform one?

16.59. Calculate the Wasserstein distance $d_{W,2}(\mu, \nu)$ for $\mu \sim \mathcal{N}(a_1, \sigma_1^2), \nu \sim \mathcal{N}(a_2, \sigma_2^2)$.

16.60. Prove Proposition 16.5.

16.61. Let $\{\lambda_k\}, \{\theta_k\}$ be a given sequences of nonnegative real numbers such that $\sum_k \lambda_k < +\infty, \sum_k \theta_k < +\infty$, and $\{\xi_k\}, \{\eta_k\}$ are sequences of independent centered Gaussian random variables with the variances $\{\lambda_k\}$ and $\{\theta_k\}$, respectively. Find the Wasserstein distance $d_{W,2}$ between the distributions of the random elements generated by these two sequences in the space ℓ_2.

16.62. Let $\{X(t), Y(t), t \in [a,b]\}$ be centered Gaussian processes, and let their covariance functions R_X, R_Y be continuous on $[a,b]^2$.

(1) Prove that the processes X, Y generate random elements in the space $L_2([a,b])$.

(2) Prove the following estimate for the Wasserstein distance between the distributions μ_X, μ_Y of the random elements generated by the processes X, Y in $L_2([a,b])$,

$$d_{W,2}(\mu_X, \mu_Y) \le \sqrt{\int_{[a,b]^2} (Q_X(t,s) - Q_Y(t,s))^2 \, ds dt},$$

where $Q_X, Q_Y \in L_2([a,b]^2)$ are arbitrary kernels satisfying

$$\int_a^b Q_X(t,r) Q_X(s,r) \, dr = R_X(t,s), \quad \int_a^b Q_Y(t,r) Q_Y(s,r) \, dr = R_Y(t,s),$$

$t, s \in [a,b]$.

16.63. Prove that the Wasserstein distance $d_{W,2}$ between the distributions of the random elements generated by the Wiener process and the Brownian bridge in $L_2([0,1])$ is bounded by $1/\sqrt{3} - (\sqrt{2})/3$ from below and by $1/\sqrt{3}$ from above.

16.64. Prove that the creating of convex combinations of probability measures does not enlarge the Wasserstein distance $d_{W,p}$; that is, for any $\mu_1, \dots, \mu_m, \nu_1, \dots, \nu_m \in \mathcal{P}(\mathbb{X})$ and $\alpha_1, \dots, \alpha_m \ge 0$ with $\sum_{k=1}^m \alpha_k = 1$,

$$d_{W,p}\left(\sum_{k=1}^m \alpha_k \mu_k, \sum_{k=1}^m \alpha_k \nu_k \right) \le \max_{k=1,\dots,m} d_{W,p}(\mu_k, \nu_k).$$

Does this property hold for other coupling metrics?

16.65. A σ-finite measure λ is said to dominate measure $\mu \in \mathcal{P}(\mathbb{X})$ if $\mu \ll \lambda$. In Definitions 16.11—16.13, the values of the Hellinger metric $d_H(\mu, \nu)$, Hellinger integrals $H_\theta(\mu, \nu), \theta \in [0,1]$, and relative entropy $\mathcal{E}(\mu \| \nu)$ are defined in terms of a measure λ that dominates both μ and ν. Prove that, for a given $\mu, \nu \in \mathcal{P}(\mathbb{X})$:

(1) There exists at least one such a measure λ.

(2) The values of $d_H(\mu, \nu)$, $H_\theta(\mu, \nu), \theta \in [0,1]$, and $\mathcal{E}(\mu \| \nu)$ do not depend on the choice of λ.

16.66. Verify that

$$H_\theta(\mu_1 \times \mu_2, \nu_1 \times \nu_2) = H_\theta(\mu_1, \nu_1) H_\theta(\mu_2, \nu_2), \quad \theta \in [0,1].$$

Use this relation for proving Proposition 16.6.

16.67. Prove Proposition 16.3.

16.68. Let $\mu, \nu \in \mathcal{P}(\mathbb{X})$. Prove the following statements.

(1) $H_0(\mu, \nu) = H_1(\mu, \nu) = 1$ and $H_\theta(\mu, \nu) \le 1$ for every $\theta \in (0,1)$. If $H_\theta(\mu, \nu) = 1$ for at least one $\theta \in (0,1)$, then $\mu = \nu$.

(2) The function $H_{\mu,\nu} : [0,1] \ni \theta \mapsto H_\theta(\mu, \nu)$ is log-convex; that is,

$$H_{\mu,\nu}(\alpha\theta_1 + (1-\alpha)\theta_2) \le H^{\alpha}_{\mu,\nu}(\theta_1) H^{1-\alpha}_{\mu,\nu}(\theta_2), \quad \theta_1, \theta_2 \in [0,1], \alpha \in (0,1).$$

(3) The function $H_{\mu,\nu}$ is continuous on the interval $(0,1)$.

(4) The measure μ is absolutely continuous w.r.t. ν if and only if the function $H_{\mu,\nu}$ is continuous at the point 1.

(5) In order for the measures μ and ν to be mutually singular it is necessary that for every $\theta \in (0,1)$, and it is sufficient that for some $\theta \in (0,1)$, the Hellinger integral $H_\theta(\mu, \nu)$ equals 0.

16.69. (*Kakutani alternative*). Let $(\mathbb{X}, \mathfrak{X}) = (\prod_{n\in\mathbb{N}} \mathbb{X}_n, \bigotimes_{n\in\mathbb{N}} \mathfrak{X}_n)$ be a countable product of a measurable spaces $(\mathbb{X}_n, \mathfrak{X}_n), n \in \mathbb{N}$, and let μ and ν be the product measures on this space: $\mu = \prod_{n\in\mathbb{N}} \nu_n, \mu = \prod_{n\in\mathbb{N}} \nu_n$, where $\mu_n, \nu_n \in \mathcal{P}(\mathbb{X}_n), n \in \mathbb{N}$. Assuming that for every $n \in \mathbb{N}$ the measure μ_n is absolutely continuous w.r.t. ν_n, prove that for the measures μ, ν only the two following relations are possible.

(a) $\mu \ll \nu$;

(b) $\mu \perp \nu$.

Prove that the second relation holds if and only if $\prod_{n\in\mathbb{N}} H_{1/2}(\mu_n, \nu_n) = 0$.

16.70. Prove that the Hellinger intregrals are continuous w.r.t. the total variation convergence; that is, as soon as $\mu_n \overset{\text{var}}{\to} \mu, \nu_n \overset{\text{var}}{\to} \nu, n \to \infty$, one has

$$H_\theta(\mu_n, \nu_n) \to H_\theta(\mu, \nu), \quad n \to \infty, \quad \theta \in [0,1].$$

16.71. Calculate $H_\theta(\mu, \nu), \theta \in [0,1]$ for

(a) $\mu \sim \mathcal{N}(a_1, \sigma^2), \nu \sim \mathcal{N}(a_2, \sigma^2)$.

(b) $\mu \sim \mathcal{N}(a, \sigma_1^2), \nu \sim \mathcal{N}(a, \sigma_2^2)$.

(c) μ is the uniform distribution on $[a_1, b_1]$; ν is the uniform distribution on $[a_2, b_2]$ $(a_1 < b_1, a_2 < b_2)$.

16.72. Let μ, ν be the distributions of Poisson random variables with the parameters λ and ρ, respectively. Find $H_\theta(\mu, \nu), \theta \in [0,1]$. In the case $\lambda \ne \rho$, find θ_* such that $H_{\theta_*}(\mu, \nu) = \min_{\theta\in[0,1]} H_\theta(\mu, \nu)$.

16.73. Let μ and ν be the distributions of the random vectors $(\xi_1, \dots, \xi_m)$ and $(\eta_1, \dots, \eta_m)$. Assuming that the components of the vectors are independent and $\xi_k \sim \text{Pois}(\lambda_k), \eta_k \sim \text{Pois}(\rho_k), k = 1, \dots, m$, find $H_\theta(\mu, \nu), \theta \in [0,1]$.

16.74. Let μ and ν be the distributions of the random elements in $L_2([0,T])$ defined by the Poisson processes with the intensity measures κ_1 and κ_2, respectively (see Definition 5.3). Find $H_\theta(\mu,\nu), \theta \in [0,1]$.

16.75. Prove that the relative entropy $\mathcal{E}(\mu\|\nu)$ is equal to the left derivative at the point 1 of the function $\theta \mapsto H_\theta(\mu,\nu)$:

$$\mathcal{E}(\mu\|\nu) = \lim_{\theta\to 1-} \frac{H_\theta(\mu,\nu)-1}{\theta-1}.$$

16.76. Prove that

$$\mathcal{E}(\mu_1\times\mu_2\|\nu_1\times\nu_2) = \mathcal{E}(\mu_1\|\nu_1) + \mathcal{E}(\mu_2\|\nu_2).$$

16.77. (*Variational formula for the entropy*). Prove that, for arbitrary measure $\nu \in \mathcal{P}(\mathbb{X})$ and arbitrary nonnegative function $h \in L_1(\mathbb{X},\nu)$,

$$\ln \int_{\mathbb{X}} h\,d\nu = \max_{\mu\in\mathcal{P}(\mathbb{X})} \left(\int_{\mathbb{X}} \ln h\,d\mu - \mathcal{E}(\mu\|\nu) \right).$$

Hints

16.1–16.3. Prove that $\{\omega : \widetilde{X}(\cdot,\omega) \in B\} \in \mathcal{F}$ for arbitrary open (or closed) ball $B \subset \mathbb{X}$ and respective modification $\widetilde{X}$ of the process X. Use Lemma 16.1.
16.4. (1) Use the "principle of the fitting sets". Prove that the class of the sets described in the formulation of the problem is a σ-algebra that contains all open sets.
(2) If $F_{\varepsilon/2}$ is a closed set from the previous statement and $K_{\varepsilon/2}$ is a compact set from the statement of the Ulam theorem, then $\widetilde{K}_\varepsilon = F_{\varepsilon/2} \cap K_{\varepsilon/2}$ is the required compact.
(3) Consider the following classes of functions: $\mathcal{K}_0 = C_{b,u}(\mathbb{X})$; $\mathcal{K}_1 = \{$the functions of the form $f = \mathbb{I}_G$, G is an open set$\}$; $\mathcal{K}_2 = \{$the functions of the form $f = \mathbb{I}_A$, A is a Borel set$\}$; $\mathcal{K}_3 = L_p(\mathbb{X},\mu)$. Prove that every function from the class $\mathcal{K}_i$ $(i = 1,2,3)$ can be obtained as an L_p-limit of a sequence of linear combinations of the functions from $\mathcal{K}_{i-1}$.
16.7. (1) Use statement (a) of Problem 1.16.
(2), (3) Use reasoning analogous to that given in the proof of Problem 16.6.
16.10. Use that the image of a compact set under a continuous mapping is also a compact set.
16.11. Use the previous problem and the following two statements. (1) The functions $\pi_{\mathbb{X}} : \mathbb{X}\times\mathbb{Y} \ni (x,y) \mapsto x \in \mathbb{X}$, $\pi_{\mathbb{Y}} : \mathbb{X}\times\mathbb{Y} \ni (x,y) \mapsto y \in \mathbb{Y}$ are continuous; (2) if K_1, K_2 are the compact sets in $\mathbb{X}, \mathbb{Y}$ then $K_1 \times K_2$ is a compact set in $\mathbb{X}\times\mathbb{Y}$.
16.13. (a) For a given $t_1,\dots,t_m$ and n greater than some $n_0 = n_0(t_1,\dots,t_m)$, $\mathsf{P}^{X_n}_{t_1,\dots,t_m} = \mathsf{P}^{X}_{t_1,\dots,t_m}$.
(b) For an open set $G = \{y \mid \sup_t |y(t)| < \frac{1}{2}\}$, $\lim_n \mathsf{P}(X_n \in G) = 0 < 1 = \mathsf{P}(X \in G)$.
16.14. If the function λ does not satisfy conditions $\lambda(c) = a, \lambda(d) = b$, then $\sup_t |x(\lambda(t)) - y(t)| = 1$.

16.15. Use the previous problem.
16.17. Let $x_a = \mathbb{I}_{t\in[a,1]} \in \mathbb{D}([0,1]), a \in [0,1]$. Then, for every $a_1 \neq a_2$, the uniform distance between x_{a_1} and x_{a_2} is equal to 1.
16.22, 16.23. Verify that $x \in C([0,1])$ is a continuity point for the functional I_{ab} if and only if $\int_0^1 \mathbb{I}_{\{a\}\cup\{b\}}(x(t))\,dt = 0$. Use the hint to Problem 3.21.
16.24. Verify that $x \in \mathbb{D}([0,1])$ is a continuity point for the functional I_{ab} if and only if $\int_0^1 \mathbb{I}_{\{a\}\cup\{b\}}(x(t))\,dt = 0$.
16.25. Verify that $x \in C([0,1])$ is a discontinuity point for the functional $\tau(\cdot,z)$ in the following cases.
(1) $\{x(t) = z\} \neq \varnothing$ and at least one of the sets $\{x(t) < z\}, \{x(t) > z\}$ is not empty.
(2) There exists a nonempty interval $(a,b) \subset \mathbb{R}^+$ such that $x(t) = z, t \in (a,b)$. Prove that otherwise $x \in C([0,1])$ is a continuity point for $\tau(\cdot,z)$ and use Problem 3.23.
16.26. (a) $|\max_t x(t) - \max_t y(t)| \le \sup_t |x(t) - y(t)|$.
(b) Verify that $x \in C([0,1])$ is a continuity point for the functional ϑ if and only if the function x takes its maximum value on $[0,1]$ in a unique point, and use Problem 3.22.
16.27. Describe explicitly the sets of discontinuity points for the functionals $\varkappa, \chi$ (see the Hint to Problem 16.25) and use Problem 3.23.
16.33. Use the invariance principle, Theorem 16.3, and Problem 16.22.
16.34. Use Problem 10.32 and the strong Markov property for the random walk (see also [22], Vol. 1, Chapter III, §4).
16.35. (1) Use the reflection principle (see Problem 10.32 or [22], Vol. 1, Chapter III, §1).
(2), (3) Use the invariance principle and item 1). In item (2), you can also use the reflection principle for the Wiener process; see Problem 7.109.
16.36. (1) $\mathsf{P}(\tau(z,W) \le x) = \mathsf{P}(\max_{s\le x} W(s) \ge z)$.
(2) Use item (1).
16.37. Use the strong Markov property for the Wiener process (see Definition 12.9 and Theorem 12.5).
16.38. Let $n \in \mathbb{N}$, and denote $\eta = \tau(1,W)$, $\eta_n = \tau(1/n,W)$. By Problem 16.36, $\eta \stackrel{d}{=} n^2\eta_n$. Thus, for every $n \ge 1$, $\eta_n \stackrel{d}{=} n^{-2}(\eta_1' + \cdots + \eta_n')$, where $\eta_1', \ldots, \eta_n'$ are the independent random variables identically distributed with η. Therefore, η has a *stable distribution with the parameter* $\alpha = \frac{1}{2}$. In addition, $\eta \ge 0$. For the description of a characteristic function of a stable distribution, see [22] Vol. 2, Chapter XVII, §§3,5.
16.39. Prove the relation

$$\mathsf{P}(H_S^N(z) = m) = \sum_{k=1}^{N} \mathsf{P}(H_S^{N-k}(0) = m)\mathsf{P}(\tau = k),$$

where $\tau = \min\{l : S_l \ge z \cdot \sqrt{N}\}$. Use this relation and Problems 16.34, 16.36.
16.40. See [22], Vol. 1, Chapter III, §7.
16.41–16.44. See [4], Chapter 2, §7.
16.46. Use Problem 6.5, item (e).
16.47. Use Problems 16.27, 16.40, and the invariance principle.

16.48.

$$\begin{aligned}\mathsf{P}(\chi(W)<x)&=\mathsf{P}(W(s)\neq 0, s\in[x,1])\\&=\mathsf{P}(W(x)>0,\ \min_{s\in[x,1]} W(s)-W(x)\geq -W(x))\\&+\mathsf{P}(W(x)<0,\ \max_{s\in[x,1]} W(s)-W(x)\leq -W(x))\\&=\int_{\mathbb{R}}\frac{e^{-y^2/(2x)}}{\sqrt{2\pi x}}\left[1-\int_{|y|}^{\infty}\frac{2}{\sqrt{2\pi(1-x)}}e^{-z^2/(2(1-x))}\,dz\right]dy.\end{aligned}$$

16.49. Use Theorem 16.9 and Problem 16.28.
16.50. Use Theorem 16.9 and Problem 16.24.
16.51. (1) If $\varepsilon\geq d_U(\mu,\nu)$, then $F_\mu(x)\leq F_\nu(x)+\varepsilon\leq F_\nu(x+\varepsilon)+\varepsilon$ for every $x\in\mathbb{R}$.
(2) If $\varepsilon>d_L(\mu,\nu)$, then

$$F_\mu(x)\leq F_\nu(x+\varepsilon)+\varepsilon=F_\nu(x)+\int_x^{x+\varepsilon}p_\nu(y)\,dy+\varepsilon,\quad x\in\mathbb{R}.$$

16.53. For random elements taking values in a Polish space $\mathbb{X}$, prove the statement analogous to the one given in Problem 1.11. Then use this statement in order to solve the problem.
16.54. See Problem 16.11.
16.55. Use Problem 16.54 and the Fatou lemma.
16.56. Use the triangle inequality for the Wasserstein metric and, analogously to the solution of Problem 16.55, the Fatou lemma.
16.57. If X is a random variable and $X_n=[nX]/n, n\in\mathbb{N}$ is its discrete approximation, then $\mathsf{E}|X-X_n|^p\leq n^{-p}\to 0, n\to\infty$, and therefore the distributions of the variables X_n converge to the distribution of X in the metric $d_{W,p}$. Therefore the statement of item (b) can be proved using item (a) and Problem 16.56.
16.58. Use formula (16.5).
16.59. Use formula (16.5) and the fact that the distribution function for $\mathcal{N}(a,\sigma^2)$ has the form $F(x)=\Phi((x-a)/\sigma)$, where Φ denotes the distribution function for $\mathcal{N}(0,1)$.
16.61. Use Problem 16.59 and Proposition 16.34.
16.62. Use Problem 6.13 in item (1) and Problems 6.28, 6.30 in item (2).
16.63. Use Problems 6.35 and 16.59 in order to obtain the upper and the lower estimates, respectively.
16.64. Let $X_1,\dots,X_m$ be the random elements with distributions $\mu_1,\dots,\mu_m$, respectively, and θ be a random variable, independent of $X_1,\dots,X_m$ and taking values $1,\dots,m$ with probabilities $\alpha_1,\dots,\alpha_m$. Then the random element

$$X_\theta=\begin{cases}X_1, & \theta=1\\ \dots & \\ X_m, & \theta=m\end{cases}$$

has the distribution $\alpha_1\mu_1+\dots+\alpha_m\mu_m$.

16.66.

$$\frac{d(\mu_1 \times \mu_2)}{d(\lambda_1 \times \lambda_2)} = \frac{d\mu_1}{d\lambda_1}\frac{d\mu_2}{d\lambda_2}.$$

The Hellinger metric and the Hellinger affinity satisfy relation $1 - d_H^2(\mu,\nu) = H_{1/2}(\mu,\nu)$.
16.69. Use Problem 16.68.
16.75. Use item (4) of Problem 16.68, the Fatou lemma, and the Lebesgue dominated convergence theorem.
16.77. Use Jensen's inequality.

Answers and Solutions

16.1. Let B be a closed ball with the center $x \in C([0,T])$ and the radius r. Then $\{\omega \mid \widetilde{X}(\cdot,\omega) \in B\} = \bigcap_{t\in\mathbb{Q}\cap[0,T]}\{\omega \mid |\widetilde{X}(t,\omega) - x(t)| \le r\} \in \mathcal{F}$.
16.2. Let B be a closed ball with the center $x \in L_2([0,T])$ and the radius r. Then $\{\omega \mid \widetilde{X}(\cdot,\omega) \in B\} = \{\omega \mid \int_0^T (\widetilde{X}(t,\omega) - x(t))^2\,dt \le r\}$. Because the process $\widetilde{X}(t) - x(t)$ is measurable, $\int_0^T (\widetilde{X}(t) - x(t))^2\,dt$ is a random variable and thus $\{\omega \mid \widetilde{X}(\cdot,\omega) \in B\} \in \mathcal{F}$.
16.3. See [4], Theorem 14.5.
16.6. (1) Let B be a closed ball with the center x and radius r, then $\{\xi \in B\} = \bigcap_k\{|\xi_k - x_k| \le r\} \in \mathcal{F}$.
(2) Every compact set can be represented as an intersection of a countable family of sets, each one being a finite union of the open balls. Therefore, every compact set belongs to $\mathbb{X}$. Let us prove that every open ball B with the radius 1 has zero measure w.r.t. distribution of the element ξ; because every compact set is covered by a finite union of such balls, this will provide the required statement. Let the center of the ball B be a sequence $x = (x_k)_{k\in\mathbb{N}}$. Then, for every $k \in \mathbb{N}$, at least one of the inequalities holds true: $|x_k - 1| \ge 1, |x_k + 1| \ge 1$. Consider the following sequence $(y_k)_{k\in\mathbb{N}}$: if for the given k the first relation holds, then $y_k = -1$; otherwise $y_k = 1$. Then $\{\xi \in B\} \subset \bigcap_k\{\xi_k = y_k\}$ and $\mathsf{P}(\xi \in B) \le \prod_{k\in\mathbb{N}} \frac{1}{2} = 0$.
16.9. Consider the points $x_{jk}^n = (x_{jk}^{1,n}, x_{jk}^{2,n}) \in [0,1]^2$, $j,k = 1,\dots,n$ such that

$$x_{jk}^{1,n} \in \left(\frac{j-1}{n}, \frac{j}{n}\right), \quad x_{jk}^{2,n} \in \left(\frac{k-1}{n}, \frac{k}{n}\right), \quad k,j = 1,\dots,n,$$

and $x_{jk}^{r,n} \ne x_{il}^{r,n}, r = 1,2$ for every j,k,i,l such that $(j,k) \ne (i,l)$. There exists a Borel (and even a continuous) function $f_n : [0,1] \to [0,1]$ such that $f_n(x_{kj}^{1,n}) = x_{kj}^{2,n}, k,j = 1,\dots,n$. Define the distribution of the random vector $X_n = (X_n^1, X_n^2)$ in the following way. X_n takes values x_{jk}^n, $j,k = 1,\dots,n$ with the probabilities n^{-2}. By the definition, $f_n(X_n^1) = X_n^2$. On the other hand, (X_n^1, X_n^2) weakly converges to the vector $X = (X^1, X^2)$ with independent components uniformly distributed on $[0,1]$. For arbitrary Borel function $f : [0,1] \to [0,1]$, one has $\mathrm{cov}(f(X^1), X^2) = 0$ and therefore relation $f(X^1) = X^2$ does not hold.
16.12. Both the spaces $\mathrm{Lip}([0,1])$ and $\mathrm{H}_\gamma([0,1])$ with arbitrary $\gamma \in (0,1)$ are Banach. None of these spaces is separable.

16.14. $d(x,y)=\max\Big[|a-c|,|b-d|\Big]$,

$$d_0(x,y)=\min\left(1,\max\left[\left|\ln\frac{a}{c}\right|,\left|\ln\frac{b-a}{d-c}\right|,\left|\ln\frac{1-b}{1-d}\right|\right]\right).$$

16.16. Yes, it is.

16.18. If $x_n\to x$ in the metric d of the space $\mathbb{D}([0,1])$, then there exists a sequence $\lambda_n\in\Lambda$ such that $\sup_t|\lambda_n(t)-t|\to 0, \sup_t|x_n(t)-x(\lambda_n(t))|\to 0$. As soon as x is continuous, it is uniformly continuous, and thus $\sup_t|x(t)-x(\lambda_n(t))|\to 0$. This gives the required convergence $\sup_t|x_n(t)-x(t)|\to 0$.

16.20. $x_n(t)=\mathbb{I}_{t\in[1/2-1/(2n),1]}, y_n(t)=\mathbb{I}_{t\in[1/2-1/(3n),1]}$.

16.21. No, it is not. Consider the functions $\hat{x}_n=x_n,\hat{y}_n=-y_n$, where x_n,y_n are the functions from the previous solution. Then $\hat{x}_n\to\mathbb{I}_{t\in[1/2,1]},\hat{y}_n\to-\mathbb{I}_{t\in[1/2,1]}$, but $\hat{x}_n+\hat{y}_n\not\to 0$.

16.28. If $x_n\to x$ in the metric d of the space $\mathbb{D}([0,1])$ then there exists a sequence $\lambda_n\in\Lambda,n\geq 1$ such that $\sup_t|\lambda_n(t)-t|\to 0,\sup_t|x_n(t)-x(\lambda_n(t))|\to 0$. Because a is continuous, $\sup_t|a(\lambda_n(t))-a(t)|\to 0$. Thus

$$\left|\sup_t\frac{x_n(t)}{a(t)}-\sup_t\frac{x(t)}{a(t)}\right|=\left|\sup_t\frac{x_n(t)}{a(t)}-\sup_t\frac{x(\lambda_n(t))}{a(\lambda_n(t))}\right|\to 0.$$

16.29. No, it is not.

16.31. For $a\notin\mathbb{Z}^+$.

16.36. (a)

$$p(x)=\frac{d}{dx}\left[\frac{2}{\sqrt{2\pi x}}\int_z^\infty e^{-y^2/(2x)}\,dy\right]=\frac{1}{\sqrt{2\pi}}x^{-3/2}e^{-z^2/(2x)}.$$

16.38. $\Pi(du)=(1/\sqrt{2\pi})\mathbb{I}_{u>0}u^{-3/2}\,du,\ \psi(z)=-\sqrt{2|z|}$.

16.39. For $z>0,\alpha\in(0,1)$,

$$\mathsf{P}\left(\frac{1}{N}H_S^N(z)\leq\alpha\right)\to\int_0^{1-\alpha}\sqrt{\frac{2}{\pi^3}}\cdot\frac{z}{s^{3/2}}e^{-z^2/(2s)}\cdot\arcsin\sqrt{\frac{\alpha}{1-s}}\,ds+$$
$$+\int_{1-\alpha}^\infty\frac{1}{\sqrt{2\pi}}\cdot\frac{z}{s^{3/2}}e^{-z^2/(2s)}\,ds,\quad N\to\infty.$$

16.52. Statement (c) (the triangle inequality) follows immediately from the relation $(A^\varepsilon)^\delta\subset A^{\varepsilon+\delta},\varepsilon,\delta>0$ (we leave details for the reader). It is obvious that $d_{LP}(\mu,\mu)=0$. Because $\bigcap_{\varepsilon>0}A^\varepsilon=A$ for any closed set A, it follows from the relation $d_{LP}(\mu,\nu)=0$ that $\mu(A)\leq\nu(A)$ for every closed A. Then for every continuous function f taking values in $(0,1)$ one has

$$\int_{\mathbb{X}}f(x)\,\mu(dx)=\lim_{n\to+\infty}\sum_{k=1}^n f(t_{n,k})\mu\Big(\{x|\,f(x)\in[t_{n,k-1},t_{n,k}]\}\Big)$$
$$\leq\lim_{n\to+\infty}\sum_{k=1}^n f(t_{n,k})\nu\Big(\{x|\,f(x)\in[t_{n,k-1},t_{n,k}]\}\Big)=\int_{\mathbb{X}}f(x)\,\nu(dx),$$

where the sequence of partitions $\pi_n = \{0 = t_{n,0} < \cdots < t_{n,n} = 1\}$ is chosen in such a way that $|\pi_n| \to \infty, n \to \infty$ and $\mu(f = t_{n,k}) = \nu(f = t_{n,k}) = 0,\ 1 \le k \le n$. This inequality and the analogous inequality for $\widetilde{f} = 1 - f$ provide that $\int_{\mathbb{X}} f d\mu = \int_{\mathbb{X}} f d\nu$. Because f is arbitrary, this yields $\mu = \nu$ and completes the proof of statement (a).

Let us prove (b). It follows from (a) that $d_{LP}(\mu,\nu) = 0$ if and only if $d_{LP}(\nu,\mu) = 0$. Assume that $d_{LP}(\mu,\nu) > 0$ and take $t \in (0, d_{LP}(\mu,\nu))$. By definition, there exists a set $A \in \mathcal{B}(\mathbb{X})$ such that $\mu(A) > \nu(A^t) + t$. Denote $B = \mathbb{X}\backslash A^t$; then the latter inequality can be written as $\mu(A) > 1 - \nu(B) + t$ or, equivalently, $\nu(B) > \mu(\mathbb{X}\backslash A) + t$. If $x \in B^t$ then there exists some $y \in \mathbb{X}\backslash A^t$ such that $\rho(x,y) < t$ and thus $x \notin A$. Therefore, $B^t \subset \mathbb{X}\backslash A$ and we have the inequality

$$\nu(B) > \mu(\mathbb{X}\backslash A) + t \ge \mu(B^t) + t,$$

which yields $t < d_{LP}(\nu,\mu)$. Because $t \in (0, d_{LP}(\mu,\nu))$ is arbitrary, we obtain that $d_{LP}(\mu,\nu) \le d_{LP}(\nu,\mu)$. By the same arguments applied to the pair (ν,μ) instead of (μ,ν), we have $d_{LP}(\mu,\nu) \ge d_{LP}(\nu,\mu)$ and thus $d_{LP}(\mu,\nu) = d_{LP}(\nu,\mu)$.

16.55. If $d_{W,p}(\mu,\nu) = +\infty$ then one can take arbitrary coupling Z, thus only the case $d_{W,p}(\mu,\nu) < +\infty$ needs detailed consideration. Take a sequence $Z^n = (X^n, Y^n) \in C(\mu,\nu)$ such that $\mathsf{E}\rho^p(X^n,Y^n) \to d^p_{W,p}(\mu,\nu), n \to \infty$. By Problem 16.54 the family of distributions of $Z^n, n \ge 1$ is tight. Using the Prokhorov theorem and passing to a limit, we may assume that $Z^n, n \ge 1$ converges weakly to some $Z^* = (X^*, Y^*)$. Because the projection in $\mathbb{X} \times \mathbb{X}$ on one component is a continuous function, by Theorem 16.3 we have $X^n \Rightarrow X^*, Y^n \Rightarrow Y^*$. Therefore, X^* and Y^* have distributions μ and ν, respectively; that is, $Z^* \in C(\mu,\nu)$.

Consider a sequence of continuous functions $f_k : \mathbb{R}^+ \to \mathbb{R}^+, k \in \mathbb{N}$ such that $\sum_{k=1}^\infty f_k \equiv 1$ and $f_k(t) = 0\ t \notin [k-1,k]$. Every function $\phi_k(z) = \rho^p(x,y) f_k(\rho(x,y)), z = (x,y) \in \mathbb{X} \times \mathbb{X}$ is continuous and bounded, and thus

$$\mathsf{E}\phi_k(Z^n) \to \mathsf{E}\phi_k(Z^*), \quad n \to \infty, \quad k \in \mathbb{N}.$$

Therefore, by the Fatou lemma,

$$\begin{aligned} \mathsf{E}\rho^p(X^*,Y^*) &= \sum_{k=1}^\infty \mathsf{E}\phi_k(Z^*) \le \limsup_{n\to\infty} \sum_{k=1}^\infty \mathsf{E}\phi_k(Z^n) \\ &= \limsup_{n\to\infty} \mathsf{E}\rho^p(X^n,Y^n) = d^p_{W,p}(\mu,\nu). \end{aligned}$$

Hence Z^* is an optimal coupling.

16.57. (a) Consider the case $p > 1$. Denote by $T = \{t_k, k \in \mathbb{N}\}$ the set of all the points that have a positive measure μ or ν. Then the distribution of any vector $Z \in C(\mu,\nu)$ is defined by the matrix $\{z_{jk} = \mathsf{P}(Z = (t_j,t_k))\}_{j,k\in\mathbb{N}}$ that satisfies relations

$$\sum_i z_{ik} = \nu(\{t_k\}) =: y_k, \quad \sum_i z_{ji} = \mu(\{t_j\}) =: x_j, \quad k,j \in \mathbb{N}.$$

We denote the class of such matrices also by $C(\mu,\nu)$. By the definition of the Wasserstein metric and Problem 16.55,

$$d_{W,p}^p(\mu,\nu) = \inf_{\{z_{jk}\}\in C(\mu,\nu)} \sum_{j,k\in\mathbb{N}} z_{jk}c_{jk}$$
$$= \sum_{j,k\in\mathbb{N}} z_{jk}^* c_{jk}, \quad c_{jk} := |t_j - t_k|^p, \quad k,j\in\mathbb{N},$$

where the matrix $\{z_{jk}^*\}$ corresponds to the distribution of an optimal coupling $Z^* \in C(\mu,\nu)$. We write $j \prec k$, if $t_j < t_k$. Let us show that for arbitrary $j_1 \prec j_2, k_1 \prec j_k$ at least one number $z_{j_1k_2}^*, z_{j_2k_1}^*$ is equal to 0. This would be enough for solving the problem, because this condition on the matrix $\{z_{jk}^*\}$, together with the conditions

$$\sum_i z_{ik}^* = y_k, \quad \sum_i z_{ji}^* = x_j, \quad k,j\in\mathbb{N},$$

defines this matrix uniquely, and, on the other hand, this condition is satisfied for the matrix corresponding to the coupling described in Proposition 16.2.

Assume that for some $j_1 \prec j_2, k_1 \prec k_2$ the required condition fails, put $a = \min(z_{j_1k_2}^*, z_{j_2k_1}^*) > 0$ and define the new matrix $\{\widetilde{z}_{jk}\}$ by

$$\widetilde{z}_{jk} = \begin{cases} z_{jk}^*, & (j,k) \notin \{(j_l,k_r), l,r=1,2\}, \\ z_{jk}^* + a, & (j,k) = (j_1,k_1) \text{ or } (j_2,k_2), \\ z_{jk}^* - a, & (j,k) = (j_1,k_2) \text{ or } (j_2,k_1). \end{cases}$$

By the construction, $\{\widetilde{z}_{jk}\} \in C(\mu,\nu)$ and

$$\sum_{j,k\in\mathbb{N}} \widetilde{z}_{jk}c_{jk} = a[c_{j_1k_1} + c_{j_2k_2} - c_{j_1k_2} - c_{j_2k_1}] + \sum_{j,k\in\mathbb{N}} z_{jk}^* c_{jk}.$$

It can be checked directly that for every $s_1 < s_2, r_1 < r_2$,

$$|r_1 - s_1|^p + |r_2 - s_2|^p < |r_1 - s_2|^p + |r_2 - s_1|^p,$$

whence $c_{j_1k_1} + c_{j_2k_2} - c_{j_1k_2} - c_{j_2k_1} < 0$. This means that

$$\sum_{j,k\in\mathbb{N}} \widetilde{z}_{jk}c_{jk} < \sum_{j,k\in\mathbb{N}} z_{jk}^* c_{jk};$$

consequently, $\{z_{jk}^*\}$ does not correspond to an optimal coupling. This contradiction shows that the above assumption is impossible; that is, the matrix $\{z_{jk}^*\}$ satisfies the required condition. The case $p = 1$ can be obtained by an appropriate limit procedure for $p \to 1+$.

16.58. $d_{W,2}^2(\mu,\nu) = \int_0^1 \left(x + \lambda^{-1}\ln(1-x)\right)^2 dx = 1/3 - (3/2)\lambda^{-1} + 2\lambda^{-2}$. $\lambda_{\min} = 8/3$.

16.59. Because $F_{1,2}(x) = \Phi\left((x - a_{1,2})/\sigma_{1,2}\right)$, the optimal coupling for μ,ν has the form $(X,Y) = (a_1 + \sigma_1\eta, a_2 + \sigma_2\eta)$, where $\eta \sim \mathcal{N}(0,1)$. Therefore,

$$d_{W,2}(\mu,\nu) = \sqrt{\mathsf{E}[a_1 - a_2 + (\sigma_1 - \sigma_2)\eta]^2} = \sqrt{(a_1 - a_2)^2 + (\sigma_1 - \sigma_2)^2}.$$

16.60. Consider a random element $Z = (Z_1, Z_2, Z_3, Z_4)$ taking values in $\mathbb{X}\times\mathbb{X} = \mathbb{X}_1 \times \mathbb{X}_2 \times \mathbb{X}_1 \times \mathbb{X}_2$. If $Z \in C(\mu, \nu)$, then the components Z_1, Z_2, Z_3, Z_4 have distributions μ_1, ν_1, μ_2, and ν_2, respectively. Thus

$$\mathsf{E}\rho^p\Big((Z_1,Z_2),(Z_3,Z_4)\Big) = \mathsf{E}\rho_1^p(Z_1,Z_3) + \mathsf{E}\rho_2^p(Z_2,Z_4)$$
$$\geq \inf_{(X_1,Y_1)\in C(\mu_1,\nu_1)} \mathsf{E}\rho_1^p(X_1,Y_1) + \inf_{(X_2,Y_2)\in C(\mu_2,\nu_2)} \mathsf{E}\rho_1^p(X_2,Y_2),$$

and, by the definition of the Wasserstein metric, we have $d_{W,p}^p(\mu,\nu) \geq d_{W,p}^p(\mu_1,\nu_1) + d_{W,p}^p(\mu_2,\nu_2)$. Next, let $\varepsilon > 0$ be fixed. Take such random elements $(X_1^\varepsilon, Y_1^\varepsilon) \in C(\mu_1,\nu_1), (X_2^\varepsilon, Y_2^\varepsilon) \in C(\mu_2,\nu_2)$ that

$$\mathsf{E}\rho_1^p(X_1^\varepsilon,Y_1^\varepsilon) \leq d_{W,p}^p(\mu_1,\nu_1) + \varepsilon, \quad \mathsf{E}\rho_2^p(X_2^\varepsilon,Y_2^\varepsilon) \leq d_{W,p}^p(\mu_2,\nu_2) + \varepsilon.$$

Such elements exist by the definition of the Wasserstein metric. Construct the random element $Z^\varepsilon = (Z_1^\varepsilon, Z_2^\varepsilon, Z_3^\varepsilon, Z_4^\varepsilon)$ with $(Z_1^\varepsilon, Z_3^\varepsilon) \stackrel{d}{=} (X_1^\varepsilon, Y_1^\varepsilon)$, $(Z_2^\varepsilon, Z_4^\varepsilon) \stackrel{d}{=} (X_2^\varepsilon, Y_2^\varepsilon)$, and elements $(Z_1^\varepsilon, Z_3^\varepsilon), (Z_2^\varepsilon, Z_4^\varepsilon)$ being independent. By construction, Z_1^ε and Z_2^ε are independent. In addition, Z_1^ε has distribution μ_1 and Z_2^ε has distribution μ_2. Thus, $(Z_1^\varepsilon, Z_2^\varepsilon)$ has distribution μ. Analogously, $(Z_3^\varepsilon, Z_4^\varepsilon)$ has distribution ν. Therefore $Z^\varepsilon \in C(\mu,\nu)$ and

$$d_{W,p}^p(\mu,\nu) \leq \mathsf{E}\rho^p\Big((Z_1^\varepsilon,Z_2^\varepsilon),(Z_3^\varepsilon,Z_4^\varepsilon)\Big) \leq d_{W,p}^p(\mu_1,\nu_1) + d_{W,p}^p(\mu_2,\nu_2) + 2\varepsilon.$$

This gives the required statement because $\varepsilon > 0$ is arbitrary.

16.61. $\sqrt{\sum_k (\lambda_k^{1/2} - \theta_k^{1/2})^2}$.

16.62. (1) The processes X, Y are mean square continuous (Theorem 4.1), and thus have measurable modification (Theorem 3.1). Hence these processes generate random elements in $L_2([a,b])$ (Problem 16.2).

(2) Let W be the Wiener process. Put

$$X(t) = \int_a^b Q_X(t,s)\,dW(s), \quad Y(t) = \int_a^b Q_Y(t,s)\,dW(s), \quad t \in [a,b].$$

The processes X, Y are centered and their covariance functions equal R_X and R_Y, respectively (Problem 6.35). Therefore $(X,Y) \in C(\mu,\nu)$ and

$$d_{W,2}^2(\mu,\nu) \leq \mathsf{E}\|X-Y\|_{L_2([a,b])}^2 = \mathsf{E}\int_a^b (X(t)-Y(t))^2\,dt$$

$$= \int_a^b \mathsf{E}\left(\int_a^b (Q_X(t,s) - Q_Y(t,s))\,dW(s)\right)^2 dt = \int_{[a,b]^2} (Q_X(t,s) - Q_Y(t,s))^2\,ds\,dt.$$

16.63. It follows from Problem 6.13 that the pair of the processes $W(t), W(t)-t, t \in [0,1]$ is a coupling for μ, ν (here W is the Wiener process). Then

$$d_{W,2}^2(\mu,\nu) \leq \int_0^1 t^2\,dt = \frac{1}{3}.$$

On the other hand, for arbitrary coupling $(X,Y) \in C(\mu,\nu)$,

$$\mathsf{E}\|X-Y\|^2_{L_2([0,1])} \ge \mathsf{E}\left[\int_0^1 (X(t)-Y(t))\,dt\right]^2 = \mathsf{E}[\xi-\eta]^2,$$

where we have used the notation $\xi = \int_0^1 X(t)\,dt, \eta = \int_0^1 Y(t)\,dt$. The variables ξ and η are centered Gaussian ones with the variances $a = \int_0^1\int_0^1 (t\wedge s)\,dtds = \frac{1}{3}$ and $b = \int_0^1\int_0^1 (t\wedge s - ts)\,dtds = \frac{2}{9}$, respectively. Then Problem 16.59 yields that

$$\sqrt{\mathsf{E}\|X-Y\|^2_{L_2([0,1])}} \ge \sqrt{a}-\sqrt{b} = \frac{1}{\sqrt{3}} - \frac{\sqrt{2}}{3}.$$

16.64. Let $X_1,\dots,X_m,Y_1,\dots,Y_m$ be random elements with distributions $\mu_1,\dots,\mu_m$, $\nu_1,\dots,\nu_m$, and θ be an independent random variable that takes values $1,\dots,m$ with probabilities $\alpha_1,\dots,\alpha_m$. Then the variables X_θ, Y_θ (see the Hint for the notation) have distributions $\sum_k \alpha_k\mu_k, \sum_k \alpha_k\nu_k$. For every $\varepsilon > 0$ the elements $X_1,\dots,X_m,Y_1,\dots,Y_m$ can be constructed in such a way that

$$\mathsf{E}\rho^p(X_j,Y_j) \le \varepsilon + \max_k d^p_{W,p}(\mu_k,\nu_k), \quad j=1,\dots,m.$$

Then

$$d^p_{W,p}(\mu,\nu) \le \mathsf{E}\rho^p(X,Y) = \sum_j \alpha_j \mathsf{E}\rho^p(X_j,Y_j) \le \varepsilon + \max_k d^p_{W,p}(\mu_k,\nu_k).$$

Because $\varepsilon > 0$ is arbitrary, this finishes the proof.

Literally the same arguments show that, for arbitrary H from (16.3) (not necessarily a metric), the function $d_{H,\min} := \inf_{(X,Y)\in C(\mu,\nu)} H(X,Y)$ has the same property

$$d_{H,\min}\left(\sum_{k=1}^m \alpha_k\mu_k, \sum_{k=1}^m \alpha_k\nu_k\right) \le \max_{k=1,\dots,m} d_{H,\min}(\mu_k,\nu_k)$$

as soon as, in the previous notation,

$$H(X_\theta,Y_\theta) \le \sum_k \alpha_k H(X_k,Y_k).$$

16.65. (a) $\lambda = \frac{1}{2}(\mu+\nu)$ is a probability measure that dominates both μ and ν.
(b) Let λ_1,λ_2 dominate μ,ν simultaneously. Assume first that $\lambda_1 \ll \lambda_2$. Then

$$\frac{d\mu}{d\lambda_2} = \frac{d\mu}{d\lambda_1}\frac{d\lambda_1}{d\lambda_2} \quad \lambda_2-\text{a.s.},$$

and thus

$$\int_{\mathbb{X}} \left(\frac{d\mu}{d\lambda_2}\right)^\theta \left(\frac{d\mu}{d\lambda_2}\right)^{1-\theta} d\lambda_2 = \int_{\mathbb{X}} \left(\frac{d\mu}{d\lambda_1}\right)^\theta \left(\frac{d\mu}{d\lambda_1}\right)^{1-\theta} \left(\frac{d\lambda_1}{d\lambda_2}\right)^\theta \left(\frac{d\lambda_1}{d\lambda_2}\right)^{1-\theta} d\lambda_2$$

$$= \int_{\mathbb{X}} \left(\frac{d\mu}{d\lambda_1}\right)^{\theta} \left(\frac{d\mu}{d\lambda_1}\right)^{1-\theta} \left(\frac{d\lambda_1}{d\lambda_2}\right) d\lambda_2 = \int_{\mathbb{X}} \left(\frac{d\mu}{d\lambda_1}\right)^{\theta} \left(\frac{d\mu}{d\lambda_1}\right)^{1-\theta} d\lambda_1.$$

Now, let λ_1, λ_2 do not satisfy any additional assumption. Then the measure $\lambda_3 = \frac{1}{2}(\lambda_1 + \lambda_2)$ dominates both λ_1 and λ_2, and therefore, taking into account previous considerations, we get

$$\int_{\mathbb{X}} \left(\frac{d\mu}{d\lambda_1}\right)^{\theta} \left(\frac{d\mu}{d\lambda_1}\right)^{1-\theta} d\lambda_1 = \int_{\mathbb{X}} \left(\frac{d\mu}{d\lambda_3}\right)^{\theta} \left(\frac{d\mu}{d\lambda_3}\right)^{1-\theta} d\lambda_3$$
$$= \int_{\mathbb{X}} \left(\frac{d\mu}{d\lambda_2}\right)^{\theta} \left(\frac{d\mu}{d\lambda_2}\right)^{1-\theta} d\lambda_2.$$

Invariance of Definitions 16.11, 16.13 with respect to the choice of λ can be proved analogously.

16.67. Denote $f = d\mu/d\lambda, g = d\nu/d\lambda$, then $(\mu-\nu)(A) = \int_A (f-g)\,d\lambda, A \in \mathcal{X}$, and thus $\|\mu-\nu\|_{\mathrm{var}} = \int_{\mathbb{X}} |f-g|\,d\lambda$. Hence

$$d_H^2(\mu,\nu) = \int_{\mathbb{X}} (\sqrt{f}-\sqrt{g})^2\,d\lambda \le \int_{\mathbb{X}} |\sqrt{f}-\sqrt{g}|(\sqrt{f}+\sqrt{g})\,d\lambda$$
$$= d_{TV}(\mu,\nu) \le \left[\int_{\mathbb{X}} (\sqrt{f}-\sqrt{g})^2\,d\lambda\right]^{1/2} \left[\int_{\mathbb{X}} (\sqrt{f}+\sqrt{g})^2\,d\lambda\right]^{1/2}$$
$$= d_H(\mu,\nu)\sqrt{2+2H_{1/2}(\mu,\nu)} \le 2d_H(\mu,\nu).$$

16.68. Denote $f = d\mu/d\lambda, g = d\nu/d\lambda$. Then $H_0(\mu,\nu) = \int_{\mathbb{X}} g\,d\lambda = 1, H_1(\mu,\nu) = \int_{\mathbb{X}} f\,d\lambda = 1$. Inequality $H_\theta(\mu,\nu) \le 1$ comes from the Hölder inequality with $p = 1/\theta$ applied to $f^\theta, g^{1-\theta}$. Log-convexity of the function $H_{\mu,\nu}$ is provided by the relation

$$H_{\mu,\nu}(\alpha\theta_1 + (1-\alpha)\theta_2) = \int_{\mathbb{X}} f^{\alpha\theta_1+(1-\alpha)\theta_2} g^{1-\alpha\theta_1-(1-\alpha)\theta_2}\,d\lambda$$
$$= \int_{\mathbb{X}} \left(f^{\theta_1} g^{1-\theta_1}\right)^{\alpha} \left(f^{\theta_2} g^{1-\theta_2}\right)^{1-\alpha} d\lambda \le H_{\mu,\nu}^{\alpha}(\theta_1) H_{\mu,\nu}^{1-\alpha}(\theta_2);$$

in the last inequality we have used the Hölder inequality with $p = 1/\alpha$. This proves the statements (1), (2).

The measures μ and ν are mutually singular if and only if $\mu(A) = 1, \nu(A) = 0$ for some set $A \in \mathcal{X}$. This condition is equivalent to the condition for the product fg to be equal to zero λ-a.s. (verify this!), whence the statement (5) follows.

Statement (3) follows from the statements (2) and (5): if $H_{\mu,\nu}$ takes value 0 at some point, then $H_{\mu,\nu}$ is zero identically (and thus is continuous) on $(0,1)$. If, otherwise, all the values of $H_{\mu,\nu}$ are positive, then $\theta \mapsto \ln H_{\mu,\nu}(\theta)$ is a bounded convex function on $[0,1]$, and thus is continuous on $(0,1)$.

Consider statement (4). If $\mu \ll \nu$ then one can assume $\lambda = \nu$ and $H_\theta(\mu,\nu) = \int_{\mathbb{X}} f^\theta\,d\nu$. We have $f^\theta \to f, \theta \to 1-$ pointwise. In addition, $f^\theta \le f \vee 1 \in L_1(\mathbb{X},\nu)$ for every $\theta \in (0,1)$, and, combined with the dominated convergence theorem, it implies that $H_\theta(\mu,\nu) \to \int_{\mathbb{X}} f\,d\nu = 1 = H_1(\mu,\nu), \theta \to 1-$. On the other hand, if $\mu \not\ll \nu$, then there exists $A \in \mathcal{X}$ with $\mu(A) > 0, \nu(A) = 0$. Hence, for every $\theta \in (0,1)$

$$H_\theta(\mu,\nu)=\int_{\mathbb{X}\setminus A} f^\theta g^{1-\theta}\,d\lambda \le [\mu(\mathbb{X}\setminus A)]^\theta.$$

Taking a limit as $\theta \to 1-$, we get

$$\limsup_{\theta\to 1-} H_\theta(\mu,\nu) \le \mu(\mathbb{X}\setminus A) < 1 = H_1(\mu,\nu);$$

that is, the function $H_{\mu,\nu}$ is discontinuous at the point 1.

16.69. Denote $h_n(\theta)=H_\theta(\mu_n,\nu_n), H_N(\theta)=\prod_{n=N}^\infty h_n(\theta)$; then $H_1(\theta)=H_\theta(\mu,\nu)$ (verify this!). By the assumption, $\mu_n \ll \nu_n$, and thus every function h_n is continuous at the point 1 (Problem 16.68, item 4)). Assume $\mu \not\ll \nu$. Then H_1 is discontinuous at the point 1 and $\gamma := \liminf_{\theta\to 1-} H_1(\theta) < 1$. Therefore, by statement (2) from the same problem, $H_1(\frac12) \le \gamma^{1/2}$. Then there exists $N_1 \in \mathbb{N}$ such that $\prod_{n=1}^{N_1-1} h_n(\frac12) \le \gamma^{1/3}$. Because $h_n(\theta) \to 1, \theta \to 1-$ for any $n \in \mathbb{N}$, we have the relation $\liminf_{\theta\to 1-} H_{N_1}(\theta)=\gamma$, which, together with statement (2) of Problem 16.68, provides an estimate $H_{N_1}\left(\frac12\right) \le \gamma^{1/2}$. Therefore, there exists $N_2 \in \mathbb{N}$ such that $\prod_{n=N_1}^{N_2-1} h_n\left(\frac12\right) \le \gamma^{1/3}$. Repeating these arguments, we obtain a sequence $N_k, k \ge 1$ such that $\prod_{n=N_{k-1}}^{N_k-1} h_n\left(\frac12\right) \le \gamma^{1/3} < 1, k \in \mathbb{N}$ (here we denote $N_0=1$). Thus

$$H_{\frac12}(\mu,\nu)=\prod_{k=1}^\infty \left(\prod_{n=N_{k-1}}^{N_k-1} h_n\left(\frac12\right)\right) \le \prod_{k=1}^\infty \gamma^{1/3}=0,$$

and therefore $\mu \perp \nu$.

16.70. Take $\lambda=\frac14\left(\mu+\nu+\sum_{n=1}^\infty 2^{-n}(\mu_n+\nu_n)\right)$; then the measure λ dominates every measure $\mu,\nu,\mu_n,\nu_n, n\ge 1$. Denote

$$f=\frac{d\mu}{d\lambda},\quad g=\frac{d\nu}{d\lambda},\quad f_n=\frac{d\mu_n}{d\lambda},\quad g_n=\frac{d\nu_n}{d\lambda}.$$

Then

$$\int_X |f_n-f|\,d\lambda=\|\mu_n-\mu\|_{\mathrm{var}}\to 0,\quad \int_X |g_n-g|\,d\lambda=\|\nu_n-\nu\|_{\mathrm{var}}\to 0,\quad n\to\infty.$$

By the Hölder and Minkowski inequalities,

$$\begin{aligned}|H_\theta(\mu,\nu)-H_\theta(\mu_n,\nu_n)| &= \left|\int_{\mathbb{X}} f^\theta g^{1-\theta}\,d\lambda-\int_{\mathbb{X}} f_n^\theta g_n^{1-\theta}\,d\lambda\right|\\ &\le \left(\int_X |f-f_n|\,d\lambda\right)^\theta \left(\int_X g\,d\lambda\right)^{1-\theta}\\ &+\left(\int_X |g-g_n|\,d\lambda\right)^{1-\theta}\left(\int_X f_n\,d\lambda\right)^{1-\theta}\to 0,\end{aligned}$$

$n\to\infty, \theta\in(0,1)$.

16.71. (a) $e^{-(\theta(1-\theta)(a_1-a_2)^2)/(2\sigma^2)}$.

(b)
$$\frac{\sigma_1^{1-\theta}\sigma_2^{\theta}}{\sqrt{(1-\theta)\sigma_1^2+\theta\sigma_2^2}}.$$

(c) One can assume that $a_1 \le a_2$. In this case,

$$H_\theta(\mu,\nu) = \frac{(b_1-a_2)_+}{(b_1-a_1)^\theta(b_2-a_2)^{1-\theta}}, \quad \theta \in (0,1).$$

16.72. $H_\theta(\mu,\nu) = \exp[\lambda^\theta\rho^{1-\theta} - \theta\lambda - (1-\theta)\rho]$. $\theta_* = \log_{\lambda/\rho}[(\lambda/\rho-1)/\ln(\lambda/\rho)]$.
16.73. $H_\theta(\mu,\nu) = \exp\left[\sum_{k=1}^m \left(\lambda_k^\theta\rho_k^{1-\theta} - \theta\lambda_k - (1-\theta)\rho_k\right)\right]$.
16.74. Denote $\kappa_i(T) = \kappa_i([0,T])$, $\hat{\kappa}_i = \kappa_i/(\kappa_i(T))$, $i = 1,2$. Then

$$H_\theta(\mu,\nu) = \exp\left[\kappa_1^\theta(T)\kappa_2^{1-\theta}(T)H_\theta(\hat{\kappa}_1,\hat{\kappa}_2) - \theta\kappa_1^\theta(T) - (1-\theta)\kappa_2^\theta(T)\right].$$

16.77. Let us prove that $\int_{\mathbb{X}} \ln h\, d\mu - \mathcal{E}(\mu\|\nu) \le \ln\int_{\mathbb{X}} h\, d\nu$ for arbitrary measure $\mu \in \mathcal{P}(\mathbb{X})$. If $\mu \not\ll \nu$, then $\mathcal{E}(\mu\|\nu) = +\infty$ and the required inequality holds true. If $\mu \ll \nu$, denote $f = d\mu/d\nu$. Applying Jensen's inequality to the concave function $\ln(\cdot)$, we get

$$\begin{aligned}\int_{\mathbb{X}} \ln h\, d\mu - \mathcal{E}(\mu\|\nu) &= \int_{\mathbb{X}} (\ln h - \ln f)\, d\mu = \int_{\mathbb{X}} (\ln h - \ln f)\, d\mu \\ &\le \ln\int_{\mathbb{X}} \frac{h}{f}\, d\mu = \ln\int_{\mathbb{X}} h\, d\nu.\end{aligned}$$

On the other hand, if $h/f =$const, then this inequality becomes an equality. Put $d\mu = f d\nu$, where $f = h/(\int_{\mathbb{X}} h\, d\nu)$ if $\int_{\mathbb{X}} h\, d\mu > 0$ and $f \equiv 1$ otherwise. Then μ is a probability measure with $\int_{\mathbb{X}} \ln h\, d\mu - \mathcal{E}(\mu\|\nu) = \ln\int_{\mathbb{X}} h\, d\nu$.

17

Statistics of stochastic processes

Theoretical grounds

General statement of the problem of testing two hypotheses
Let the trajectory $x(\cdot)$ of a stochastic process $\{X(t),\ t \in [0,T]\}$ be observed. It is known that the paths of the process belong to a metric space of functions $F_{[0,T]}$ defined on $[0,T]$. For example, it can be the space of continuous functions $C([0,T])$ or Skorokhod space $\mathbb{D}([0,T])$; see Chapter 16. On $F_{[0,T]}$ the *Borel σ-field* $\mathcal{B}$ is considered.

Two hypotheses concerning finite-dimensional distributions of the process X are given. According to the hypothesis H_k, $k = 1,2$, a probability measure μ_k on the σ-field $\mathcal{B}$ corresponds to the process $\{X(t)\}$; that is, under the hypothesis H_k the equality holds $P(X(\cdot) \in A) = \mu_k(A)$, $A \in \mathcal{B}$. Based on the observations, one has to select one of the hypotheses. It can be done on the basis of either randomized or nonrandomized decision rule, as we show below.

It is said that a *randomized decision rule R* is given, if for each possible trajectory $x(\cdot) \in F_{[0,T]}$ the probability $p(x(\cdot))$ is defined (here p is a measurable functional on $F_{[0,T]}$) to accept H_1 if the path $x(\cdot)$ is observed, and $1 - p(x(\cdot))$ is the probability to accept the alternative hypothesis H_2. The rule is characterized by the *probabilities of Type I and Type II errors*: $\alpha_{12} = P(H_1|H_2)$ to accept H_1 when H_2 is true, and $\alpha_{21} = P(H_2|H_1)$ to accept H_2 when H_1 is true. The error probabilities are expressed by integrals

$$\alpha_{12} = \int_{F_{[0,T]}} p(x)\mu_1(dx), \quad \alpha_{21} = \int_{F_{[0,T]}} (1 - p(x))\mu_1(dx).$$

It is natural to look for the rules minimizing the error probabilities.

In many cases it is enough to content oneself with *nonrandomized decision rules* for which $p(x(\cdot))$ takes only values 0 and 1. Then $F_{[0,T]}$ is partitioned into two measurable sets G_1 and $G_2 := F_{[0,T]} \setminus G_1$; if $x(\cdot) \in G_1$ then H_1 is accepted, whereas if $x(\cdot) \in G_2$ then H_2 is accepted. The set G_1 is called *the critical region for testing H_1*. The error probabilities are calculated as $\alpha_{ij} = \mu_j(G_i)$, $i \neq j$.

D. Gusak et al., *Theory of Stochastic Processes*, Problem Books in Mathematics,
DOI 10.1007/978-0-387-87862-1_17, © Springer Science+Business Media, LLC 2010

Absolutely continuous measures on function spaces
Let μ_1 and μ_2 be two finite measures on the σ-field $\mathcal{B}$ in $F_{[0,T]}$.

Definition 17.1. *The measures μ_1 and μ_2 are* singular *if there exists a partition of the total space $F_{[0,T]}$ into two sets Q_1 and Q_2 such that $\mu_1(Q_2)=0$ and $\mu_2(Q_1)=0$. Notation: $\mu_1 \perp \mu_2$.*

Definition 17.2. *The measure μ_2 is* absolutely continuous *with respect to μ_1 if for each $A \in \mathcal{B}$ such that $\mu_1(A)=0$, it holds $\mu_2(A)=0$. Notation: $\mu_2 \ll \mu_1$.*

If $\mu_2 \ll \mu_1$ then by the Radon–Nikodim theorem there exists a $\mathcal{B}$-measurable nonnegative function $\rho(x)$ such that for all $A \in \mathcal{B}$

$$\mu_2(A) = \int_A \rho(x)\mu_1(dx). \tag{17.1}$$

The function is called *the density of the measure μ_2 with respect to μ_1*. Notation:

$$\rho(x) = \frac{d\mu_2}{d\mu_1}(x).$$

Definition 17.3. *If $\mu_2 \ll \mu_1$ and $\mu_1 \ll \mu_2$ simultaneously, then the measures μ_1 and μ_2 are called* equivalent. *Notation: $\mu_1 \sim \mu_2$.*

Measures μ_1 and μ_2 are equivalent if and only if the function ρ from the equality (17.1) is positive a.e. with respect to μ_1. In this case

$$\mu_1(A) = \int_A \frac{1}{\rho(x)}\mu_2(dx), \quad A \in \mathcal{B}.$$

For any finite measures μ_1 and μ_2 one can find pairwise disjoint sets Δ_1, Δ_2, and Δ such that $\mu_1(\Delta_2)=0$ and $\mu_2(\Delta_1)=0$, and on Δ the measures are equivalent; that is, there exists a measurable function $\rho:\ \Delta \to (0,+\infty)$ such that for all $A \in \mathcal{B}$,

$$\mu_2(A\cap\Delta) = \int_{A\cap\Delta} \rho(x)\mu_1(dx), \quad \mu_1(A\cap\Delta) = \int_{A\cap\Delta} \frac{1}{\rho(x)}\mu_2(dx). \tag{17.2}$$

Let H be a real separable infinite-dimensional Hilbert space. Consider a finite measure μ on the Borel σ-field $\mathcal{B}(H)$.

Definition 17.4. *The mean value of a measure μ is called a vector $m_\mu \in H$ such that*

$$(m_\mu, x) = \int_H (z,x)d\mu(z), \quad x \in H.$$

The correlation operator of a measure μ is called a linear operator S_μ in H such that

$$(S_\mu x, y) = \int_H (z-m_\mu, x)(z-m_\mu, y)d\mu(z), \quad x,y \in H.$$

It is known that the correlation operator S_μ of a measure μ, if it exists, is a continuous self-adjoint operator. Moreover it is positive; that is, $(S_\mu x,x) \geq 0$, $x \in H$.

Definition 17.5. *A measure μ is called a Gaussian measure on H if for each linear continuous functional f on H, the induced measure $\mu \circ f^{-1}$ is a Gaussian measure on the real line.*

The correlation operator S_μ of a Gaussian measure μ always exists, moreover $\sum_{i=1}^{\infty} \lambda_i(S_\mu) < \infty$ where $\lambda_i(S_\mu)$ are eigenvalues of S_μ that are counted according to their multiplicity. Let $\{e_i,\ i \geq 1\}$ be the corresponding orthonormal eigenvectors. They form a basis in H. Define the operator $\sqrt{S_\mu}$ in H,

$$\sqrt{S_\mu}x = \sum_{i=1}^{\infty} \sqrt{\lambda_i(S_\mu)}(x,e_i)e_i, \quad x \in H.$$

Theorem 17.1. *(Hajek–Feldman theorem) Let μ and ν be two Gaussian measures in H, with common correlation operator S and mean values $m_\mu = 0$ and $m_\nu = a$. If $a \in \sqrt{S}(H)$ then $\mu \sim \nu$, and if $a \notin \sqrt{S}(H)$ then $\mu \perp \nu$. In the case $\mu \sim \nu$, the Radon–Nikodim derivative is*

$$\frac{d\nu}{d\mu}(x) = \exp\{-\frac{1}{2}\left\|\left(\sqrt{S}\right)^{-1}a\right\|^2 + \sum_{k=1}^{\infty}\frac{x_k a_k}{\lambda_k}\}.$$

Here λ_k are positive eigenvalues of the operator S, and φ_k are the corresponding eigenvectors, the coefficients $x_k = (x,\varphi_k)$, $a_k = (a,\varphi_k)$, $k \geq 1$, and the series converges for μ–almost all x.

The Neyman–Pearson criterion

Fix $\varepsilon \in (0,1)$. The Neyman–Pearson criterion presents a randomized rule for hypothesis testing, which for a given upper bound ε of a Type I error (that is, when $\alpha_{12} \leq \varepsilon$), minimizes a Type II error α_{21}.

Consider three cases concerning the measures μ_1 and μ_2 related to the hypotheses H_1 and H_2.

(1) $\mu_1 \perp \mu_2$. Then there exists a set G_1 such that $\mu_1(G_1) = 1$ and $\mu_2(G_1) = 0$. If $x(\cdot) \in G_1$ then we accept H_1, otherwise if $x(\cdot) \notin G_1$ then H_2 is accepted. We have

$$\alpha_{12} = \mu_2(G_1) = 0, \quad \alpha_{21} = \mu_1(F_{[0,T]} \setminus G_1) = 0.$$

Thus, in this case one can test the hypotheses without error.

(2) $\mu_1 \sim \mu_2$. Let ρ be the density of μ_2 with respect to μ_1. For $\lambda > 0$ denote

$$R^\lambda = \{x \in F_{[0,T]} : \rho(x) < \lambda\}, \quad \Gamma^\lambda = \{x \in F_{[0,T]} : \rho(x) = \lambda\}.$$

Then there exists $\bar{\lambda}$ such that

$$\mu_2(R^{\bar{\lambda}}) \leq \varepsilon, \quad \mu_2(R^{\bar{\lambda}} \cup \Gamma^{\bar{\lambda}}) \geq \varepsilon.$$

Consider three options.

(2a) $\mu_2(R^{\bar{\lambda}}) = \varepsilon$. Then set $G_1 = R^{\bar{\lambda}}$. We have

$$\alpha_{12} = \varepsilon, \quad \alpha_{21} = 1 - \mu_1(R^{\bar{\lambda}}).$$

(2b) $\mu_2(R^{\bar{\lambda}}) < \varepsilon$ and $\mu_2(R^{\bar{\lambda}} \cup \Gamma^{\bar{\lambda}}) = \varepsilon$. Then set $G_1 = R^{\bar{\lambda}} \cup \Gamma^{\bar{\lambda}}$. We have

$$\alpha_{12} = \varepsilon, \quad \alpha_{21} = 1 - \mu_1(R^{\bar{\lambda}}) - \mu_1(\Gamma^{\bar{\lambda}}).$$

(2c) $\mu_2(R^{\bar{\lambda}}) < \varepsilon$ and $\mu_2(R^{\bar{\lambda}} \cup \Gamma^{\bar{\lambda}}) > \varepsilon$. Then we construct a randomized rule by means of the probability functional $p(x)$:

$$\begin{aligned} p(x) &= 1 && \text{if } x \in R^{\bar{\lambda}}, \\ p(x) &= 0 && \text{if } x \in F_{[0,T]} \setminus (R^{\bar{\lambda}} \cup \Gamma^{\bar{\lambda}}), \\ p(x) &= \tfrac{\varepsilon - \mu_2(R^{\bar{\lambda}})}{\mu_2(\Gamma^{\bar{\lambda}})} && \text{if } x \in \Gamma^{\bar{\lambda}}. \end{aligned}$$

If the measure μ_2 of any single path equals 0, then in case (2c) we define a nonrandomized rule as follows. There exists $D \subset \Gamma^{\bar{\lambda}}$ such that $\mu_2(D) = \varepsilon - \mu_2(R^{\bar{\lambda}})$; we set $G_1 = R^{\bar{\lambda}} \cup D$ and obtain the decision rule with $\alpha_{12} = \varepsilon$ and minimal α_{21}.

(3) Now, let μ_1 and μ_2 be neither singular nor equivalent. There exist pairwise disjoint sets Δ_1, Δ_2, and Δ such that (17.2) holds and $\mu_2(\Delta_1) = \mu_1(\Delta_2) = 0$. Let

$$R^{\lambda} = \{x \in \Delta : \rho(x) < \lambda\} \cup \Delta_1, \quad \Gamma^{\lambda} = \{x \in \Delta : \rho(x) = \lambda\}.$$

Consider two options.

(3a) $\varepsilon \geq 1 - \mu_2(\Delta_2)$. We set $G_1 = \Delta_1 \cup \Delta$, then

$$\alpha_{12} = 1 - \mu_2(\Delta_2) \leq \varepsilon, \quad \alpha_{21} = 0.$$

(3b) $\varepsilon < 1 - \mu_2(\Delta_2)$. Choose $\bar{\lambda}$ such that

$$\mu_2(R^{\bar{\lambda}}) \leq \varepsilon, \quad \mu_2(R^{\bar{\lambda}} \cup \Gamma^{\bar{\lambda}}) \geq \varepsilon,$$

and construct the rule as in case (2).

Therefore, in order to construct an optimal criterion one has to find the sets Δ_k on which the singular measures are concentrated, or in the case of equivalent measures to find the relative density of measures, wherein the probability law of $\rho(x(\cdot))$ is needed for each hypothesis.

Hypothesis testing for diffusion processes

The case of different diffusion matrices

Let $x(t)$, $t \in [0,T]$, be a path of a diffusion process in $\mathbb{R}^m$, and under a hypothesis H_k the drift vector of the diffusion process is $a_k(t,x)$, and its diffusion matrix is $B_k(t,x)$ (all the functions are continuous in both arguments); $k = 1,2$. This means that under H_k the observed diffusion process is a weak solution to the stochastic integral equation

$$x(t) = x_0 + \int_0^t a_k(s,x(s))ds + \int_0^t B_k^{1/2}(s,x(s))dW(s),\ t \in [0,T]. \tag{17.3}$$

Here W is an m–dimensional Wiener process; that is, $W(t) = (W_1(t),\ldots,W_m(t))^\top$, $t \in [0,T]$, where W_i, $i = \overline{1,m}$ are independent scalar Wiener processes, and $B_k^{1/2}$ is a positive semidefinite matrix such that its square is a positive semidefinite matrix B_k. For the equation (17.3) the analogue of Theorem 14.5 about the existence of a weak solution holds true.

Having the path $x(t)$ one can find $B_k(t,x(t))$, $t \in [0,T]$, provided the hypothesis H_k is true. This can be done as follows.

For $z \in \mathbb{R}^m$ we set

$$\lambda(t,z) = \lim_{n\to\infty} \sum_{k=0}^{2^n-1} \left(x\left(\frac{k+1}{2^n}t\right) - x\left(\frac{k}{2^n}t\right), z\right)^2. \tag{17.4}$$

The limit in (17.4) exists a.s. under each hypothesis $H_k, k = 1,2$, and

$$\lambda(t,z) = \int_0^t (B_k(s,x(s))z,z)\,ds,\ t \in [0,T], \tag{17.5}$$

if H_k is true.

If on the observed path for some $t \in [0,T]$ and $z \in \mathbb{R}^m$,

$$\int_0^t (B_1(s,x(s))z,z)ds \neq \int_0^t (B_2(s,x(s))z,z)ds, \tag{17.6}$$

then the equality (17.5) is correct only for a single value of k. Due to the continuity of the integrand functions, (17.6) holds true if and only if for some $t \in [0,T]$ and $z \in \mathbb{R}^m$,

$$(B_1(t,x(t))z,z) \neq (B_2(t,x(t))z,z).$$

Thus, under this condition we accept H_k if (17.5) holds for that k, and finally obtain the error-free decision rule.

Condition for equivalence of measures, and distribution of density under various hypotheses

Now, let along the observed path

$$\forall z \in \mathbb{R}^m: \quad (B_1(t,x(t))z,z) = (B_2(t,x(t))z,z).$$

Then $B_1(t,x(t)) \equiv B_2(t,x(t))$. Therefore, one can assume that

$$\forall t \in [0,T], \quad x \in \mathbb{R}^m: \quad B_1(t,x) = B_2(t,x).$$

Assume that the distribution of $x(0)$ is given and does not depend on the choice of the hypothesis. Denote

$$B(t,x) = B_1(t,x) = B_2(t,x), \quad a(t,x) = a_2(t,x) - a_1(t,x).$$

Let μ_k be a measure generated by the observed process on the space $C([0,T])$ under the hypothesis H_k; $k = 1,2$.

Theorem 17.2. *For the equivalence of measures $\mu_1 \sim \mu_2$, the next condition is sufficient: for each t,x there exists $b(t,x) \in \mathbb{R}^m$ such that the next two conditions hold:*

(1) $$a(t,x) = B(t,x)b(t,x).$$

(2) $$\int_0^T (a(t,x(t)), b(t,x(t)))dt < \infty$$

for almost every $x(\cdot)$ with respect to the measure μ_2.

Therein the density of μ_2 with respect to μ_1 is

$$\rho(x(\cdot)) = \exp\{\int_0^T (b(t,x(t)), dx(t))$$

$$-\frac{1}{2}\int_0^T (b(t,x(t)), a_1(t,x(t)) + a_2(t,x(t)))dt\}. \qquad (17.7)$$

Here the differential $dx(t)$ is written based on the stochastic equation (17.3), and the first integral on the right-hand side of (17.7) is understood, respectively, as a sum of the Lebesgue integral and the stochastic Ito integral.

Homogeneous in space processes

Let $a_k(t,x) = a_k(t)$ and $B_k(t,x) = B_k(t)$, $k = 1,2$; that is, the coefficients of the diffusion process do not depend on the spatial variable. As above we assume that all the coefficients are continuous functions. Then the process $\{x(t),\ t \in [0,T]\}$ has independent increments. From (17.5) it follows that

$$\lambda(t,z) = \int_0^t (B_k(s)z,z)ds$$

if H_k is true. Therefore, the hypotheses are tested without error if there exists such t that $B_1(t) \neq B_2(t)$.

Let $B_1(t) = B_2(t) = B(t)$ and $a(t) = a_2(t) - a_1(t)$. Denote by L_t the range $\{B(t)z:\ z \in \mathbb{R}^m\}$ and by E the set of such $t \in [0,T]$ that $a(t)$ does not belong to L_t. Let $P(t)$ be the projection operator on L_t. If $\lambda^1(E) > 0$ then the hypotheses are tested without error:

$$I(x(\cdot)) := \int_0^T \|P(t)(x(t) - a_1(t))\|^2 dt = 0$$

under the hypothesis H_1, and $I(x(\cdot)) > 0$ under the hypothesis H_2.

Now, let $a(t) \in L_t$, $t \in [0,T]$; that is,

$$\forall t \ \exists b(t) \in \mathbb{R}^m : a(t) = B(t)b(t). \tag{17.8}$$

In order for the vector $b(t)$ in (17.8) to be uniquely defined, we select it from the subspace L_t; this is possible because the matrix $B(t)$ is symmetric. Note that then $(a(t), b(t)) \geq 0$. Under condition (17.8), the necessary and sufficient condition for the absolute continuity of the measures μ_1 and μ_2 is the condition

$$\int_0^T (a(t), b(t))dt < \infty. \tag{17.9}$$

Under the conditions (17.8) and (17.9), the density of the measure μ_2 with respect to μ_1 in the space $C([0,T])$ is

$$\rho(x(\cdot)) = \exp\{\int_0^T (b(t), dx(t)) - \frac{1}{2}\int_0^T (b(t), a_1(t) + a_2(t))dt\}. \tag{17.10}$$

Under the hypothesis H_k it holds $\log \rho(x(\cdot)) \sim \mathcal{N}(m_k, \sigma^2)$ with

$$\sigma^2 = \int_0^T (a(t), b(t))dt, \quad m_k = (-1)^k \frac{\sigma^2}{2}; \quad k = 1, 2.$$

This makes it possible to construct the Neyman–Pearson criterion.

Next, we construct an error-free test under the assumption (17.8) and the condition

$$\int_0^T (a(t), b(t))dt = +\infty. \tag{17.11}$$

Select a sequence of continuous functions $b_n(t)$ such that

$$n \leq \int_0^T (B(t)b_n(t), b_n(t))dt = \int_0^T (a(t), b_n(t))dt < \infty$$

(this is possible due to the imposed assumptions). Then we accept the hypothesis H_1 if

$$\lim_{n\to\infty} \left(\int_0^T (B(t)b_n(t), b_n(t))dt \right)^{-1} \int_0^T b_n(t)d(x(t) - a_1(t)) = 0,$$

otherwise the hypothesis H_2 is accepted.

Hypothesis testing about the mean of Gaussian process
Let $x(t), t \in [0,T]$, be a path of a scalar Gaussian process with given continuous correlation function $R(t,s)$. Under the hypothesis H_1 the mean of the process equals 0, whereas under the alternative hypothesis H_2 the mean is equal to a given continuous function $a(t)$.
Condition for singularity of measures
Introduce a linear operator R in $X := L_2([0,T])$,

$$(Rg)(t) = \int_0^T R(t,s)g(s)ds, \quad g \in X, \quad t \in [0,T].$$

This is a Hilbert–Schmidt integral operator. Its eigenspace which corresponds to a zero eigenvalue is the kernel of the operator R. Also the operator has a sequence of positive eigenvalues and corresponding normalized eigenfunctions $\{\lambda_k, \varphi_k;\ k \geq 1\}$ with $\sum_{k\geq 1} \lambda_k < \infty$, the functions φ_k are pairwise orthogonal, and their linear combinations are dense in the range of the operator R.

Consider two cases.

(1) In the space X the function $a(\cdot)$ has no series expansion in the functions φ_k. Then the measures μ_1 and μ_2 on the space X that correspond to the hypotheses H_1 and H_2, are singular. We describe a decision rule. Let

$$\hat{a}(\cdot) = a(\cdot) - \sum_{k\geq 1} (a, \varphi_k)\varphi_k(\cdot). \tag{17.12}$$

Hereafter $(\cdot,\cdot)$ is the inner product in X, and in the case of an infinite number of φ_k, the series in (17.12) converges in the norm in X. If

$$\hat{l}(x(\cdot)) := \int_0^T x(t)\hat{a}(t)dt = 0$$

then we accept H_1; otherwise we accept H_2.

(2) In the space X the function $a(\cdot)$ has a Fourier expansion in the functions φ_k:

$$a(t) = \sum_{k\geq 1} a_k \varphi_k(t), \quad a_k := (a, \varphi_k). \tag{17.13}$$

(In the case of an infinite number of φ_k, the series in (17.13) converges in the norm in X). Assume additionally that

$$\sum_{k=1}^{\infty} \frac{a_k^2}{\lambda_k} = \infty. \tag{17.14}$$

Then $\mu_1 \perp \mu_2$. A decision rule is constructed as follows. Select a sequence $\{m_n\}$ such that

$$\forall n \geq 1: \ \sum_{k=1}^{m_n} \frac{a_k^2}{\lambda_k} \geq n.$$

If

$$\lim_{n\to\infty}\left(\sum_{k=1}^{m_n}\frac{a_k^2}{\lambda_k}\right)^{-1}\sum_{k=1}^{m_n}\frac{a_k}{\lambda_k}\int_0^T x(t)\varphi_k(t)dt=0 \tag{17.15}$$

then we accept the hypothesis H_1; otherwise we accept H_2.

Condition for equivalence of measures

In the notations of the previous subsection, the criterion of the equivalence of the measures μ_1 and μ_2 is the condition

$$\sum_{k\geq 1}\frac{a_k^2}{\lambda_k}<\infty. \tag{17.16}$$

Under this condition the density of μ_2 with respect to μ_1 is

$$\rho(x)=\exp\left\{\sum_{k\geq 1}\frac{x_k a_k}{\lambda_k}-\frac{1}{2}\sum_{k\geq 1}\frac{a_k^2}{\lambda_k}\right\}, \tag{17.17}$$

where $x_k:=(x,\varphi_k)$, and the first series under the exponent converges for μ_1–almost all x. Under the hypothesis H_k,

$$\log\rho(x(\cdot))\sim\mathcal{N}(m_k,\sigma^2),\quad \sigma^2=\sum_{k\geq 1}\frac{a_k^2}{\lambda_k},\quad m_k=(-1)^k\frac{1}{2}\sigma^2;\quad k\geq 1.$$

For the condition (17.16) it is sufficient that in $X=L_2([0,T])$ there exists a solution $b(\cdot)$ to the Fredholm Type I equation

$$a(t)=\int_0^T R(t,s)b(s)ds,\quad 0\leq t\leq T. \tag{17.18}$$

Via this solution the density ρ can be written differently:

$$\rho(x)=\exp\left\{\int_0^T x(s)b(s)ds-\frac{1}{2}\int_0^T a(s)b(s)ds\right\}. \tag{17.19}$$

Parameter estimation of distributions of stochastic process

Let $x(t)$, $0\leq t\leq T$, be the observed path of a stochastic process that generates a probability measure μ_θ on the function space $F_{[0,T]}$. A parameter θ is to be estimated and belongs to a parameter set Θ, which is a complete separable metric space or a Borel subset in such a space.

Definition 17.6. *A function* $\theta(x):F_{[0,T]}\to\Theta$ *which is* $\mathcal{B}-\mathscr{B}(\Theta)$ *measurable is called* the estimator $\theta(x)$ of the parameter θ *for any family of measures* μ_θ.

Assume that there exists a σ-finite measure ν on Borel σ-field $\mathcal{B}$ in $F_{[0,T]}$, with respect to which all the measures μ_θ are absolutely continuous and

$$\frac{d\mu_\theta}{d\nu}(x) = \rho(\theta,x), \quad \theta \in \Theta, \quad x \in F_{[0,T]}.$$

Then the family of measures $\{\mu_\theta,\ \theta \in \Theta\}$ is called *regular*.

Definition 17.7. *An estimator $\theta(x)$ of the parameter θ for a regular family of measures $\{\mu_\theta\}$ is called* strictly consistent under increasing T, *if $\theta(x) \to \theta$ as $T \to \infty$, a.s.*

For a regular family of measures, the estimator can be found by *the maximum likelihood method* via maximization of the function $\rho(\theta,x(\cdot))$ on Θ.

A real parameter θ for a regular family of measures $\{\mu_\theta,\ \theta \in \Theta\}$ can be estimated by *the Bayes method* as well. Let Θ be a finite or infinite interval on the real line, on which a pdf is given. We call it *the prior density* of the parameter θ. Based on the path $x = x(\cdot)$ one can compute *the posterior density*

$$\rho(\theta|x) := \frac{\rho(\theta,x)\rho(\theta)}{\int_\Theta \rho(\theta,x)\rho(\theta)d\theta}, \qquad \theta \in \Theta.$$

It is correctly defined if for the observed path it holds $\rho(\theta,x) > 0$, for a.e. $\theta \in \Theta$.

For an estimator $\theta(x)$, we introduce two loss functions: quadratic $L^2(\theta(x),\theta) := (\theta(x)-\theta)^2$ and all-or-nothing loss function $L_0(\theta(x),\theta) := \mathbb{I}_{\theta(x)\neq\theta}$. The latter is approximated by the functions $L_\varepsilon(\theta(x),\theta) := \mathbb{I}_{|\theta(x)-\theta|>\varepsilon}$ as $\varepsilon \to 0+$.

Under the *quadratic* loss function, *the Bayes estimator* $\hat{\theta}_2(x)$ of the parameter θ is defined as a minimum point of the next function (we suppose that the posterior density possesses a finite second moment):

$$Q(\hat{\theta}) := \int_\Theta L^2(\hat{\theta},\theta)\rho(\theta|x)d\theta, \quad \hat{\theta} \in \Theta.$$

This implies that $\hat{\theta}_2(x)$ coincides with the expectation of the posterior distribution; that is,

$$\hat{\theta}_2(x) = \int_\Theta \theta\rho(\theta|x)d\theta.$$

Under the *all-or-nothing* loss function, we have the approximating cost functions

$$Q_\varepsilon(\hat{\theta}) := \int_\Theta L_\varepsilon(\hat{\theta},\theta)\rho(\theta|x)d\theta, \quad \hat{\theta} \in \Theta.$$

Their minimum points, under unimodal and smooth posterior density, tend to the mode of this density. Therefore, the mode is taken as *the Bayes estimator* $\hat{\theta}_0(x)$ under a given loss function,

$$\hat{\theta}_0(x) = \operatorname*{argmax}_{\theta \in \Theta} \rho(\theta|x).$$

The case of a pairwise singular family $\{\mu_\theta,\ \theta \in \Theta\}$ is more specific for statistics of stochastic processes, in contrast to classical mathematical statistics. It is natural to expect in this case, that the parameter θ can be estimated without error by a single path $x(t),\ 0 \le t \le T$.

Definition 17.8. *An estimator* $\theta(x)$ *of the parameter* θ *for a pairwise singular family of measures* $\{\mu_\theta\}$ *is called* consistent *if*

$$\forall \theta \in \Theta: \quad \mu_\theta\{x \in F_{[0,T]}:\ \theta(x) = \theta\} = 1.$$

Thus, the consistent estimator makes it possible to find the parameter without error for a singular family of measures, which is impossible for a regular family of measures.

Bibliography

[51], Chapter 24; [57], Chapters 7, 17; [31], Chapter 4; [37], Chapters 2–4.

Problems

17.1. On $[0,T]$ a process is observed which is the Wiener process $\{W(t)\}$ under the hypothesis H_1, and is the process $\{\gamma t + W(t)\}$ with given $\gamma \ne 0$ under the hypothesis H_2. Construct the Neyman–Pearson test.

17.2. On $[0,T]$ a process is observed which is a homogeneous Poisson process with intensity λ_k under the hypothesis H_k; $k = 1,2$.

(a) Prove that the corresponding measures in the space $E = \mathbb{D}([0,T])$ are equivalent with density

$$\frac{d\mu_2}{d\mu_1}(x) = \rho(x(\cdot)) = \left(\frac{\lambda_2}{\lambda_1}\right)^{x(T)} e^{(\lambda_1 - \lambda_2)T}, \quad x \in E.$$

(b) Construct the Neyman–Pearson criterion to test the hypotheses.

17.3. Let $\{x(t),\ t \in [0,T]\}$ and $\mu_1,\ \mu_2$ be the objects described in the subsection of Theoretical grounds *Hypothesis testing about the mean of Gaussian process. Condition for singularity of measures*. Prove that:

(a) In cases (1) and (2) of the above-mentioned subsection the measures μ_1 and μ_2 are singular.

(b) Under condition (17.16) it holds $\mu_1 \sim \mu_2$, and the density of μ_2 with respect to μ_1 is given in (17.17).

17.4. Prove that the decision rule described in the subsection of Theoretical grounds *Hypothesis testing about the mean of Gaussian process. Condition for singularity of measures*, case (1), tests the hypotheses without error.

17.5. Prove that the decision rule described in the subsection of Theoretical grounds *Hypothesis testing about the mean of Gaussian process. Condition for singularity of measures*, case (2), tests the hypotheses without error.

17.6. On $[0,1]$ a path $x(\cdot)$ of a Gaussian process with correlation function $e^{-|t-s|}$ is observed. Under the hypothesis H_1 the mean of the process equals 0, whereas under the hypothesis H_2 it is equal to a given function $a \in C^2([0,1])$ with $a'(0) = a(0)$, $a'(1) = -a(1)$. Prove that the Neyman–Pearson criterion is constructed as follows. If

$$\int_0^1 x(s)\frac{a(s)-a''(s)}{2}ds < \sigma\Phi^{-1}(\varepsilon)+\sigma^2$$

then H_1 is accepted; otherwise H_2 is accepted. Here $\sigma > 0$ and

$$2\sigma^2 = \|a\|_{L_2}^2 + \|a'\|_{L_2}^2 + a(0)^2 + a(1)^2,$$

Φ is the cdf of standard normal law, and Φ^{-1} is the inverse function to the function Φ.

17.7. On $[0,T]$ a path $x(\cdot)$ of a zero mean Gaussion process is observed. Under the hypothesis H_1 the correlation function of the process is $R(t,s)$, whereas under the hypothesis H_2 it is equal to $\sigma^2 R(t,x)$ with unknown positive $\sigma^2 \neq 1$. Because σ^2 is unknown, the hypothesis H_2 is composite. Here $R(t,s)$ is a given continuous function such that the integral operator in $L_2([0,T])$,

$$(Ag)(t) = \int_0^T R(t,s)g(s)ds, \quad t \in [0,T], \quad g \in L_2([0,T]),$$

has an infinite number of positive eigenvalues. Construct an error-free criterion to test the hypotheses.

17.8. On $[0,2]$ a path $x(\cdot)$ of a scalar diffusion process is observed. Under the hypothesis H_1 the diffusion coefficient $b_1(t,x) = 1$, whereas under the hypothesis H_2 the diffusion coefficient $b_2(t,x) = t$. Under each hypothesis H_k the drift coefficient is the unknown continuous function $a_k(t,x)$, $k = 1,2$. Construct an error-free test.

17.9. On $[0,T]$ a path $x(\cdot)$ of a scalar diffusion process starting from 0 is observed. Under both hypotheses H_1 and H_2 its diffusion coefficient is t, $t \in [0,T]$, and under the hypothesis H_1 its drift coefficient is 0.

(a) Let $T < 1$ and under H_2 the drift coefficient is $|\log t|^{-1/2}$ for $t \in (0,T]$ and 0 for $t = 0$. Construct an error-free test.

(b) Let under H_2 the drift coefficient be $\sqrt{t}$, $t \in [0,T]$. Construct the Neyman–Pearson criterion to test the hypotheses.

17.10. On $[0,T]$ a path $x(\cdot)$ of a two-dimensional diffusion process starting from the origin is observed. Under the hypotheses H_1 and H_2 its diffusion matrix is diagonal with entries 1 and t on the diagonal. Under H_1 its drift vector is 0, whereas under H_2 the drift vector is $(\sqrt[4]{t};\sqrt[4]{t})^\top$. Construct the Neyman–Pearson criterion to test the hypotheses.

17.11. On $[0,T]$ a path $N(\cdot)$ of a homogeneous Poisson process with intensity λ is observed.

(a) Based on Problem 17.2 (a), show that the maximum likelihood estimator of the parameter λ is $\hat{\lambda}_T = N(T)/T$ (more precisely the maximum in $\lambda > 0$ of the density at the observed path is attained if $N(T) > 0$, and in the case $N(T) = 0$ we set $\hat{\lambda}_T = 0$).

(b) Prove the next: it is an unbiased estimator; that is, $\mathsf{E}\hat{\lambda}_T = \lambda$; it is a strongly consistent estimator under increasing T; the normalized estimator $\sqrt{T}(\hat{\lambda}_T - \lambda)$ converges in distribution to the normal law as $T \to \infty$; that is, $\hat{\lambda}_T$ is an asymptotically normal estimator.

17.12. On $[0,T]$ a path $N(\cdot)$ is observed of a nonhomogeneous Poisson process with intensity function λt (it is a density of the intensity measure with respect to Lebesgue measure).

(a) Show that the maximum likelihood estimator of the parameter λ is $\hat{\lambda}_T = 2N(T)/T^2$ (more precisely, the maximum in $\lambda > 0$ of the density on the observed path is attained if $N(T) > 0$, and in the case $N(T) = 0$ we set $\hat{\lambda}_T = 0$).

(b) Prove that $\hat{\lambda}$ is an unbiased and strongly consistent estimator (see the corresponding definitions in Problem 17.11).

(c) Prove that $T(\hat{\lambda}_T - \lambda)$ converges in distribution to the normal law as $T \to \infty$; that is, $\hat{\lambda}_T$ is an asymptotically normal estimator.

17.13. Let $f,g \in C(\mathbb{R}^+)$; $f(t) \geq 0$, $t > 0$; $g(t) > 0$, $t > 0$. Nonhomogeneous Poisson processes $\{N_f(t), N_g(t),\ t \geq 0\}$ are given with intensity functions f and g (these functions are the densities of the intensity measures with respect to Lebesgue measure). Let μ_1 be the measure generated by the process N_g on $\mathbb{D}([0,T])$, and μ_2 be the similar measure for N_f. Prove that $\mu_2 \ll \mu_1$ and

$$\frac{d\mu_2}{d\mu_1}(x) = \prod_i \frac{f(t_i)}{g(t_i)} \mathbb{I}_{x(t_i)-x(t_i-)=1} \exp\{\int_0^T (g(t) - f(t))dt\}, \tag{17.20}$$

$x \in \mathbb{D}([0,T])$. Here t_i are jump points of the function x, and if $x \in C([0,T])$ then the product in (17.20) is set to be equal to 1.

17.14. On $[0,T]$ a path $N(\cdot)$ is observed of a nonhomogeneous Poisson process with intensity function $1 + \lambda_0 t$, $\lambda_0 > 0$ (it is a density of the intensity measure with respect to Lebesgue measure).

(a) Write an equation for the maximum likelihood estimator $\hat{\lambda}_T$ of the parameter λ_0 and show that with probability 1 this equation has a unique positive root for all $T \geq T_0(\omega)$.

(b) Prove that $\hat{\lambda}_T$ is strongly consistent; that is, $\hat{\lambda}_T \to \lambda$ as $T \to \infty$, a.s.

17.15. On $[0,T]$ a path $x(\cdot)$ is observed of a mean square continuous stochastic process with given correlation function $r(s,t)$. For the integral operator J on $L_2([0,T])$ with the kernel $r(t,s)$ it holds that $\operatorname{Ker} J = \{0\}$. A mean value m of the process is estimated, and m does not depend on t. Let

$$M = \left\{ \hat{m} = \int_0^T f(t)x(t)dt \;\middle|\; f \in C([0,T]);\ \forall\, m \in \mathbb{R}:\ \mathsf{E}_m\hat{m} = m \right\}$$

(that is, M is a certain class of linear unbiased estimators). Here the integral is a mean square limit of integral Riemann sums, and E_m is a standard notation for the expectation with respect to the distribution μ_m of the observed process with mean m. Prove that:

(a)

$$\inf_{\hat{m}\in M} \mathsf{D}\hat{m} = \left(\sum_{n=1}^{\infty} \lambda_n^{-1} a_n^2 \right)^{-1},$$

where $\{\lambda_n, \varphi_n,\ n \geq 1\}$ are all the eigenvalues of J and corresponding orthonormal eigenfunctions, and $a_n = \int_0^T \varphi_n(t)dt,\ n \geq 1$.

(b) In particular if the series in (a) diverges then $\exists\ \{\hat{m}_k,\ k \geq 1\} \subset M:\ \hat{m}_k \to m$ as $k \to \infty$, a.s.; that is, then $\hat{m}_k$ is strictly consistent in the sense of Definition 17.7.

17.16. On $[0,T]$ a path $x(\cdot)$ is observed of a mean square continuous stochastic process with given correlation function $r(s,t)$. A mean value m of the process is estimated, and m does not depend on t. Let

$$M = \{\hat{m}_F = \int_0^T x(t)dF(t) \mid F \text{ is a function of bounded variation};$$
$$\forall\, m \in \mathbb{R}:\ \mathsf{E}_m\hat{m}_F = m\}.$$

Here the integral is a mean square limit of integral Riemann sums. For the notation E_m see the previous problem.

Suppose that there exists an estimator $\hat{m}_{F_0} \in M$ such that for all $s \in [0,T]$, $\int_0^T r(s,t)dF_0(t) = C$. Prove that

$$\min_{\hat{m}_H \in M} \mathsf{D}\hat{m}_H = \mathsf{D}\hat{m}_{F_0} = C.$$

17.17. On $[0,T]$ a path $x(\cdot)$ of the process with given correlation function $r(s,t)$ is observed . A mean value m of the process is estimated, and m does not depend on t. Let M be the class of estimators from Problem 17.16. Prove that the next estimator has the least variance in M.

(a) $\hat{m}_1 = (2+\beta T)^{-1}\left(x(0)+x(T)+\beta\int_0^T x(t)dt\right)$, if $r(s,t) = \exp\{-\beta|t-s|\}$ with $\beta > 0$.

(b) $\hat{m}_2 = x(0)$ if $r(s,t) = \min(s+1,t+1)$.

(c) In cases (a) and (b) prove that $\mathsf{D}\hat{m}_G > \min_{\hat{m}\in M}\mathsf{D}\hat{m}$, where $\hat{m}_G \in M$ and G is an absolutely continuous function of bounded variation (that is, $G(t) = G(0) + \int_0^t f(s)ds,\ t \in [0,T]$, with $f \in L_1([0,T])$).

17.18. On $[0,T]$ a path $x(\cdot)$ of the process $\{\mu t + \sigma W(t)\}$ is observed where W is a separable Wiener process with unknown parameters $\mu \in \mathbb{R}$ and $\sigma^2 > 0$.

(a) Construct an error-free estimate of the parameter σ^2.

(b) For a fixed σ^2, prove that the maximum likelihood estimator of the parameter μ is $\hat{\mu}_T = x(T)/T$.

(c) Prove that the expectation of $\hat{\mu}_T$ is μ, and $\hat{\mu}_T \to \mu$ as $T \to \infty$, a.s., and $\sqrt{T}(\hat{\mu}_T - \mu) \sim \mathcal{N}(0,\sigma^2)$.

17.19. Assume the conditions of Problem 17.18 and let the prior distribution $\mathcal{N}(\mu_0, \sigma_0^2)$ of the parameter μ be given.

(a) Find the posterior distribution of the parameter μ.

(b) Construct the Bayes estimator of the parameter under the quadratic loss function.

17.20. On $[0,T]$ a path $N(\cdot)$ is observed of a homogeneous Poisson process with intensity λ that has the prior gamma distribution $\Gamma(\alpha,\beta)$.

(a) Find the posterior distribution of the parameter λ.

(b) Construct the Bayes estimator λ under the quadratic loss function and under the all-or-nothing loss function.

17.21. On $[0,T]$ the process is observed

$$\left\{ x(t) = \int_0^t \left(\varphi(s) + \sum_{i=1}^m \theta_i g_i(s) \right) ds + \sigma W(t) \right\},$$

where W is a Wiener process, the unknown function φ belongs to a fixed subspace $K \subset L_2([0,T])$; $\{g_i\}$ are given functions from $L_2([0,T])$ that are linearly independent modulus K; that is, a linear combination of these functions which belongs to K is always a combination with zero coefficients; and $\theta = (\theta_1,\ldots,\theta_m)^\top \in \mathbb{R}^m$ and $\sigma > 0$ are unknown parameters. Let $M =$

$$\left\{ \hat{\theta} = \int_0^T f(t)dx(t) \;\middle|\; f \in L_2([0,T],\mathbb{R}^m);\ \forall\, \theta \in \mathbb{R}^m\ \forall\, \varphi \in K:\ \mathsf{E}_{\varphi,\theta}\hat{\theta} = \theta \right\}.$$

Prove that there exists a unique estimator $\hat{\theta}^* \in M$ such that for any estimate $\hat{\theta} \in M$ the matrix $S - S^*$ is positive semidefinite. Here S and S^* are covariance matrices of the estimators $\hat{\theta}$ and $\hat{\theta}^*$.

17.22. Let $X = [0,1]^2$, and $\Theta = [0,1] \cup [2,3]$, and for $\theta \in \Theta$ μ_θ be a measure on $\mathcal{B}(X)$. If $\theta \in [0,1]$ then μ_θ is Lebesgue measure on $[0,1] \times \{\theta\}$, whereas for $\theta \in [2,3]$, μ_θ is Lebesgue measure on $\{\theta - 2\} \times [0,1]$.

(a) Check that the measures $\{\mu_\theta\}$ are pairwise singular.

(b) Prove that there is no consistent estimator $\theta(x)$, $x \in X$, of the parameter θ.

Hints

17.1. Both processes are diffusion ones with continuous paths, and the density is

$$\rho(x) = \frac{d\mu_2}{d\mu_1}(x), \quad x \in C([0,T]).$$

17.2. (a) Use a representation of a homogeneous Poisson process given in Problem 5.17.
(b) The Neyman–Pearson criterion for equivalent measures can be applied.
17.3. The μ_1 and μ_2 are Gaussian measures on Hilbert space $X = L_2([0,T])$. Use the Hajek–Feldman theorem.
17.4. Let L be a closure of the set $R(X)$, where R is an integral operator with kernel $R(t,s)$. If a vector h is orthogonal to $R(X)$ then under the hypothesis H_1 the variance of r.v. $(x(\cdot),h(\cdot))$ is 0, therefore, the r.v. is equal to 0, a.s. Then $\mu_1(L) = 1$.
17.5. Under the hypothesis H_1, $\{(x(\cdot),\varphi_k(\cdot)),\ k \geq 1\}$ is a sequence of independent Gaussian random variables with distributions $\mathcal{N}(0,\lambda_k)$, $k \geq 1$.
17.6. Solve an integral equation (17.18) where $T = 1$, $a(\cdot)$ is the function from the problem situation, and $R(t,s) = e^{-|t-s|}$, $t,s \in [0,1]$.
17.7. Let $\{\varphi_n,\ n \in \mathbb{N}\}$ be an orthonormal system of eigenfunctions of the operator A with corresponding eigenvalues λ_n, $n \in \mathbb{N}$. Then under both hypotheses $x_n := \int_0^T x(t)\varphi_n(t)dt$, $n \geq 1$ is a sequence of centered independent Gaussian random variables.
17.8. Because the diffusion coefficients are different, singular measures on $C([0,2])$ correspond to the hypotheses.
17.9. (a) The equality (17.11) holds true.
(b) The density of μ_2 with respect to μ_1 can be found by the formula (17.10).
17.10. The condition (17.9) holds.
17.11. (a) Let

$$\rho_\lambda(x) = \frac{d\mu_\lambda}{d\mu_1}(x), \quad x \in \mathbb{D}([0,T]).$$

Here μ_λ, $\lambda > 0$ is a measure generated by a homogeneous Poisson process with intensity λ. Then $\hat{\lambda}_T$ is a point of maximum in $\lambda > 0$ of the log-density $L(\lambda;N) := \log \rho_\lambda(N)$.
(b) For $T \in \mathbb{N}$ use the SLLN and CLT.
17.12. (a) Let ν_λ be a measure on $\mathbb{D}([0,T])$ generated by given process. The formula for the density

$$\rho_\lambda(x) := \frac{d\nu_\lambda}{d\nu_1}(x)$$

is derived similarly to Problem 17.2 (a).

(b), (c) The process $\{N_1(t) := N(\sqrt{2t}),\ t \geq 0\}$ is a homogeneous process with intensity λ.
17.13. Use Problem 5.17 and generalize the solution of Problem 17.2 (a).
17.14. (a) Use Problem 17.13. The derivative in λ of the log-density $L(\lambda,N)$ at the observed path is a strictly decreasing function in λ.
(b) Investigate the behavior of the function

$$\varphi(\lambda,T) := T^{-2}\frac{\partial L(\lambda,N)}{\partial\lambda}$$

as $T\to\infty$, when λ is from the complement to the fixed neighborhood of λ_0.
17.15. (a) Expand f in Fourier series by the basis $\{\varphi_n\}$.
(b) Use the Riesz lemma about a subsequence of random variables that converges a.s.
17.16. Let the minimum of the variance be attained at $\hat{m}_F \in M$. For $\alpha,\beta\in[0,T]$ introduce $G(t) = \mathbb{I}_{t\geq\alpha} - \mathbb{I}_{t\geq\beta}$, $t\in[0,T]$. Then for all $\delta\in\mathbb{R}$ it holds $\hat{m}_{F+\delta G}\in M$.
17.17. Use Problem 17.16.
17.18. (a) Use Problem (17.4).
(b) Let the measure μ_1 correspond to the process $\{\sigma W(t)\}$, and the measure μ_2 be generated by the given process. Use the formula (17.7).
(c) Use Problem 3.18.
17.19. Use a density $\rho(x)$ from the solution of Problem 17.18.
17.20. The density of the distribution of the process is derived in Problem 17.2.
17.21. Reformulate this problem in terms of vectors in the space $H = L_2([0,T])$.
17.22. (b) Prove to the contrary. Let $\theta(x)$ be a consistent estimator. Introduce $A_1 = \{x:\ \theta(x)\in[0,1]\}$. Then for all $x_2\in[0,1]$ it holds $\lambda^1(\{x_1\in[0,1]:\ (x_1,x_2)\in A_1\}) = 1$.

Answers and Solutions

17.1. Under the hypothesis H_k the observed process $\{x(t),\ t\in[0,T]\}$ generates a measure μ_k on the space $E = C([0,T])$; $k=1,2$. By Theorem 17.2 we have $\mu_1\sim\mu_2$, and by formula (17.7) it holds

$$\rho(x) = \exp\{\gamma x(T) - \frac{\gamma^2}{2}T\},\quad x\in E.$$

Without loss of generality we can assume that $\gamma>0$ (for $\gamma<0$ one should consider the process $y(t) := -x(t)$). Then for $\lambda>0$

$$R^\lambda := \{x\in E:\ \rho(x)<\lambda\} = \left\{x\in E:\ x(T) < \frac{1}{\gamma}(\log\lambda + \frac{1}{2}\gamma^2 T)\right\}.$$

Under the hypothesis H_1, $x(T)\sim\mathcal{N}(0,T)$, whereas under the hypothesis H_2, $x(T)\sim\mathcal{N}(\gamma T,T)$. Fix $\varepsilon\in(0,1)$. We are looking for $c=c(\varepsilon)$ such that $\mu_2(\{x:\ x(T)<c\}) = \varepsilon$. We have

$$\mu_2(\{x:\ x(T)<c\}) = \mu_2\left(\left\{x:\ \frac{x(T)-\gamma T}{\sqrt{T}} < \frac{c-\gamma T}{\sqrt{T}}\right\}\right)$$
$$= \Phi\left(\frac{c-\gamma T}{\sqrt{T}}\right) = \varepsilon$$

for $c(\varepsilon) = \Phi^{-1}(\varepsilon)\sqrt{T} + \gamma T$. According to the Neyman–Pearson criterion we accept H_1 if $x(T) < c(\varepsilon)$; otherwise we accept H_2. At that

$$\alpha_{21} = 1 - \mu_1(\{x\colon x(T) < c(\varepsilon)\}) = \bar{\Phi}\left(\frac{c(\varepsilon)}{\sqrt{T}}\right),$$

where $\bar{\Phi} := 1 - \Phi$.

17.2. (a) Let $\{\xi_n,\ n \geq 1\}$ be independent random variables, uniformly distributed on $[0,T]$, and $\nu_i \sim \mathrm{Pois}(\lambda_i T)$, ν_i is independent of $\{\xi_n\}$, $i = 1,2$. According to Problem 5.17 the process

$$\left\{X_i(t) = \sum_{n=1}^{\nu_i} \mathbb{I}_{\xi_n \leq t},\quad t \in [0,T]\right\}$$

is a homogeneous Poisson process on $[0,T]$ with intensity λ_i. A measure μ_i on E that is generated by the process X_i, is concentrated on the set of functions of the form

$$f_0(t) = 0,\quad f_k(x,t) = \sum_{n=1}^{k} \mathbb{I}_{x_n \leq t},\quad t \in [0,T],$$

where $k \geq 1$ and $x = (x_1,\ldots,x_k)$ is a vector of k distinct points from the interval $(0,T)$. Let $F_k = \{f_k(x,\cdot) \mid x = (x_1,\ldots,x_k) \in A_k\}$ where A_k is a symmetric Borel set in $(0,T)^k$, and $F_0 = \{f_0\}$. Then

$$\mu_i(F_k) = \mathsf{P}\left\{\sum_{n=1}^{\nu_i} \mathbb{I}_{\xi_n \leq t} \in F_k,\ \nu_i = k\right\} = \mathsf{P}\{(\xi_1,\ldots,\xi_k) \in A_k\} \cdot \mathsf{P}\{\nu_i = k\},$$

where $k \geq 1$ and $i = 1,2$. Hence for $k \geq 0$ we have

$$\frac{\mu_2(F_k)}{\mu_1(F_k)} = \frac{\mathsf{P}\{\nu_2 = k\}}{\mathsf{P}\{\nu_1 = k\}} = \left(\frac{\lambda_2}{\lambda_1}\right)^k e^{(\lambda_1 - \lambda_2)T}.$$

Therefore, for any Borel set $B \subset \mathbb{D}([0,T])$,

$$\int_B \left(\frac{\lambda_2}{\lambda_1}\right)^{x(T)} e^{(\lambda_1-\lambda_2)T} d\mu_1(x) =$$

$$= \sum_{k=0}^{\infty} \int_{\{x \in B\colon x(T)=k\}} \left(\frac{\lambda_2}{\lambda_1}\right)^{x(T)} e^{(\lambda_1-\lambda_2)T} d\mu_1(x) =$$

$$= \sum_{k=0}^{\infty} \mu_2(\{x \in B\colon x(T) = k\}) = \mu_2(B).$$

(b) Suppose that $\lambda_2 > \lambda_1$. For $\lambda > 0$ we have

$$R^\lambda := \{x \in E\colon \rho(x) < \lambda\} = \{x \in E\colon x(T) < y\},\ y = \frac{\log\lambda + (\lambda_2 - \lambda_1)T}{\log\lambda_2 - \log\lambda_1}.$$

Given $\varepsilon \in (0,1)$ we are looking for $y = n_\varepsilon \in \mathbb{Z}^+$ such that $\mu_2(R^\lambda) < \varepsilon$ and $\mu_2(\mathbb{R}^\lambda \cup \Gamma^\lambda) \geq \varepsilon$. Under the hypothesis H_2 we have $x(T) \sim \mathrm{Pois}(\lambda_2 T)$. The desired n_ε can be found uniquely from the condition

$$\sum_{0\leq k<n_\varepsilon} \frac{(T\lambda_2)^k}{k!} e^{-T\lambda_2} < \varepsilon \leq \sum_{0\leq k\leq n_\varepsilon} \frac{(T\lambda_2)^k}{k!} e^{-T\lambda_2}.$$

If $x(T) < n_\varepsilon$ then we accept the hypothesis H_1, and if $x(T) > n_\varepsilon$ then accept H_2. In the case $x(T) = n_\varepsilon$ we accept H_2 with probability p and accept H_1 with probability $1-p$. Here

$$p = \frac{\varepsilon - \mu_2(R^\lambda)}{\mu_2(\Gamma^\lambda)} = \left(\varepsilon - \sum_{k<n_\varepsilon} \frac{(T\lambda_2)^k}{k!} e^{-T\lambda_2}\right) \left(\frac{(T\lambda_2)^{n_\varepsilon}}{n_\varepsilon !} e^{-T\lambda_2}\right)^{-1}.$$

It can happen that $p = 1$; then we obtain a nonrandomized rule: for $x(T) \geq n_\varepsilon$ accept H_2, and for $x(T) < n_\varepsilon$ accept H_1.

17.3. Let $\{\lambda_k, \varphi_k;\ k \geq 1\}$ be a sequence of positive eigenvalues and corresponding normalized eigenfunctions of the integral operator R on X with kernel $R(t,s)$. Introduce an operator $R^{1/2}$ on X,

$$R^{1/2} g = \sum_{k\geq 1} \lambda_k^{1/2} (g, \varphi_k)\varphi_k, \quad g \in X.$$

The measure μ_1 is Gaussian with zero mean and the correlation operator R, and the measure μ_2 is Gaussian as well with the same correlation operator and the mean $a = a(t) \in X$. By the Hajek–Feldman theorem, in the case $a(\cdot) \in R^{1/2}(X)$ the measures are equivalent; otherwise they are singular. In the case (1) from the above-mentioned subsection, $a(\cdot)$ does not belong to the closure L of the set $R^{1/2}(X)$, and in the case (2) it holds $a(\cdot) \in L \setminus R^{1/2}(X)$. Therefore, in both cases the measures are singular. Under the condition (17.16) it holds $a(\cdot) \in R^{1/2}(X)$, and then $\mu_1 \sim \mu_2$. The desired density is found by the Hajek–Feldman theorem.

17.4. The function $\hat{a}(\cdot)$ from (17.12) is a nonzero vector orthogonal to L. Under the hypothesis H_1 it holds $x(\cdot) \in L$ a.s. (see Hints), hence $\hat{l}(x(\cdot)) = (x, \hat{a}) = 0$ a.s. Under the hypothesis H_2 we have $x(t) = a(t) + x_0(t)$ where $x_0(\cdot) \in L$ a.s., and then $\hat{l}(x(\cdot)) = (a, \hat{a}) + (x_0, \hat{a}) = (a, \hat{a})$ a.s. Moreover

$$(a, \hat{a}) = \|a\|^2 - \sum_{k\geq 1} (a, \varphi_k)^2 > 0$$

because $a \notin L$ by the problem situation.

17.5. Let z_n be a Gaussian r.v. under the limit in (17.15). Under the hypothesis H_1 we have $\mathsf{E} z_n = 0$,

$$\mathsf{D} z_n = \left(\sum_{k=1}^{m_n} \frac{a_k^2}{\lambda_k}\right)^{-1} \leq \frac{1}{n}, \quad \mathsf{E} z_n^4 \leq \frac{3}{n^2},$$

therefore,

$$\sum_{n=1}^{\infty} \mathsf{E} z_n^4 < \infty.$$

For $\varepsilon > 0$ by the Chebyshev inequality we have $\mathsf{P}\{|z_n| \geq \varepsilon\} \leq \varepsilon^{-4}\mathsf{E} z_n^4$, thus

$$\sum_{n=1}^{\infty} \mathsf{P}\{|z_n| \geq \varepsilon\} < \infty,$$

and by the Borel–Cantelli lemma with probability 1 for $n \geq n_0(\varepsilon, \omega)$ it holds $|z_n| < \varepsilon$. Therefore, under the hypothesis H_1 it holds $z_n \to 0$, a.s. Next, under the hypothesis H_2,

$$z_n = 1 + \left(\sum_{k=1}^{m_n} \frac{a_k^2}{\lambda_k} \right)^{-1} \sum_{k=1}^{m_n} \frac{a_k}{\lambda_k} (x - a, \varphi_k) \to 1, \text{ a.s.}$$

17.6. We are looking for a continuous solution $b(\cdot)$ to the integral equation

$$a(t) = \int_0^1 e^{-|t-s|} b(s) ds, \quad t \in [0,1].$$

Rewrite the equation in a form

$$a(t) = \int_0^t e^{s-t} b(s) ds + \int_t^1 e^{t-s} b(s) ds,$$

whence

$$a'(t) = -\int_0^t e^{s-t} b(s) ds + \int_t^1 e^{t-s} b(s) ds,$$
$$a''(t) = \int_0^t e^{s-t} b(s) ds + \int_t^1 e^{t-s} b(s) ds - 2b(t),$$

$$b(t) = \frac{a(t) - a''(t)}{2}, \quad t \in [0,1]. \tag{17.21}$$

Integrating by parts we verify that this continuous function does satisfy the given integral equation. It is essential that $a \in C^2([0,1])$, $a'(0) = a(0)$, and $a'(1) = -a(1)$.

Then the density $\rho = \frac{d\mu_2}{d\mu_1}$ can be written in the form (17.19), with the function $b(\cdot)$ given in (17.21). We have

$$\int_0^1 a(s) b(s) ds = \frac{1}{2} \left(\int_0^1 a^2(s) ds - \int_0^1 a(s) a''(s) ds \right)$$
$$= \frac{1}{2} \left(\|a\|_{L_2}^2 + \|a'\|_{L_2}^2 + a^2(0) + a^2(1) \right),$$

that is denoted by σ^2 in the problem situation. Then

$$\log \rho(x) = \int_0^1 x(s)b(s)ds - \frac{\sigma^2}{2}, \quad x \in C([0,1]), \tag{17.22}$$

and under both hypotheses

$$\mathsf{D}\log \rho(x(\cdot)) = \int_0^1 \left(\int_0^1 e^{-|t-s|} b(s)ds \right) b(t)dt = \int_0^1 a(t)b(t)dt = \sigma^2.$$

Next, under H_1,

$$\mathsf{E}\log \rho(x(\cdot)) = -\frac{\sigma^2}{2} \text{ and } \log \rho(x(\cdot)) \sim \mathcal{N}\left(-\frac{\sigma^2}{2}, \sigma^2\right).$$

According to representation (17.22) we have under H_2, that

$$\mathsf{E}\log \rho(x(\cdot)) = \int_0^1 a(s)b(s)ds - \frac{\sigma^2}{2} = \frac{\sigma^2}{2} \text{ and } \log \rho(x(\cdot)) \sim \mathcal{N}\left(\frac{\sigma^2}{2}, \sigma^2\right).$$

We use the Neyman–Pearson criterion to construct the decision rule. Fix $\varepsilon \in (0,1)$. We accept H_1 if $\log \rho(x(\cdot)) < C$, where a threshold C is found from the equation

$$\mu_2(\{x:\ \log \rho(x) < C\}) = \mathsf{P}\left\{\mathcal{N}\left(\frac{\sigma^2}{2}, \sigma^2\right) < C\right\} = \varepsilon,$$

and $C = \sigma \Phi^{-1}(\varepsilon) + \sigma^2/2$. Thus, H_1 is accepted if

$$\int_0^1 x(s)b(s)ds < \sigma \Phi^{-1}(\varepsilon) + \sigma^2;$$

otherwise we accept H_2. There the Type II error is

$$\alpha_{21} = \mu_1(\{x: \log \rho(x) \geq C\}) = \mathsf{P}\left\{\mathcal{N}\left(-\frac{\sigma^2}{2}, \sigma^2\right) \geq C\right\} = \bar{\Phi}\left(\Phi^{-1}(\varepsilon) + \sigma\right).$$

17.7. Under the hypothesis H_1 we have $\mathsf{D}x_n = \lambda_n$, whereas under the hypothesis H_2 it holds $\mathsf{D}x_n = \sigma^2 \lambda_n$. By the SLLN, under H_1 we have

$$y_n := \frac{1}{n} \sum_{k=1}^{n} \frac{x_k^2}{\lambda_k} \to 1, \text{ a.s.},$$

whereas under hypothesis H_2 it holds $y_n \to \sigma^2 \neq 1$, a.s. Thus, we accept H_1 if

$$\lim_{n\to\infty} \frac{1}{n} \sum_{k=1}^{n} \frac{1}{\lambda_k} \left(\int_0^T x(t)\varphi_k(t)dt \right)^2 = 1;$$

otherwise we accept H_2.

17.8. For $t \in [0,2]$ we set

$$\lambda(t) = \lim_{n\to\infty} \sum_{k=0}^{2^n-1} \left(x\left(\frac{k+1}{2^n}t\right) - x\left(\frac{k}{2^n}t\right) \right)^2 .$$

Under the hypothesis H_k this limit exists a.s., and by the formula (17.5)

$$\lambda(t) = \int_0^t b_k(s,x(s))ds; \quad k = 1,2.$$

In particular $\lambda(t) = t$ for $k = 1$, and $\lambda(t) = t^2/2$ for $k = 2$. The values of these functions differ, for example, at $t = 1$. Therefore, we calculate

$$\lambda(1) = \lim_{n\to\infty} \sum_{k=0}^{2^n-1} \left(x\left(\frac{k+1}{2^n}\right) - x\left(\frac{k}{2^n}\right) \right)^2 .$$

We accept the hypothesis H_1 if $\lambda(1) = 1$; otherwise we accept H_2.

17.9. (a) In the notations of subsection *Homogeneous in space processes* from Theoretical grounds, we have $a_1(t) = 0$, $B(t) = t$; $a_2(t) = |\log t|^{-1/2}$ for $t \in (0,T]$, $a_2(0) = 0$; $a(t) = a_2(t)$. There exists a solution $b(t)$ to the equation (17.8) and it is equal to

$$b(t) = \begin{cases} 0, & t = 0 \\ \dfrac{1}{t\sqrt{|\log t|}}, & 0 < t \le T. \end{cases}$$

Integral (17.11) is divergent:

$$\int_0^T a(t)b(t)dt = \int_0^T \frac{dt}{t|\log t|} = +\infty.$$

That is why measures in the space $C([0,T])$ that correspond to the distribution of the process under the hypotheses H_1 and H_2 are singular.

In order to construct an error-free criterion we have to find a sequence of continuous functions $b_n(t)$, $n \ge 1$, $t \in [0,T]$, such that

$$\lim_{n\to\infty} \int_0^T a(t)b_n(t)dt = +\infty.$$

The functions can be defined as follows:

$$b_n(t) = \begin{cases} b(t), & \frac{T}{n} \le t \le T, \\ b\left(\frac{T}{n}\right), & 0 \le T < \frac{T}{n}. \end{cases}$$

If

$$\lim_{n\to\infty}\Big(\int_0^T tb_n^2(t)dt\Big)^{-1}\int_0^T b_n(t)dx(t)=0$$

then we accept H_1; otherwise we accept H_2.

(b) In the notations of subsection *Homogeneous in space processes* from Theoretical grounds, we have $a_2(t)=\sqrt{t}=a(t)$; $b(t)=t^{-1/2}$ for $t\in(0,T]$ and $b(0)=0$. Condition (17.9) holds, therefore, a density of μ_2 with respect to μ_1 in the space $C([0,T])$ is equal to

$$\rho(x)=\exp\{\int_0^T\frac{dx(t)}{\sqrt{t}}-\frac{1}{2}\int_0^T\frac{1}{\sqrt{t}}\sqrt{t}dt\},$$

$$\log\rho(x)=\int_0^T\frac{dx(t)}{\sqrt{t}}-\frac{T}{2}.$$

Under the hypothesis H_k we have

$$\log\rho(x)\sim\mathcal{N}(m_k,\sigma^2),\quad \sigma^2=T,\quad m_k=(-1)^k\frac{1}{2}\sigma^2=(-1)^k\frac{T}{2};\quad k=1,2.$$

Fix a bound ε for the Type I error, $\varepsilon\in(0,1)$. We accept the hypothesis H_1 if

$$\log\rho(x(\cdot))<L:=\sigma\Phi^{-1}(\varepsilon)+\frac{T}{2}$$

which is equivalent to

$$\int_0^T\frac{dx(t)}{\sqrt{t}}<\sqrt{T}\Phi^{-1}(\varepsilon)+T.$$

Otherwise we accept H_2. Then $\alpha_{12}=\varepsilon$,

$$\alpha_{21}=\mathsf{P}\{\mathcal{N}(m_1,\sigma^2)\ge L\}=\bar{\Phi}\left(\frac{L-m_1}{\sigma}\right),$$

$$\alpha_{21}=\bar{\Phi}\left(\Phi^{-1}(\varepsilon)+\sqrt{T}\right).$$

17.10. In the notations of subsection *Homogeneous in space processes* from Theoretical grounds we have $a_1(t)=0$, $B(t)=\operatorname{diag}(1,t)$, $a_2(t)=a(t)=(\sqrt[4]{t};\sqrt[4]{t})^\top$. From equation (17.8) we find $b(t)=(\sqrt[4]{t};b_2(t))^\top$ with $b_2(t)=t^{-3/4}$ for $t\in(0,T]$ and $b_2(0)=0$. Check the condition (17.9):

$$\int_0^T(a(t),b(t))dt=\int_0^T\left(\sqrt{t}+\frac{1}{\sqrt{t}}\right)dt=\frac{2}{3}T^{3/2}+2T^{1/2}<\infty.$$

Then by the formula (17.10)

$$\log \rho(x) = \int_0^T \sqrt[4]{t}\, dx_1(t) + \int_0^T \frac{dx_2(t)}{t^{3/4}} - \left(\frac{T^{3/2}}{3} + T^{1/2} \right),$$

where $(x_1(t), x_2(t))^\top = x(t)$ is the observed vector path. Under the hypothesis H_k, $k = 1, 2$,

$$\log \rho(x(\cdot)) \sim \mathcal{N}(m_k, \sigma^2), \quad \sigma^2 = 2\left(\frac{T^{3/2}}{3} + T^{1/2} \right), \quad m_k = (-1)^k \frac{\sigma^2}{2}.$$

Let the bound ε of the Type I error be given, $\varepsilon \in (0,1)$. We accept the hypothesis H_1 if

$$\log \rho(x(\cdot)) < L := \sigma \Phi^{-1}(\varepsilon) + \frac{\sigma^2}{2},$$

which is equivalent to

$$\int_0^T \sqrt[4]{t}\, dx_1(t) + \int_0^T \frac{dx_2(t)}{t^{3/4}} < \sigma^2;$$

otherwise we accept H_2. Then $\alpha_{12} = \varepsilon$ and

$$\alpha_{21} = \mathsf{P}\{\mathcal{N}(m_1, \sigma^2) \ge L\} = \bar{\Phi}\left(\frac{L - m_1}{\sigma} \right) = \bar{\Phi}\left(\Phi^{-1}(\varepsilon) + \sigma \right).$$

17.11. (a) The log-density is $L(\lambda; N) = N(T) \log \lambda + (1 - \lambda)T$, $\lambda > 0$. In the case $N(T) > 0$ it attains its maximum at $\lambda = \hat{\lambda}_T := T^{-1} N(T)$.

(c) Introduce an i.i.d. sequence $n_i = N(i) - N(i-1)$, $i \ge 1$; $n_1 \sim \mathrm{Pois}(\lambda)$. For $N = k \in \mathbb{N}$ consider as $k \to \infty$:

$$\hat{\lambda}_k = \frac{N(k)}{k} = \frac{n_1 + \cdots + n_k}{k} \to \mathsf{E} n_1 = \lambda, \text{ a.s.} \tag{17.23}$$

Next, for any real $T \ge 1$ consider

$$\hat{\lambda}_T = \frac{N([T])}{[T]} \cdot \frac{[T]}{T} + \delta_T, \quad \delta_T := \frac{N(T) - N([T])}{T}. \tag{17.24}$$

As a result of (17.23) the first summand in (17.24) tends to λ, a.s. We have

$$0 \le \delta_T \le \frac{N([T]+1) - N([T])}{[T]} = \frac{n_{m+1}}{m}, \quad m := [T].$$

It remains to prove that as $m \to \infty$,

$$\frac{n_{m+1}}{m} \to 0, \text{ a.s.} \tag{17.25}$$

Consider

$$\sum_{m=1}^{\infty} \mathsf{E}\left(\frac{n_{m+1}}{m}\right)^2 = \sum_{m=1}^{\infty} \frac{\lambda+\lambda^2}{m^2} < \infty.$$

For $\varepsilon > 0$ we have by the Chebyshev inequality that

$$\mathsf{P}\left\{\left|\frac{n_{m+1}}{m}\right| > \varepsilon\right\} \le \frac{1}{\varepsilon^2}\left(\frac{n_{m+1}}{m}\right)^2,$$

therefore, a series of these probabilities converges. Then by the Borel–Cantelli lemma $|n_{m+1}/m| \le \varepsilon$ for all $m \ge m_0(\omega)$, a.s. This implies (17.25).

(d) For $T = k \in \mathbb{N}$ we have as $k \to \infty$:

$$\sqrt{k}(\hat{\lambda}_k - \lambda) = \frac{(n_1-\lambda)+\cdots+(n_k-\lambda)}{\sqrt{k}} \xrightarrow{d} \mathcal{N}(0, \mathsf{D}n_1) = \mathcal{N}(0,\lambda). \tag{17.26}$$

According to the expansion (17.24), for any $T \ge 1$ we have

$$\sqrt{T}(\hat{\lambda}_T - \lambda) = \frac{N([T]) - \lambda T}{T} + \sqrt{T}\delta_T.$$

Now, (17.26) implies that the first summand converges in distribution to $N(0,\lambda)$ as $T \to \infty$. The second summand is estimated as

$$0 \le \sqrt{T}\delta_T \le \frac{n_{m+1}}{\sqrt{T}},$$

where $m = [T]$. Then $\mathsf{E}|\sqrt{T}\delta_T| \to 0$ as $T \to \infty$, and $\sqrt{T}\delta_T \xrightarrow{P} 0$ as $T \to \infty$. Finally the Slutzky lemma implies that

$$\sqrt{T}(\hat{\lambda}_T - \lambda) \xrightarrow{d} \mathcal{N}(0,\lambda), \quad T \to \infty.$$

17.12. (a) The density mentioned in Hints is equal to

$$\rho_\lambda(x) = \lambda^{x(T)} e^{(1-\lambda)T^2/2}, \quad x \in \mathbb{D}([0,T]),$$

and this implies the desired relation.

(b), (c) For the process N_1 introduced in Hints, we have

$$\hat{\lambda}_T = \frac{N_1(T^2/2)}{T^2/2},$$

hence the desired relation follows from Problem 17.1. In particular

$$\sqrt{\frac{T^2}{2}}(\hat{\lambda}_T - \lambda) \xrightarrow{d} \mathcal{N}(0,\lambda), \quad T(\hat{\lambda}_T - \lambda) \xrightarrow{d} \mathcal{N}(0,2\lambda),$$

as $T \to \infty$.

17.13. Let $\{\xi_{n1},\ n \ge 1\}$ be independent random variables distributed on $[0,T]$ with a density $g_T(t) = g(t)\left(\int_0^T g(s)ds\right)^{-1}$, and $\{\xi_{n2},\ n \ge 1\}$ be an i.i.d. sequence with similar density f_T generated by the function f; $\nu_1 \sim \text{Pois}\left(\int_0^T g(t)dt\right)$ and $\nu_2 \sim \text{Pois}\left(\int_0^T f(t)dt\right)$, and ν_i is independent of $\{\xi_{ni},\ n \geq 1\}$, $i = 1,2$. According to Problem 5.17 the processes

$$\left\{X_i(t) = \sum_{n=1}^{\nu_i} \mathrm{1\!I}_{\xi_{ni} \geq t},\ t \in [0,T]\right\}$$

are nonhomogeneous Poisson processes on $[0,T]$ with intensity functions g (for $i = 1$) and f (for $i = 2$). In the notations from the solution to Problem 17.2 (a), we have for the measures μ_1 and μ_2 for $k \geq 1$:

$$\begin{aligned}\mu_2(F_k) &= \mathsf{P}\{(\xi_{12},\dots,\xi_{k2}) \in A_k\} \cdot \mathsf{P}\{\nu_2 = k\} \\ &= \int_{A_k} \prod_{i=1}^{k} \frac{f_T(t_i)}{g_T(t_i)} \cdot \prod_{k=1}^{k} g_T(t_i)dt_1 \dots dt_k \cdot \mathsf{P}\{\nu_1 = k\} \\ &\times \left(\frac{\int_0^T g(t)dt}{\int_0^T f(t)dt}\right)^k \exp\{\int_0^T (g(t) - f(t))dt\}.\end{aligned}$$

Here $\prod_{i=1}^k g_T(t_i)$ is a density of random vector $(\xi_{11},\dots,\xi_{k1})$. Then

$$\mu_2(F_k) = \int_{A_k} \prod_{i=1}^{k} \frac{f(t_i)}{g(t_i)} \times \exp\{\int_0^T (g(t) - f(t))dt\} d\mu_1(x),$$

where $t_i = t_i(x)$ are jump points of a step-function $x(\cdot)$. As in the solution to Problem 17.2 (a), this implies that for any Borel set $B \subset \mathbb{D}([0,T])$,

$$\mu_2(B) = \int_B \prod_i \frac{f(t_i)}{g(t_i)} \mathrm{1\!I}_{x(t_i)-x(t_i-)=1} \exp\{\int_0^T (g(t) - f(t))dt\} d\mu_1(x).$$

17.14. (a) Based on Problem 17.13 the estimator $\hat{\lambda}_T$ is found as a maximum point of the function

$$L_0(\lambda, N) = \sum_{i=1}^{N(T)} \log(1 + \lambda t_i) - \frac{\lambda T^2}{2}, \quad \lambda > 0,$$

or (for $N(T) \geq 1$) as a solution to the equation

$$h_T(\lambda) := \frac{1}{T^2} \sum_{i=1}^{N(T)} \frac{t_i}{1 + \lambda t_i} = \frac{1}{2}, \quad \lambda > 0.$$

The function h_T is strictly increasing and continuous in $\lambda \geq 0$. We have $h_T(+\infty) = 0 < \frac{1}{2}$. For the existence of a unique solution one has to ensure that

$$h_T(0) = \frac{1}{T^2} \sum_{i=1}^{N(T)} t_i > \frac{1}{2}. \tag{17.27}$$

We have $\mathsf{E}h_T(0) = T^{-2}\mathsf{E}N(T)\cdot\mathsf{E}(t_i \mid t_i \le T)$. Here t_i is any jump point of the observed path, and under the condition $t_i \le T$ its density is equal to

$$\frac{1+\lambda_0 t}{T+\frac{\lambda_0 T^2}{2}}.$$

Then

$$\mathsf{E}h_T(0) = \frac{1}{T^2}\int_0^T t(1+\lambda_0 t)dt = \frac{1}{2} + \frac{\lambda_0 T}{3} > \frac{1}{2}. \tag{17.28}$$

It is straightforward to check that $h_T(0) - \mathsf{E}h_T(0) \to 0$ as $T \to \infty$, a.s. Therefore, (17.28) implies that the inequality (17.27) holds with probability 1 for all $T \ge T_0(\omega)$. For such $T \ge T_0(\omega)$ there exists a unique maximum point of the function $L_0(\lambda, N)$.

(b) Notice that

$$\mathsf{E}h_T(\lambda_0) = \frac{1}{T^2}\int_0^T \frac{t(1+\lambda_0 t)}{1+\lambda_0 t}dt = \frac{1}{2}.$$

Fix $0 < \varepsilon < \lambda_0$. For $0 < \lambda \le \lambda_0 - \varepsilon$ we have

$$h_T(\lambda) \ge h_T(\lambda_0 - \varepsilon) = \mathsf{E}h_T(\lambda_0 - \varepsilon) + o(1) \ge \frac{1}{2} + \delta_1(\varepsilon) + o(1).$$

Here $\delta_1(\varepsilon) > 0$ and $o(1)$ is a r.v. tending to 0 as $T \to \infty$, a.s. In a similar way for $\lambda \ge \lambda_0 + \varepsilon$ we have

$$h_T(\lambda) \le h_T(\lambda_0 + \varepsilon) = \mathsf{E}h_T(\lambda_0 + \varepsilon) + o(1) \le \frac{1}{2} - \delta_2(\varepsilon) + o(1).$$

Because $h_T(\hat{\lambda}_T) = \frac{1}{2}$, then with probability 1 there exists $T_\varepsilon(\omega)$ such that for all $T \ge T_\varepsilon(\omega)$ it holds $|\hat{\lambda}_T - \lambda| < \varepsilon$. This proves the strong consistency of $\hat{\lambda}_T$.

17.15. (a) Let f generate $\hat{m} \in M$. The unbiasedness of $\hat{m}$ is equivalent to the condition $\int_0^T f(t)dt = 1$. Then

$$\mathsf{D}\hat{m} = (Jf, f) = \sum_{i=1}^{\infty} \lambda_n c_n^2, \quad c_n := (f, \varphi_n), \quad n \ge 1,$$

at that $\sum_{n=1}^{\infty} c_n a_n = 1$. By the Cauchy–Schwartz inequality

$$1 = \left(\sum_{n=1}^{\infty} c_n a_n\right)^2 \le \sum_{n=1}^{\infty} \lambda_n c_n^2 \cdot \sum_{n=1}^{\infty} \lambda_n^{-1} a_n^2, \tag{17.29}$$

and $\mathsf{D}\hat{m} \ge \left(\sum_{n=1}^{\infty} \lambda_n^{-1} a_n^2\right)^{-1}$.

Let the latter series converge (the divergency case is treated in solution (b) below). The equality in (17.29) is attained if c_n is proportional to $\lambda_n^{-1} a_n$ (though $\sum_{n=1}^{\infty} \lambda_n^{-1} a_n \varphi_n$ is not necessarily a continuous function). Introduce a continuous function

$$f_N(t) = \left(\sum_{n=1}^{N} \lambda_n^{-1} a_n^2\right)^{-1} \cdot \sum_{n=1}^{N} \lambda_n^{-1} a_n \varphi_n(t), \quad N \geq 1, \quad t \in [0,T],$$

and the corresponding estimator

$$\hat{m}_N = \int_0^T f_N(t)x(t)dt = m + \left(\sum_{n=1}^{N} \lambda_n^{-1} a_n^2\right)^{-1} \sum_{n=1}^{N} \sqrt{\lambda_n^{-1}} a_n x_n.$$

Here

$$x_n = \frac{1}{\sqrt{\lambda_n}} \int_0^T \varphi_n(t)(x(t)-m)\,dt, \quad n \geq 1,$$

is a sequence of uncorrelated random variables with zero mean and unit variance. Then

$$\lim_{N\to\infty} \mathsf{D}\hat{m}_N = \lim_{N\to\infty} \left(\sum_{n=1}^{N} \lambda_n^{-1} a_n^2\right)^{-1} = \left(\sum_{n=1}^{\infty} \lambda_n^{-1} a_n^2\right)^{-1}.$$

Moreover there exists the mean square limit of $\hat{m}_N$ as $N \to \infty$, and this is an unbiased estimator $\hat{m}^*$ with variance $\left(\sum_{n=1}^{\infty} \lambda_n^{-1} a_n^2\right)^{-1}$. This estimator can be out of M.

(b) In this case $\mathsf{D}\hat{m}_N \to 0$ as $N \to \infty$. Now, from the unbiasedness of $\hat{m}_N$ it follows that $\hat{m}_N \xrightarrow{P} m$ as $N \to \infty$, and by the Riesz lemma there exists a subsequence of estimators $\{\hat{m}_{N(k)},\, k \geq 1\}$ that converges to m, a.s.

17.16. Let $\hat{m}_F \in M$. Then $\int_0^T dF(t) = m$ and

$$\mathsf{D}\hat{m}_F = \int_0^T \int_0^T r(s,t)dF(s)dF(t) =: \Phi(F).$$

Suppose that the minimum of the variance is attained at $\hat{m}_F$, and G is a function introduced in Hints. For each $\delta \in \mathbb{R}$ we have

$$\Phi(F+\delta G) = \Phi(F) + \delta^2 \Phi(G) + 2\delta \int_0^T R(t)dG(t) \geq \Phi(F)$$

where $R(s) = \int_0^T r(s,t)dF(t)$, $s \in [0,T]$. This implies that

$$\int_\alpha^\beta R(t)dG(t) = R(\alpha) - R(\beta) = 0.$$

Therefore, $R(s) \equiv C$ and $\mathsf{D}\hat{m}_F = \int_0^T R(s)dF(s) = C$.

Vice versa, let F be a function of bounded variation such that $\hat{m}_F \in M$ and $R(s) \equiv C$. Let $\hat{m}_H \in M$; then for $G := H - F$ we have $\int_0^T dG(t) = 0$ and

$$\mathsf{D}\hat{m}_H = \Phi(F+G) = \Phi(F) + \Phi(G) + 2\int_0^T R(t)dG(t)$$

$$= \Phi(F) + \Phi(G) \geq \Phi(F) = \mathsf{D}\hat{m}_F.$$

17.17. (a) $\hat{m}_1 = \int_0^T x(t)dF(t)$, $F(t) = (2+\beta T)^{-1}\left(\mathrm{I\!I}_{t>0} + \mathrm{I\!I}_{t\geq T} + \beta t\right)$. The equality holds $\int_0^T dF(t) = F(T) - F(0) = 1$, therefore, $\hat{m}_1 \in M$. Next, it is straightforward that for each $s \in [0,T]$,

$$\int_0^T e^{-\beta|t-s|}dF(t) = \frac{2}{2+\beta T} = const,$$

thus, $\hat{m}_1$ has the least variance in M.

(b) $\hat{m}_2 = \int_0^T x(t)dF(t)$, $F(t) = \mathrm{I\!I}_{t>0}$. This estimator belongs to M, and for each $s \in [0,T]$ it holds

$$\int_0^T \min(s+1, t+1)dF(t) = 1 = const.$$

That is why $\hat{m}_2$ has the least variance in M.

(c) Content ourself with case (a). Based on Problem 17.16 it is enough to show there is no such function $f \in L_1([0,T])$ that

$$\int_0^T e^{-\beta|t-s|}f(t)dt \equiv 1, \quad s \in [0,T]. \tag{17.30}$$

To the contrary, suppose that (17.30) holds. Differentiating we obtain that for almost every $s \in [0,T]$,

$$-\beta e^{-\beta s}\int_0^s e^{\beta t}f(t)dt + \beta e^{\beta s}\int_s^T e^{-\beta t}f(t)dt = 0.$$

The last two equalities imply that

$$2e^{-\beta s}\int_0^s e^{\beta t}f(t)dt = 1$$

for almost every $s \in [0,T]$. Both parts of the equation are continuous in s; then $\int_0^s e^{\beta t}f(t)dt \equiv e^{\beta s}/2$. Differentiating this identity we obtain $f(s) = \beta/2$ for almost every $s \in [0,T]$. But this function does not satisfy (17.30), because the integral

$$I(s) = \int_0^T e^{-\beta|t-s|}dt = \int_{-s}^{T-s} e^{-\beta|u|}du, \quad s \in [0,T],$$

is not a constant. We came to contradiction.

17.18. (a)

$$\sigma^2 = T^{-1} \lim_{n\to\infty} \sum_{k=0}^{2^n-1} \left(x\left(\frac{k+1}{2^n}T\right) - x\left(\frac{k}{2^n}T\right)\right)^2, \quad \text{a.s.}$$

(b)

$$\rho(x) = \frac{d\mu_2}{d\mu_1}(x) = \exp\left\{\sigma^{-2}\left(\mu x(T) - \frac{\mu^2 T}{2}\right)\right\}, \quad x \in C([0,T]).$$

Hence, $\hat{\mu}_T = \operatorname{argmax}_{\mu>0} \log \rho(x(\cdot)) = T^{-1}x(T)$.

(c) It holds $\hat{\mu}_T = \mu + \sigma W(T)/T$, which implies the desired relations. For that Problem 3.18 is used.

17.19. (a) Write down the prior density up to multipliers that do not depend on μ:

$$\rho(\mu) \sim \exp\{-\frac{\mu^2}{2\sigma_0^2} + \frac{\mu\mu_0}{\sigma_0^2}\}.$$

Then the posterior density $\rho(\mu|x)$ is proportional to the expression:

$$\rho(\mu|x) \sim \rho(\mu)\rho(x|\mu) \sim \exp\{-\frac{A\mu^2}{2} + B\mu\},$$

with

$$A := \frac{T}{\sigma^2} + \frac{1}{\sigma_0^2} \text{ and } B := \frac{\mu_0}{\sigma_0^2} + \frac{x(T)}{\sigma^2}.$$

Therefore,

$$\rho(\mu|x) \sim \exp\{-\frac{(\mu - BA^{-1})^2}{2A^{-1}}\}.$$

The posterior distribution will be the normal law $\mathcal{N}(\mu_T, \sigma_T^2)$ with parameters

$$\mu_T = BA^{-1} = \frac{\mu_0\sigma^2 + \sigma_0^2 x(T)}{T\sigma_0^2 + \sigma^2} \text{ and } \sigma_T^2 = \frac{\sigma_0^2\sigma^2}{T\sigma_0^2 + \sigma^2}.$$

(b) The Bayes estimator is

$$\mu_T = K_T\mu_0 + (1 - K_T)\frac{x(T)}{T} \text{ with } K_T = \frac{\sigma^2}{\sigma^2 + T\sigma_0^2}.$$

Thus, the estimator is a convex combination of the prior estimator μ_0 and the maximum likelihood estimator $T^{-1}x(T)$. The coefficient K_T is called the confidence factor. As $T \to \infty$ it tends to 0 (that is, for large T we give credence to the maximum likelihood estimator), while as $T \to 0$ it tends to 1 (that is, for small T we give more credence to the prior information rather than to the data).

Answer: the posterior distribution is $\mathcal{N}(\mu_T, \sigma_T^2)$ with

$$\mu_T = K_T\mu_0 + (1 - K_T)\frac{x(T)}{T}, \quad K_T = \frac{\sigma^2}{\sigma^2 + T\sigma_0^2},$$

$$\sigma_T^2 = \frac{\sigma_0^2 \sigma^2}{T\sigma_0^2 + \sigma^2},$$

and μ_T is the Bayes estimator of the parameter μ.

17.20. (a) Up to multipliers that do not depend on λ, $\rho(\lambda) \sim \lambda^{\alpha-1} e^{-\beta\lambda}$. According to Problem 17.2, the density of the distribution of the process $\rho(N|\lambda) \sim \lambda^{N(T)} e^{-\lambda T}$. Then

$$\rho(\lambda|N) \sim \rho(\lambda)\rho(N|\lambda) \sim \lambda^{\alpha+N(T)-1} e^{-(\beta+T)\lambda}.$$

The posterior distribution is gamma distribution $\Gamma(\alpha+N(T),\, \beta+T)$.

(b) In the first case the Bayes estimator is

$$\hat{\lambda}_{T1} = \frac{\alpha+N(T)}{\beta+T} = K_1 \frac{\alpha}{\beta} + (1-K_1)\frac{N(T)}{T}.$$

Here $K_1 = \beta/(\beta+T)$ is the confidence factor (see the discussion in the solution of Problem 17.19).

In the second case the Bayes estimator is

$$\hat{\lambda}_{T2} = \frac{\alpha+N(T)-1}{\beta+T}.$$

It exists if $\alpha+N(T) > 1$. Under the additional constraint $\alpha > 1$ it holds

$$\hat{\lambda}_{T2} = K_2 \frac{\alpha-1}{\beta} + (1-K_2)\frac{N(T)}{T} \text{ with } K_2 = K_1.$$

This estimator is a convex combination of the prior estimator $(\alpha-1)/\beta$ (under the same loss function) and the maximum likelihood function.

Answer: gamma distribution $\Gamma(\alpha+N(T),\, \beta+T)$,

$$\hat{\lambda}_{T1} = K\frac{\alpha}{\beta} + (1-K)\frac{N(T)}{T} \text{ with } K = \frac{\beta}{\beta+T},$$

$\hat{\lambda}_{T2} = K\frac{\alpha-1}{\beta} + (1-K)\frac{N(T)}{T}$.

17.21. The unbiasedness condition $\mathsf{E}_{\varphi,\theta}\hat{\theta} = \theta$ means the following:

$$\forall\, \varphi \in K: \quad <f,\varphi> = \bar{0}; \quad <f, g^\top> = I_m.$$

Here $g = (g_1,\dots,g_m)^\top$; $<f,\varphi>$ is a column vector with components (f,φ_i), and $<f,g^\top>$ is a matrix with entries (f_i,g_j); I_m stands for the unit matrix of size m.

Looking for an optimal estimator in the class M is reduced to the following problem. Find a vector function $f = (f_1,\dots,f_m)^\top$ such that:

(1) $\{f_1,\dots,f_m\} \subset K^\perp$.
(2) $<f,g^\top> = I_m$.
(3) For each vector function h that satisfies the conditions (1) and (2), the matrix $<h,h^\top> - <f,f^\top>$ is positive semidefinite.

Let P be the projection operator on K, and $Pg = (Pg_1, \dots, Pg_m)^\top$, $\Phi = < Pg, (Pg)^\top >$. The matrix Φ is nonsingular as a result of linear independence of the functions $\{Pg_i\}$. The desired vector is unique and has a form $f = f^* = \Phi^{-1} Pg$. The covariance matrix S^* of the corresponding estimator is

$$S^* = \mathsf{E}[(\hat{\theta}^* - \theta)(\hat{\theta}^* - \theta)^\top] = \sigma^2 \Phi^{-1}.$$

17.22. (b) Continue the reasoning from Hints. By the Fubini theorem

$$\lambda^2(A_1) = \int_{[0,1]} \lambda^1(\{x_1 : (x_1, x_2) \in A_1\}) d\lambda^1(x_2) = 1.$$

Here λ^2 is Lebesgue measure on the plane. Next,

$$A_2 := X \setminus A_1 = \{x : \theta(x) \in [2,3]\}$$

and the consistency of the estimator implies that for all $x_1 \in [0,1]$ it holds

$$\lambda^1(\{x_2 \in [0,1] : (x_1, x_2) \in A_2\}) = 1.$$

Therefore,

$$\lambda^2(A_2) = \int_{[0,1]} 1 \cdot d\lambda^1(x_1) = 1.$$

But due to additivity of a measure, $1 = \lambda^2(X) = \lambda^2(A_1) + \lambda^2(A_2) = 2$. We came to a contradiction.

18

Stochastic processes in financial mathematics (discrete time)

Theoretical grounds

Consider a model of a *financial market* with a finite number of periods (i.e., of the moments of time) at which it is possible to trade, consume, spend, or receive money or other valuables. The model consists of the following components. There exist $d+1$ financial assets, $d \geq 1$, and the prices of these assets are available at moments $t \in \mathbb{T} = \{0, 1, \ldots, T\}$. The price of the ith asset at moment t is a nonnegative r.v. $S^i(t)$ defined on the fixed probability space $(\Omega, \mathcal{F}, \mathsf{P})$. This space is assumed to support some filtration $\{\mathcal{F}_t\}_{t \in \mathbb{T}}$, and we suppose that the random vector $\overline{S}_t = (S^0(t), S(t)) = (S^0(t), S^1(t), \ldots, S^d(t))$ is measurable with respect to the σ-field $\mathcal{F}_t$. With the purpose of technical simplifying we assume in what follows that $\mathcal{F}_0 = \{\varnothing, \Omega\}$ and $\mathcal{F}_T = \mathcal{F}$. In the most applications the asset $S^0(t)$ is considered as a risk-free (riskless) bond (numéraire), and sometimes it is supposed that $S^0(t) = (1+r)^t$, where $r > -1$ is a risk-free interest rate. In real situations $r > 0$, but it is not obligatory. Other assets are considered as risky ones, for example, stocks, property, currency, and so on.

Definition 18.1. *A predictable $d+1$-dimensional stochastic process $\overline{\xi} = (\xi^0, \xi) = \{(\xi^0(t), \xi^1(t), \ldots, \xi^d(t)), t \in \mathbb{T}\}$ is called the* trading strategy *(portfolio) of a financial investor.*

A coordinate $\xi^i(t)$ of the strategy ξ corresponds to the quantity of units of the ith asset during the tth trading period between the moments $t-1$ and t. Therefore, $\xi^i(t)S^i(t-1)$ is the sum invested into the ith asset at the moment $t-1$, and $\xi^i(t)S^i(t)$ is the corresponding sum at the moment t. The total value of the portfolio at the moment $t-1$ equals $(\overline{\xi}(t), \overline{S}(t-1)) = \sum_{i=0}^{d} \xi^i(t)S^i(t-1)$, and at the moment t this value can be equated to $(\overline{\xi}(t), \overline{S}(t)) = \sum_{i=0}^{d} \xi^i(t)S^i(t)$ ($(\cdot, \cdot)$ is, as always, the symbol of the inner product in Euclidean space). The predictability of the strategy reflects the fact that the distribution of resources happens at the beginning of each trading period, when the future prices are unknown.

Definition 18.2. *The strategy $\overline{\xi}$ is called* self-financing, *if the investor's capital $V(t)$ satisfies the equality $V(t) = (\overline{\xi}(t), \overline{S}(t)) = (\overline{\xi}(t+1), \overline{S}(t))$, $t \in \{1, 2, \ldots, T-1\}$.*

D. Gusak et al., *Theory of Stochastic Processes*, Problem Books in Mathematics, DOI 10.1007/978-0-387-87862-1_18,

The self-financing property of the trading strategy means that the portfolio is always redistributed in such a way that its total value is preserved. The strategy is self-financing if and only if for any $t \in \{1,2,\ldots,T\}$,

$$V(t) = (\overline{\xi}(t),\overline{S}(t)) = (\overline{\xi}(1),\overline{S}(0)) + \sum_{k=1}^{t}(\overline{\xi}(k),(\overline{S}(k)-\overline{S}(k-1))).$$

The value $(\overline{\xi}(1),\overline{S}(0))$ is the initial investment that is necessary for the purchasing of the portfolio $\overline{\xi}(1)$. Below we assume that $S^0(t) > 0$ P-a.s., for all $t \in \mathbb{T}$. In this case it is possible to define the discounted prices of the assets $X^i(t) := (S^i(t))/(S^0(t))$, $t \in \mathbb{T}$, $i = 0,1,\ldots,d$. Evidently, after the discounting we obtain that $X^0(t) \equiv 1$, and $X(t) = (X^1(t),\ldots,X^d(t))$ is the value of the vector of risk assets in terms of units of the asset $S^0(t)$, which is the discounting factor. Despite the fact that the asset $S^0(t)$ is called risk-free and the vector $X(t)$ is called the vector of risk assets, these notions are relative, to some extent. Introduce the extended vector of discounted prices $\overline{X}(t) = (1,X^1(t),\ldots,X^d(t))$.

Definition 18.3. *A stochastic process of the form* $\{V(t),\mathcal{F}_t,t \in \mathbb{T}\}$ *where*

$$V(t) = (\overline{\xi}(t),\overline{X}(t)) = \frac{(\overline{\xi}(t),\overline{S}(t))}{S^0(t)}$$

is called the discounted capital of investor.

If a strategy is self-financing, the equality

$$V(t) = (\overline{\xi}(1),\overline{X}(0)) + \sum_{k=1}^{t}(\xi(k),(X(k)-X(k-1)))$$

holds true for all $t \in \mathbb{T}$. Here $\sum_{k=1}^{0} := 0$.

Definition 18.4. *A self-financing strategy is called* the arbitrage possibility *if its capital* V *satisfies inequalities* $V(0) \le 0$, $V(T) \ge 0$ *P-a.s., and* $\mathsf{P}(V(T) > 0) > 0$.

Definition 18.5. *A probability measure* Q *on* $(\Omega,\mathcal{F})$ *is called* the martingale measure, *if the vector-valued discounted price process* $\{X(t),\mathcal{F}_t,t \in \mathbb{T}\}$ *is a d-dimensional* Q*-martingale; that is,* $\mathsf{E}_{\mathsf{Q}}X^i(t) < \infty$ *and* $X^i(s) = \mathsf{E}_{\mathsf{Q}}(X^i(t)/\mathcal{F}_s)$, $0 \le s \le t \le T$, $1 \le i \le d$.

Theorem 18.1. *A financial market is free of arbitrage if and only if the set* $\mathcal{P}$ *of all martingale measures, which is equivalent to measure* P, *is nonempty. In this case there exists a measure* $\mathsf{P}^* \in \mathcal{P}$ *with bounded density* $d\mathsf{P}^*/d\mathsf{P}$.

Definition 18.6. *A nonnegative r.v.* C *on* $(\Omega,\mathcal{F},\mathsf{P})$ *is called* the European contingent claim (payoff).

The European contingent claim can be interpreted as an asset that guarantees to its owner the payment $C(\omega)$ at moment T. The moment T is called the expiration date, or maturity date of the claim C. The corresponding discounted contingent claim has a form $H = C/S_T^0$. If this discounted contingent claim can be presented in a functional form, namely, $H = f(X(\cdot)$, where $f : \mathbb{R}^{d(T+1)} \to \mathbb{R}^+$ is a measurable function, then it is called the derivative, or derivative security, of the vector of primary financial assets $X(t), t \in \mathbb{T}$.

Definition 18.7. *A contingent claim C is called* attainable (replicable, redundant), *if there exists a self-financing strategy ξ such that the value of portfolio at the maturity date equals C; that is, $C = (\overline{\xi}_T, \overline{S}_T)$ P-a.s.*

In this case we say that a strategy ξ creates a replicating portfolio for C (replicates C, is a hedging strategy for C).

A contingent claim is attainable if and only if the corresponding discounted contingent claim has a form

$$H = (\overline{\xi}_T, \overline{X}_T) = V_T = V_0 + \sum_{k=1}^{T} (\xi_k, (X_k - X_{k-1})),$$

for some self-financing strategy ξ.

Definition 18.8. *An arbitrage-free financial market is called* complete *if on this market any contingent claim is attainable.*

Theorem 18.2. *A financial market is complete if and only if there exists and is unique the equivalent martingale measure.*

Definition 18.9. *A number $\pi(H)$ is called* the arbitrage-free price (fair price) *of discounted European contingent claim H if there exists a nonnegative adapted stochastic process $X^{d+1} = \{X^{d+1}(t), \mathcal{F}_t, t \in \mathbb{T}\}$ such that $X^d(0) = \pi(H)$, $X^d(T) = H$, and the extended financial market $(X^0(t), \dots, X^{d+1}(t))$ is arbitrage-free.*

Theorem 18.3. *If a contingent claim is attainable then it has a unique arbitrage-free price $\mathsf{E}_{\mathsf{P}^*}(H)$. If a contingent claim is not attainable, then the set of its arbitrage-free prices is an interval of the form $(\pi^{\downarrow}(H), \pi^{\uparrow}(H))$ on nonnegative axis (possibly $\pi^{\uparrow}(H) = \infty$).*

Definition 18.10. *A nonnegative adapted stochastic process $C = \{C(t), \mathcal{F}_t, t \in \mathbb{T}\}$ is called* the American contingent claim.

An American contingent claim is a contract that is issued at moment $t = 0$ and obliges the writer to pay a certain amount $C(t)$, provided the buyer decides at moment t to exercise this contract. The contract is exercised only once. If the buyer has not decided to exercise the contract till the maturity date T, then at this moment the contract is automatically exercised. The buyer has a possibility to exercise the contract not only at nonrandom moment $t \in \mathbb{T}$, but at any stopping time $\tau \in \mathbb{T}$. The aim of the buyer is to find an optimal stopping time τ_0 in the sense that $\mathsf{E}C(\tau_0) = \sup_{0 \le \tau \le T} \mathsf{E}C(\tau)$, where τ are stopping times, or, in terms of the corresponding discounted contingent claim $H(t) = (C(t))/(S^0(t))$, $\mathsf{E}H(\tau_0) = \sup_{0 \le \tau \le T} \mathsf{E}H(\tau)$.

Definition 18.11. An American call option *on an asset S is the derivative that can be exercised at any stopping time $\tau \in \mathbb{T}$, and in this case the payment is $(S(\tau) - K)^+$, where $S(t)$ is the price at moment t of the underlying asset. An American put option is defined similarly.*

The strategies of the buyer and writer of an American option are different: the buyer wants to exercise the option at that moment τ_0 where the mean value of the payment is the biggest, and the writer wants to create his portfolio in order to have a possibility to exercise the option whenever the buyer comes.

Bibliography

[55], Chapter IV; [23], Chapters 1,5,6; [84], Volume 2, Chapters V and VI; [21], Chapters I and II; [46], Chapters 4–8; [54], Chapter IX; [62], Chapters 1–4.

Problems

18.1. The owner of a European call option has a right, but not an obligation, to buy some asset, for example, some stock S, at moment T at price K, which is fixed initially. This price is called the strike price. Similarly, the owner of a European put option has a right, but not an obligation, to buy some asset, for example, the same stock S, at moment T at the price K, which is fixed initially.

(1) Prove that the value of the call option $C = C^{call}$ equals $C = C(T) = (S(T) - K)^+$, and the value of the put option $P = P^{put}$ equals $P = P(T) = (K - S(T))^+$.

(2) Let a finance market be arbitrage-free, $\pi(C)$ be the arbitrage-free price of a European call option, and $\pi(P)$ be the arbitrage-free price of the European put option, both with strike price K. Prove that $\pi(C) \le S(0)$ and $\pi(P) \le K$, where $S(0)$ is the initial price of the risk asset.

18.2. (1) We know that the financial market is arbitrage-free, the price of an asset at moment 0 equals $S(0)$, and at moment T the possible values of this asset are $S(\omega_i)$, $i = 1,\dots,M$. Also, let the risk-free interest rate at any moment equal r. What is the risk-free price of a European call option on this asset if the strike price equals K with $K < \min_{1\le i\le M} S(\omega_i)$?

(2) What is the risk-free price of a European call option on this asset if the strike price is zero?

18.3. We know that the financial market is arbitrage-free, the interest rate equals r at any period, and T is the expiration date.

(1) Prove the following inequalities constructing an explicitly arbitrage strategy in an opposite case :

(a) $\pi(P) \ge K(1+r)^{-T} - S(0)$.

(b) $\pi(C) \ge S(0) - K(1+r)^{-T}$.

(2) Using a definition of the martingale measure prove the following specifications of the inequalities from item (1).

(a) Prove that the arbitrage-free price of a European put option admits the bounds

$$\max(0,\, (1+r)^{-T}K - S(0)) \le \pi(P) \le (1+r)^{-T}K.$$

(b) Prove that the price of the corresponding call option admits the bounds

$$\max(0,\, S(0) - (1+r)^{-T}K] \le \pi(C) \le S(0).$$

18.4. We know that a financial market is arbitrage-free, the interest rate equals r at any period, T is the expiration date, and K is the strike price of all the options mentioned below.

(1) (a) Prove that under the conditions for absence of arbitrage, the put–call parity holds between arbitrage-free prices of call and put options: $S(0)+\pi(P)-\pi(C)=K(1+r)^{-T}$.

(b) Prove the following generalization of the put–call parity to any intermediate moment: $S(t)+P(t)-C(t)=K(1+r)^{-T+t}$, where $P(t)$ and $C(t)$ are arbitrage-free prices of put and call options at moment t, respectively.

(2) Prove that the selling one asset, selling one put option, and buying one call option yields positive profit with vanishing risk (the arbitrage) under the assumption $S(0)+\pi(P)-\pi(C)>K(1+r)^{-T}$.

(3) Prove that the buying one asset, buying one put option, and selling one call option provides the arbitrage under the assumption $S+\pi(P)-\pi(C)<K(1+r)^{-T}$.

18.5. Let the price of an asset (e.g., stock) at moment t equal $S(t)$. All the options under consideration are supposed to have the expiration date T and the strike price K, unless otherwise specified. The interest rate equals r at any period between buying and exercising options. Calculate the capital at moment T of the investor whose activity at moment t can be described as follows.

(a) She has one call option and one put option.

(b) She has one call option with strike price K_1 and sells one put option with strike price K_2.

(c) She has two call options and sells one asset.

(d) She has one asset and sells one call option.

18.6. (Law of one price) Let the financial market be arbitrage-free, C be an attainable contingent claim, and $\overline{\xi}=\{\overline{\xi}(t),t\in\mathbb{T}\}$ be any replicating portfolio for C. Prove that the initial capital $V(0)=(\overline{\xi}(1),\overline{S}(0))$ is the same for any such portfolio.

18.7. (Binomial model, or Cox–Ross–Rubinstein model) Assume that there is one riskless asset (a bond) $\{B_n=(1+r)^n,0\le n\le N\}$ with the interest rate $r>-1$ and one risky asset (a stock) $\{S_n,0\le n\le N\}$ within the financial market. The price S_n can be calculated as follows. $S_0>0$ is a given value, S_{n+1} is equal either to $S_n(1+a)$ or $S_n(1+b)$, where $-1<a<b$. Hence, $\Omega=\{1+a,1+b\}^N$. We put $\mathfrak{F}_0=\{\varnothing,\Omega\}$ and $\mathfrak{F}_n=\sigma\{S_1,\dots,S_n\},\ 1\le n\le N$. Assume that every element of Ω has positive probability. Let $R_n=S_n/S_{n-1},\ 1\le n\le N$. If $\{y_1,\dots,y_n\}$ is some element of Ω then $\mathsf{P}(\{y_1,\dots,y_n\})=\mathsf{P}(R_1=y_1,\dots,R_n=y_n)$.

(1) Show that $\mathfrak{F}_n=\sigma\{R_1,\dots,R_n\},\ 1\le n\le N$.

(2) Show that the discounted stock price $X_n:=S_n/(1+r)^n$ is a P^*-martingale if and only if

$$\mathsf{E}_{\mathsf{P}^*}(R_{n+1}/\mathfrak{F}_n)=1+r,\ 0\le n\le N-1.$$

(3) Prove that the condition $r\in(a,b)$ is necessary for the market to be arbitrage-free.

(4) Prove that under the condition $r\in(a,b)$ a random sequence $\{X_n,\mathfrak{F}_n,0\le n\le N\}$ is a P^*-martingale if and only if random variables $R_1,\dots,R_n$ are mutually independent and identically distributed and $\mathsf{P}^*(R_1=1+b)=(r-a)/(b-a)=:p^*$. Show that the market is complete in this case.

18.8. In the framework of an arbitrage-free and complete binomial model consider a discounted derivative security of the form $H = f(X_0,\dots,X_N)$, where $f : \mathbb{R}^{N+1} \to \mathbb{R}^+$ is a measurable function.

(1) Prove that H is integrable with respect to the martingale measure P^*.

(2) Prove that the capital V_n of any hedging strategy for H can be presented in the form $V_n = \mathsf{E}_{\mathsf{P}^*}(H/\mathfrak{F}_n) = v_n(X_0, X_1(\omega)\dots,X_n(\omega))$, where the function $v_n(x_0,\dots,x_n) = \mathsf{E}_{\mathsf{P}^*} f\Big(x_0,\dots,x_n,x_n(X_1/X_0),\dots,x_n((X_{N-n})/(X_0))\Big)$.

(3) Prove that a self-financing strategy $\overline{\xi} = (\xi^0, \xi)$, which is a replicating strategy for H, has the form $\xi_n(\omega) = \Delta_n(X_0, X_1(\omega),\dots,X_{n-1}(\omega))$, where

$$\Delta_n(x_0,x_1,\dots,x_{n-1}) = \frac{v_n(x_0,\dots,x_{n-1},x_{n-1}\hat{b}) - v_n(x_0,\dots,x_{n-1},x_{n-1}\hat{a})}{x_{n-1}(\hat{b}-\hat{a})},$$
$$\hat{a} = \frac{1+a}{1+r},\ \hat{b} = \frac{1+b}{1+r},$$

and

$$\xi_1^0(\omega) = \mathsf{E}_{\mathsf{P}^*}(H) - \xi_1(\omega)X_0,\ \xi_{n+1}^0(\omega) - \xi_n^0(\omega) = -(\xi_{n+1}(\omega) - \xi_n(\omega))X_n.$$

(4) Let $H = f(X_N)$. Prove that in this case the functions $v_n(x_n)$ can be presented in the form $v_n(x_n) = \sum_{k=0}^{N-n} f(x_n\hat{a}^{N-n-k}\hat{b}^k)C_{N-n}^k(p^*)^k(1-p^*)^{N-n-k}$; in particular, the unique arbitrage-free price of the contingent claim H can be presented in the form $\pi(H) = v_0(X_0) = \sum_{k=0}^{N} f(X_0\hat{a}^{N-k}\hat{b}^k)C_N^k(p^*)^k(1-p^*)^{N-k}$.

(5) Denote by C_n the price at moment n of a European call option with the expiration date N and the strike price K. Prove that under the conditions of item (4) of Problem 18.7 it holds that $C_n = c(n, S_n)$, where

$$\frac{c(n,x)}{(1+r)^{n-N}} = \mathsf{E}_{\mathsf{P}^*}\left(x\prod_{i=n+1}^{N} R_i - K\right)^+$$
$$= \sum_{j=0}^{N-n} \frac{(N-n)!}{(N-n-j)!j!}(p^*)^j(1-p^*)^{N-n-j}\left(x(1+a)^{N-n-j}(1+b)^j - K\right)^+.$$

18.9. (Trinomial model) Let a financial market consist of one risky asset (stock) $\{S_n,\ 0 \le n \le N\}$ and one riskless asset (bond) $\{B_n = (1+r)^n,\ 0 \le n \le N\}$ with the interest rate $r > -1$. The price of S_n is defined as follows. $S_0 > 0$ is a fixed number, and S_{n+1} equals either $S_n(1+a)$, or $S_n(1+b)$, or $S_n(1+c)$, where $-1 < a < b < c$. Therefore, $\Omega = \{1+a, 1+b, 1+c\}^N$; that is, any element of Ω can be presented as $\omega = \{y_1,\dots,y_N\}$, where $y_n = 1+a$, or $1+b$, or $1+c$. Similarly to the binomial model, put $\mathfrak{F}_0 = \{\varnothing, \Omega\}$ and $\mathfrak{F}_n = \sigma\{S_1,\dots,S_n\}$, $1 \le n \le N$. Suppose that any element of Ω has positive probability.

(1) How many equivalent martingale measures exist in this model? Is this model arbitrage-free? complete?

(2) If we consider two martingale measures, do they lead to the same price of attainable payoff; that is, does the law of one price hold in the trinomial model?

18.10. Denote by $C(t)$ and $Y(t)$, $t \in \mathbb{T}$ the price which will pay the buyer of a European and American option, respectively, if he buys them at moment t. It is supposed that the options are derivatives at the same asset. Prove that $Y(t) \ge C(t)$.

18.11. Consider an American option with payoffs $\{Y(t), t \in \mathbb{T}\}$. Let a European option at moment T have the same payoff $Y(T)$. Prove the following statement. If $C(t) \geq Y(t)$ for any $t \in \mathbb{T}$ and any $\omega \in \Omega$, then $C(t) = Y(t)$ and the optimal strategy for the buyer is to wait until moment T and then exercise the option.

18.12. (Hedging of American option) Assume an American option can be exercised at any moment $n = 0, 1, \dots, N$, and let $\{S_n, \mathfrak{F}_n, 0 \leq n \leq N\}$ be a stochastic process, that is equal to the profit, provided the option is exercised at moment n. Denote by $\{S_n^0, 0 \leq n \leq N\}$ the price of a risk-free asset (the discounting factor), which is supposed to be nonrandom, and let $X_n = S_n/S_n^0$. Denote also by $\{Y_n, 0 \leq n \leq N\}$ the price (value) of the option at moment n. The market is supposed to be arbitrage-free and complete, and let P^* be the unique martingale measure.

(1) Prove that $Y_N = S_N$.

(2) Prove that for any $0 \leq n \leq N-1$ the equality holds

$$Y_n = \max(S_n, S_n^0 \mathsf{E}_{\mathsf{P}^*}(\frac{Y_{n+1}}{S_{n+1}^0}/\mathfrak{F}_n)).$$

(3) Prove that in the case $S_n^0 = (1+r)^n$, the price of the American option can be presented as

$$Y_n = \max(S_n, \frac{1}{1+r}\mathsf{E}_{\mathsf{P}^*}(Y_{n+1}/\mathfrak{F}_n)).$$

(4) Prove that the discounted price of the American option $Z_n := Y_n/S_n^0$ is a P^*-supermartingale; moreover it is the smallest P^*-supermartingale dominating the sequence $\{X_n, 0 \leq n \leq N\}$. Prove that Z_n is a Snell envelope of the sequence $\{X_n\}$.

(5) Prove that the following equalities hold: $Z_n = \sup_{\tau \in \mathcal{T}_{n,N}} \mathsf{E}_{\mathsf{P}^*}(X_\tau/\mathfrak{F}_n)$, $0 \leq n \leq N$. (See Problems 7.22 and 15.20 for the corresponding definitions.)

18.13. (Price of an American put option in the context of the Cox–Ross–Rubinstein model) Let a financial market consist of one stock $\{S_n, 0 \leq n \leq N\}$, with the price defined by the Cox–Ross–Rubinstein model and one bond with the interest rate $r > -1$ (see Problem 18.7).

(1) Prove that at moment n with $0 \leq n \leq N$ the price P_n of the American put option with the maturity date N and the strike price K equals $P_n = P(n, S_n)$, where $P(n,x)$ can be found as follows. $P(N,x) = (K-x)^+$, and for $0 \leq n \leq N-1$ it holds $P(n,x) = \max((K-x)^+, ((f(n+1,x))/(1+r)))$, where $f(n+1,x) = pP(n+1, x(1+a)) + (1-p)P(n+1, x(1+b))$, with $p = (b-r)/(b-a)$.

(2) Prove that $P(0,x) = \sup_{\tau \in \mathcal{T}_{0,N}} \mathsf{E}_{\mathsf{P}^*}((1+r)^{-\tau}(K - xV_\tau)^+)$, where the sequence $\{V_n, 0 \leq n \leq N\}$ is given by $V_0 = 1, V_n = \prod_{i=1}^n U_i, 1 \leq n \leq N$, and U_i are some random variables. Determine their simultaneous distribution with respect to the measure P^*.

(3) Use item (2) and prove that the function $P(0,x) : \mathbb{R}^+ \to \mathbb{R}^+$ is convex.

(4) Let $a < 0$. Prove that there exists a real number $x^* \in [0, K]$ such that for $x \leq x^*$ it holds $P(0,x) > (K-x)^+$.

(5) Let the owner have an American put option at moment $t = 0$. What are the values of S_0 for which it is the most profitable to exercise this option at the same moment?

18.14. An American option is called attainable if for any stopping time $0 \le \tau \le T$ there exists a self-financing strategy such that the corresponding capital V satisfies the equality $V(\tau) = Y(\tau)$. Let an American option be attainable, and the corresponding discounted stochastic process $H(t) := Y(t)/S^0(t)$ be a submartingale with respect to some martingale measure P^*. Prove that the optimal stopping time τ coincides with the exercise date (i.e., $\tau = T$), and the price of this American option coincides with the price of a European option, $C(T) = Y(T)$.

18.15. Consider an American call option that can be exercised at any moment $t \in \mathbb{T}$, and in the case where it is exercised at moment $t \in \mathbb{T}$, the strike price equals $K(1+q)^t$, where q is a fixed number; that is, the payoff at moment t equals $(S(t) - K(1+q)^t)^+$. For $q \le r$, where r is the interest rate, prove that the option will not be exercised before moment $t = T$.

18.16. Let a risky asset at moment T have the price $S(T)$, where $S(T)$ is a r.v. with the distribution determined by the measure P. At moment T, let the option on this asset have the price $C(T)$. Consider a portfolio consisting of ξ_0 units of a riskless asset and ξ units of a risky asset, and the portfolio is constant during all trading periods; such a strategy is called "buy and hold"; it is not obligatory self-financing. We suppose that the interest rate is zero, and let the initial capital equal $V(0)$.

(1) Prove that the additional costs that must be invested by the owner of this portfolio with the purpose to be able at moment T to exercise the contingent claim $C(T)$, can be calculated by the formula $D := C(T) - V(0) - \xi(S(T) - S(0))$.

(2) In terms of $\mathsf{E}S(T), \mathsf{E}C(T)$, $\mathsf{D}S(T)$, and $\operatorname{cov}(S(T), C(T))$, determine those values of $V(0)$ and ξ that minimize $\mathsf{E}(D^2)$, and prove that under these values of $V(0)$ and ξ it holds $\mathsf{E}D = 0$.

(3) Prove that in the case of a complete market the option $C(T)$ linearly depends on $S(T) - S(0)$, and that $V(0)$ and ξ can be chosen in such a way that $D = 0$.

Hints

18.7. (1) Express $\{S_1, \ldots, S_n\}$ via $\{R_1, \ldots, R_n\}$ and vice versa.
(2) Write the equation $\mathsf{E}_{\mathsf{Q}}(X_{n+1}/\mathfrak{F}_n) = X_n$ in the equivalent form $\mathsf{E}_{\mathsf{Q}}(((X_{n+1})/(X_n))/\mathfrak{F}_n) = 1$.
(3) Let the market be arbitrage-free. Then there exists a measure $\mathsf{P}^* \sim \mathsf{P}$ such that X_n is a P^*-martingale. Furthermore, use the statement of item (2).
(4) If R_n are mutually independent and $\mathsf{P}^*(R_i = 1+b) = p^*$ then $\mathsf{E}_{\mathsf{P}^*}(R_{n+1}/\mathfrak{F}_n) = 1+r$. Check this and use the statement (2). And conversely, let $\mathsf{E}_{\mathsf{P}^*}(R_{n+1}/\mathfrak{F}_n) = 1+r$. Derive from here that $\mathsf{P}^*(R_{n+1} = 1+b/\mathfrak{F}_n) = p^*$ and $\mathsf{P}(R_n = 1+a/\mathfrak{F}_n) = 1-p^*$. Prove by induction that $\mathsf{P}^*(R_1 = x_1, \ldots, R_n = x_n) = \prod_{i=1}^n p_i$, where $p_i = p^*$, if $x_i = 1+b$, and $p_i = 1-p^*$, if $x_i = 1+a$. Note that the P^*-martingale property of X_n uniquely determines the distribution $(R_1, \ldots, R_N)$ with respect to the measure P^*, so, uniquely determines the measure P^* itself. That is why the market is complete.
18.8. (1) Integrability of H is evident, because all the random variables take only a finite number of values.

(2) To prove the equality $V_n = \mathsf{E}_{\mathsf{P}^*}(H/\mathcal{F}_n)$ it is necessary to prove at first, using backward induction, that all the values V_n of the capital are nonnegative. Then, with the help of the formula for the capital of a self-financing strategy, it is possible to prove that V_n is a martingale with respect to the measure P^*. To prove the second inequality you can write X_k, $n \le k \le N$ in the form $X_k = X_n(X_k/X_n)$ and use the fact that the r.v. X_k/X_n is independent of $\mathcal{F}_n$ and has the same distribution as $(X_{k-n})/X_0$.
(3) Write the equality $\xi_n(\omega)(X_n(\omega) - X_{n-1}(\omega)) = V_n(\omega) - V_{n-1}(\omega)$, in which $\xi_n(\omega)$, $X_{n-1}(\omega)$, $V_{n-1}(\omega)$ depend only on the first $n-1$ components of the vector ω. Denote $\omega^a := (y_1, \dots, y_{n-1}, 1+a, y_{n-1}, \dots, y_N)$, $\omega^b := (y_1, \dots, y_{n-1}, 1+b, y_{n-1}, \dots, y_N)$ and obtain the equalities $\xi_n(\omega)(X_{n-1}(\omega)\hat{b} - X_{n-1}(\omega)) = V_n(\omega^b) - V_{n-1}(\omega)$ and $\xi_n(\omega)(X_{n-1}(\omega)\hat{a} - X_{n-1}(\omega)) = V_n(\omega^a) - V_{n-1}(\omega)$. Derive from here the formulae for $\xi_n(\omega)$. The formulae for $\xi_n^0(\omega)$ can be derived from a definition of the self-financing strategy.
(4) Use item (2).
18.12. (1) Apply backward induction starting at moment N. Use the fact that at moment N it is necessary to pay the price for an option that equals his benefit at this moment (i.e., $Y_N = S_N$), and at moment $N-1$ it is necessary to have the capital that is sufficient for buying at that moment (i.e., S_{N-1}), and also have the sum that is sufficient to buy it at moment N, but the cost of this sum at moment $N-1$ equals $S_{N-1}^0 \mathsf{E}_{\mathsf{P}^*}(X_N/\mathcal{F}_{N-1}) = S_{N-1}^0 \mathsf{E}_{\mathsf{P}^*}\left((S_N/S_N^0)/\mathcal{F}_{N-1}\right)$.
Statements (2)–(5) follow from item (1) if using Problems 15.21–15.23.
18.13. Apply Problem 18.12.
18.15. Apply Problem 18.11.

Answers and Solutions

18.1. (1) This statement is evident.
(2) If $\pi(C) > S(0)$ then at moment $t = 0$ the seller of the call option sells it and buys the stock at the price $S(0)$. Therefore, at moment $t = T$ he can pay for the claim concerning the call option, for any market scenario and any strike price. In this case he will obtain a guaranteed profit $\pi(C) - S(0) > 0$. Next, if $\pi(P) > K$ then the seller of the put option sells it at the price $\pi(P)$ and at moment $t = T$ has a possibility to buy the stock for the stock price $K < \pi(P)$ in the case where the option will be exercised. In this case he will obtain a guaranteed profit $\pi(P) - K > 0$.
18.2. (1) Nonarbitrage price of such an option equals $\pi(C) = S(0) - (K/((1+r)^T))$.
(2) $S(0)$.
18.3. (1)(a) Suppose that $\pi(P) < K(1+r)^{-T} - S(0)$; that is, $(\pi(P) + S(0))(1+r)^T < K$. At moment $t = 0$ it is possible to borrow the sum $\pi(P) + S(0)$, buy the stock at the price $S(0)$, and buy the option with the strike price K at the price $\pi(P)$. At moment T we sell the stock at the price K (or even at a higher price, if its market price exceeds K) and return $(\pi(P) + S(0))(1+r)^T$ as a repayment for the borrowed sum. So, we will have a guaranteed profit not less than $K - (\pi(P) + S(0))(1+r)^T > 0$.
(b) Suppose that $\pi(C) < S(0) - K(1+r)^{-T}$; that is, $(S(0) - \pi(C))(1+r)^T > K$. At moment $t = 0$ it is possible to make a short sale of the stock (short sale of a

stock is an immediate sale without real ownership) at the price $S(0)$ and buy the option with the strike price K at the price $\pi(C)$. At moment T we have the sum $(S(0)-\pi(C))(1+r)^T$, so we can buy the stock at the price K (or even at a lower price if its market price is lower than K) and return the borrowed sum. So, we obtain a guaranteed profit not less than $(S(0)-\pi(C))(1+r)^T-K>0$.

18.4. (1) Both relations of the put–call parity are direct consequences of a definition of the martingale measure and the evident equality $C(T)-P(T)=S(T)-K$.

(2) At moment $t=0$ we act as proposed, obtain the sum $S(0)+\pi(P)-\pi(C)$, which we put into a bank account and obtain at moment $t=T$ the sum $(S(0)+\pi(P)-\pi(C))(1+r)^T$. If $S(T)\geq K$ we use the call option, buy the stock at the price K, and return the borrowed sum. Possible exercising of the put option will not lead to losses. If $S(T)<K$, then after exercising the put option we buy the stock at the price K, and return our debt with the help of this stock. We do not exercise the call option. In both cases we have a profit not less than $(S(0)+\pi(P)-\pi(C))(1+r)^T-K>0$.

(3) At moment $t=0$ we borrow the sum $S+\pi(P)-\pi(C)$ and act as mentioned in the problem situation. If $S(T)<K$ then we use the put option, sell the stock at the price K, and return the borrowed sum which size is now $(S(0)+\pi(P)-\pi(C))(1+r)^T$. Possible exercising of a call option will not lead to losses. If $S(T)\geq K$ then the call option will be exercised, we will sell the stock at the price K, will not exercise the call option, and also can return the borrowed sum which size is now $(S(0)+\pi(P)-\pi(C))(1+r)^T$. In both cases we have a profit at least $K-(S(0)+\pi(P)-\pi(C))(1+r)^T>0$.

18.5. (a) $((S(T)-K)^+ + (K-S(T))^+ = S(T)-K$.

(b) $(S(T)-K_1)^+ + \pi(P)(1+r)^{T-t} - (K_2-S(T))^+$.

(c) $2(S(T)-K)^+ + S(t)(1+r)^{T-t}$.

(d) $S(T)\mathbb{I}_{S(T)<K} + K\mathbb{I}_{S(T)\geq K} + \pi(C)(1+r)^{T-t}$.

18.9. (1) There will be an infinite number of martingale measures; all of them will create three-dimensional vectors of the form

$$\left(p_1, \frac{c-r-p_1(c-a)}{c-b}, \frac{c-r-p_1(b-a)}{c-b}\right), \text{ with } \frac{b-r}{b-a}<p_1<\frac{c-r}{c-a}.$$

The market will be arbitrage-free and incomplete.

(2) The law of one price holds.

18.10. The solution is based on the fact that it is possible to postpone exercising the American option until moment T.

18.11. If the buyer requires the American option and $C(t)\geq Y(t)$, it is not reasonable to exercise it at moment $t<T$ and get the sum $Y(t)$, because it is possible at that moment to guarantee instead the payoff $C(t)$. For instance, it is possible to sell the European option or take a short position with respect to the portfolio that hedges such an option. So, it is necessary to wait until moment T, but it means that the prices of both options coincide.

18.14. If $Y(t)/S^0(t)$ is a submartingale, then by Theorem 7.5 (a version of Doob's optional sampling theorem for continuous time), $\mathsf{E}_{\mathsf{P}^*}(Y(\tau)/S^0(\tau))\leq \mathsf{E}_{\mathsf{P}^*}(Y(T)/S^0(T))$, therefore,

$$\max_{0\leq\tau\leq T}\mathsf{E}_{\mathsf{P}^*}[Y(\tau)/S^0(\tau))=\mathsf{E}_{\mathsf{P}^*}(Y(T)/S^0(T)),$$

but this is the price of the corresponding European option that is exercised at moment $\tau = T$. Because the option is attainable, there exists a strategy that hedges this option; that is, the strategy where capital at moment T equals the value of the option.

19

Stochastic processes in financial mathematics (continuous time)

Theoretical grounds

Let $t \in \mathbb{T}$, where $\mathbb{T} = \mathbb{R}^+$ or $\mathbb{T} = [0,T]$. Also, let us have a filtration $\{\mathcal{F}_t, t \in \mathbb{T}\}$. Consider a financial market with one risk-free asset (bond) $B(t)$ and one risk asset (stock) $S(t)$, adapted to this filtration.

Definition 19.1. *A couple of stochastic processes $\{\varphi(t)\}$ and $\{\psi(t)\}$, where $\{\varphi(t)\}$ is a number of bonds and $\{\psi(t)\}$ is a number of stocks, is called a* portfolio. *The process φ is supposed to be $\mathcal{F}$-adapted, and the process ψ is supposed to be $\mathcal{F}$-predictable; that is, ψ is adapted to the information that comes strictly before the moment t (an exact definition of the predictable process with continuous time parameter is contained, e.g., in the book [84]; all the processes with continuous or continuous from the left trajectories are predictable).*

The investor's capital, that corresponds to the portfolio mentioned above, equals $V(t) = \psi(t)S(t) + \varphi(t)B(t)$.

Definition 19.2. *A portfolio (φ, ψ) is called* self-financing *if $dV(t) = \psi(t)\,dS(t) + \varphi(t)\,dB(t)$ (it means that the changes of capital occur only as a consequence of changes in bond and stock prices without any external entry or departure of the capital; also, we suppose that the stochastic process which describes the stock price admits a stochastic differential).*

Let $\mathbb{T} = [0,T]$ and X be an $\mathcal{F}_T$-adapted r.v. (the contingent claim). A self-financing portfolio is the replicating strategy for X if the following equality holds, $V(T) = \psi(T)S(T) + \varphi(T)B(T) = X$. A probability measure $\mathsf{P}^* \sim \mathsf{P}$ is called the martingale measure if the corresponding discounted process $\{B^{-1}(t)S(t)\}$ is a P^*-martingale. The existence of the measure P^* is equivalent to the arbitrage-free property of the market, and the uniqueness of such a measure is equivalent to the market completeness. In turn, the completeness of the market means that any $\mathcal{F}_T$-measurable integrable contingent claim X is attainable; that is, there exists a replicating portfolio for such a claim.

D. Gusak et al., *Theory of Stochastic Processes*, Problem Books in Mathematics,
DOI 10.1007/978-0-387-87862-1_19,

Let W be a Wiener process that is adapted to the filtration $\mathfrak{F}$, and a financial market consist of two assets with prices described by the formulas $B(t) = \exp\{rt\}$ (bond price) and $S(t) = S_0 \exp\{(\mu - \sigma^2/2)t + \sigma W(t)\}$ (stock price; this model is called *geometrical Brownian motion*), $t \geq 0$. Such a model is called a *(B,S)-model*, or *Black–Scholes model*. Consider a European call option with the strike price K, the exercise date T, and with the payoff of the form $C(S(T),T) = (S(T) - K)^+$. Denote by $C(S,t)$ the arbitrage-free (fair) price of the option at moment t under the condition that the stock price equals S. (Here $C(S,t)$ is called the price function, and $\pi(t) = C(S(t),t)$ is called the price process.) Then the function $C(S,t)$ satisfies *Black–Scholes equation*

$$\frac{\partial C}{\partial t} + \frac{1}{2}\sigma^2 S^2 \frac{\partial^2 C}{\partial S^2} + rS\frac{\partial C}{\partial S} - rC = 0$$

with boundary conditions $C(0,t) = 0$, $C(S,T) = (S-K)^+$. The solution to this equation has a form

$$C(S,t) = S\Phi(d_1^t) - Ke^{-r(T-t)}\Phi(d_2^t) \quad \textit{(Black–Scholes formula)},$$

where $\Phi(x) = \int_{-\infty}^{x} e^{-(u^2/2)}du/\sqrt{2\pi}$ is standard Gaussian distribution function, and

$$d_1^t = \frac{\log\frac{S}{K} + (r + \frac{1}{2}\sigma^2)(T-t)}{\sigma\sqrt{T-t}}, \quad d_2^t = \frac{\log\frac{S}{K} + (r - \frac{1}{2}\sigma^2)(T-t)}{\sigma\sqrt{T-t}}.$$

In particular, for $t = 0$ we obtain the Black–Scholes formula for the arbitrage-free (fair) price of the call option at initial moment:

$$\pi(C(S(T),T)) := C(S,0) = S\Phi(d_1^0) - Ke^{-rT}\Phi(d_2^0).$$

How can the Black–Scholes equation be deduced? If the investor's portfolio consists of one call option and of $(-\Delta)$ stocks, then its cost equals $\pi = C - \Delta \cdot S$, and the change of this cost equals $d\pi = dC - \Delta \cdot dS$, but from reasoning based on the absence of arbitrage, it must be equal to $d\pi = r\pi\, dt$, whence

$$\pi = \frac{1}{r}\left(\frac{\partial C}{\partial t} + \frac{1}{2}\sigma^2 S^2 \frac{\partial^2 C}{\partial S^2}\right),$$

and we can easily deduce the Black–Scholes equation.

The values $\Delta = \partial C(S,t)/\partial S$, $\gamma = \partial^2 C(S,t)/\partial S^2$, $\theta = -\partial C(S,t)/\partial t$, $\rho = \partial C(S,t)/\partial r$, and $\mathcal{V} = \partial C(S,t)/\partial \sigma$ are called Delta, Gamma, Theta, Rho, and Vega options, respectively, and as a whole they are called *Greeks*, although "vega" does not correspond to a letter of the Greek alphabet.

The prices of put and call options, $C(S,t)$ and $P(S,t)$, correspondingly, satisfy the following put–call parity relation, $C(S,t) - P(S,t) = S - Ke^{-r(T-t)}$.

Consider a stock, on which some dividends are paid. Suppose that they are paid with a constant rate that is proportional to the stock price, so that the owner of $a(t)$ units of this stock, $t \in [\alpha, \beta]$, obtains the total quantity of dividends that is equal to

$$r_D \int_{\alpha}^{\beta} a(t)S(t)dt.$$

The coefficient r_D is called *the dividend yield.* If $\overline{C}(S,t)$ denotes the price of the call option on the stock with dividend payments, then $\overline{C}(S,t)$ is a solution to the modified Black–Scholes equation

$$\frac{\partial C}{\partial t} + \frac{1}{2}\sigma^2 S^2 \frac{\partial^2 C}{\partial S^2} + (r - r_D)S\frac{\partial C}{\partial S} - rC = 0.$$

Bibliography

[55], Chapter IV; [84], Volume 2, Chapters VII and VIII; [21], Chapters IV-VII; [46], Chapters 11–13; [54]; [61], Chapter X; [68], Chapter 13; [85], Chapters 10–14.

Problems

19.1. Suppose that the price of a discounted asset at moment T has lognormal distribution with parameters a and σ^2; that is, $\log((S(T))/(S(0)))$ has normal distribution with parameters mentioned above. Calculate $\mathsf{E}S(T)$.

19.2. In the framework of an arbitrage-free market and the Black–Scholes model consider two European call options with the same strike price and on the same underlying asset. Is it true that the option with a longer time to maturity has a larger arbitrage-free price? What can you say in this connection concerning a European put option?

19.3. Let a market be arbitrage-free, dividends not be paid, and $C = C(S,t)$ be the price of a European call option at moment t under the condition that the stock price equals S. Prove the following inequalities.

(a) $C(S,t) \le S$.

(b) $C(S,t) \ge S - Ke^{-r(T-t)}$.

(c) If we consider two call options with the same exercise date and on the same stock but with different strike prices K_1 and K_2, then $0 \le C(S,t;K_1) - C(S,t;K_2) \le K_2 - K_1$.

(d) If we consider two call options with the same strike price and on the same stock but with different exercise dates $T_1 < T_2$, then $C(S,t;T_1) \le C(S,t;T_2)$.

19.4. Prove that the functions $C(S,t) = AS$ and $C(S,t) = Ae^{rt}$, where A is an arbitrary constant, are solutions to the Black–Scholes equation.

19.5. Find on your own the solution to the Black–Scholes equation using the following steps.

(1) Change the variables $S = Ke^x$, $t = T - 2\tau/\sigma^2$, and $C = Kv(x,\tau)$, and reduce the Black–Scholes equation to the equation of the form

$$\frac{\partial v}{\partial \tau} = \frac{\partial^2 v}{\partial x^2} + (k-1)\frac{\partial v}{\partial x} - kv,$$

where $k = 2r/\sigma^2$. A new boundary condition is $v(x,0) = (e^x - 1)^+$.

(2) Put $v(x,\tau) = e^{\alpha x + \beta\tau} u(x,\tau)$ with unknown coefficients α and β. Get an equation for v, put $\beta = \alpha^2 + (k-1)\alpha - k$, $2\alpha + (k-1) = 0$ in it, and reduce the equation for u to the form $\partial u/\partial\tau = \partial^2 u/\partial x^2$, $x \in \mathbb{R}$, $\tau > 0$ (the heat equation, or the diffusion equation), where

$$\begin{aligned} u(x,0) &= u_0(x) = \max\{e^{(k+1)x/2} - e^{(k-1)x/2}, 0\}, \\ v(x,\tau) &= \exp\{-(k-1)x/2 - (k+1)^2\tau/4\}u(x,\tau). \end{aligned}$$

Check that the function

$$u(x,\tau) = \frac{1}{2\sqrt{\pi\tau}} \int_{\mathbb{R}} u_0(y) e^{-(((x-y)^2)/4\tau)} dy$$

is the unique solution to the diffusion equation (Cauchy problem) with initial condition $u(x,0) = u_0(x)$.

(3) Obtain the Black–Scholes formula by the inverse change of variables.

19.6. In the framework of the Black–Scholes model consider the discounted capital $V(t) = \psi(t)Z(t) + \varphi(t)$, where $Z(t) = B^{-1}(t)S(t)$. Prove that the portfolio (φ,ψ) is self-financing if and only if $dV(t) = \psi(t)dZ(t)$.

19.7. Let a market consist of one stock $S(t) = S(0)\exp\{\sigma W(t) + \mu t\}$ and one bond $B(t) = \exp\{rt\}$. Also, suppose that a filtration $\mathfrak{F}$ is natural, that is, generated by a Wiener process $\{W(t), t \in [0,T]\}$. Prove that this market is arbitrage-free, and for any nonnegative $\mathfrak{F}_T$-measurable claim X such that $\mathsf{E}X^{2+\alpha} < \infty$ for some $\alpha > 0$, there exists a replicating portfolio (φ,ψ). Also, prove that the arbitrage-free price of X at moment t equals

$$\pi(X)(t) = B(t)\,\mathsf{E}_{\mathsf{P}^*}(B^{-1}(T)X|\mathfrak{F}_t) = e^{-r(T-t)}\mathsf{E}_{\mathsf{P}^*}(X|\mathfrak{F}_t),$$

where P^* is the equivalent martingale measure with respect to which the discounted process $\{B^{-1}(t)S(t)\}$ is a martingale.

19.8. Let a stochastic process X be the nominal return,

$$dX(t) = X(t)(\alpha\,dt + \sigma\,dW(t)),$$

and a stochastic process Y describe the inflation, $dY(t) = Y(t)(\gamma dt + \delta\,d\widetilde{W}(t))$, where $\widetilde{W}$ is a Wiener process, independent of W. We suppose that the coefficients α, σ, γ, and δ are constant. Derive the SDE for real return $Z(t) := X(t)/Y(t)$.

19.9. In the framework of the Black–Scholes model consider a European contingent claim of the form

$$H = \begin{cases} K, & \text{if } S(T) \le A, \\ K + A - S(T), & \text{if } A < S(T) < K + A, \\ 0, & \text{if } S(T) > K + A. \end{cases}$$

The expiration date of H is supposed to equal T. Define a portfolio consisting of bonds, stocks, and a European call option, that will be constant in time and replicates the claim H. Define the arbitrage-free price of H.

19.10. (1) Using Problem 19.5, choose such a change of variables that permits the reduction of the equation

$$\frac{\partial u}{\partial t} = \frac{\partial^2 u}{\partial x^2} + a\frac{\partial u}{\partial x} + bu,\ a,\, b \in \mathbb{R}$$

to the diffusion one.

(2) Choose such a change of time that permits the reduction of the equation

$$c(t)\frac{\partial u}{\partial t} = \frac{\partial^2 u}{\partial x^2},\ c(t) > 0,\ t > 0$$

to the diffusion one.

(3) Suppose that $\sigma^2(\cdot)$ and $r(\cdot)$ in the Black–Scholes equation are the functions of t, however, $(r(t))/(\sigma^2(t))$ does not depend on t. Rewrite the Black–Scholes formula for this case.

19.11. Suppose that in the Black–Scholes equation the functions $r(\cdot)$ and $\sigma^2(\cdot)$ are known nonrandom functions of t. Prove that the following steps reduce the Black–Scholes equation to the diffusion one.

(1) Put $S = Ke^x$, $C = Kv$, and $t = T - t'$, and obtain the equation

$$\frac{\partial v}{\partial t'} = \frac{1}{2}\sigma^2(t')\frac{\partial^2 v}{\partial x^2} + \left(r(t') - \frac{1}{2}\sigma^2(t')\right)\frac{\partial v}{\partial x} - r(t')v.$$

(2) Change the time variable as $\widehat{\tau}(t') = \int_0^{t'} \frac{1}{2}\sigma^2(s)ds$ and obtain the equation

$$\frac{\partial v}{\partial \widehat{\tau}} = \frac{\partial^2 v}{\partial x^2} + a(\widehat{\tau})\frac{\partial v}{\partial x} - b(\widehat{\tau})v,$$

where $a(\widehat{\tau}) = 2r/\sigma^2 - 1$, $b(\widehat{\tau}) = 2r/\sigma^2$.

(3) Prove that the general solution to the first-order partial differential equation of the form

$$\frac{\partial v}{\partial \widehat{\tau}} = a(\widehat{\tau})\frac{\partial v}{\partial x} - b(\widehat{\tau})v$$

can be presented as $v(x,\widehat{\tau}) = F(x + A(\widehat{\tau}))e^{-B(\widehat{\tau})}$, where $dA(\widehat{\tau})/d\widehat{\tau} = a(\widehat{\tau})$, $dB(\widehat{\tau})/d\widehat{\tau} = b(\widehat{\tau})$, and $F(\cdot)$ is an arbitrary function.

(4) Prove that the solution to the second-order partial differential equation from item (2) has a form

$$v(x,\widehat{\tau}) = e^{-B(\widehat{\tau})}V(\widehat{x},\widehat{\tau}),$$

where $\hat{x} = x + A(\hat{\tau})$, $A(\hat{\tau})$ $B(\hat{\tau})$ are the functions of $\hat{\tau}$, taken from item (3), and V is a solution to the diffusion equation $\partial V/d\hat{\tau} = \partial^2 V/\partial x^2$.

(5) Transform the initial data correspondingly to the change of variables.

19.12. Let $C(S,t)$ and $P(S,t)$ be the prices at moment t of a European call and put option, correspondingly, with the same strike price and exercise date.

(1) Prove that both P and $C-P$ satisfy the Black–Scholes equation; moreover, the boundary condition for $C-P$ is extremely simple: $C(S,T) - P(S,T) = S - K$.

(2) Deduce from the put–call parity that $S - Ke^{-r(T-t)}$ is a solution to the Black–Scholes equation with the same boundary condition.

19.13. Use the exact solution to the diffusion equation to find the Black–Scholes price $P(S,t)$ of a put option $P(S,T) = (K-S)^+$ without using the put–call parity.

19.14. (1) Prove that in the case where the initial condition of the boundary value problem for the heat equation is positive, then $u(x,\tau) > 0$ for any $\tau > 0$.

(2) Deduce from here that for any option with positive payoff its price is also positive, if it satisfies the Black–Scholes equation.

19.15. (1) In the framework of the Black–Scholes model find the arbitrage-free option price with the payoff $f(S(T))$, where the function $f \in C(\mathbb{R})$ and increases at infinity not faster than a polynomial.

(2) Find the arbitrage-free option price with the payoff of the form $B\mathscr{H}(K - S(T))$, where $\mathscr{H}(s) = \mathbb{I}_{s\geq 0}$ is a Heaviside function and $B > 0$ is some constant (the option "cash-or-nothing").

(3) The European digital call option of the kind asset-or-nothing has the payoff $S(T)$ in the case $S(T) > K$, and zero payoff in the case $S(T) \leq K$. Find its price.

19.16. What is a probability that a European call option will expire in-the-money?

19.17. Calculate the price of a European call option on an asset with dividends, if the dividend yield equals r_D on the interval $[0,T]$.

19.18. What is the put–call parity relation for options on the asset with dividends?

19.19. What is the delta for the call option with the continuous and constant dividend yield r_D?

19.20. Calculate Δ, Γ, θ, ρ, and $\mathscr{V}$ for put and call options.

19.21. On the Black–Scholes market a company issued an asset "Golden logarithm" (briefly GLO). The owner of GLO(T) with the expiration date T receives at moment T the sum $\log S(T)$ (in the case $S(T) < 1$ the owner pays the corresponding sum to the company). Define the price process for GLO(T).

19.22. Let the functions $r(x)$ and $u(x)$ denote the interest rate and the process (flow) of the cash receipt, correspondingly, under the condition that the initial value of a risk asset $X(0) = x$, where X is a time-homogeneous diffusion process with the drift coefficient $\mu = \mu(x)$ and diffusion $\sigma = \sigma(x)$; μ and σ are continuous functions, and $\sigma(x) \neq 0$ for all $x \in \mathbb{R}$. Suppose that u is bounded and continuous.

(1) Write the stochastic differential of the process X.

(2) Check that the function

$$u(t,x) := \mathsf{E}\left(e^{-\int_0^t r(X(s))ds} u(X(t)) \big| X(0) = x\right), \quad t \in \mathbb{R}^+$$

can be considered as the expected discounted cash flow at moment t under the condition that $X(0) = x$.

(3) Find a partial derivative equation for which the function $u(t,x)$ is a solution.

19.23. (When is the right time to sell the stocks?) Suppose that the stock price $\{S(t), t \in \mathbb{R}^+\}$ is a diffusion process of the form $dS(t) = rS(t)dt + \sigma S(t)dW(t)$, $S(0) = x > 0$ (for the explicit form of $S(t)$ see Problem 14.3). Here W is a one-dimensional Wiener process, $r > 0$, and $\sigma \neq 0$. Suppose that there is a fixed transaction cost $a > 0$, connected to the sale of the asset. Then, regarding inflation, the discounted asset price at moment t equals $e^{-\rho t}(S(t) - a)$. Find the optimal stopping time τ_0, for which

$$\mathsf{E}_{s,x} e^{-\rho \tau_0}(S(\tau_0) - a) = \sup_{\tau} \mathsf{E}_{s,x} e^{-\rho \tau}(S(\tau) - a) = \sup_{\tau} \mathsf{E}_{s,x} g(\tau, S(\tau)),$$

where $g(t,y) = e^{-\rho t}(y - a)$.

19.24. (Vasicek stochastic model of interest rate) According to the Vasicek model, the interest rate $r(\cdot)$ satisfies a SDE,

$$dr(t) = (b - ar(t))dt + \sigma dW(t),$$

where W is a Wiener process.

(1) Find an explicit form of $r(\cdot)$.

(2) Find the limit distribution of $r(t)$ as $t \to \infty$.

19.25. (Cox–Ingersoll–Ross stochastic model of interest rate) According to the Cox–Ingersoll–Ross model, the interest rate $r(\cdot)$ satisfies a SDE

$$dr(t) = (\alpha - \beta r(t))dt + \sigma\sqrt{r(t)}dW(t),$$

where W is a Wiener process, $\alpha > 0$, and $\beta > 0$. The process $\{r(t)\}$ is also called the square of the Bessel process. (Concerning the existence and uniqueness of the strong solution to this equation see Problem 14.16.)

(1) Define the SDE for $\{\sqrt{r(t)}\}$ in the case $\alpha = 0$.

(2) Suppose that a nonrandom function $u(\cdot)$ satisfies the ordinary differential equation $u'(t) = -\beta u(t) - (\sigma^2/2)u^2(t)$, $u(0) = \theta \in \mathbb{R}$. Fix $T > 0$ and assume that $\alpha = 0$. Find the differential equation for the function $G(t) = \mathsf{E}\exp\{-u(T-t)r(t)\}$. Calculate the mean value and variance of $r(T)$.

(3) In the general case, calculate the density and moment-generating function for the distribution of $r(t)$.

Hints

19.2. Yes, for a call option it is true. To check this statement it is necessary to prove that if we fix all other parameters, then the stochastic process $\{Y_t := e^{-rt}(S_t - K)^+, t \geq 0\}$ becomes a submartingale with respect to the natural filtration and to the risk-neutral measure. For put options the situation becomes more complicated, and the answer is negative in the general case.

19.3. *I method.* Prove that under opposite inequalities the arbitrage is possible. *II method.* Directly use the form of the solution to the Black–Scholes equation.

19.6. Verify directly the definition of the self-financing property.

19.10. (3) Choose a change of time in order to reduce the equation to the diffusion one, and then use the usual Black–Scholes formula or apply Problem 19.11.

19.15. (2), (3) Substitute the corresponding function f into the formula obtained in item (1).

19.16. This is the probability of the event $\{S(T) \geq K\}$, and the distribution of $\log S(T)$ is Gaussian.

19.18. Solve the Black–Scholes equation for $C - P$ (with dividends), using the boundary condition $C(S,T) - P(S,T) = S - K$.

19.22. Apply Problem 14.14.

19.24. (2) Use the equality (19.3) (see Answers and Solutions to this chapter) and the fact that the integral on the right-hand side has a Gaussian distribution.

Answers and Solutions

19.1.

$$\mathsf{E}S(T) = \frac{1}{\sigma\sqrt{2\pi}} e^{a+(\sigma^2/2)}.$$

19.7. It is possible to construct the martingale measure P^* by the Girsanov theorem (see Problems 14.25 and 14.22). For this purpose it is necessary to put

$$\frac{d\mathsf{P}^*}{d\mathsf{P}} := \exp\left\{-\left(\frac{\mu - r}{\sigma} - \frac{1}{2}\sigma\right) W(t) - \frac{1}{2}\left(\frac{\mu - r}{\sigma} - \frac{1}{2}\sigma\right)^2 t\right\}.$$

Then the Novikov's condition evidently holds on the finite interval $[0,T]$, stochastic process $\tilde{W}(t) := W(t) + ((\mu - r)/(\sigma) - \frac{1}{2}\sigma$ is a Wiener process with respect to the same filtration, and the discounted process $(S(t))/(B(t))$ is a martingale with respect to the measure P^* and the same filtration. Now, because $\mathsf{E}X^{2+\alpha} < \infty$ for some

$\alpha > 0$, it is easy to check with the help of the Hölder inequality that the claim X is square integrable with respect to a measure P^*. Put $V(t) = \mathsf{E}_{\mathsf{P}^*}(B^{-1}(T)X|\mathcal{F}_t)$. Because the filtration $\mathcal{F}$ is generated by a Wiener process, then, for example, by Theorem 5.13 [57], the representation holds $V(t) = \mathsf{E}V(0) + \int_0^t \beta(s)dW(s)$. Now, it is necessary to put $\psi(t) = \mu^{-1}\beta(t)B(t)S^{-1}(t)$ and $\varphi(t) = V(t) - \psi(t)S(t)B^{-1}(t)$. Furthermore you can check it on your own with the help of the Itô formula that the following equations hold: $B(t)V(t) = \varphi(t)B(t) + \psi(t)S(t)$, whence, in particular, $X = \varphi(T)B(T) + \psi(T)S(T)$; that is, our portfolio replicates X, and also $B(t)V(t) = B(t)dV(t) + V(t)dB(t) = \varphi(t)dB(t) + \psi(t)dS(t)$; it is equivalent to the self-financing property of the strategy (φ, ψ).

19.8. $dZ(t) = Z(t)((\alpha - \gamma + \delta^2)dt + \sigma dW(t) - \delta d\widetilde{W}(t))$.

19.9. Write H in the form $H = K \cdot 1 - (S(T) - A)^+ + (S(T) - A - K)^+$. Thus, the desired portfolio can be constructed from K bonds of the price 1 each, short position in option $(S(T) - A)^+$ (i.e., this option must be sold), and the long position in option $(S(T) - A - K)^+$ (i.e., this option must be bought). Hence the arbitrage-free price of the claim H at moment t equals

$$\pi(H)(t) = Ke^{-r(T-t)} - \pi\left((S(T) - A)^+\right) + \pi\left((S(T) - A - K)^+\right),$$

where the arbitrage-free prices of the options mentioned above have to be defined by the Black–Scholes formula.

19.15. (1) The required arbitrage-free price equals

$$\frac{1}{\sigma\sqrt{2\pi}} \int_{\mathbb{R}} f(e^y) \exp\left\{\frac{-(y-\mu)^2}{2\sigma^2}\right\} dy.$$

19.21. The required price process has a form $\pi(t) = \log S(t) + (r - \sigma^2/2)(T - t)$.

19.23. The infinitesimal operator $\widetilde{L}$ of the process $Y(t) = (s + t, S(t))$ is given by the formula

$$\widetilde{L}f(s,x) = \frac{\partial f}{\partial s} + rx\frac{\partial f}{\partial x} + \frac{1}{2}\sigma^2 x^2 \frac{\partial^2 f}{\partial x^2},\ f \in C^2(\mathbb{R}^2).$$

Therefore, in our case $\widetilde{L}g(s,x) = e^{-\rho s}((r - \rho)x + \rho a)$, whence

$$U := \{(s,x)|\widetilde{L}g(s,x) > 0\} = \begin{cases} \mathbb{R} \times \mathbb{R}^+, & \text{if } r \geq \rho, \\ \{(s,x)|x < \frac{a\rho}{\rho - r}\}, & \text{if } r < \rho. \end{cases}$$

So, if $r \geq \rho$, then $U = \Gamma^c = \mathbb{R} \times \mathbb{R}^+$, and there is no optimal stopping time. If $r > \rho$, then $v_g = \infty$, and for $r = \rho$ it holds $v_g(s,x) = xe^{-\rho s}$ (prove these statements). Consider the case $r < \rho$ and prove that the set Γ^c is invariant in t; that is, $\Gamma^c + (t_0, 0) = \Gamma^c$ for all t_0. Indeed,

$$\begin{aligned}\Gamma^c + (t_0, 0) &= \{(t+t_0, x) | (t,x) \in \Gamma^c\} \\ &= \{(s,x) | (s-t_0, x) \in \Gamma^c\} = \{(s,x) | g(s-t_0,x) < v_g(s-t_0,x)\} \\ &= \{(s,x) | e^{\rho t_0} g(s,x) < e^{\rho t_0} v_g(s,x)\} = \{(s,x) | g(s,x) < v_g(s,x)\} = \Gamma^c.\end{aligned}$$

Here the equalities

$$\begin{aligned}v_g(s-t_0,x) &= \sup_\tau \mathsf{E}_{s-t_0,x} e^{-\rho\tau}(S(\tau)-a) = \sup_\tau \mathsf{E} e^{-\rho(\tau+(s-t_0))}(S(\tau)-a) \\ &= e^{\rho t_0} \sup_\tau \mathsf{E} e^{-\rho(\tau+s)}(S(\tau)-a) = e^{\rho t_0} v_g(s,x)\end{aligned}$$

were used. Therefore, a connected component of the set Γ^c containing U must have the form $\Gamma^c(x_0) = \{(t,x) | 0 < x < x_0\}$, for some $x_0 > a\rho/(\rho - r)$. Note that Γ^c cannot have any other components, because another component V of the set Γ^c must satisfy the relation $\widetilde{L}g < 0$ in V, and then for $y \in V$

$$\mathsf{E}_y g(Y(\tau)) = g(y) + \mathsf{E}_y \int_0^\tau \widetilde{L}g(Y(t))dt < g(y),$$

for all stopping times bounded by the exit time from a strip in V. So, it follows from Theorem 15.2, item (2), that $v_g(y) = g(y)$, and then $V = \varnothing$. Put $\tau(x_0) = \tau_{\Gamma^c(x_0)}$ and calculate

$$\widetilde{g}(s,x) = \widetilde{g}_{x_0}(s,x) := \mathsf{E}_{s,x} g\left(Y(\tau(x_0))\right).$$

This function is a solution to the boundary value problem

$$\begin{aligned}\frac{\partial f}{\partial s} + rx\frac{\partial f}{\partial x} + \frac{1}{2}\sigma^2 x^2 \frac{\partial^2 f}{\partial x^2} &= 0,\ 0 < x < x_0, \\ f(s,x_0) &= e^{-\rho s}(x_0 - a).\end{aligned} \tag{19.1}$$

If we try a solution of (19.1) of the form

$$f(s,x) = e^{-\rho s}\varphi(x),$$

we get the following one-dimensional problem

$$\begin{aligned}-\rho\varphi + rx\varphi'(x) + \tfrac{1}{2}\sigma^2 x^2 \varphi''(x) &= 0,\ 0 < x < x_0, \\ \varphi(x_0) &= x_0 - a.\end{aligned} \tag{19.2}$$

The general solution to the equation (19.2) has a form

$$\varphi(x) = C_1 x^{\gamma_1} + C_2 x^{\gamma_2},$$

where C_i, $i = 1,2$ are arbitrary constants, and

$$\gamma_i = \sigma^{-2}\left(\frac{1}{2}\sigma^2 - r \pm \sqrt{(r - \frac{1}{2}\sigma^2)^2 + 2\rho\sigma^2}\right),\ \gamma_2 < 0 < \gamma_1.$$

Because the function $\varphi(x)$ is bounded as $x \to 0$, it should hold $C_2 = 0$, and the boundary requirement gives $C_1 = x_0^{-\gamma_1}(x_0 - a)$. Hence,

$$\widetilde{g}_{x_0}(s,x) = f(s,x) = e^{-\rho s}(x_0 - a)\left(\frac{x}{x_0}\right)^{\gamma_1}.$$

If we fix (s,x), then the maximal value of $\widetilde{g}_{x_0}(s,x)$ is attained at $x_0 = x_{\max} = a\gamma_1/(\gamma_1 - 1)$ (here $\gamma_1 > 1$ if and only if $r < \rho$). At last,

$$v_g(s,x) = \sup_{x_0} \mathsf{E}_{s,x} g\Big(\tau(x_0), X(\tau(x_0))\Big) = \sup_{x_0} \widetilde{g}_{x_0}(s,x) = \widetilde{g}_{x_{\max}}(s,x).$$

The conclusion is that one should sell the stock at the first moment when the price of it reaches the value $x_{\max} = a\gamma_1/(\gamma_1 - 1)$. The expected discounted profit obtained from this strategy equals

$$v_g(s,x) = e^{-\rho s}\left(\frac{\gamma_1 - 1}{a}\right)^{\gamma_1 - 1}\left(\frac{x}{\gamma_1}\right)^{\gamma_1}.$$

19.24. (1)

$$r(t) = \frac{b}{a} + \left(r(0) - \frac{b}{a}\right)e^{-at} + \sigma\int_0^t e^{-a(t-s)}dW(s). \tag{19.3}$$

19.25. (1) Denote $q(t) = \sqrt{r(t)}$. Then

$$dq(t) = -\left(\frac{\beta}{2}q(t) + \frac{1}{8}\frac{\sigma^2}{q(t)}\right)dt + \frac{\sigma}{2}dW(t).$$

(2) The function G satisfies the differential equation (in the integral form) $G(t) = \exp\{-\theta r(T)\} - \int_t^T G(s)r(s)(u'(T-s) + \beta u(T-s) + (\sigma^2/2)u^2(T-s))ds = \mathsf{E}\exp\{-\theta r(T)\}$; that is, in fact, it does not depend on t. Now, it is easy to prove (please do it yourself) that

$$u(t) = \frac{\theta\beta e^{-\beta t}}{\beta + \frac{\sigma^2}{2}\theta(1 - e^{-\beta t})}.$$

Now, it is necessary to write the equality $G(0) = \mathsf{E}\exp\{-u(T)r(0)\} = \mathsf{E}\exp\{-\theta r(T)\}$ and to take the derivative of the left-hand and right-hand sides of this equality in θ to find the corresponding moments.
(3) The density of distribution of $r(t)$ equals

$$f_t(x) = ce^{-c(u+x)}\left(\tfrac{x}{u}\right)^{q/2} I_q(2c\sqrt{xr}),\ x \geq 0,\ c = \tfrac{2\beta}{\sigma^2(1-e^{-\beta t})},\ u = r(0)e^{-\beta t},$$
$$q = \tfrac{2\alpha}{\sigma^2} - 1,$$

and the function

$$I_q(x) = \sum_{k=0}^{\infty} \frac{(x/2)^{2k+q}}{k!\Gamma(k+q+1)}$$

is a modified Bessel function of the first kind and order q. This is a noncentral χ^2 distribution with $2(q+1)$ degrees of freedom and the skew coefficient $2cu$. The density of distribution of $r(t)$ can be also presented in the form

$$f_t(x) = \sum_{k=0}^{\infty} (((cu)^k)/k!)e^{-cu} g_{k+q+1,c}(x),$$

where the function $g_{\gamma,\lambda}(x) = (1/(\Gamma(\gamma)))\lambda^{\gamma} x^{\gamma-1} e^{-\lambda x}$ is the density of Gamma distribution $\Gamma(\gamma, \lambda)$. The moment generating function equals

$$m(v) = \left(\frac{c}{c-v}\right)^{q+1} \exp\left\{\frac{cuv}{c-v}\right\}.$$

(Noncentral χ^2 distributions are considered in detail in [42].)

20

Basic functionals of the risk theory

Theoretical grounds

Mathematical foundations of investigating of the risk process in insurance were created by Swedish mathematician Filip Lundberg in 1903–1909. For a long time this theory had been developed by mostly Nordic mathematicians, such as Cramér, Segerdal, Teklind, and others. Later on risk theory started to develop not only with connection to insurance but also as the method of solving different problems in actuarial and financial mathematics, econometrics. In the second half of the twentieth century the applied area of risk theory was expanded significantly.

Processes with independent increments play a very important role in risk theory. The definitions and the characteristics of the homogeneous processes with independent increments are presented in Chapter 5. Not general multidimensional processes but rather real-valued ones with independent increments and with the jumps of the same sign are used in queueing and risk theory. In particular, stepwise processes $\xi(t)$ with jumps ξ_k satisfying one of the following conditions,

$$\mathsf{E}\left(e^{i\alpha\xi_1}/\xi_1>0\right)=\frac{c}{c-i\alpha},\quad \mathsf{E}\left(e^{i\alpha\xi_1}/\xi_1<0\right)=\frac{b}{b+i\alpha}, \tag{20.1}$$

have a range of application.

Theorem 20.1. *The Lévy process $\{\xi(t),\, t\ge 0\}$ is piecewise constant (stepwise) with probability one if and only if its Lévy measure Π satisfies the condition $\Pi(\mathbb{R}\setminus\{0\})<\infty$ and the characteristic function in the Lévy–Khinchin formula (Theorem 5.2) of an increment $\xi(t)-\xi(0)$ is as follows,*

$$\mathsf{E}e^{i\alpha(\xi(t)-\xi(0))}=e^{t\psi(\alpha)}, \tag{20.2}$$

with the cumulant function $\psi(\alpha)$ *determined by the relation*

$$\psi(\alpha)=\int_{-\infty}^{\infty}(e^{i\alpha x}-1)\,\Pi(dx). \tag{20.3}$$

D. Gusak et al., *Theory of Stochastic Processes*, Problem Books in Mathematics,
DOI 10.1007/978-0-387-87862-1_20, © Springer Science+Business Media, LLC 2010

Theorem 20.2. *The Lévy process $\{\xi(t), t \geq 0\}$ is a nondecreasing function of time with probability one if and only if*

$$\mathsf{E}e^{i\alpha(\xi(t)-\xi(0))} = e^{t\psi(\alpha)}, \quad \psi(\alpha) = i\alpha a + \int_0^\infty \left(e^{i\alpha x} - 1\right) \Pi(dx), \tag{20.4}$$

where $a \geq 0$ and the measure Π satisfies the condition $\int_0^1 x\Pi(dx) < \infty$.

Theorem 20.3. *The process $\{\xi(t), t \geq 0\}$ with independent increments has a bounded variation with probability one on any bounded interval if and only if*

$$\mathsf{E}e^{i\alpha(\xi(t)-\xi(0))} = e^{t\psi(\alpha)}, \quad \psi(\alpha) = i\alpha a + \int_{\mathbb{R}} \left(e^{i\alpha x} - 1\right) \Pi(dx), \tag{20.5}$$

where $a \in \mathbb{R}$, $\int_{-1}^1 |x|\Pi(dx) < \infty$.

Theorem 20.4. *The characteristic function of the compound Poisson process*

$$\xi(t) = at + \sum_{k=0}^{N(t)} \xi_k, \quad (\xi_0 = \xi(0) = 0), \tag{20.6}$$

where $\{\xi_k, k \geq 1\}$ are i.i.d. random variables independent of the simple Poisson process $N(t)$ with intensity $\lambda > 0$, can be expressed in the form of identity (20.2) *with the following cumulant function,*

$$\psi(\alpha) = i\alpha a + \lambda \int_{\mathbb{R}} \left(e^{i\alpha x} - 1\right) dF(x), \quad F(x) = \mathsf{P}(\xi_1 < x), \quad x \in \mathbb{R}. \tag{20.7}$$

Definition 20.1. *The compound Poisson* (20.6) *process with jumps of the same sign is said to be:*

(1) Upper continuous *if $a > 0$, $\mathsf{P}(\xi_k < 0) = 1$.*
(2) Lower continuous *if $a < 0$, $\mathsf{P}(\xi_k > 0) = 1$.*

We call such kinds of processes *semicontinuous.*

Definition 20.2. *The compound Poisson process* (20.6) *with $a \leq 0$ is said to be* almost upper semicontinuous *if the first condition in* (20.1) *is satisfied with $c > 0$. The processes* (20.6) *with $a \geq 0$ are said* almost lower semicontinuous *if the second condition holds true in* (20.1) *with $b > 0$. If for these processes $a = 0$, then they are called stepwise almost upper or lower semicontinuous.*

Let us introduce the basic notions connected with *risk processes* and their basic characteristics.

Definition 20.3. The classic risk process *(or* the reserve process*) is the process*

$$R_u(t) = \xi_u(t) = u + Ct - \sum_{k=0}^{N(t)} \xi_k, \quad C > 0, \quad u > 0, \quad \mathsf{P}(\xi_k > 0) = 1, \tag{20.8}$$

which describes the reserved capital of an insurance company at time t. Here u is an initial capital, C is a gross risk premium rate. The process $S(t) = \sum_{k=0}^{N(t)} \xi_k$ ($\xi_0 = S(0) = 0$) determines the outpayments of claims with mean value $0 < \mu = \mathsf{E}\xi_1 < \infty$.

Definition 20.4. *The* safety security loading *is the number*

$$\delta = \frac{\mathsf{E}\xi(1)}{\mathsf{E}S(1)} = \frac{C-\lambda\mu}{\lambda\mu} > 0, \quad \xi(t) = \xi_0(t), \tag{20.9}$$

(Here $\mathsf{E}\xi(1) = m := C - \lambda\mu$.*)*

Definition 20.5. *The* claim surplus process *is the process*

$$\zeta(t) = u - \xi_u(t) = S(t) - Ct. \tag{20.10}$$

Let us denote the extremums of the processes $\xi(t)$ and $\zeta(t)$ as

$$\begin{aligned} \xi^{\pm}(t) &= \sup_{0\le t'\le t} (\inf)\, \xi(t'); & \zeta^{\pm}(t) &= \sup_{0\le t'\le t} (\inf)\, \zeta(t'); \\ \xi^{\pm} &= \sup_{0\le t<\infty} (\inf)\, \xi(t); & \zeta^{\pm} &= \sup_{0\le t<\infty} (\inf)\, \zeta(t); \end{aligned} \tag{20.11}$$

$$\tau^+(u) = \inf\{t \ge 0 \mid \zeta(t) > u\} = \inf\{t \ge 0 \mid \xi_u(t) < 0\}.$$

Definition 20.6. *The* ultimate ruin probability *is*

$$\Psi(u) = \mathsf{P}(\xi_u(t) < 0 \quad \textit{for some} \quad t > 0). \tag{20.12}$$

It can be written in the terms of distributions of the extremums as follows:

$$\Psi(u) = \mathsf{P}(\zeta^+ > u) = \mathsf{P}(\tau^+(u) < \infty) = \mathsf{P}(\xi^- < -u). \tag{20.13}$$

Definition 20.7. *The* ruin probability with finite horizon $[0,T]$ *is the probability*

$$\Psi(u,T) = \mathsf{P}(\zeta^+(T) > u) = \mathsf{P}(\tau^+(u) < T). \tag{20.14}$$

Besides the extreme values (20.11), other boundary functionals are also used in risk theory. In particular, the following overjump functionals are used:

$$\begin{aligned} \gamma^+(u) &= \zeta(\tau^+(u)) - u \; - \text{ the value of overjump;} \\ \gamma_+(u) &= u - \zeta(\tau^+(u) - 0) \; - \text{ the value of lowerjump;} \\ \gamma_u^+ &= \gamma^+(u) + \gamma_+(u), \quad u > 0; \end{aligned} \tag{20.15}$$

here γ_u^+ is the value of the jump covering the level u. Let $\gamma_+(u)$ be the lowerjump of the stepwise process $\zeta(t)$ with $a = 0$ under the condition that the first jump $\xi_1 > u$ crossed the level $u \ge 0$. Then $\gamma_+(u)$ takes up the fixed value u with positive probability $\mathsf{P}(\gamma_+(u) = u,\ \zeta^+ > u) = \overline{F}(u) = \mathsf{P}(\xi_1 > u) > 0$.

Let us also mention that all boundary functionals (20.11), (20.15) have their own interpretation in risk theory, namely:

$\tau^+(u)$ is the ruin time.

$\gamma^+(u)$ is the security of ruin.

$\gamma_+(u)$ is the surplus $\zeta(t)$ prior to ruin.

γ_u^+ is the claim causing ruin.

(Figures 2 and 3, page 360 contain the graphs of the process $\xi_u(t)$ and $\zeta_u(t)$, and of the functionals mentioned above). Denote by $\tau'(u) = \inf\{t > \tau^+(u) \mid \xi_u(t) > 0\}$ the time of returning of $\xi_u(t)$ after the ruin into the half-plane $\Pi^+ = \{y \geq 0\}$,

$$T(u) = \begin{cases} \tau'(u) - \tau^+(u), & \tau^+(u) < \infty, \\ \infty, & \tau^+(u) = \infty. \end{cases} \tag{20.16}$$

$T(u)$ is said to be the first "red period", determining the first duration of $\xi_u(t)$ being in the risk zone $\Pi^- = \{y < 0\}$ or $\zeta(t)$ being in the risk zone $\Pi_u^+ = \{x > u\}$. (Figure 6, page 363 contains the graphs of these functionals for the classic risk process). The risk zone $\Pi_u^+ = \{x > u\}$ and the survival zone $\{x \leq u\}$ for the process $\zeta(t)$ are divided by the "critical" boundary $x = u$. Let

$$\begin{aligned} Z^+(u) &= \sup_{\tau^+(u) \leq t < \infty} \zeta(t), \\ Z_1^+(u) &= \sup_{\tau^+(u) \leq t < \tau'(u)} \zeta(t). \end{aligned} \tag{20.17}$$

$Z^+(u)$ determines the the total maximal deficit; $Z_1^+(u)$ is the maximal deficit during a period $T(u)$.

The duration of the process $\zeta(t)$ being over "critical" level $u > 0$ is defined by the integral functional

$$Q_u(t) = \int_0^t \mathbb{I}_{\zeta(s) > u}\, ds, \tag{20.18}$$

which determines for $t \to \infty$ the total duration of the "red period"

$$Q_u(\infty) = \int_0^\infty \mathbb{I}_{\zeta(s) > u}\, ds.$$

Let us note that the functionals (20.16)–(20.18) are needed to study the behavior of the risk processes after the ruin. This need is explained by the possibility for the insurance agency to function even after the ruin. It can borrow some capital. In order to estimate the predicted loan, it is important to know the distribution of these functionals.

To study the distributions of the functionals from (20.11), (20.15)–(20.18), we need the results from the theory of boundary problems for the processes with independent increments which can be found in [7, 33, 50].

Let $\{\xi(t), t \geq 0\}$ $(\xi(0) = 0)$ be a general real-valued homogeneous process with independent increments that has the following characteristic function in the Lévy–Khinchin form:

$$\begin{aligned} &\mathsf{E}e^{i\alpha\xi(t)} = e^{t\psi(\alpha)}, \\ &\psi(\alpha) = i\alpha\gamma - \frac{\sigma^2}{2}\alpha^2 + \int_{-\infty}^{\infty} \left(e^{i\alpha x} - 1 - \frac{i\alpha x}{1+x^2}\right) \Pi(dx), \end{aligned} \tag{20.19}$$

$$\int_{0<|x|\le 1} x^2 \Pi(dx) < \infty.$$

We denote as θ_s an exponentially distributed random variable which is independent of $\xi(t)$,

$$\mathsf{P}(\theta_s > t) = e^{-st}, \qquad s > 0, \quad t > 0,$$
$$\varphi(s,\alpha) = \mathsf{E}e^{i\alpha\xi(\theta_s)}, \qquad \varphi_\pm(s,\alpha) = \mathsf{E}e^{i\alpha\xi^\pm(\theta_s)},$$

and consider a randomly stopped process $\xi(\theta_s)$. The introduction of θ_s allows us to write in the short form the Laplace–Karson transform of the distributions of $\xi(t)$, $\xi^\pm(t)$ and their characteristic function. In particular,

$$P(s,x) := \mathsf{P}(\xi(\theta_s) < x) = s\int_0^\infty e^{-st}\mathsf{P}(\xi(t) < x)\,dt, \quad x \in \mathbb{R},$$

$$\mathsf{E}e^{i\alpha\xi(\theta_s)} = s\int_0^\infty e^{-st}\mathsf{E}e^{i\alpha\xi(t)}\,dt, \quad \mathsf{E}e^{i\alpha\xi^\pm(\theta_s)} = s\int_0^\infty e^{-st}\mathsf{E}e^{i\alpha\xi^\pm(t)}\,dt.$$

It is easy to prove that

$$\varphi(s,\alpha) = \mathsf{E}e^{i\alpha\xi(\theta_s)} = \frac{s}{s-\psi(\alpha)}. \tag{20.20}$$

Theorem 20.5. *The following main factorization identity holds true for the characteristic function of* $\xi(\theta_s)$*(see Theorem* 2.2 in [33]):

$$\varphi(s,\alpha) = \varphi_+(s,\alpha)\varphi_-(s,\alpha), \tag{20.21}$$
$$\varphi_\pm(s,\alpha) = \exp\left\{\pm\int_0^{\pm\infty}\left(e^{i\alpha x}-1\right)dN_s^\pm(x)\right\}, \tag{20.22}$$
$$N_s^+(x) = -\int_0^\infty e^{-st}t^{-1}\mathsf{P}(\xi(t) > x)\,dt, \quad x > 0,$$
$$N_s^-(x) = \int_0^\infty e^{-st}t^{-1}\mathsf{P}(\xi(t) < x)\,dt, \quad x < 0.$$

The relations (20.22) are called the Spitzer–Rogozin identities. The characteristic functions $\varphi_\pm(s,\alpha)$ in them are expressed via the complicated transformations of the distributions for the positive (negative) values of $\xi(\cdot)$.

Let us mention that the following supplements to the extremums of the process

$$\hat{\xi}^\pm(t) = \xi(t) - \xi^\mp(t), \quad \hat{\xi}^\pm(\theta_s) = \xi(\theta_s) - \xi^\mp(\theta_s)$$

satisfy the following relations,

$$\hat{\xi}^\pm(\theta_s) \stackrel{d}{=} \xi^\pm(\theta_s);$$

that is,

$$\mathsf{E}\exp\{i\alpha\hat{\xi}^\pm(\theta_s)\} = \varphi_\pm(s,\alpha).$$

It means that the components of the main factorization identity (20.22) can be interpreted as the characteristic functions of the supplements $\hat{\xi}^{\pm}(\theta_s)$.

Later in the text we use the following notations,

$$P_+(s,x) := \mathsf{P}(\xi^+(\theta_s) < x), \quad x > 0;$$

$$P_-(s,x) := \mathsf{P}(\xi^-(\theta_s) < x), \quad x < 0.$$

We denote as θ'_v the exponentially distributed random variable (independent of θ_s and $\xi(t)$) with a parameter $v > 0$. The following statement on the second factorization identity holds true.

Theorem 20.6. *The joint distribution of the pair* $\{\tau^+(\cdot), \gamma^+(\cdot)\}$ *is determined by the moment generating function*

$$\mathsf{E}e^{-s\tau^+(\theta'_v)-z\gamma^+(\theta'_v)}\mathrm{I\!I}_{\tau^+(\theta'_v)<\infty} \tag{20.23}$$

$$= \frac{v}{v-z}\left(1 - \frac{\varphi_+(s,iv)}{\varphi_+(s,iz)}\right), \quad (s,z,v > 0).$$

It is easy to determine the moment generating functions inverting (20.23) *in* v*:*

$$\mathsf{E}e^{-s\tau^+(x)-z\gamma_+(x)}\mathrm{I\!I}_{\tau^+(x)<\infty} = \frac{1}{\varphi_+(s,iz)}\int_x^{\infty} e^{z(x-y)}dP_+(s,y), \quad m \in \mathbb{R},$$

$$\mathsf{E}e^{-z\gamma_+(x)}\mathrm{I\!I}_{\tau^+(x)<\infty} = \frac{1}{\varphi_+(iz)}\int_x^{\infty} e^{z(x-y)}dP_+(y), \quad m < 0,$$

where $P_+(y) = \mathsf{P}(\zeta^+ < y)$.

Theorem 20.7. *If the pair* $\{\tau^+(0), \gamma^+(0)\}$ *satisfies the condition* $\mathsf{P}(\tau^+(0) = \gamma^+(0) = 0) = 0$*, then the joint moment generating function* $\{\tau^+(0), \gamma^+(0)\}$ *is as follows*

$$f_+(s,z) = \mathsf{E}e^{-z\gamma^+(0)-s\tau^+(0)}\mathrm{I\!I}_{\tau^+(0)<\infty} = 1 - \frac{p_+(s)}{\varphi_+(s,iz)}, \tag{20.24}$$

$$p_+(s) = \mathsf{P}(\xi^+(\theta_s) = 0), \quad q_+(s) = 1 - p_+(s),$$

$$\mathsf{E}e^{-z\xi^+(\theta_s)} = \frac{p_+(s)}{1 - f_+(s,z)} = \frac{p_+(s)}{1 - q_+(s)\tilde{g}_s(z)}, \tag{20.25}$$

$$f_+(s,z) = \mathsf{E}e^{-z\gamma^+(0)}\mathrm{I\!I}_{\xi^+(\theta_s)>0} = q_+(s)\tilde{g}_s(z),$$

$$\tilde{g}_s(z) = \mathsf{E}\left(e^{-z\gamma^+(0)}/\xi^+(\theta_s) > 0\right).$$

If $m = \mathsf{E}\xi(1) \geq 0$*, then* $f_+(s,z) \underset{s\to 0}{\to} \mathsf{E}e^{-z\gamma^+(0)}\mathrm{I\!I}_{\xi^+>0} = \mathsf{E}e^{-z\gamma^+(0)}$ *(because* $\mathsf{P}(\xi^+ > 0) = \mathsf{P}(\xi^+ = +\infty) = 1$*), and the distributions* $\gamma^+(0)$ *and* $\gamma^+(\infty) = \lim_{u\to\infty}\gamma^+(u)$ *are connected by the relation*

$$\mathsf{E}e^{-z\gamma^+(\infty)} = \frac{1}{z\mathsf{E}\gamma^+(0)}\left(1 - \mathsf{E}e^{-z\gamma^+(0)}\right). \tag{20.26}$$

If $m<0$, then the moment generating function of the absolute maximum can be expressed by the generalized Pollaczek–Khinchin formula

$$\mathsf{E}e^{-z\xi^+}=\frac{p_+}{1-q_+\tilde{g}_0(z)},\quad \tilde{g}_0(z)=\mathsf{E}\left(e^{-z\gamma^+(0)}/\xi^+>0\right), \tag{20.27}$$

$$q_+(s)=\mathsf{P}(\xi^+(\theta_s)>0)\underset{s\to 0}{\to} q_+=\mathsf{P}(\xi^+>0),\quad 0<q_+<1,\quad p_+=1-q_+.$$

The complicated dependence of the distribution of the positive (negative) values of the process for the positive (negative) components $\varphi_\pm(s,\alpha)$ of the main factorization identity becomes considerably simpler for the semicontinuous and almost semicontinuous processes. Later we consider only the processes $\xi(t)$, the jumping part of which has the bounded variation. For such kind of processes $\int_{|x|\le 1}|x|\Pi(dx)<\infty$ and the cumulant is as follows.

$$\psi(\alpha)=i\alpha a-\frac{\sigma^2}{2}\alpha^2+\int_{\mathbb{R}}\left(e^{i\alpha x}-1\right)\Pi(dx),\quad \sigma^2\ge 0. \tag{20.28}$$

Thus, the drift coefficient $a\neq 0$ and jumps of the process have different signs for the semicontinuous processes with $\sigma^2=0$. For almost semicontinuous processes with $\sigma^2=0$ only the exponentially distributed jumps and drift coefficient a have different signs.

Let us denote $k(r):=\psi(-ir)$ and write the Lundberg equation

$$k(r)-s=0,\quad s\ge 0. \tag{20.29}$$

For upper (lower) semicontinuous and almost semicontinuous processes due to the convexity of $k(r)$ in the neighborhood of $r=0$ the equation (20.29) has only one positive (negative) root $r_s=\pm\rho_\pm(s)$ which completely determines $\varphi_\pm(s,\alpha)$.

Theorem 20.8. *The following relations hold for the upper continuous nonmonotonic process $\xi(t)$ with the cumulant (20.28), where $\Pi(dx)=0$ for $x>0$ and $m=k'(0)$:*

$$\varphi_+(s,\alpha)=\frac{\rho_+(s)}{\rho_+(s)-i\alpha},\quad k(\rho_+(s))=s,$$

$$\overline{P}_+(s,x):=\mathsf{P}(\xi^+(\theta_s)>x)=e^{-\rho_+(s)x},\quad x\ge 0, \tag{20.30}$$
$$\rho_+(s)\underset{s\to 0}{\to} 0\ \mathit{for}\quad m\ge 0;\quad \rho_+'(0)=m^{-1}\ \mathit{for}\quad m>0.$$

The distribution of $\xi^-(\theta_s)$ can be determined by the relation

$$P_-(s,x)=\frac{1}{\rho_+(s)}P'(s,x)+P(s,x),\quad x<0. \tag{20.31}$$

If $\sigma^2>0$, then the derivative $P'(s,x)$ (in x) exists for all $x\in\mathbb{R}^1$. If $\sigma^2=0$, $a>0$, and $\xi(t)$ have the bounded variation, then the derivative $P'(s,x)$ exists only for $x\neq 0$ and

$$P'(s,+0)-P'(s,-0)=\frac{s}{a},$$

$$p_-(s):=\mathsf{P}(\xi^-(\theta_s)=0)=\frac{s}{a\rho_+(s)}>0, \tag{20.32}$$

$$\rho_+'(0)=\frac{1}{m},\ p_-(0)=\frac{m}{a},\quad m>0,$$

where a is a constant drift from (20.28).

Theorem 20.9. *The following relations hold for the lower continuous nonmonotonic process $\xi(t)$ with the cumulant* (20.28), *where $\Pi(dx)=0$ for $x<0$ and $m=k'(0)$:*

$$\varphi_-(s,\alpha)=\frac{\rho_-(s)}{\rho_-(s)+i\alpha},\quad k(-\rho_-(s))=s,$$

$$\rho_-(s)\underset{s\to 0}{\to} 0 \text{ for } \ m\le 0;\ \ \rho_-'(0)=m^{-1} \text{ for } m<0, \tag{20.33}$$

$$P_-(s,x)=e^{\rho_-(s)x},\quad x\le 0.$$

The distribution of $\xi^+(\theta_s)$ is determined by the relation

$$\overline{P}_+(s,x)=\frac{1}{\rho_-(s)}P'(s,x)+\overline{P}(s,x),\quad x>0.$$

If $\sigma^2=0$, $a<0$, then

$$\begin{aligned} p_+(s)&=\mathsf{P}(\xi^+(\theta_s)=0)=\frac{s}{|a|\rho_-(s)}>0,\\ p_+(s)&\underset{s\to 0}{\to} p_+=\frac{|m|}{|a|},\quad q_+=\frac{\lambda\mu}{|a|},\quad m<0. \end{aligned} \tag{20.34}$$

Theorem 20.10. *For the almost upper semicontinuous process $\xi(t)$ satisfying the first condition in* (20.1) *with $c>0$ and with cumulant function of the form*

$$\psi(\alpha)=c\lambda_1\int_0^\infty(e^{i\alpha x}-1)\,e^{-cx}dx+\int_{-\infty}^0(e^{i\alpha x}-1)\Pi(dx),\,\lambda_1>0,$$

the following relations holds:

$$\varphi_+(s,\alpha)=\frac{p_+(s)(c-i\alpha)}{\rho_+(s)-i\alpha},\quad \rho_+(s)=cp_+(s),\quad k(\rho_+(s))=s,$$

$$\overline{P}_+(s,x)=q_+(s)e^{-\rho_+(s)x},\quad x\ge 0, \tag{20.35}$$

$$\rho_+(s)\underset{s\to 0}{\to} 0 \text{ if } \ m\ge 0;\ \ \rho_+'(0)=m^{-1},\ \ p_+'(0)=(cm)^{-1},\, m=\mathsf{E}\xi(1)=k'(0)>0.$$

$$P_-(s,x)=\frac{1}{p_+(s)}\left(P(s,x)-cq_+(s)\int_0^\infty e^{-cy}P(s,x-y)\,dy\right),\quad x<0. \tag{20.36}$$

Theorem 20.11. *For the almost lower semicontinuous process $\xi(t)$ satisfying the second condition in (20.1) with $b>0$ and with cumulant function of the form*

$$\psi(\alpha)=b\lambda_2\int_{-\infty}^{0}\left(e^{i\alpha x}-1\right)e^{bx}dx+\int_{0}^{\infty}(e^{i\alpha x}-1)\Pi(dx),\,\lambda_2>0,$$

the following relations hold true.

$$\begin{aligned}\varphi_-(s,\alpha)&=\frac{p_-(s)(b+i\alpha)}{\rho_-(s)+i\alpha},\quad \rho_-(s)=bp_-(s),\quad k(-\rho_-(s))=s,\\ \rho_-'(0)&=|m|^{-1},\quad p_-'(0)=(b|m|)^{-1},\quad m=k'(0)<0; \\ P_-(s,x)&=q_-(s)e^{\rho_-(s)x},\quad x<0,\quad \rho_-(s)\underset{s\to0}{\to}0\ \text{if}\quad m\le 0.\end{aligned}\tag{20.37}$$

$$P_+(s,x)=\frac{1}{p_-(s)}\left(\overline{P}(s,x)-bq_-(s)\int_{-\infty}^{0}e^{by}\overline{P}(s,x-y)\,dy\right),\quad x>0.\tag{20.38}$$

The corresponding results for the distributions of the absolute extremums are simple corollaries from Theorems 20.8–20.11.

Corollary 20.1. *The following relations are true for the upper continuous processes $\xi(t)$ in accordance to the sign of $m=\mathsf{E}\xi(1)$ (or to the sign of safety security loading $\delta=m/(\lambda\mu)$):*

(1) If $m>0$ then $\rho_+(s)\underset{s\to0}{\to}0$, $\rho_+(s)s^{-1}\underset{s\to0}{\to}m^{-1}$, $\mathsf{P}(\xi^+=+\infty)=1$. It follows from (20.31) that for $s\to0$ we obtain

$$\mathsf{P}(\xi^-<x)=m\left(\int_0^\infty \mathsf{P}(\xi(t)<x)\,dt\right)_x',\quad x<0.\tag{20.39}$$

(2) If $m=0$ then we have for $s\to0$ that $\rho_+(s)\approx\sqrt{2s}\sigma_1^{-1}$, $\sigma_1=\sqrt{\mathsf{D}\xi(1)}$, $\mathsf{P}(\xi^\pm=\pm\infty)=1$.

(3) If $m<0$ then $\rho_+(s)\underset{s\to0}{\to}\rho_+>0$ and, according to (20.30), we obtain the following relation as $s\to0$,

$$\mathsf{P}(\xi^+>x)=e^{-\rho_+x},\quad x\ge0,\quad \mathsf{P}(\xi^-=-\infty)=1.\tag{20.40}$$

Corollary 20.2. *The following is true for the lower continuous processes in accordance with the sign of m (or with the sign of δ).*

(1) If $m<0$, then $\rho_-(s)\underset{s\to0}{\to}0$, $\rho_-(s)s^{-1}\underset{s\to0}{\to}|m|^{-1}$ and $\mathsf{P}(\xi^-=-\infty)=1$. According to (20.34) we obtain the following relations as $s\to0$:

$$\mathsf{P}(\xi^+>x)=m\left(\int_0^\infty \mathsf{P}(\xi(t)>x)\,dt\right)_x',\quad x>0.\tag{20.41}$$

(2) If $m=0$, then we obtain the following relations as $s\to0$: $\rho_\pm(s)\approx\sqrt{2s}\sigma_1^{-1}$, $\sigma_1=\sqrt{\mathsf{D}\xi(1)}$, $\mathsf{P}(\xi^\pm=\pm\infty)=1$.

(3) If $m>0$, then $\rho_-(s)\underset{s\to0}{\to}\rho_->0$ and according to (20.33) we obtain the following relations as $s\to0$:

$$\mathsf{P}(\xi^-<x)=e^{\rho_-x},\quad x\le0,\quad \mathsf{P}(\xi^+=+\infty)=1.\tag{20.42}$$

Note that the process $\xi(t) = at + \sigma W(t)$ is both upper and lower continuous. Thus, its characteristic functions $\xi^{\pm}(\theta_s)$ are determined by the first formulas in (20.30) and (20.33).

Corollary 20.3. *The following relations hold true for the almost upper semicontinuous process $\xi(t)$ with $a = 0$ in accordance with the sign of m (or with the sign of δ).*

(1) If $m > 0$ then $\rho_+(s) \underset{s\to 0}{\to} 0$, $\rho_+(s)s^{-1} \underset{s\to 0}{\to} \rho_+'(0) = m^{-1}$, $\mathsf{P}(\xi^+ = +\infty) = 1$.

According to (20.36) for $x < 0$

$$\begin{aligned}\lim_{s\to 0} P_-(s,x) &= \mathsf{P}(\xi^- < x)\\ &= cm\left(\int_0^\infty \mathsf{P}(\xi(t) < x)\,dt - c\int_0^\infty\int_0^\infty e^{-cy}\mathsf{P}(\xi(t) < x - y)\,dt\,dy\right).\end{aligned} \tag{20.43}$$

(2) If $m = 0$ then tending $s \to 0$ we obtain that $\rho_+(s) \approx \sqrt{2s}\sigma_1^{-1}$, $\mathsf{P}(\xi^{\pm} = \pm\infty) = 1$.

(3) If $m < 0$ then $\rho_+(s) \underset{s\to 0}{\to} \rho_+ = cp_+, p_+ = \mathsf{P}(\xi^+ = 0) = c|m|/\lambda > 0$, $q_+ = 1 - p_+$ and according to (20.35)

$$\mathsf{P}(\xi^+ > x) = q_+ e^{-\rho_+ x}, \quad x > 0, \quad \mathsf{P}(\xi^- = -\infty) = 1. \tag{20.44}$$

Corollary 20.4. *The following relations are true for the almost lower semicontinuous process $\xi(t)$ with $a = 0$.*

(1) If $m < 0$ then $\rho_-(s) \underset{s\to 0}{\to} 0$, $\rho_-(s)s^{-1} \underset{s\to 0}{\to} |m|^{-1}$, $\mathsf{P}(\xi^- = -\infty) = 1$, $p_+ = b|m|/|\lambda|$, $q_+ = 1 - p_+$.

According to (20.38) we have for $x > 0$:

$$\begin{aligned}\lim_{s\to 0} \overline{P}_+(s,x) &= \mathsf{P}(\xi^+ > x)\\ &= b|m|\left(\int_0^\infty \mathsf{P}(\xi(t) > x)\,dt - b\int_0^\infty\int_x^\infty e^{b(x-y)}\mathsf{P}(\xi(t) > y)\,dy\,dt\right).\end{aligned} \tag{20.45}$$

(2) If $m = 0$ then given $s \to 0$, $\rho_-(s) \approx \sqrt{2s}\sigma_1^{-1}$, $\mathsf{P}(\xi^{\pm} = \pm\infty) = 1$.

(3) If $m > 0$ then $\rho_-(s) \underset{s\to 0}{\to} \rho_- = bp_-$, $p_- = \mathsf{P}(\xi^- = 0) = (bm)/\lambda > 0$ and according to (20.37) for $x < 0$

$$\mathsf{P}(\xi^- < x) = q_- e^{\rho_- x}, \quad q_- = 1 - p_-, \quad \mathsf{P}(\xi^+ = +\infty) = 1. \tag{20.46}$$

Let us define the Laplace–Karson transform of the ruin probability with finite horizon (20.14):

$$\Psi_s(u) = s\int_0^\infty e^{-st}\Psi(u,t)\,dt.$$

Then, according to (20.34),

$$\Psi_s(u) = \frac{1}{\rho_-(s)} P'(s,-u) + \overline{P}(s,-u), \quad u > 0 \tag{20.47}$$

for the upper continuous classic ruin process $\xi_u(t)$ assigned by the formula (20.8). And it follows from (20.39) that for $m > 0$ and $s \to 0$,

$$\Psi(u) = \lim_{s\to 0} \Psi_s(u) = m \left(\int_0^\infty \mathsf{P}(\xi(t) < x)\,dt \right)'_{x=-u} = \mathsf{P}(\xi^- < -u). \tag{20.48}$$

We should take into account that the jumps ξ_k are positive for the lower continuous claim surplus process $\zeta(t) = S(t) - Ct$ (see (20.10)). On the other hand, for the almost lower semicontinuous risk process we have that

$$\zeta(t) = S(t) - C(t), \quad C(t) = \sum_{k=0}^{N_2(t)} \xi_k'', \quad \mathsf{E}e^{i\alpha\xi_k''} = \frac{b}{b - i\alpha}, \quad b > 0, \tag{20.49}$$

the jumps of the process $S(t) = \sum_{k=0}^{N_1(t)} \xi_k'$, that is, the claims ξ_k' are also positive (the Poisson processes $N_{1,2}(t)$ with intensities $\lambda_1 > 0$, $\lambda_2 > 0$ are independent). The cumulant of $\zeta(t)$ can be written in two ways:

$$\psi(\alpha) = \lambda_1(\varphi_1(\alpha) - 1) + \lambda_2(\frac{b}{b + i\alpha} - 1), \quad \varphi_1(\alpha) = \mathsf{E}e^{i\alpha\xi_1'}$$

or

$$\psi(\alpha) = \lambda(\varphi(\alpha) - 1), \quad \varphi(\alpha) = \lambda \left(p\varphi_1(\alpha) + q\frac{b}{b + i\alpha} \right),$$

where $\lambda = \lambda_1 + \lambda_2$, $p = \lambda_1/\lambda$, $q = 1 - p$, $\delta = (\lambda|\mu|)/(\lambda_1\mu_1) > 0$, $\lambda\mu = \lambda_1\mu_1 - \lambda_2 b^{-1}$, $\mu_1 = \mathsf{E}\xi_1'$. Let also $\mathsf{P}(\xi_1' > 0) = \overline{F}_1(0) = 1$, $\overline{F}_1(x) = \mathsf{P}(\xi_1' > x)$, $\overline{F}(x) = p\overline{F}_1(x)$, $x > 0$. For both processes (20.10) and (20.49) the conditional moment generating function for $\gamma^+(0)$ in Pollaczek–Khinchin formula (see (20.27)) is as follows.

$$\mathsf{E}\left(e^{-z\gamma^+(0)}/\zeta^+ > 0\right) = \\ = \begin{cases} \widetilde{\varphi}_0(z) = \left(\int_0^\infty \overline{F}(x)\,dx\right)^{-1} \int_0^\infty e^{-zx}\overline{F}(x)\,dx, & \text{for } \zeta(t) \text{ in (20.10)} \\ \widetilde{g}_0(z) = \frac{1}{q_+}\int_0^\infty e^{-yz}(dF(y) + b\overline{F}(y)dy), & \text{for } \zeta(t) \text{ in (20.49).} \end{cases} \tag{20.50}$$

Furthermore, the moment generating function for ζ^+ is determined by the classic Pollaczek–Khinchin formula. This implies the following decomposition,

$$\mathsf{E}e^{-z\zeta^+} = \frac{p_+}{1 - q_+\widetilde{\varphi}_0(z)} = p_+ \sum_{n=0}^\infty (q_+\widetilde{\varphi}_0(z))^n. \tag{20.51}$$

Let us denote $\widetilde{S}_n = \sum_{k\le n} \widetilde{\xi}_k$, where $\widetilde{\xi}_k$ are i.i.d. random variables with the moment generating function $\widetilde{\varphi}_0(z)$. Inverting (20.51) in the variable z, we obtain

$$\Psi(u) = \mathsf{P}(\zeta^+ > u) = p_+ \sum_{n=1}^\infty q_+^n \mathsf{P}(\widetilde{S}_n > u). \tag{20.52}$$

It means that $\zeta^+ \stackrel{d}{=} \widetilde{S}_{\widetilde{\nu}}$. Here the random variable $\widetilde{\nu}$ follows the geometric distribution with parameter $q_+ = \mathsf{P}(\zeta^+ > 0) < 1$.

Definition 20.8. *The process* (20.49) *is called the* risk process with random (exponentially distributed) claims *(Figures 4 and 5, page 362 contain the graphical images of the functionals of the process* (20.49)*).*

Let us mention that if $\mathsf{E}e^{i\alpha\xi_k'} = c/(c-i\alpha)$, $\mathsf{E}e^{i\alpha\xi_k''} = b/(b-i\alpha)$ then the process (20.49) is both the almost upper and almost lower semicontinuous.

The joint moment generating function of the overjump functionals $\{\tau^+(x), \gamma_k(x), k=\overline{1,3}\}$, where $\gamma_1(x) = \gamma^+(x)$, $\gamma_2(x) = \gamma_+(x)$, $\gamma_3(x) = \gamma_x^+$, that is,

$$V(s,x,u_1,u_2,u_3) = \mathsf{E}e^{-s\tau^+(x)-\sum_{k=1}^{3} u_k\gamma_k(x)}\mathrm{I\!I}_{\tau^+(x)<\infty},$$

can be determined by some integral equation on the semi-axis $x > 0$.

For the lower continuous (almost lower semicontinuous) processes $\zeta(t)$ this equation is as follows.

$$(s+\lambda)V(s,x,u_1,u_2,u_3) - \lambda\int_{-\infty}^{x} V(s,x-y,u_1,u_2,u_3)dF(y) = A(x,u_1,u_2,u_3),$$

where

$$A(x,u_1,u_2,u_3) = \lambda\int_{x}^{\infty} e^{(u_1-u_2)x-(u_1+u_2)z}dF(z), \quad x>0.$$

Let us denote the convolution $A(\cdot)$ with $dP_-(\cdot)$ (which is the exponential distribution of $\zeta^-(\theta_s)$; see (20.33), (20.37)), as $G(\cdot)$. Then

$$G(s,x,u_1,u_2,u_3) = \int_{-\infty}^{+0} A(x-y,u_1,u_2,u_3)\,dP_-(s,y) \tag{20.53}$$

$$= p_-(s)A(x,u_1,u_2,u_3) + q_-(s)\rho_-(s)\int_{-\infty}^{0} A(x-y,u_1,u_2,u_3)\,dP_-(s,y),$$

$p_-(s) \geq 0$. Note that for the lower continuous processes $\zeta(t)$ in (20.53) $p_-(s) = 0$, $q_-(s) = 1$.

Theorem 20.12. *The joint moment generating function of the overjump functionals for the lower semicontinuous (almost semicontinuous) risk processes $\zeta(t)$ is determined by the relation*

$$V(s,u,u_1,u_2,u_3) = s^{-1}\int_{-0}^{u} G(s,u-y,u_1,u_2,u_3)\,dP_+(s,y). \tag{20.54}$$

The moment generating function of the pair $\{\tau^+(u), \gamma_k(u)\}$, $(k=\overline{1,3})$ is determined by the relation

$$\begin{aligned} V_k(s,u,u_k) :&= \mathsf{E}e^{-s\tau^+(u)-u_k\gamma_k(u)}\mathrm{I\!I}_{\tau^+(u)<\infty} \\ &= s^{-1}\int_{-0}^{u} G_k(s,u-y,u_k)\,dP_+(s,y), \quad k=\overline{1,3}, \end{aligned} \tag{20.55}$$

where $p_+(s) > 0$ *for the processes* (20.10) *and* (20.49),

$$\begin{aligned} P_{\pm}(s,x) &= \mathsf{P}(\zeta^{\pm}(\theta_s) < x) \quad (\pm x > 0), \\ G_1(s,u,u_1) &= G(s,u,u_1,0,0), \\ G_2(s,u,u_2) &= G(s,u,0,u_2,0), \\ G_3(s,u,u_3) &= G(s,u,0,0,u_3). \end{aligned}$$

In turn, functions G_k corresponding to the process $\zeta(t)$ from (20.10) have a form

$$\begin{aligned} G_1(s,u,u_1) &= \frac{\lambda\rho_-(s)}{\rho_-(s)-u_1}\int_0^{\infty}\left(e^{-u_1 y} - e^{-\rho_-(s)y}\right) dF(u+y), \\ G_2(s,u,u_2) &= \lambda\rho_-(s)\int_0^{\infty} e^{-u_2(u+z)-\rho_-(s)z}\overline{F}(u+z)\,dz, \qquad (20.56) \\ G_3(s,u,u_3) &= \lambda\int_u^{\infty} e^{-u_3 z}\left(1-e^{\rho_-(s)(u-z)}\right) dF(z). \end{aligned}$$

The relations (20.54) and (20.55) can be easily inverted in u_k $(k=\overline{1,3})$. Denote

$$\phi(s,k,u,x) = \frac{\partial}{\partial x}\mathsf{P}(\gamma_k(u) < x, \zeta^+(\theta_s) > u).$$

This value tends to a limit as $s \to 0$. We denote this limit $\phi_k(u,x)$.

The limit distributions of the overjump functionals and marginal ruin functions can be found using these relations.

Theorem 20.13. *The first two ruin functions for the lower continuous ruin processes* $\zeta(t)$ *in the case when* $m < 0$ *are determined by the relations*

$$\Phi_1(u,x) := \mathsf{P}(\zeta^+ > u, \gamma^+(u) > x) = \int_x^{\infty}\phi_1(u,z)dz \qquad (20.57)$$

$$= \frac{\lambda}{c}\overline{\overline{F}}(u+x) + \frac{\lambda}{|m|}\int_0^u \overline{\overline{F}}(u+x-z)\,dP_+(z), \quad \overline{\overline{F}}(y) = \int_y^{\infty}\overline{F}(z)\,dz,$$

$$\Phi_2(u,y) := \mathsf{P}(\zeta^+ > u, \gamma_+(u) > y) = \int_y^{\infty}\phi_2(u,z)dz \qquad (20.58)$$

$$= \begin{cases} \frac{\lambda}{|m|}P_+(u)\overline{\overline{F}}(y), & y > u, \\ \frac{\lambda}{|m|}\left(p_+\overline{\overline{F}}(u) + \overline{\overline{F}}(y)\int_{u-y}^{u} dP_+(z) + \int_0^{u-y}\overline{\overline{F}}(u-z)\,dP_+(z)\right), & 0 < y < u, \end{cases}$$

where $P_+(u) = \mathsf{P}(\zeta^+ < u)$, $p_+ = |m|/C$.

The distribution density of the claim γ_u^+ *that caused the ruin (i.e., of the third ruin function) is as follows.*

$$\phi_3(u,y) := \frac{\partial}{\partial z}\mathsf{P}(\gamma_u^+ < y, \zeta^+ > u) = -\frac{\partial}{\partial y}\Phi_3(u,y)$$

$$= \begin{cases} \frac{\lambda}{|m|}F'(y)\int_0^u (y-u+z)\,dP_+(z), & y > u, \\ \frac{\lambda}{|m|}F'(y)\int_{-y}^0 (z+y)\,dP_+(y+u), & 0 < y < u. \end{cases} \qquad (20.59)$$

In order to study the distribution of the first duration and the total duration of "red period" it should be taken into account that

$$T(u) \stackrel{d}{=} \tau^-(-\gamma^+(u)), \qquad u > 0, \tag{20.60}$$
$$\tau^-(x) = \inf\{t > 0 |\, \zeta(t) < x\}, \qquad x < 0.$$

The statement below follows from (20.55), (20.56), and (20.60) (if $k = 1$, $s \to 0$).

Theorem 20.14. *Consider the lower continuous risk process $\zeta(t)$ (see (20.10)). The following relations hold true for $m = \mathsf{E}\zeta(1) < 0$.*

$$g_+(u,z) := \mathsf{E}e^{-z\gamma^+(u)}\mathrm{I\!I}_{\tau^+(u)<\infty}$$

$$= \frac{\lambda}{c}\int_0^\infty e^{-zx}\overline{F}(u+x)\,dx + \frac{\lambda}{|m|}\int_{0+}^u\int_0^\infty e^{-zx}\overline{F}(u-y+x)\,dx\,dP_+(y). \tag{20.61}$$

The following formula is a consequence of (20.60). *It is called the corrected dos Reis formula.*

$$g_u(s) := \mathsf{E}e^{-sT'(u)}\mathrm{I\!I}_{\tau^+(u)<\infty} = g_+(u,\rho_-(s)). \tag{20.62}$$

The integral transform of the moment generating function for the total duration of the "red period" is determined by the following relation ($m < 0$).

$$d_+(u,\mu) := \int_0^\infty e^{i\alpha u}d_u\mathsf{E}e^{-\mu Q_u(\infty)} = \frac{\varphi_+(\alpha)}{\varphi_+(\mu,\alpha)}, \quad \varphi_+(\alpha) = \mathsf{E}e^{i\alpha\zeta^+}. \tag{20.63}$$

The following analogue of (20.61) is true for the almost lower semicontinuous risk process (20.49) with arbitrary distributed claims and exponentially distributed premiums in the case when $b > 0$, $m = \lambda(p\mu_1 - qb^{-1}) < 0$, $\lambda = \lambda_1 + \lambda_2$.

$$g_+(u,z) := \int_0^\infty e^{-zx}\phi_1(u,x)dx = \int_0^\infty e^{-yz}\,dF_*(u+y) \tag{20.64}$$
$$+ \frac{\lambda}{b|m|}\int_{0+}^u\int_0^\infty e^{-yz}F_*(u-y+x)\,dx\,dP_+(y) \underset{u\to 0}{\to} \int_0^\infty e^{-yz}dF_*(y) = q_+\tilde{g}_0(z),$$

where $\tilde{g}_0(z) = \frac{1}{q_+}\int_0^\infty e^{-yz}\,dF_*(y) = (\overline{F}_*(0))^{-1}\int_0^\infty e^{-yz}\,dF_*(y)$.

Moreover,

$$F_*'(y) := F'(y) + b\overline{F}(y), \quad y > 0, \; p_+ = \frac{b|m|}{\lambda}.$$

The moment generating function of $T(u)$ is determined by the generalization of the formula (20.62):

$$g_u(s) = q_-(s)g_+(u,\rho_-(s)). \tag{20.65}$$

The integral Fourier–Stieltjes transform for $Q_u(\infty)$ is determined by the relation (20.63) the same way as for lower continuous processes. The moment generating function for the total deficit maximum $Z^+(u)$ (see (20.17)) is determined for the processes (20.10) and (20.49) by the relation

$$\varkappa_u(z) := \mathsf{E}e^{-zZ^+(u)}\mathrm{I\!I}_{\zeta^+>u} = e^{-uz}\mathsf{E}e^{-z\zeta^+}. \tag{20.66}$$

Theorem 20.15. *Let $\zeta(t)$ be the lower almost semicontinuous stepwise risk process. The limiting ruin densities for $m<0$ are determined by the relations*

$$\phi_1(u,x) = F'_*(u+x) + \frac{\lambda}{b|m|}\int_0^u F'_*(u+x-y)dP_+(y), \quad x \geq 0, \tag{20.67}$$

$$F'_*(x) = F'(x) + b\overline{F}(x), \quad x \geq 0, \quad \overline{F}(0) = p > 0.$$

If $y=u\geq 0$ then $\mathsf{P}(\gamma_2(u) = \gamma_+(u) = u) = \overline{F}(u) = p\overline{F}_1(u)$, and for $y \neq u$

$$\phi_2(u,y) = \begin{cases} \frac{\lambda}{|m|}\overline{F}(y)P_+(u), & y>u, \\ \frac{\lambda}{|m|}\overline{F}(y)\left(\int_{u-y}^u dP_+(z) + b^{-1}P'_+(u-y)\right), & 0<y<u, \end{cases} \tag{20.68}$$

$$\phi_3(u,y) = \begin{cases} \frac{\lambda}{|m|}F'(y)\int_{0-}^u (z+y-u+b^{-1})dP_+(z), & y>u, \\ \frac{\lambda}{|m|}F'(y)\int_{-y}^0 (z+y+b^{-1})dP_+(z+u), & 0<y<u. \end{cases} \tag{20.69}$$

Let us remark that (20.67) follows from (20.64) inverting it on $u_1 = z$ and (20.68) and (20.69) follow from (20.55) using the boundary transition as $s\to 0$ and inversion on $u_{2,3}$. At that time the integral in the first row in (20.69) can be written as

$$\int_{0-}^u (z+y-u+b^{-1})dP_+(y)$$

$$= p_+(z-u+b^{-1}) + \int_{0+}^u (z+y-u+b^{-1})dP_+(y), \quad z>u.$$

The relations (20.57) and (20.58) as $u\to 0$ imply the following.

Corollary 20.5. *Let $\zeta(t)$ be the lower continuous risk process (see (20.10)). Then for $m<0$ the following relations are true for $\mathsf{E}e^{-s\tau^+(0)}\mathrm{I\!I}_{\gamma_k(0)>z} = \mathsf{P}(\zeta^+(\theta_s)>\theta,\ \gamma_k(0) > z)$, $k=\overline{1,3}$.*

$$\begin{aligned} \mathsf{P}(\gamma^+(0)>z, \zeta^+(\theta_s)>0) &= \frac{\lambda}{C}\int_z^\infty e^{(z-y)\rho_-(s)}\overline{F}(y)\,dy, \\ \mathsf{P}(\gamma_+(0)>z, \zeta^+(\theta_s)>0) &= \frac{\lambda}{C}\int_z^\infty e^{-y\rho_-(s)}\overline{F}(y)\,dy, \\ \mathsf{P}(\gamma_0^+>z, \zeta^+(\theta_s)>0) &= \frac{\lambda}{C\rho_-(s)}\int_z^\infty \left(1-e^{-y\rho_-(s)}\right)dF(y). \end{aligned} \tag{20.70}$$

If $m<0$ then for $s\to 0$ and the lower continuous process $\zeta(t)$ the formula (20.70) implies $p_+ = |m|/C$,

$$\begin{aligned} \mathsf{P}(\gamma^+(0)>z, \zeta^+>0) &= \mathsf{P}(\gamma_+(0)>z, \zeta^+>0) = \frac{\lambda}{C}\overline{\overline{F}}(z), \\ \mathsf{P}(\gamma_0^+>z, \zeta^+>0) &= \frac{\lambda}{C}\int_z^\infty y\,dF(y), \quad \overline{\overline{F}}(z) = \int_z^\infty \overline{F}(y)\,dy, \\ \mathsf{P}(\gamma_k(0)>0, \zeta^+>0) &= \mathsf{P}(\zeta^+>0) = q_+ = \frac{\lambda\mu}{C}, \quad \mu = \overline{\overline{F}}(0). \end{aligned} \tag{20.71}$$

Corollary 20.6. *If $\zeta(t)$ is the lower almost semicontinuous risk process (see (20.49)) then the following relations are true $(\overline{F}(x) = p\overline{F}_1(x), x > 0, p = \overline{F}(0))$*

$$p_\pm(s) = \mathsf{P}(\zeta^\pm(\theta_s) = 0) > 0, \quad p_+(s)\rho_-(s) = \frac{sb}{s+\lambda},$$

$$\begin{aligned}
\mathsf{P}(\gamma^+(0) > x, \zeta^+(\theta_s) > 0) &= \frac{\lambda}{s+\lambda}\left(\overline{F}(x) + q_-(s)b\int_x^\infty e^{(x-y)\rho_-(s)}\overline{F}(y)\,dy\right),\\
\mathsf{P}(\gamma_+(0) > y, \zeta^+(\theta_s) > 0) &= \frac{\lambda b}{s+\lambda}q_-(s)\int_y^\infty e^{-v\rho_-(s)}\overline{F}(v)\,dv, \qquad (20.72)\\
\mathsf{P}(\gamma_0^+ > z, \zeta^+(\theta_s) > 0) &= \frac{\lambda}{s+\lambda}\left(\overline{F}(z) + \frac{bq_-(s)}{\rho_-(s)}\int_z^\infty \left(1 - e^{-y\rho_-(s)}\right)dF(y)\right).
\end{aligned}$$

If $s \to 0$ and $m < 0$ then it follows from (20.72) for the lower almost semicontinuous process $\zeta(t)$ that $p_+ = (b|m|)/\lambda$ and

$$\begin{aligned}
\mathsf{P}(\gamma^+(0) > x, \zeta^+ > 0) &= p\left(\overline{F}_1(x) + b\overline{\overline{F}}_1(x)\right) \underset{x\to 0}{\to} q_+ = p(1 + b\mu_1),\\
\mathsf{P}(\gamma_+(0) > y, \zeta^+ > 0) &= bp\overline{\overline{F}}_1(y), \quad \overline{\overline{F}}_1(y) = \int_y^\infty \overline{F}_1(x)\,dx, \qquad (20.73)
\end{aligned}$$

$$\mathsf{P}(\gamma_+(0) > 0, \zeta^+ > 0) < q_+, \quad \mathsf{P}(\gamma_+(0) = 0, \zeta^+ > 0) = p\overline{F}_1(0),$$
$$\mathsf{P}(\gamma_0^+ > z, \zeta^+ > 0) = p\left(\overline{F}_1(z) + b\int_z^\infty y\,dF_1(y)\right) \underset{z\to 0}{\to} q_+.$$

In order to calculate the moments $m_k = \mathsf{E}\zeta(1)^k$ (given $m_k < \infty$, $k = \overline{1,4}$) of the risk process $\zeta(t) = S(t) - Ct$ (or $\zeta(t) = S(t) - C(t)$) the derivatives of its cumulant $k(r) = \ln \mathsf{E}e^{r\zeta(1)} = \psi(-ir)$ at zero point ($r = 0$) are used. They are said to be semiinvariants

$$\begin{aligned}
k'(0) &=: \varkappa_1 = m_1\\
k''(0) &=: \varkappa_2 = \mathsf{D}\zeta(1) = m_2 - m_1^2, \qquad (20.74)\\
k'''(0) &=: \varkappa_3 = m_3 - 3m_1m_2 + 2m_1^3,\\
k^{(4)}(0) &=: \varkappa_4 = m_4 - 3m_1 - 4m_1m_3 + 12m_1^2m_2 - 6m_1^4.
\end{aligned}$$

This implies that

$$m_1 = \varkappa_1, \quad m_2 = \varkappa_2 + \varkappa_1^2, \quad m_3 = \varkappa_3 + 3\varkappa_1\varkappa_2 + \varkappa_1^3, \qquad (20.75)$$
$$m_4 = \varkappa_4 + 3\varkappa_2^2 + 4\varkappa_1\varkappa_3 + 6\varkappa_1^2\varkappa_2 + \varkappa_1^4.$$

And let us finally mention that it is not always possible to find the ruin probabilities $\Psi(T,u)$ and $\Psi(u)$ in an explicit form from the integro-differential equation derived for $\Psi(T,u)$. Most often the Laplace–Karson transform on T or Laplace or Fourier transform on u are used. That is why the approximating estimates of these probabilities are often used in risk theory. They could be found in [30, 33].

Bibliography

[7, 33, 47, 50]; [55] Chapter III; [1, 10, 30, 68].

Problems

20.1. Let us consider the process $\xi(t) = at + \sigma W(t)$ with a characteristic function $\mathsf{E}e^{i\alpha\xi(t)} = e^{t\psi(\alpha)}$, $\psi(\alpha) = ia\alpha - \frac{1}{2}\sigma^2\alpha^2$. Write the characteristic function for $\xi(\theta_s)$, express the components of the main factorization identity (characteristic functions of $\xi^{\pm}(\theta_s)$) in terms of roots $r_s = \pm\rho_{\pm}(s)$ of the Lundberg equation. Find the distributions of the extremums $P_{\pm}(s,x) = \mathsf{P}(\xi^{\pm}(\theta_s) < x)$, $(\pm x > 0)$. For $\mathsf{E}\xi(1) = a < 0$, $(a > 0)$ find the distributions of the absolute extremums. Find out the shape of the distribution for $\gamma_k(x)$ $(\gamma_1(x) = \gamma^+(x), \gamma_2(x) = \gamma_+(x))$ and for the first duration of $T(x)$ being over the level x. If $a < 0$ find the moment generating function for the total duration of being over the level $x > 0$. Show that $Q_0(t)$ satisfies the arcsine law for $a = 0$.

20.2. Let $\zeta(t) = S(t) - C(t)$ be a risk process with claims ξ_k' and premiums ξ_k'' both following the exponential distributions, ξ_k' with parameter c and ξ_k'' with parameter 1. Furthermore, let

$$S(t) = \sum_{k=0}^{N_1(t)} \xi_k', \quad \mathsf{E}e^{i\alpha\xi_k'} = \frac{c}{c - i\alpha},$$

$$C(t) = \sum_{k=0}^{N_2(t)} \xi_k'', \quad \mathsf{E}e^{i\alpha\xi_k''} = \frac{1}{1 - i\alpha},$$

where $N_{1,2}(t)$ are independent Poisson processes with $\lambda_1 = \lambda_2 = 1$.

Find $\psi(\alpha)$, $\varphi(s,\alpha)$, and $\varphi_{\pm}(s,\alpha)$ using the roots of the Lundberg equation. If $\pm m < 0$ $(m = \mathsf{E}\zeta(1))$, find the characteristic function of $\zeta^{\pm}$. Find the joint moment generating function of $\{\tau^+(x), \gamma^+(x)\}$ relying on the second factorization identity (20.23). If $m < 0$, find the moment generating function for the distribution of the first duration of "red period" $T(x)$ and the moment generating function for the total duration $Q_x(\infty)$ of being over the level $x > 0$.

20.3. Let $\zeta(t) = -t + \sum_{k=0}^{N(t)} \xi_k$ be a claim surplus process following the exponential distribution $\varphi(\alpha) = \mathsf{E}e^{i\alpha\xi_k} = c(c - i\alpha)^{-1}, c > 0$, and $N(t)$ be a Poisson process with intensity $\lambda > 0$. Find $\psi(\alpha)$, $\varphi(s,\alpha)$ and express $\varphi_{\pm}(s,\alpha)$ via roots of the Lundberg equation. Find the characteristic function of $\zeta^{\pm}$ and the ruin probability for $\pm m < 0$ $(m = \mathsf{E}\zeta(1) = (\lambda - c)/c)$ if the initial capital $u > 0$ and $m < 0$. Write the formulas for the densities of the ruin functions $\phi_k(u,z) = (\partial/\partial z)\mathsf{P}(\gamma^k(u) < z, \zeta^+ > u)$, $k = 1,2,3$.

20.4. Calculate all three densities of the ruin functions for the process $\zeta(t)$ from Problem 20.3 taking into account that

$$P_+(y) = \mathsf{P}(\zeta^+ < y) = 1 - q_+ e^{-\rho_+ y}, \quad y > 0.$$

Prove also that the first density function in the solution of Problem 20.3 can be simplified (see (20.93) below):

$$\frac{\partial}{\partial x}\mathsf{P}(\gamma^+(u) < x, \zeta^+ > u) = \lambda e^{-\rho_+ u - cx},$$

$$\mathsf{P}(\gamma^+(u) > x, \zeta^+ > u) = \frac{\lambda}{c} e^{-\rho_+ u - cx}.$$

Find the moment generating function for $\gamma^+(x)$ and the moment generating function for the first duration of the "red period" $T(u)$.

20.5. Let us consider the risk processes with random premiums from Problem 20.2 and the classical risk process with linear premium function from Problem 20.4 the claims of which follow the exponential distribution with parameter $c > 0$. It was shown that the moment generating functions of the distribution $\gamma^+(u)$ have the same form (see the expressions for $g_+(u,z)$ in the end of the solutions of Problems 20.2 and 20.4). Find the moment generating function of the total deficit $Z^+(u)$.

20.6. Let $\zeta(t) = (\sum_{k=0}^{N(t)} \xi_k) - t$ be a claim surplus process where ξ_k have the characteristic function

$$\varphi(\alpha) = \mathsf{E}e^{i\alpha\xi_k} = \frac{1}{2}\left(\frac{3}{3-i\alpha} + \frac{7}{7-i\alpha}\right),$$

where $N(t)$ is a Poisson process with intensity $\lambda = 3$, $(m = \mathsf{E}\zeta(1) = -2/7)$. Prove that

$$\psi(\alpha) = \frac{-(i\alpha)^3 + 7(i\alpha)^2 - 6i\alpha}{(3-i\alpha)(7-i\alpha)},$$

$$\varphi(s,\alpha) = \frac{s}{s - \psi(\alpha)} = \frac{s(3-i\alpha)(7-i\alpha)}{P_3(s,i\alpha)},$$

where $\varphi(s,\alpha)$ is a fractional rational function, and $P_3(s,r)$ is a cubic polynomial:

$$P_3(s,r) = r^3 + r^2(s-7) + r(6-10s) + 21, \quad s > 0.$$

Find the roots of the equation $P_3(0,r) = 0$ and show that the negative root $r_1(s) = -\rho_-(s) \underset{s\to 0}{\longrightarrow} 0$ and the positive roots $r_2(s) = \rho_+(s) < r_3(s)$ stay positive as $s \to 0$: $\rho_+(s) = r_2(s) \to 1$, $r_3(s) \to 6$.

Express $\varphi_\pm(s,\alpha)$ via roots found. Find the distribution of ζ^+ and the ruin probability $\Psi(u)$.

20.7. Find the first two ruin functions $\Phi_{1,2}(u,x)$ for the process from Problem 20.6 taking into account that

$$\overline{F}(x) = \frac{1}{2}\left(e^{-3x} + e^{-7x}\right), \quad x > 0,$$

$$\mathsf{P}(\zeta^+ > x) = \frac{24}{35}e^{-x} + \frac{1}{35}e^{-6x}, \quad x > 0.$$

20.8. Find the characteristic function of the absolute maximum ζ^+ for the process $\zeta(t)$ from Problem 20.6 using the Pollaczek–Khinchin formula (see (20.50) and (20.51)) and show that the denominator of the characteristic

$$\mathsf{E}e^{i\alpha\zeta^+} = \frac{2}{7}\frac{(3-i\alpha)(7-i\alpha)}{(3-i\alpha)(7-i\alpha)-3(5-i\alpha)}$$

coincides with $P_2(0,r)$ after the substitution $r = i\alpha$.

As a result the identity of the last characteristic function obtained by the Pollaczek–Khinchin formula and the characteristic function for ζ^+ obtained from Problem 20.6 using factorization is assigned.

20.9. Let $\zeta(t) = S(t) - C(t)$ be a risk process with exponentially distributed premiums

$$S(t) = \sum_{k=0}^{N_1(t)} \xi_k', \quad C(t) = \sum_{k=0}^{N_2(t)} \xi_k'',$$

where $N_{1,2}(t)$ are independent Poisson processes with $\lambda_1 = \lambda_2 = 1$,

$$\varphi_1(\alpha) = \mathsf{E}e^{i\alpha\xi_k'} = \frac{1}{(1-i\alpha)^2}, \quad \varphi_2(\alpha) = \mathsf{E}e^{i\alpha\xi_k''} = \frac{b}{b-i\alpha}.$$

Prove that $\psi(\alpha) = \lambda_1(\varphi_1(\alpha)-1) + \lambda_2(\varphi_2(-\alpha)-1)$,

$$\varphi(s,\alpha) = \frac{s}{s-\psi(\alpha)}$$

and find

$$\varphi_\pm(s,\alpha) = \mathsf{E}e^{i\alpha\zeta^\pm(\theta_s)}.$$

20.10. Consider the process from Problem 20.9 (given that $b = 1/14$, $m = 2 - b^{-1} = -12 < 0$) taking into account that $s^{-1}\rho_-(s) \underset{s\to 0}{\to} |m|^{-1} = 1/12$, $s^{-1}\rho_-(s) \underset{s\to 0}{\to} (b|m|)^{-1} = 7/6$. Show that

$$\varphi_+(\alpha) = \lim_{s\to 0}\frac{s}{\rho_-(s)}\frac{(1-i\alpha)^2}{P_2(s,i\alpha)} = b|m|\frac{(1-i\alpha)^2}{P_2(0,i\alpha)}.$$

Show that the fractional rational characteristic function $\varphi_+(\alpha)$ allows the decomposition

$$\varphi_+(\alpha) = \frac{3}{7} + \frac{27}{41}\frac{1}{1-4i\alpha} + \frac{25}{287}\frac{1}{7i\alpha-12}. \tag{20.76}$$

Invert (20.76) in α and find the distribution for ζ^+ and $\Psi(u)$. Find the densities $\phi_{1,2}(x)$, using the formula (20.67) which follows from (20.64) after inversion on z, and (20.68).

20.11. For the above problem find the moment generating function ζ^+ using the Pollaczek–Khinchin formula (20.51), taking into account the equalities $\overline{F}_1(x) = (1+x)e^{-x}$, $\mu_1 = \int_0^\infty \overline{F}_1(x)\,dx = 2$.

20.12. Calculate the ruin functions for the process from Problem 20.9 for $u=0$ and $m=2-b^{-1}<0$, using formula (20.73) and the relations $\overline{F}_1(x)=(1+x)e^{-x}$ and $\overline{\overline{F}}_1(x)=(2+x)e^{-x}$.

20.13. Find the moment generating function $T(0)$ for the duration of the first "red period" for the ruin process $\zeta(t)$ from Problem 20.6 (given $u=0$, $m<0$) using the formula (20.50) for the moment generating function of $\gamma^+(0)$:

$$g_+(0,z)=\mathsf{E}e^{-z\gamma^+(0)}\mathrm{I\!I}_{\zeta^+>0}=q_+\widetilde{\varphi}_0(z).$$

Use formula (20.62).

20.14. Consider the process $\zeta(t)$ from Problem 20.9. The moment generating function for $\gamma^+(0)$ is determined by the formula (20.50). Find the moment generating function of the first duration of "red period" given $u=0$, $b=1/14$ $(m=-12)$, using the formula (20.65). For the calculation of the moment generating function $\widetilde{\varphi}_0(z)$ it should be taken into account that $\overline{F}_1(x)=(1+x)e^{-x}$, $x>0$, $\mu=2$,

$$\widetilde{\varphi}_0(z)=\frac{1}{2}\int_0^\infty e^{-zx}\overline{F}_1(x)\,dx.$$

20.15. Find the moment generating function of $\gamma^+(u)$ for the risk process from Problem 20.6 (given $u>0$, $\lambda=3$, $m=-2/7<0$) . Use the solution of Problem 20.7 for $\Phi_1(u,x)$.

Use the obtained expression

$$g_+(u,z)=\mathsf{E}e^{-z\gamma^+(u)}\mathrm{I\!I}_{\tau^+(u)<\infty}=\int_0^\infty e^{-xz}d\Phi_1(u,x)$$

to determine the moment generating function of the total deficit $Z^+(u)$ by the formula (20.66).

20.16. Consider the risk process with random premiums from Problem 20.9 (given $u>0, b=1/14, m=-12<0$). Find the moment generating function of $\gamma^+(u)$ using (20.64) and the relations

$$\overline{F}(x)=(1+x)e^{-x},\quad (x>0),\quad \overline{P}_+(y)=\frac{27}{41}e^{-y/4}-\frac{25}{287}e^{-12y/7},\quad (y>0).$$

Or, in order to do it, you can use the density $\phi_1(u,x)$ found in Problem 20.10 and calculate

$$g_+(u,z)=\int_0^\infty e^{-zx}\phi_1(u,x)dx.$$

Use this relation for $g_+(u,z)$ in order to determine the moment generating function of the total deficit $Z^+(u)$ by the formula (20.66).

20.17. The risk process $\zeta(t)=S(t)-Ct$ $\left(\delta=(\lambda\mu)^{-1}(C-\lambda\mu)>0\right)$ has the cumulant

$$k(z) := \frac{1}{t}\ln \mathsf{E}e^{-z\zeta(t)} = Cz + \lambda(f(z)-1), \quad f(z) = \mathsf{E}e^{-z\xi_1} = \int_0^\infty e^{-zx}\,dF(x).$$

Rewrite it as the queueing process $\eta(t)$ with cumulant

$$k_1(z) = z + \lambda_1(f(z)-1), \quad \lambda_1 = \lambda C^{-1}, \quad 1 - f(z) = z\int_0^\infty e^{-zx}\overline{F}(x)\,dx.$$

Investigate the virtual time waiting process $w(t)$ ($w(0)=0$) for $\eta(t)$ using formula (11.3). Find $\widetilde{\omega}(z,s) := \mathsf{E}e^{-zw(\theta_s)}$ by rewriting this expectation via the probability $\widetilde{p}_0(s) = s\int_0^\infty e^{-su}P_0(u)\,du$ of the system to be free of claims in the exponentially distributed moment of time θ_s. Using the boundary transition as $z\to\infty$ show that the atomic probability of $w(\theta_s)$ being in 0 is positive:

$$\mathsf{P}(w(\theta_s)=0) = p_+(s) = \lim_{z\to 0}\widetilde{\omega}(z,s) = \widetilde{p}_0(s) > 0.$$

20.18. It can be identified based on the Figure 1 (page 360) that

$$\widetilde{\theta}_1 \overset{d}{=} \tau_\eta^-(-\xi_1), \quad \tau_\eta^-(-x) = \sup\{t \mid \eta(t) < x\}, \quad x<0.$$

for the queueing process $\eta(t)$ from the previous problem. Find the moment generating function $\widetilde{\theta}_1$ using the average in ξ_1 and prove that

$$\pi(s) = \mathsf{E}e^{-s\widetilde{\theta}_1}\,\mathbb{I}_{\widetilde{\theta}_1<\infty} = \mathsf{E}e^{-s\tau_\eta^-(-\xi_1)}\,\mathbb{I}_{\tau_\eta^-(-\xi_1)<\infty} = f(\rho(s)),$$

where $\rho(s)$ is the positive root of the Lundberg equation $k_1(\rho)=s$.

20.19. Consider the process $\eta(t)$ from Problem 20.17 given $\delta>0$. Find the moment generating function for the distribution of stationary waiting time $w_* = \lim_{t\to\infty} w(t)$; that is, find $\omega_*(z) = \mathsf{E}e^{-zw_*}$.

20.20. Prove the identity of the moment generating functions for $w(\theta_s)$, $\eta^+(\theta_s)$, and $-\widehat{\xi}_1^-(\theta_s)$, and also for w_*, η^+ and $-\widehat{\xi}_1^-$ using the remark of Theorem 20.5 and notations $\xi(t) = -\zeta(t)$ ($\xi_1(t) = -\eta(t)$) within Problem 20.17. The Pollaczek–Khinchin formula can be established in such a way for the moment generating function of the virtual waiting time:

$$\mathsf{E}e^{-zw_*} = \frac{p_+}{1 - q_+\mu^{-1}\int_0^\infty e^{-zx}\overline{F}(x)\,dx}.$$

Show the following relations for the risk process $\zeta(t)$.

$$\mathsf{E}e^{-z\zeta^+(\theta_s)} = \widetilde{\omega}(Cz,s), \quad \mathsf{E}e^{-z\zeta^+} = \omega_*(Cz), \quad \widetilde{\omega}(z,s) = \mathsf{E}e^{-zw(\theta_1)}.$$

20.21. Prove that the cumulant for the risk process $\zeta(t)$ from Problem 20.9 is the fractional rational function of the form

$$k(r) = \psi(\alpha)|_{i\alpha=r} = r\cdot\frac{\lambda_1(2-r)(b+r) - \lambda_2(1-r)^2}{(1-r)^2(b+r)}.$$

Calculate the first three moments $m_k = \mathsf{E}\zeta(1)^k$ ($k=\overline{1,3}$) using (20.74) and (20.75) for $\lambda_1=\lambda_2=1$, $b=1/14$.

20.22. Consider the risk process $\zeta(t) = \sum_{k=1}^{N(t)} \xi_k - t$ $(c = 1)$ with characteristic function of claims

$$\varphi(\alpha) = \mathsf{E}e^{i\alpha\xi_k} = \frac{\delta^2}{(\delta - i\alpha)^2}, \quad \delta > 0,$$

Find $m = \mathsf{E}\xi(1)$, $\psi(\alpha)$, $\varphi(s,\alpha)$, $\varphi_\pm(s,\alpha)$, and write the main factorization identity. Using the second factorization identity (20.23) find the moment generating function for pairs $\{\tau^+(\theta'_\mu), \gamma^+(\theta'_\mu)\}$, $\{\tau^+(x), \gamma^+(x)\}$. If $m = 2\lambda\delta^{-1} - 1 < 0(\lambda < \delta/2)$ find the characteristic function for $\varphi_+(\alpha)$ and compare it with one determined by the Pollaczek–Khinchin formula. Find the distribution function of ζ^+ if $\lambda = \delta/4$.

20.23. Consider the process $\zeta(t)$ from the previous problem with $\lambda = \delta/4$. Find $g_+(u,z)$ with help of formula (20.61), $d_+(\alpha,\mu)$ using formula (20.63), $g_u(s)$ with help of formula (20.62), and $\varkappa_u(z)$ using formula (20.66).

20.24. Using the equality (20.57) with $m < 0$, prove for the process $\zeta(t)$ from the formula (20.10) the following equality

$$\mathsf{E}\gamma^+(u)\,\mathbb{I}_{\zeta^+ > u} = \int_0^\infty \Phi_1(u,x)dx = \frac{\lambda}{c}\widehat{F}_3(u) + \frac{\lambda}{|m|}\int_0^u \widehat{F}_3(u-z)dP_+(z), \quad u \geq 0,$$

where $\widehat{F}_3(u) = \int_u^\infty \overline{\overline{F}}(x)dx$ is the tail of the third order of the d.f. $F(x), x > 0$.

20.25. Using the equality (20.67) with $m < 0$, prove for the process $\zeta(t)$ from the formula (20.49) the following equality

$$\mathsf{E}\gamma^+(u)\,\mathbb{I}_{\zeta^+ > u} = \int_0^\infty x\phi_1(u,x)dx = \overline{\overline{F}}_*(u) + \frac{\lambda}{b|m|}\int_0^u \overline{\overline{F}}_*(u-z)dP_+(z), \quad u \geq 0,$$

where $\overline{\overline{F}}_*(u) = \int_u^\infty \overline{F}_*(x)dx, \quad \overline{F}_*(x) = \overline{F}(x) + b\overline{\overline{F}}(x), x > 0.$

Hints

20.4. To obtain the simplified relation for the first duration of "red period" it is sufficient to calculate the corresponding integral taking into account that for $m < 0$, $\rho_+ = c|m|$, $p_+ = (c-\lambda)/c$, $\rho_+ = cp_+$,

$$\frac{\lambda}{|m|}\int_0^u e^{-c(u+x-z)}\,dP_+(z) = \frac{\lambda}{|m|}q_+\rho_+\int_0^\infty e^{-c(u+x-z)-\rho_+ z}dz$$

$$= \lambda c q_+ p_+ e^{-c(u+x)}\int_0^u e^{cq_+ z}dz = \lambda\left(e^{-cp_+ u - cx} - e^{-c(u+x)}\right).$$

The negative part in the mentioned integral compensates the first term of the first density $\mathsf{P}(\gamma^+(u) < x, \zeta^+ > u)$ and, thus, its simple exponential expression is fulfilled. It implies that the moment generating function for the security of ruin $\gamma^+(u)$ has a form

$$g_+(u,z) = \mathsf{E}e^{-z\gamma^+(u)}\mathbb{I}_{\zeta^+>u} = \frac{cq_+}{c+z}e^{-\rho_+ u}.$$

Thus, the moment generating function $T(u)$ according to (20.62) is determined by

$$g_u(s) = g_+(u,\rho_-(s)).$$

The dual relations for the second and third densities can be simplified in a similar way.

20.5. Use the moment generating function $g_+(u,z)$ from Problem 20.4 and the formula (20.66).

20.6. The process $\zeta(t)$ is lower continuous. The fractional rational expressions for $\psi(\alpha)$ and $\varphi(s,\alpha)$ can be found calculating the cumulant $\psi(\alpha) = -i\alpha + \lambda(\varphi(\alpha) - 1)$. The lower continuity of $\zeta(t)$ implies that

$$\varphi_-(s,\alpha) = \frac{\rho_-(s)}{\rho_-(s)+i\alpha}, \quad \mathsf{P}(\zeta^-(\theta_s) < x) = e^{\rho_-(s)x}, \quad x < 0.$$

Dividing $P_3(s,r)$ by $(r+\rho_-(s))$ we obtain that

$$\begin{aligned} P_2(s,r) &= r^2 + r(s-7-\rho_-(s)) + 21s\rho_-^{-1}(s) \\ &= (\rho_+(s)-r)(r_+(s)-r). \end{aligned}$$

Furthermore, the main factorization identity implies that

$$\varphi_+(s,\alpha) = \frac{s}{\rho_-(s)}\frac{(3-i\alpha)(7-i\alpha)}{P_2(s,i\alpha)}.$$

Because $m<0$, $s^{-1}\rho_-(s) \underset{s\to 0}{\to} |m|^{-1} = 7/2$, $r_2(s) = \rho_+(s) \underset{s\to 0}{\to} \rho_+ = 1$, $r_3(s) \underset{s\to 0}{\to} 6$ then

$$\varphi_+(\alpha) = \lim_{s\to 0}\varphi_+(s,\alpha) = \frac{2}{7}\frac{(3-i\alpha)(7-i\alpha)}{P_2(0,i\alpha)}.$$

Decompose the fractional rational function of the second order into fractionally linear parts (because $P_2(0,r)$ can be decomposed as $(r-1)(r-6)$) and invert $\varphi_+(\alpha)$.

20.7. Before calculating the first two ruin functions

$$\Phi_1(u,x) := \mathsf{P}(\gamma^+(u) > x, \zeta^+ > u), \quad x \ge 0,$$

$$\Phi_2(u,x) := \mathsf{P}(\gamma_+(u) > x, \zeta^+ > u), \quad x \ge 0,$$

use the formulas (20.57) and (20.58) for $x=0$ and corresponding conditions from the previous problem: $\lambda = 3$, $c=1$, $m=-2/7$, $p_+ = 2/7$,

$$\overline{F}(x) = \frac{1}{2}\left(e^{-3x}+e^{-7x}\right), \quad \overline{\overline{F}}(x) = \frac{1}{2}\left(\frac{1}{3}e^{-3x}+\frac{1}{7}e^{-7x}\right), \quad x>0,$$

$$P_+'(z) = \frac{24}{35}e^{-z} + \frac{6}{35}e^{-6z}, \quad z>0,$$

and show that

$$\Phi_1(u,0) = \Phi_2(u,0) = \overline{P}_+(u) = \frac{24}{35}e^{-u} + \frac{1}{35}e^{-6u}, \quad u > 0.$$

The ruin functions $\Phi_1(u,x)$ and $\Phi_2(u,x)$ can be found by formulas (20.57) and (20.58).

20.9. Find $\psi(\alpha)$ and show (for $m = 2 - b^{-1} = (2b-1)/b$) that

$$\varphi(s,\alpha) = \frac{s(b+i\alpha)(1-i\alpha)^2}{P_3(s,i\alpha)},$$

$$P_3(s,r) = r^3(s+2) + r^2(s(b-2)+b-4) - r(s+1)mb + bs = 0.$$

The negative root $r_1(s) = -\rho_-(s)$ of the cubic equation $P_3(s,r) = 0$ can be used for the determination of $\varphi_-(s,\alpha)$, and

$$\begin{aligned} P_2(s,r) &= P_3(s,r)(r+\rho_-(s))^{-1} \\ &= r^2(s+2) + (b-4+s(b-2)-(2+s)\rho_-(s))r + bs\rho_-^{-1}(s) \end{aligned}$$

for $\varphi_+(s,\alpha)$.

20.10. For calculation $\phi_{1,2}(u,x)$ by the formulas (20.67)–(20.68) it should be taken into account that under conditions of the problem

$$\lambda = \lambda_1 + \lambda_2 = 1,\ p = \overline{F}(0) = \frac{1}{2}, \quad b = \frac{1}{14}, \quad m = -12, \quad \frac{\lambda}{b|m|} = \frac{7}{31},$$

$$\overline{F}(x) = \frac{1}{2}(1+x)e^{-x}, \quad F'(x) = \frac{1}{2}xe^{-x}, \quad x > 0,$$

$$F_*'(x) = F'(x) + b\overline{F}(x) = \frac{15x+1}{28}e^{-x}, \quad x > 0,$$

$$P_+(y) = 1 - \frac{1}{41}(27e^{-y/4} - \frac{25}{7}e^{-12y/7}), \quad y > 0.$$

Notice that for $y = u$ (see (16.47) in [33])

$$\mathsf{P}(\gamma_+(u) = u,\ \zeta_+ > u) = p\bar{F}_1(u) > 0$$

in order to calculate $\phi_2(u,y)$.

Answers and Solutions

20.1. It was mentioned before that $\xi(t)$ is both an upper and lower continuous process for which

$$\varphi(s,\alpha) = \mathsf{E}e^{i\alpha\xi(\theta_s)} = \frac{2s}{2s - 2i\alpha a - (i\alpha)^2\sigma^2}, \tag{20.77}$$

and the characteristic functions of $\xi^{\pm}(\theta_s)$ can be expressed by formulas (20.30) and (20.33). According to (20.77), the Lundberg equation can be reduced to a quadratic one:

$$\sigma^2 r^2 + 2ar - 2s = 0, \quad r_{1,2}(s) = \pm\rho_{\pm}(s),$$

$$\rho_{\pm}(s) = \frac{\sqrt{2s\sigma^2 + a^2} \mp a}{\sigma^2}, \quad s \geq 0.$$

Thus, the main factorization identity and its components have the form:

$$\varphi(s,\alpha) = \varphi_+(s,\alpha)\varphi_-(s,\alpha),$$

$$\varphi_{\pm}(s,\alpha) = \frac{\rho_{\pm}(s)}{\rho_{\pm}(s) \mp i\alpha}, \quad s \geq 0. \tag{20.78}$$

The characteristic functions from (20.78) can be easily inverted in α and the densities of $\xi^{\pm}(\theta_s)$ can be found:

$$p_{\pm}(s,x) = \frac{\partial}{\partial x}\mathsf{P}(\xi^{\pm} < x) = \rho_{\pm}(s)e^{\mp\rho_{\pm}(s)x}, \quad (\pm x > 0). \tag{20.79}$$

(1) For $a < 0$ we have that $\rho_+(s) \underset{s\to 0}{\to} \rho_+ = 2|a|\sigma^{-2} > 0$, $\rho_-(s) \underset{s\to 0}{\to} 0$. So, the characteristic function and the distribution of ξ^+ are

$$\varphi_+(\alpha) := \mathsf{E}e^{i\alpha\xi^+} = \frac{\rho_+}{\rho_+ - i\alpha}, \quad \mathsf{P}(\xi^+ > x) = e^{-\rho_+ x}, \quad x \geq 0, \tag{20.80}$$

$$\mathsf{P}(\xi^- = -\infty) = 1.$$

(2) For $a = 0$ we have that $\rho_{\pm}(s) = \sqrt{2s}/\sigma \underset{s\to 0}{\to} 0$, $\mathsf{P}(\xi^{\pm} = \pm\infty) = 1$.

(3) For $a > 0$ we have that $\rho_-(s) \underset{s\to 0}{\to} \rho_- = 2a\sigma^{-2} > 0$, $\rho_+(s) \underset{s\to 0}{\to} 0$ therefore $\mathsf{P}(\xi^+ = +\infty) = 1$,

$$\varphi_-(\alpha) := \mathsf{E}e^{i\alpha\xi^-} = \frac{\rho_-}{\rho_- + i\alpha}, \quad \mathsf{P}(\xi^- < x) = e^{\rho_- x}, \quad x \leq 0. \tag{20.81}$$

Because the process $\xi(t) = at + \sigma W(t)$ is continuous then $\mathsf{P}(\gamma_k(x) = 0) = 1$, $(k = \overline{1,3})$. The formula (20.62) implies

$$\mathsf{E}e^{-sT(x)}\mathbb{I}_{\tau^+(x)<\infty} = \mathsf{E}e^{-\rho_+(s)\gamma^+(x)}\mathbb{I}_{\tau^+(x)<\infty} = \mathsf{P}(\tau^+(x) < \infty).$$

Thus, $\mathsf{P}(T = 0/\tau^+(x) < \infty) = 1$; that is, T is a degenerating random variable.

The moment generating function of the total duration of being over the level x, that is,

$$D_x(\mu) = \mathsf{E}e^{-\mu Q_x(\infty)}, \quad Q_x(\infty) = \int_0^{\infty} \mathbb{I}_{\xi(t)>x}\,dt,$$

for $m = a < 0$ is determined, according to (20.63), by the integral transform

$$d_+(\alpha,\mu) = \int_0^{\infty} e^{i\alpha x}\,d_x D_x(\mu) = \frac{\varphi_+(\alpha)}{\varphi_+(\mu,\alpha)}. \tag{20.82}$$

Furthermore, the following relation holds true

$$d_+(\alpha,\mu) = \frac{\rho_+}{\rho_+(\mu)}\frac{\rho_+(\mu) - i\alpha}{\rho_+ - i\alpha}, \quad \rho_+ = \frac{2|a|}{\sigma^2}, \quad \rho_+(\mu) = \frac{\sqrt{2\mu\sigma^2 + a^2} - a}{\sigma^2}.$$

Inverting it in α we can find the moment generating function

$$D_x(\mu) = 1 - \frac{\rho_+(\mu) - \rho_+}{\rho_+(\mu)} e^{-\rho_+ x}, \quad x > 0. \tag{20.83}$$

The integral transform for $D_x(s,\mu) = \mathsf{E}e^{-\mu Q_x(\theta_s)}$ $(x \geq 0)$ can be defined in a similar way to (20.82), according to (2.70) in [7]:

$$\begin{gathered} d_+(s,\alpha,\mu) = \int_0^\infty e^{i\alpha x} d_x D_x(s,\mu) \\ = \frac{\varphi_+(s,\alpha)}{\varphi_+(s+\mu,\alpha)} = \frac{\rho_+(s)}{\rho_+(s+\mu)} \frac{\rho_+(s+\mu) - i\alpha}{\rho_+(s) - i\alpha}. \end{gathered} \tag{20.84}$$

After inversion in α we can find, similarly to (20.83), that

$$D_x(s,\mu) = 1 - \frac{\rho_+(s+\mu) - \rho_+(s)}{\rho_+(s+\mu)} e^{-\rho_+(s)x}, \quad x \geq 0. \tag{20.85}$$

For $x = 0$

$$D_0(s,\mu) = \mathsf{E}e^{-\mu Q_0(\theta_s)} = \frac{\rho_+(s)}{\rho_+(s+\mu)}.$$

If $a = 0$, then $\rho_\pm(s) = \sqrt{2s}/\sigma$. So, for $\xi(t) = \sigma w(t)$ we get

$$s \int_0^\infty e^{-st} \mathsf{E}e^{-\mu Q_0(t)} dt = \mathsf{E}e^{-\mu Q_0(\theta_s)} = \sqrt{\frac{s}{s+\mu}}.$$

After inversion in s we obtain the well-known result for the distribution $Q_0(t)$:

$$\mathsf{P}(Q_0(t) < x) = \frac{2}{\pi} \arcsin \sqrt{\frac{x}{t}}, \quad (0 \leq x \leq t).$$

20.2. It was mentioned above that the considering process is both upper and lower almost semicontinuous. We have

$$\begin{aligned} -\psi(\alpha) &= \left(1 - \frac{1}{1+i\alpha}\right) + \left(1 - \frac{c}{c - i\alpha}\right) = \frac{(c-1)i\alpha - (i\alpha)^2}{(1+i\alpha)(c-i\alpha)}, \\ \varphi(s,\alpha) &= \frac{s}{s - \psi(\alpha)} = \frac{(c-i\alpha)(1+i\alpha)}{(s-2)\alpha^2 + (1-c)(1-s)i\alpha + sc}. \end{aligned} \tag{20.86}$$

Hence, the characteristic function $\varphi(s,\alpha)$ is a rational function of the second order. Let us decompose $\varphi(s,\alpha)$ into a product of a fractional linear multipliers that determine $\varphi_\pm(s,\alpha)$. After substitution $r = i\alpha$ (making the denominator in (20.86) equal to 0) we obtain the Lundberg equation which is quadratic:

$$-(2+s)r^2 + rcm(1-s) - sc = 0, \quad m = (1-c)c^{-1}. \tag{20.87}$$

It follows for $s = 0$ that $2r^2 + rcm = 0$ and the roots are $r_0 = 0$, $r_1^0 = -cm/2$, ($r_1^0 > 0$ if $m < 0$, $r_1^0 < 0$ if $m > 0$).

If $s > 0$ it is possible to find the roots of the equation (20.87). In particular, for $m = 0$ the roots are of very simple form: $r_{1,2}(s) = \pm\sqrt{s/(2+s)} = \pm\rho_\pm(s)$. If $m \neq 0$

then the roots are $r_1(s) \neq |r_2(s)|$, $\rho_+(s) = r_1(s)$, $\rho_-(s) = -r_2(s)$. For $s > 0$ these roots determine the characteristic function for the distribution of $\zeta^\pm(\theta_s)$, according to (20.35) and (20.37),

$$\varphi_+(s,\alpha) = \frac{p_+(s)(c-i\alpha)}{\rho_+(s)-i\alpha}, \quad \rho_+ = cp_+(s),$$
$$\mathsf{P}(\zeta^+(\theta_s)=0) = p_+(s), \quad p_+(s)p_-(s) = \frac{s}{s+\lambda}, \tag{20.88}$$

$$\varphi_-(s,\alpha) = \frac{p_-(s)(1+i\alpha)}{\rho_-(s)+i\alpha}, \quad \rho_-(s) = p_-(s) = \mathsf{P}(\zeta^-(\theta_s)=0).$$

If $m = (1-c)/c < 0$, that is equivalent to $c > 1$, then $\rho_+ = c|m|/2 = (c-1)/2 > 0$, thus

$$\varphi_+(\alpha) = \mathsf{E}e^{i\alpha\zeta^+} = \frac{p_+(c-i\alpha)}{\rho_+ - i\alpha}, \quad \rho_+ = cp_+. \tag{20.89}$$

If $m > 0$, that is equivalent to $c < 1$, then $\rho_- = cm/2 = (1-c)/2 > 0$, thus

$$\varphi_-(\alpha) = \mathsf{E}e^{i\alpha\zeta^-} = \frac{p_-(1+i\alpha)}{\rho_- + i\alpha}, \quad \rho_- = p_-. \tag{20.90}$$

According to (20.23) it is easy to calculate the joint moment generating function of $\{\tau^+(x), \gamma^+(x)\}$:

$$\mathsf{E}e^{-s\tau^+(\theta'_\mu)-u\gamma^+(\theta'_\mu)}\mathbb{I}_{\tau^+(\theta'_\mu)<\infty} = \frac{\mu q_+(s)}{\rho_+(s)+\mu}\frac{c}{c+u},$$

which after inversion in μ is

$$\mathsf{E}e^{-s\tau^+(x)-u\gamma^+(x)}\mathbb{I}_{\tau^+(x)<\infty} = q_+(s)e^{-\rho_+(s)x}\frac{c}{c+u}. \tag{20.91}$$

This implies the following relation (after the inversion in u),

$$\mathsf{E}e^{-s\tau^+(x)}\mathbb{I}_{\tau^+(x)<\infty}\mathsf{P}(\gamma^+(x) > z) = q_+(s)e^{-\rho_+(s)x}e^{-cz}, \quad z > 0, \tag{20.92}$$

Thus, the overjump $\gamma^+(x)$ follows the exponential distribution with the same parameter $c > 0$ as claims ξ_k. If $s \to 0$ then it follows from (20.90) that

$$g_+(x,z) = \mathsf{E}e^{-z\gamma^+(x)}\mathbb{I}_{\tau^+(x)<\infty} = \frac{cq_+}{c+z}e^{-\rho_+ x},$$

and according to the formula (20.65) we could find

$$\mathsf{E}e^{-sT(x)}\mathbb{I}_{\tau^+(x)<\infty} = q_-(s)g_+(x,\rho_-(s)) = q_-(s)\frac{cq_+}{c+\rho_-(s)}e^{-\rho_+ x}.$$

If $m < 0$ then the moment generating function $Q_x(\infty)$ is determined in a similar way to the (20.79) and (20.80) relation but with other values for the roots ρ_+ and $\rho_+(\mu)$. They can be determined by the equation (20.87) for $s = 0$ and $s = \mu$.

20.3. Let us mention that the process $\zeta(t)$ is lower continuous and almost upper semicontinuous. Its cumulants $\psi(\alpha)$ and $\varphi(s,\alpha)$ can be easily calculated:

$$\psi(\alpha)=\lambda(\varphi(\alpha)-1)-i\alpha=\frac{\lambda i\alpha}{c-i\alpha}-i\alpha=\frac{(i\alpha)^2+cmi\alpha}{c-i\alpha},$$
$$\varphi(s,\alpha)=\frac{s}{s-\psi(\alpha)}=\frac{s(c-i\alpha)}{cs-i\alpha(s+cm)-(i\alpha)^2}.$$

The Lundberg equation can be reduced to a quadratic one

$$r^2+(s+cm)r-cs=0,\quad (D_s=(s+cm)^2+4cs>0),$$

with roots $\pm r_{1,2}(s)>0$

$$r_{1,2}(s)=\frac{1}{2}\left(-(s+cm)^2\pm\sqrt{D_s}\right),\quad \rho_+(s)=r_1(s),\ \rho_-(s)=-r_2(s).$$

These roots determine the components of the main factorization identity ($\rho_+(s)=cp_+(s)<c$)

$$\varphi_\pm(s,\alpha)=\frac{p_+(s)(c-i\alpha)}{\rho_+(s)-i\alpha},\quad \varphi_-(s,\alpha)=\frac{\rho_-(s)}{\rho_-(s)+i\alpha}.$$

(1) For $m<0$ we have that $\rho_+(s)\underset{s\to 0}{\to}\rho_+=c|m|>0$, $\rho_-(s)\underset{s\to 0}{\to}0$, thus $\mathsf{P}(\zeta^-=-\infty)=1$ and

$$\varphi_+(\alpha)=\mathsf{E}e^{i\alpha\zeta^+}=\frac{p_+(c-i\alpha)}{\rho_+-i\alpha},\quad \mathsf{P}(\zeta^+>x)=q_+e^{-\rho_+x},\quad x>0.$$

So, the ruin probability for $m<0$ and $u>0$ is equal to

$$\Psi(u)=\mathsf{P}(\zeta^+>u)=q_+e^{-\rho_+u}.$$

(2) For $m>0$ we have that $\rho_+(s)\underset{s\to 0}{\to}0$, $\rho_-(s)\underset{s\to 0}{\to}\rho_-=cm=cp_-$, thus $\mathsf{P}(\xi^+=\infty)=1$, and

$$\varphi_-(\alpha)=\mathsf{E}e^{i\alpha\zeta^-}=\frac{\rho_-}{\rho_-+i\alpha},\quad \mathsf{P}(\zeta^-<x)=e^{\rho_-x},\quad x\leq 0.$$

(3) For $m=0$ we have that $\rho_\pm(s)\underset{s\to 0}{\to}0$, thus $\mathsf{P}(\zeta^\pm=\pm\infty)=1$.

Because $\overline{F}(x)=e^{-cx}$, then the following relations take place according to the formulas (20.57)–(20.58) for the marginal densities of the ruin functions (for $\gamma_1(x)=\gamma^+(x)$, $\gamma_2(x)=\gamma_+(x)$ and $\gamma_3(x)=\gamma_x^+$, respectively):

$$\phi_1(u,x) := \frac{\partial}{\partial x}\mathsf{P}(\gamma^+(u) < x, \zeta^+ > u)$$
$$= \lambda e^{-c(x+u)} + \frac{\lambda}{|m|}\int_0^u e^{-c(u+x-z)}\,dP_+(z),$$
$$\phi_2(u,y) := \frac{\partial}{\partial y}\mathsf{P}(\gamma_+(u) < y, \zeta^+ > u)$$
$$= \begin{cases} \frac{\lambda}{|m|}e^{-cy}P_+(u), \quad P_+(u) = \mathsf{P}(\zeta^+ < u), \quad y > u, \\ \frac{\lambda}{|m|}e^{-cy}\mathsf{P}(u-y < \zeta^+ < y), \quad 0 < y < u, \end{cases} \tag{20.93}$$
$$\phi_3(u,z) := \frac{\partial}{\partial z}\mathsf{P}(\gamma_u^+ < z, \zeta^+ > u)$$
$$= \begin{cases} \frac{\lambda c}{|m|}e^{-cz}\int_0^u (z-u+y)\,dP_+(y), \quad z > u, \\ \frac{\lambda c}{|m|}e^{-cz}\int_{-z}^0 (z+y)\,dP_+(y+u), \quad 0 < z < u. \end{cases}$$

20.4. $\phi_1(u,x) = (\partial/\partial x)\mathsf{P}(\gamma^+(u) < x, \zeta^+ > u) = \lambda e^{-\rho_+ u - cx}, \quad x > 0;$

$$\phi_2(u,y) = \frac{\partial}{\partial y}\mathsf{P}(\gamma_+(u) < y, \zeta^+ > u)$$
$$= \begin{cases} \frac{\lambda}{|m|}e^{-cy}\left(1-e^{-\rho_+ u}\right), & y > u, \\ \frac{\lambda}{|m|}q_+ e^{-cy}\left(e^{-\rho_+(u-y)} - e^{-\rho_+ u}\right), & 0 < y < u; \end{cases}$$
$$\phi_3(u,z) = \frac{\partial}{\partial z}\mathsf{P}(\gamma_u^+ < z, \zeta^+ > u)$$
$$= \begin{cases} \frac{\lambda c q_+}{|m|}e^{-cz}\left((1-e^{-\rho_+ u})(z+\rho_+^{-1}) - u\right), & z > u, \\ \frac{\lambda c q_+}{|m|}e^{-cz}\left(\rho_+^{-1}e^{-\rho_+(u-z)} - e^{-\rho_+ u}(z+\rho_+^{-1})\right) & 0 < z < u; \end{cases}$$
$$g_+(u,z) = \frac{cq_+}{c+z}e^{-\rho_+ z}, \quad g_u(s) = \frac{cq_+}{c+\rho_-(s)}e^{-\rho_+ u}.$$

20.5.

$$\mathsf{E}e^{-zZ^+(u)}\mathbb{I}_{\zeta^+ > u} = \frac{c\rho_+ q_+}{\rho_+ + z}e^{-(\rho_+ + z)u}.$$

20.6.

$$\varphi_-(s,\alpha) = \frac{\rho_-(s)}{\rho_-(s)+i\alpha}, \quad \varphi_+(s,\alpha) = \frac{s}{\rho_-(s)}\frac{(3-i\alpha)(7-i\alpha)}{(\rho_+(s)-i\alpha)(r_2(s)-i\alpha)}.$$

$$\varphi_+(\alpha) = \frac{24}{35}\frac{1}{1-i\alpha} + \frac{1}{35}\frac{6}{6-i\alpha}, \quad \rho_-(s)s^{-1} \underset{s\to 0}{\to} \rho_-'(0) = \frac{1}{|m|};$$
$$\Psi(u) = \mathsf{P}(\zeta^+ > u) = \frac{24}{35}e^{-u} + \frac{1}{35}e^{-6u}.$$

20.7.

$$\Phi_1(u,x) = \frac{3}{5}e^{-u}\left(e^{-3x} + \frac{1}{7}e^{-7x}\right) + \frac{3}{10}e^{-6u}\left(\frac{3}{7}e^{-7x} - \frac{1}{3}e^{-3x}\right), \quad x \geq 0.$$

$$\Phi_2(u,y) = \begin{cases} \frac{3}{20}\left(\frac{1}{3}e^{-3y} + \frac{1}{7}e^{-7y}\right)\left(24e^{-u} - e^{-6u}\right), y > u, \\ \frac{3}{10}\left(e^{-u}\left(6e^{-2y} + 2e^{-6y} - 4e^{-3y} - \frac{12}{7}e^{-7y}\right)\right. \\ \left. + e^{-6u}\left(\frac{1}{2}e^{-y} - \frac{1}{6}\left(e^{3y} + e^{-3y}\right) - \frac{1}{14}e^{-7y}\right)\right), \ 0 < y < u. \end{cases}$$

20.9.

$$\varphi_-(s,\alpha) = \frac{p_-(s)(b+i\alpha)}{\rho_-(s)+i\alpha}, \quad \varphi_+(s,\alpha) = \frac{s}{\rho_-(s)}\frac{(1-i\alpha)^2}{P_2(s,i\alpha)}.$$

20.10.

$$\Psi(u) = \mathsf{P}(\zeta^+ > u) = \frac{1}{41}\left(27e^{-u/4} - \frac{25}{7}e^{-12u/7}\right),$$

$$\phi_1(u,x) = \frac{e^{-x}}{41}\left(\frac{25}{7}(3x-4)e^{--12u/7} + \frac{9}{4}(5x+7)e^{-u/4}\right), \quad x \geq 0;$$

If $y = u$ $\mathsf{P}(\gamma_2(u) = u, \zeta^+ > u) = \frac{1}{2}(1+u)e^{-u}$, if $y \neq u$

$$\phi_2(u,y) = \begin{cases} \frac{1}{12}(1+y)e^{-y}\left(1 - \frac{27}{41}e^{-u/4} + \frac{25}{287}e^{-12u/7}\right), \quad y > u; \\ \frac{1}{12}(1+y)\frac{e^{-y}}{41}\left(\frac{25}{7}e^{-12u/7} - 27e^{-u/4} + \frac{243}{2}e^{-(u-y)/4}\right. \\ \left. - \frac{625}{7}e^{-12(u-y)/7}\right), \quad 0 < y < u. \end{cases}$$

20.11.

$$\mathsf{E}e^{-z\zeta^+} = \frac{2p_+(1+z)^2}{2p_+ + 2z^2 + (4-q_+)z}.$$

20.12.

$$\begin{aligned} \Phi_1(0,x) &= p(\overline{F}_1(x) + b\overline{\overline{F}}_1(x)) \\ &= \frac{1}{2}e^{-x}(1+x+b(2+x)) \underset{x\to 0}{\to} \frac{1}{2}(1+2b)\,|_{b=1/14} = \frac{4}{7}; \\ \Phi_2(0,y) &= pb\overline{\overline{F}}_1(y) = \frac{1}{2}b(2+y)e^{-y} \underset{y\to 0}{\to} \frac{1}{2}b\,|_{b=1/14} = \frac{1}{14}; \end{aligned}$$

$$\mathsf{P}(\zeta^+ > 0, \gamma_+(0) = 0) = \frac{1}{2}, \quad \phi_2(0,0) = \frac{1}{14}.$$

$$\begin{aligned} \Phi_3(0,z) &= p\left(\overline{F}_1(z) + b\int_z^\infty y\,dF_1(y)\right) \\ &= \frac{1}{2}e^{-z}(1+z+b(z^2+2z+2)) \underset{z\to 0}{\to} \frac{1}{2}(1+2b)\,|_{b=1/14} = \frac{4}{7}. \end{aligned}$$

20.13.

$$g_0(s) = \mathsf{E}e^{-zT(0)}\mathbb{I}_{\zeta^+>0} = \frac{3}{2}\left(\frac{1}{3+\rho_-(s)} + \frac{1}{7+\rho_-(s)}\right).$$

20.14.

$$g_0(s) = \mathsf{E}e^{-sT(0)}\mathbb{I}_{\zeta^+>0} = q_+q_-(s)\frac{2+\rho_-(s)}{2(1+\rho_-(s))^2}.$$

20.15.

$$g_+(u,z) = \frac{3}{10}\frac{4(12+3z)e^{-u}+(2-z)e^{-6u}}{(3+z)(7+z)}.$$

20.16.

$$g_+(u,z) = \frac{1}{41}\frac{1}{1+z}\left(\frac{63}{4}e^{-u/4} - \frac{100}{7}e^{-12u/7} + \frac{15}{1+z}\left(\frac{5}{7}e^{-12u/7}+\frac{3}{4}e^{-u/4}\right)\right).$$

20.17. $\widetilde{\omega}(z,s) = (s-k_1(z))^{-1}(s-z\widetilde{p}_0(s))$, where $\widetilde{p}_0(s)$ is defined in (11.5).
20.20. $\omega_*(z) = \lim_{s\to 0}\widetilde{\omega}(z,s) = p_+\left(1-\lambda_1\int_0^\infty e^{-zx}\overline{F}(x)\,dx\right)^{-1}$,

$$p_+ = \mathsf{P}(w_*=0) = 1-\lambda_1\mu, \quad q_+ = 1-p_+ = \lambda_1\mu.$$

20.22. $m = \lambda\mu - 1 = 2\lambda\delta^{-1}-1$,
$\psi(\alpha) = \frac{\lambda\delta^2}{(\delta-i\alpha)^2} - i\alpha$; $\varphi(s,\alpha) = \frac{s(\delta-i\alpha)^2}{P_3(s,\alpha)}$, $P_3(s,r) = P_2(s,r)(\rho_-(s)+r)$, $P_2(s,r) = r^2+(s-2\delta+\lambda-\rho_-(s))r+s\delta^2\rho_-^{-1}(s)$, $\varphi(s,\alpha) = \varphi_+(s,\alpha)\varphi_-(s,\alpha)$; $\varphi_-(s,\alpha) = \frac{\rho_-(s)}{\rho_-(s)+i\alpha}$, $\varphi_+(s,\alpha) = \frac{s}{\rho_-(s)}\frac{(\delta-i\alpha)^2}{P_2(s,i\alpha)}$.
For $\lambda < \delta/2 \quad (m<0)$

$$\begin{aligned}
\varphi_+(\alpha) &= \lim_{s\to 0}\varphi_+(s,\alpha) = |m|\frac{(\delta-i\alpha)^2}{P_2(0,i\alpha)},\\
P_2(0,r) &= r^2+(\lambda-2\delta)r+|m|\delta^2 = (r-r_1)(r-r_2),\\
r_{1,2} &= \frac{1}{2}(2\delta-\lambda\mp\sqrt{\lambda(4\delta+\lambda)}) > 0.
\end{aligned}$$

After decomposition of $\varphi_+(\alpha)$ into linear-fractional functions we obtain

$$\varphi_+(\alpha) = |m|\left(1+\frac{(r_1-\delta)^2}{r_2-r_1}\frac{1}{r_1-i\alpha}+\frac{(r_2-\delta)^2}{r_1-r_2}\frac{1}{r_2-i\alpha}\right).$$

Thus, the distribution of ζ^+ can be expressed via $e^{-xr_{1,2}}$ inverting the previous relation on α. According to (20.52) and using the Pollacek–Khinchin formula we can obtain the similar result for $\varphi_+(\alpha)$ if $\lambda = \delta/4$ $\left(p_+ = q_+ = \frac{1}{2}, r_{1,2} = (7\mp\sqrt{17})\delta/8\right)$. Inverting it on α we obtain

$$\mathsf{P}\{\zeta^+>x\} = \frac{1}{4}\left[\left(1+\frac{5}{\sqrt{17}}\right)e^{-(7-\sqrt{17})\delta x/8}+\left(1-\frac{5}{\sqrt{17}}\right)e^{-(7+\sqrt{17})\delta x/8}\right].$$

A

Appendix

1. Main processes of queuing theory

$\xi(t)$ $\xi(t) = t - s(t)$, controlling process

$\eta(t) = s(t) - t = -\xi(t)$

$\alpha(t)$, non-busy process

$\alpha(t) = \xi^{+}(t)$

$\omega(t) = \alpha(t) - \xi(t)$, virtual waiting process $(\omega(0) = 0)$

$\theta_k = \tau_\eta^{-}(-\xi_1) = \tau_\xi^{+}(\xi_1)$, duration of busy period

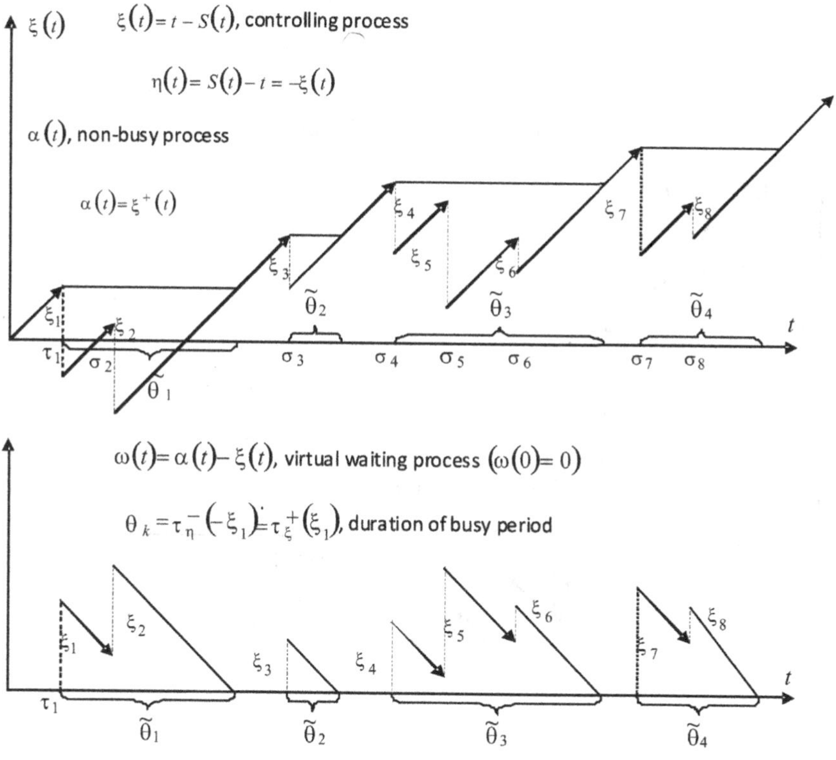

D. Gusak et al., *Theory of Stochastic Processes*, Problem Books in Mathematics,   359
DOI 10.1007/978-0-387-87862-1, © Springer Science+Business Media, LLC 2010

$\beta(t) = t - \alpha(t)$, busy process

$\tau^{+}(x) = \inf\{t > 0 : \xi(t) > x\}, x > 0$

$\tau^{+}(\xi_1) = \tau_\xi^{+}(\xi_1) = \tau_\eta^{-}(-\xi_1)$

$\tau_\eta^{-}(-x) = \inf\{t > 0 : \eta(t) < -x\}$

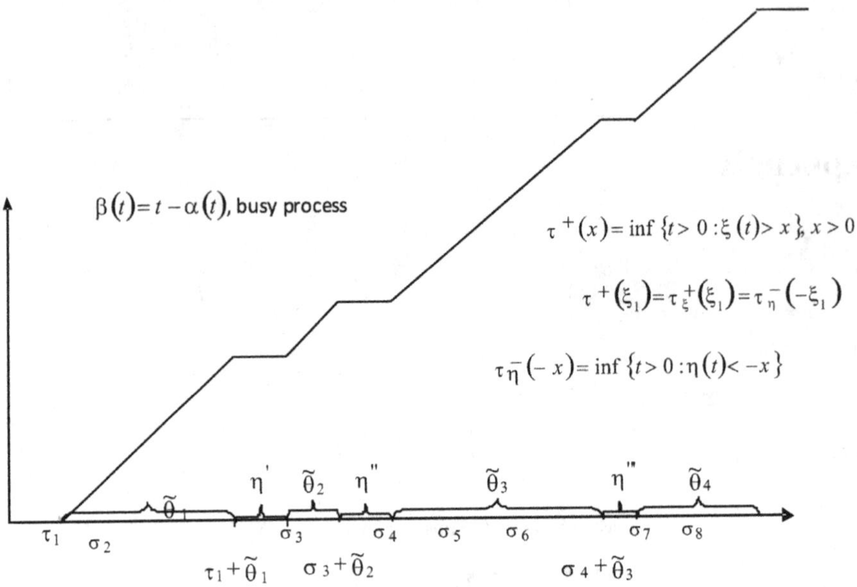

2. Classical reserve risk process and its functionals

$R_u(t) = u + ct - S(t), c > 0$

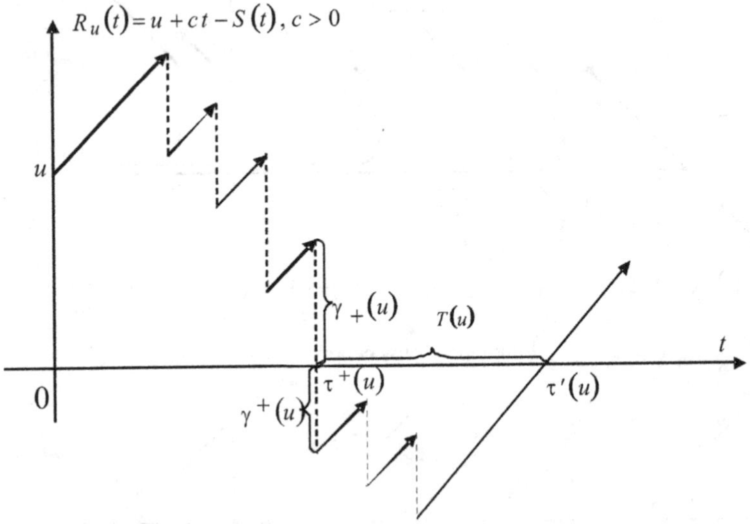

3. Claim surplus process and its functionals

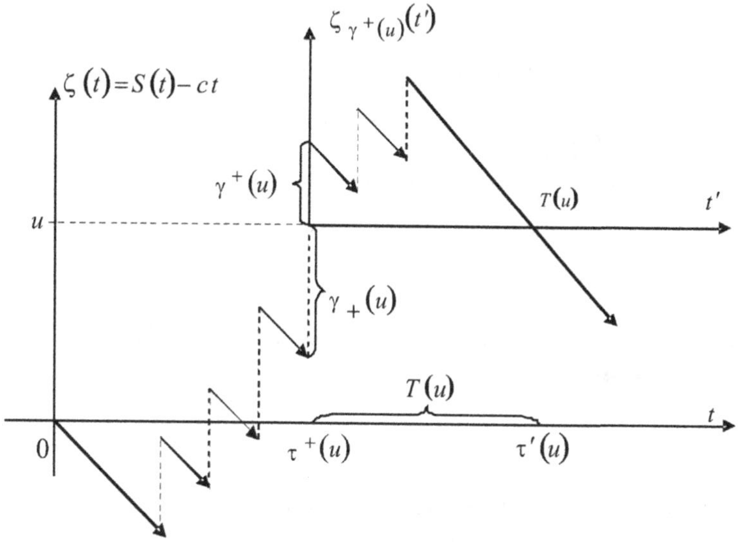

4. Reserve risk process with random premiums

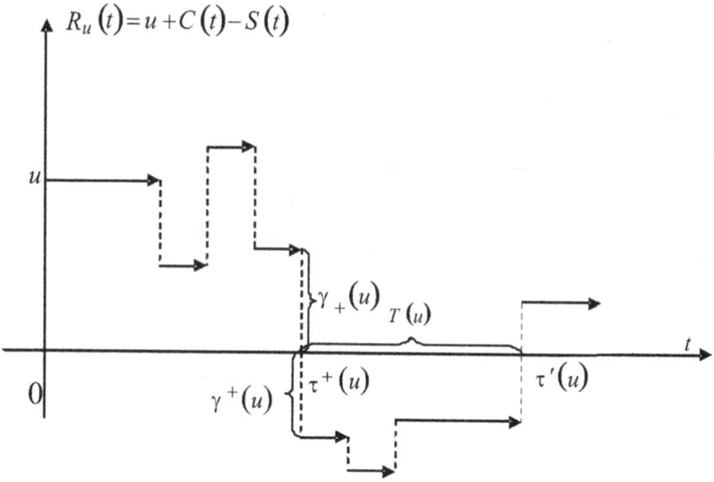

5. Claim surplus process with random premiums

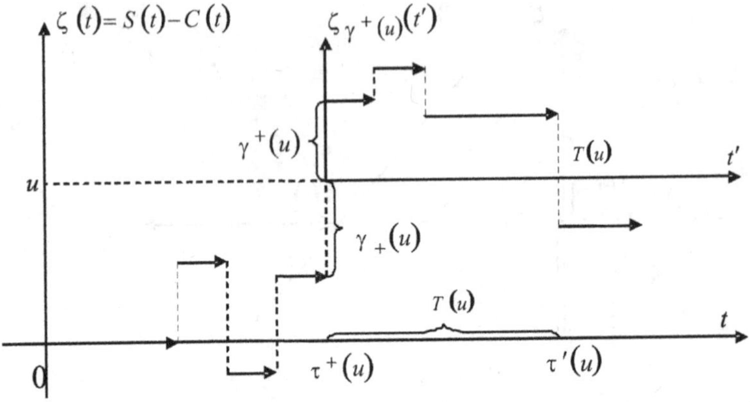

6. Risk and survival zones for $\varsigma(t)$ in risk zone $\{y > u\}$ and survival zone $\{y \le u\}$

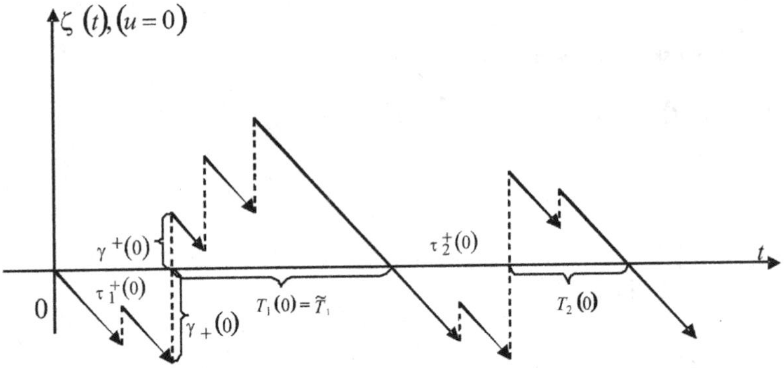

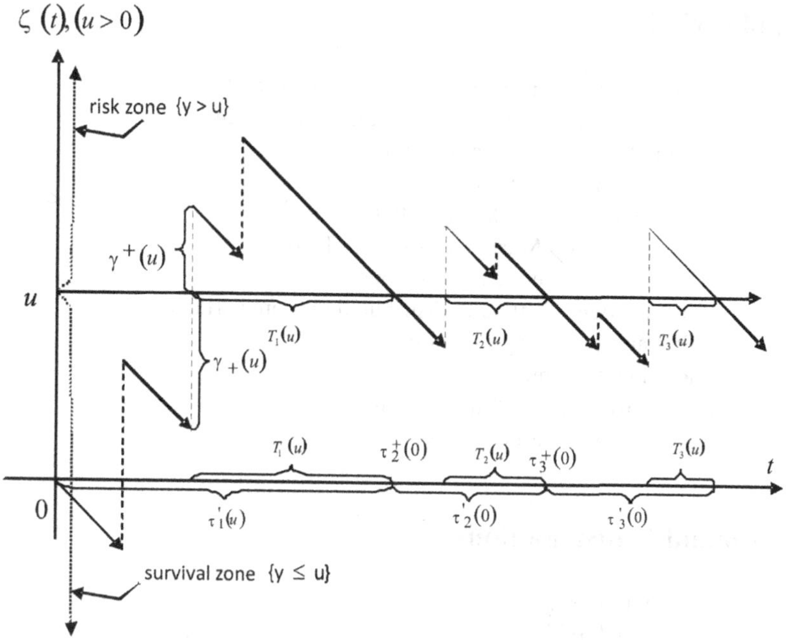

ζ (t), (u > 0)
risk zone {y > u}
γ⁺(u)
u
γ₊(u)
T₁(u)
T₂(u)
T₃(u)
τ₂⁺(0)
τ₃⁺(0)
t
0
τ'₁(u)
τ'₂(0)
τ'₃(0)
survival zone {y ≤ u}

List of abbreviations

càdlàg	Right continuous having left-hand limits (p. 24)
càglàd	Left continuous having right-hand limits (p. 24)
cdf	Cumulant distribution function
CLT	Central limit theorem
HMF	Homogeneous Markov family (p. 179)
HMP	Homogeneous Markov process (p. 176)
i.i.d.	Independent identically distributed
i.i.d.r.v.	Independent identically distributed random variables
pdf	Probability density function
r.v.	Random variable
SDE	Stochastic differential equation
SLLN	Strong law of large numbers

List of probability distributions

$\mathrm{Be}(p)$	Bernoulli, $\mathsf{P}(\xi = 1) = p, \mathsf{P}(\xi = 0) = 1 - p$
$\mathrm{Bi}(n,p)$	Binomial, $\mathsf{P}(\xi = k) = C_n^k p^k (1-p)^{n-k}, k = 0, \dots, n$
$\mathrm{Geom}(p)$	Geometric, $\mathsf{P}(\xi = k) = p^{k-1}(1-p), k \in \mathbb{N}$
$\mathrm{Pois}(\lambda)$	Poisson, $\mathsf{P}(\xi = k) = (\lambda^k / k!) e^{-\lambda}, k \in \mathbb{Z}^+$
$\mathrm{U}(a,b)$	Uniform on (a,b), $\mathsf{P}(\xi \le x) = 1 \wedge ((x-a)/(b-a))_+, x \in \mathbb{R}$
$\mathcal{N}(a, \sigma^2)$	Normal (Gaussian), $\mathsf{P}(\xi \le x) = (2\pi\sigma^2)^{-1/2} \int_{-\infty}^x e^{-(y-a)^2/2\sigma^2}\, dy$, $x \in \mathbb{R}$
$\mathrm{Exp}(\lambda)$	Exponential, $\mathsf{P}(\xi \le x) = [1 - e^{-\lambda x}]_+, x \in \mathbb{R}$
$\Gamma(\alpha, \beta)$	Gamma, $\mathsf{P}(\xi \le x) = (\beta^\alpha / \Gamma(\alpha)) \int_0^x y^{\alpha-1} e^{-\beta y}\, dy, x \in \mathbb{R}^+$

List of symbols

References

1. Asmussen S (2000) Ruin Probability. World Scientist, Singapore
2. Bartlett MS (1978) An Introduction to Stochastic Processes with Special Reference to Methods and Applications. Cambridge University Press, Cambridge, UK.
3. Bertoin J (1996) Levy Processes. Cambridge University Press, Cambridge, UK.
4. Billingsley P (1968) Convergence of Probability Measures. Wiley Series in Probability and Mathematical Statistics, John Wiley, New York
5. Bogachev VI (1998) Gaussian Measures. Mathematical Surveys and Monographs, vol. 62, American Mathematical Society, Providence, RI
6. Borovkov AA (1976) Stochastic Processes in Queueing Theory. Springer-Verlag, Berlin
7. Bratijchuk NS, Gusak DV (1990) Boundary problems for processes with independent increments [in Russian]. Naukova Dumka, Kiev
8. Brzezniak Z, Zastawniak T (1999) Basic Stochastic Processes. Springer-Verlag, Berlin
9. Bulinski AV, Shirjaev AN (2003) Theory of Random Processes [in Russian]. Fizmatgiz, Laboratorija Bazovych Znanij, Moscow
10. Bühlmann H (1970) Mathematical Methods in Risk Theory Springer-Verlag, New-York
11. Chaumont L, Yor M (2003) Exercises in Probability: A Guided Tour from Measure Theory to Random Processes, Via Conditioning. Cambridge University Press, Cambridge, UK
12. Chung KL (1960) Markov Chains with Stationary Transition Probabilities. Springer, Berlin
13. Chung KL, Williams RJ, (1990) Introduction to Stochastic Integration. Springer-Verlag New York, LLC
14. Cramér H, Leadbetter MR (1967) Stationary and Related Stochastic Processes. Sample Function Properties and Their Applications. John Wiley, New York
15. Doob JL (1990) Stochastic Processes. Wiley-Interscience, New York
16. Dorogovtsev AY, Silvesrov DS, Skorokhod AV, Yadrenko MI (1997) Probability Theory: Collection of Problems. American Mathematical society, Providence, RI
17. Dudley, RM (1989) Real Analysis and Probability. Wadsworth & Brooks/Cole, Belmont, CA
18. Dynkin EB (1965) Markov processes. Vols. I, II. Grundlehren der Mathematischen Wissenschaften, vol. 121, 122, Springer-Verlag, Berlin
19. Dynkin EB, Yushkevich AA (1969) Markov Processes-Theorems and Problems. Plenum Press, New York
20. Elliot RJ (1982) Stochastic Calculus and Applications. Applications of Mathematics 18, Springer-Verlag, New York
21. Etheridge A (2006) Financial Calculus. Cambridge University Press, Cambridge, UK
22. Feller W (1970) An Introduction to Probability Theory and Its Applications (3rd ed.). Wiley, New York
23. Föllmer H, Schied A (2004) Stochastic Finance: An Introduction in Discrete Time. Walter de Gruyter, Hawthorne, NY
24. Gikhman II, Skorokhod AV (2004) The Theory of Stochastic Processes: Iosif I. Gikhman, Anatoli V. Skorokhod. In 3 volumes, Classics in Mathematics Series, Springer, Berlin
25. Gikhman II, Skorokhod AV (1996) Introduction to the Theory of Random Processes. Courier Dover, Mineola
26. Gikhman II, Skorokhod AV (1982) Stochastic Differential Equations and Their Applications [in Russian]. Naukova dumka, Kiev
27. Gikhman II, Skorokhod AV, Yadrenko MI (1988) Probability Theory and Mathematical Statistics [in Russian] Vyshcha Shkola, Kiev

28. Gnedenko BV (1973) Priority queueing systems [in Russian]. MSU, Moscow
29. Gnedenko BV, Kovalenko IN (1989) Introduction to Queueing Theory. Birkhauser Boston, Cambridge, MA
30. Grandell J (1993) Aspects of Risk Theory. Springer-Verlag, New York
31. Grenander U (1950) Stochastic Processes and Statistical Inference. Arkiv fur Matematik, Vol. 1, no. 3:1871-2487, Springer, Netherlands
32. Gross D, Shortle JF, Thompson JM, Harris CM (2008) Fundamentals of Queueing Theory (4th ed.). Wiley Series in Probability and Statistics, Hoboken, NJ
33. Gusak DV (2007) Boundary Value Problems for Processes with Independent Increments in the Risk Theory. Pratsi Instytutu Matematyky Natsional'noï Akademiï Nauk Ukraïny. Matematyka ta ïï Zastosuvannya 65. Instytut Matematyky NAN Ukraïny, Kyïv
34. Hida T (1980) Brownian Motion. Applications of Mathematics, 11, Springer-Verlag, New York
35. Ibragimov IA, Linnik YuV (1971) Independent and Stationary Sequences of Random Variables. Wolters-Noordhoff Series of Monographs and Textbooks on Pure and Applied Mathematics, Wolters-Noordhoff, Groningen
36. Ibragimov IA, Rozanov YuA (1978) Gaussian Random Processes. Applications of Math., vol. 9, Springer-Verlag, New York
37. Ibramkhalilov IS, Skorokhod AV (1980) Consistent Estimates of Parameters of Random Processes [in Russian]. Naukova dumka, Kyiv
38. Ikeda N, and Watanabe S (1989) Stochastic Differential Equations and Diffusion Processes, Second edition. North-Holland/Kodansya, Tokyo
39. Ito K (1961) Lectures on Stochastic Processes. Tata Institute of Fundamental Research, Bombay
40. Ito K, McKean H (1996) Diffusion Processes and Their Sample Paths. Springer-Verlag, New York
41. Jacod J, Shiryaev AN (1987) Limit Theorems for Stochastic Processes. Grundlehren der Mathematischen Wissenschaften, vol. 288, Springer-Verlag, Berlin
42. Johnson NL, Kotz S (1970) Distributions in Statistics: Continuous Univariate Distributions. Wiley, New York
43. Kakutani S (1944) Two-dimensional Brownian motion and harmonic functions Proc. Imp. Acad., Tokyo 20:706–714
44. Karlin S (1975) A First Course in Stochastic Processes. Second edition, Academic Press, New York
45. Karlin S (1966) Stochastic Service Systems. Nauka, Moscow
46. Kijima M (2003) Stochastic Processes with Application to Finance. Second edition. Chapman and Hall/CRC, London
47. Klimov GP (1966) Stochastic Service Systems [in Russian]. Nauka, Moscow
48. Kolmogorov AN (1992) Selected Works of A.N. Kolmogorov, Volume II: Probability theory and Mathematical statistics. Kluwer, Dordrecht
49. Koralov LB, Sinai YG (2007) Theory of Probability and Random Processes, Second edition. Springer-Verlag, Berlin
50. Korolyuk VS (1974) Boundary Problems for a Compound Poisson Process. Theory of Probability and its Applications 19, 1-14, SIAM, Philadelphia
51. Korolyuk VS, Portenko NI, Skorokhod AV, Turbin AF (1985) The Reference Book on Probability Theory and Mathematical Statistics [in Russian]. Nauka, Moscow
52. Krylov NV (2002) Introduction to the Theory of Random Processes. American Mathematical Society Bookstore, Providence, RI
53. Lamperti J (1977) Stochastic Processes. Applied Mathematical Sciences, vol. 23, Springer-Verlag, New York

54. Lamberton D, Lapeyre B (1996) Introduction to Stochastic Calculus Applied to Finance. Chapman and Hall/CRC, London
55. Leonenko MM, Mishura YuS, Parkhomenko VM, Yadrenko MI (1995) Probabilistic and Statistical Methods in Ecomometrics and Financial Mathematics. [in Ukrainian] Informtechnika, Kyiv
56. Lévy P (1948) Processus Stochastiques et Mouvement Brownien. Gauthier-Villars, Paris
57. Liptser RS, Shiryaev AN (2008) Statistics Of Random Processes, Vol. 1. Springer-Verlag New York
58. Liptser RS, Shiryaev AN (1989) Theory of Martingales. Mathematics and Its Applications (Soviet Series), 49, Kluwer Academic, Dordrecht
59. Lifshits MA (1995) Gaussian Random Functions. Springer-Verlag, New York
60. Meyer PA (1966) Probability and Potentials. Blaisdell, New York
61. Øksendal B (2000) Stochastic Differential Equations, Fifth edition. Springer-Verlag, Berlin
62. Pliska SR (1997) Introduction to Mathematical Finance. Discrete Time Models. Blackwell, Oxford
63. Port S, Stone C (1978) Brownian Motion and Classical Potential Theory. Academic Press, New York
64. Protter P (1990) Stochastic Integration and Differential Equations. A New Approach Springer-Verlag, Berlin
65. Prokhorov AV, Ushakov VG, Ushakov NG (1986) Problems in Probability Theory. [in Russian] Nauka, Moscow
66. Revuz D, Yor M (1999) Continuous martingales and Brownian Motion. Third edition. Springer-Verlag, Berlin
67. Robbins H, Sigmund D, Chow Y (1971) Great Expectations: The Theory of Optimal Stopping. Houghton Mifflin, Boston
68. Rolski T, Schmidli H, Schmidt V, Teugels J (1998) Stochastic Processes for Insurance and Finance. John Wiley and Sons, Chichester
69. Rozanov YuA (1977) Probability Theory: A Concise Course. Dover, New York
70. Rozanov YuA (1982) Markov Random Fields. Springer-Verlag, New York
71. Rozanov YuA (1995) Probability Theory, Random Processes and Mathematical Statistics. Kluwer Academic, Boston
72. Rozanov YuA (1967) Stationary Random Processes. Holden-Day, Inc., San Francisco
73. Sato K, Ito K (editor), Barndorff-Nielsen OE (editor) (2004) Stochastic Processes. Springer, New York
74. Sevast'yanov BA (1968) Branching Processes. Mathematical Notes, Volume 4, Number 2 / August, Springer Science+Business Media, New York
75. Sevastyanov BA, Zubkov AM, Chistyakov VP (1988) Collected Problems in Probability Theory. Nauka, Moscow
76. Skorokhod AV (1982) Studies in the Theory of Random Processes. Dover, New York
77. Skorokhod AV (1980) Elements of the Probability Theory and Random Processes [in Russian]. Vyshcha Shkola Publ., Kyiv
78. Skorohod AV (1991) Random Processes with Independent Increments. Mathematics and Its Applications, Soviet Series, 47 Kluwer Academic, Dordrecht
79. Skorohod AV (1996) Lectures on the Theory of Stochastic Processes. VSP, Utrecht
80. Spitzer F (2001) Principles of Random Walk. Springer-Verlag New York
81. Shiryaev AN (1969) Sequential Statistical Analysis. Translations of Mathematical Monographs 38, American Mathematical Society, Providence, RI
82. Shiryaev AN (1995) Probability. Vol 95. Graduate Texts in Mathematics, Springer-Verlag New York

83. Shiryaev AN (2004) Problems in Probability Theory. MCCME, Moscow
84. Shiryaev AN (1999) Essentials of Stochastic Finance, in 2 vol. World Scientific, River Edge, NJ
85. Steele JM (2001) Stochastic Calculus and Financial Applications. Springer-Verlag, New York
86. Striker C, Yor M (1978) Calcul stochastique dependant d'un parametre. Z. Wahrsch. Verw. Gebiete, 45: no. 2: 109–133.
87. Stroock DW, Varadhan SRS (1979) Multidimensional Diffusion Processes Springer-Verlag, New York
88. Vakhania NN, Tarieladze VI, Chobanjan SA (1987) Probability Distributions on Banach Spaces. Mathematics and Its Applications (Soviet Series), 14. D. Reidel, Dordrecht
89. Ventsel' ES and Ovcharov LA (1988). Probability Theory and Its Engineering Applications [in Russian]. Nauka, Moscow
90. Wentzell AD (1981) A Course in the Theory of Stochastic Processes. McGraw-Hill, New York
91. Yamada T, Watanabe S (1971) On the uniquenes of solutions of stochastic differential equations J. Math. Kyoto Univ., 11: 155–167
92. Zolotarev VM (1997) Modern Theory of Summation of Random Variables. VSP, Utrecht

Index